METHODS IN MOLECULAR BIOLOGY™

Series Editor
John M. Walker
School of Life Sciences
University of Hertfordshire
Hatfield, Hertfordshire, AL10 9AB, UK

For further volumes:
http://www.springer.com/series/7651

Glycosyltransferases

Methods and Protocols

Edited by

Inka Brockhausen

Department of Medicine, Division of Rheumatology, and Department of Biomedical and Molecular Sciences, Queen's University, Kingston, Ontario, Canada

Editor
Inka Brockhausen
Department of Medicine, Division of Rheumatology
and Department of Biomedical and Molecular Sciences
Queen's University
Kingston, ON, Canada

ISSN 1064-3745 ISSN 1940-6029 (electronic)
ISBN 978-1-62703-464-7 ISBN 978-1-62703-465-4 (eBook)
DOI 10.1007/978-1-62703-465-4
Springer New York Heidelberg Dordrecht London

Library of Congress Control Number: 2013938604

Printed on acid-free paper

Humana Press is a brand of Springer
Springer is part of Springer Science+Business Media (www.springer.com)

Preface

This issue on glycosyltransferases (GTs) contains a wide range of studies, methods, and protocols which form a solid basis for investigations of the role and mechanisms, biology, and pathology involving GTs. Complex glycoconjugates, such as glycoproteins, proteoglycans, and glycolipids, are assembled by GTs which synthesize specific linkages between sugars or sugars and protein. This is in contrast to the nonspecific or less-specific chemical glycation reactions, transglycosylation, and reverse glycosylation reactions. GTs are essential for the biosynthesis of complex glycoconjugates and are powerful tools to study the functions of complex glycans in health, development, and disease.

For each of the glycosylation steps, a family of similar GT proteins exists. GTs are found in all prokaryotic and eukaryotic cells. The CAZy Webpage lists about 100 different families of related proteins (GT families) that catalyze the synthesis of many different sugar linkages, with thousands of GT genes discovered so far. However, only a small percentage of GTs have been biochemically identified and characterized. This issue will help to advance the field by providing detailed methods for the study of GTs.

GTs in mammals and microbes may catalyze similar reactions but their sequences often have a low degree of similarity. Yet it is assumed that substrate binding, metal ion activation, and catalysis follow similar mechanisms. Therefore it is important to define the three-dimensional architecture of the protein and the catalytic site in order to determine the mechanisms by which these enzymes act. Thus not only amino acid sequences but also protein folds determine enzyme function. Two major classes of protein folds, GTA and GTB, have been identified, and there are several variants. Only a few GTs have been crystallized but it is hoped that the structures of many more GTs will soon be elucidated. In silico studies of the relationships of GTs and specific domains and sequences will also contribute to knowledge of their mechanism. The creation of site-specific mutants is a powerful tool to determine the role of specific amino acids in substrate binding and catalysis which complements the crystallographic studies.

GTs have been classified as inverting and retaining enzymes, depending on the anomeric configuration of the sugar linkage formed, in comparison to the sugar donor structure. For example, the sugar in UDP-Gal is in the α-configuration, and an inverting transferase catalyzes a Gal-β-linkage. In these transferases, often Mn^{2+} or a similar metal ion, as well as a D×D sequence, are required for activity. One of the acidic residues of the DxD sequence provides a catalytic base. The mechanism of retaining transferases (that synthesize Gal-α linkages and often do not require divalent metal ions) is not so clear.

With few exceptions, most GTs are very specific for their acceptor substrates and the linkage they form. An example of a GT that catalyzes the formation of two different linkages is the α3/4 Fucosyltransferase. GTs may also have two different GT domains and catalyze two different successive reactions; for example, in the synthesis of hyaluronic acid, bacterial O-antigens, and capsule polysaccharides.

GTs can be assayed by a number of different methods, including assays using radioactive or fluorescent substrates or microarrays. These GT assays help to develop the tools for the in vitro synthesis of functional carbohydrates. Although GT reactions are very specific for both the nucleotide sugar donor and the acceptor substrate, their characteristics may be altered by in vitro conditions. Many factors control sugar transfer, including cofactors, possible chaperone proteins, and the availability of sugar acceptor substrates, localization and topology of enzymes. The analysis of the intracellular organization of glycosylation and the factors controlling the activities of the participating enzymes in the cell are important areas that need more research efforts. It is also critical to understand the glycodynamics of a cell, i.e., how the cell responds to stimuli leading to biological and pathological changes in terms of alterations in glycosylation, and how this affects the biology of the cell.

In eukaryotic cells, GTs can exist as soluble enzymes in the cytosol. For example, the O-GlcNAc-transferase (OGT) glycosylates proteins with a GlcNAc residue linked to Ser/Thr, affecting the structure and function of a great variety of intracellular and nuclear proteins. However, most GTs are membrane-bound enzymes. Signals that localize GTs in the ER involve ER recognition sequences. The ER is rich in reactions involving lipid-like intermediates and membrane-bound enzymes. In contrast, enzymes localized in the Golgi are subject to control by the conserved oligomeric Golgi complex (COG). The discovery of COG has opened up another dimension of control of glycosylation. The role of the membrane in the activity of GTs is still not well known, although it is clear that lipid rafts have an influence on the organization and function of proteins in membranes.

The biosynthesis of GlcNAc-Asn-linked and Man-O-Ser/Thr-linked oligosaccharides involves dolichol-phospho-sugars as both sugar donor and acceptor substrates. Since both the enzymes involved and their substrates are in a lipid phase, they are difficult to study, and this area has therefore been neglected in the past. Useful reagents and methods are suggested here to overcome the problem inherent in assays of water-insoluble biomolecules.

Bacteria are rich in GTs that transfer either a single sugar, a polysaccharide, oligosaccharide chain, or a sugar phosphate to a lipid-linked intermediate. GTs can be expressed in bacteria, but in these artificial systems, proteins may not be folded properly, and have to be refolded.

Although the structures of the GT enzyme products are found abundantly in nature, some GTs have very low activity in in vitro assays. An example is the poly-N-acetyllactosamine structure of *N*-glycans. These ubiquitous glycan chains are assembled by a β1,4Gal-transferase which is highly active and a β1,3-GlcNAc-transferase which is poorly active in vitro. Thus, activating factors for GTs should be present in vivo but remain to be identified. It has been shown that GTs have other GTs as binding partners, and it is possible that associations with other proteins or membrane components help GTs to be highly efficient in vivo. In a unique association, lactalbumin is known to change the acceptor substrate specificity of β1,4-Gal-transferase to convert it to a lactose synthase. Another unique association has been identified for the β1,3-Gal-transferase that synthesizes the cancer-associated T antigen. Cosmc is a specific chaperone for this Gal-transferase and ensures its activity by promoting proper protein folding and thus preventing enzyme degradation.

Carbohydrates that are synthesized by GTs in the ER and Golgi and then are transported to the cell surface can be recognized by lectins, anti-carbohydrate antibodies and other carbohydrate-binding molecules. These methods help to define the abnormalities of glycan structures in disease and develop specific anticancer diagnostic and therapeutic tools.

Sensitive tools to detect and distinguish specific GT proteins are essential. Thus, photoaffinity labeling, use of antibodies, and lectins help to distinguish different members of a GT family. Abnormalities in the activities or expression of GTs are relevant to most diseases, where they can have an impact on the pathology, for example in cancer, tumor metastasis and invasiveness, inflammation, and immune functions. The cloning of GT genes allowed the studies of gene knockout mice and cells, or transfections and overexpression of GTs that can make important contributions to our understanding of glycan functions.

While endogenous natural GT inhibitors remain to be discovered, many synthetic inhibitors are known and have been tested in single assays or by robot-high-throughput assays. Inhibitors may be donor or acceptor substrate analogs, discovered through rational design requiring knowledge of the substrate recognition by GTs. They have also been discovered by testing chemical libraries and by serendipity successes.

I hope that this book of glycosyltransferase protocols will help students, postdoctoral fellows, and senior scientists to carry on the research of GTs that have been intensified during the last years. This will make a vital contribution to glycobiology and glycopathology and to applications of these enzymes in biotechnology and drug development. I am grateful to all contributors for sharing their valuable expertise.

Kingston, ON, Canada *Inka Brockhausen*

Contents

Contributors

ANGEL ASHIKOV • *Department of Cellular Chemistry, Hannover Medical School, Hannover, Germany*

BRUCE A. BAGGENSTOSS • *Department of Biochemistry and Molecular Biology, The Oklahoma Center for Medical Glycobiology, The University of Oklahoma Health Sciences Center, Oklahoma City, OK, USA*

HANS BAKKER • *Department of Cellular Chemistry, Hannover Medical School, Hannover, Germany*

ERIC PAUL BENNETT • *Department of Cellular and Molecular Medicine, Copenhagen Center for Glycomics, University of Copenhagen, Copenhagen N, Denmark*

INKA BROCKHAUSEN • *Department of Medicine, Division of Rheumatology, and Department of Biomedical and Molecular Sciences, Queen's University, Kingston, ON, Canada*

FALK F.R. BUETTNER • *Department of Cellular Chemistry, Hannover Medical School, Hannover, Germany*

HENRIK CLAUSEN • *Department of Cellular and Molecular Medicine, Copenhagen Center for Glycomics, University of Copenhagen, Copenhagen N, Denmark*

RICHARD D. CUMMINGS • *Department of Biochemistry, Emory University School of Medicine, Atlanta, GA, USA*

PAUL L. DEANGELIS • *Department of Biochemistry and Molecular Biology, The Oklahoma Center for Medical Glycobiology, The University of Oklahoma Health Sciences Center, Oklahoma City, OK, USA*

LUDIVINE DROUGAT • *Unit of Structural and Functional Glycobiology, University of Lille 1, Villeneuve d'Ascq, France*

MARIO F. FELDMAN • *Department of Biological Sciences, Alberta Glycomics Centre, University of Alberta, Edmonton, AB, Canada*

MICHAEL A. J. FERGUSON • *Division of Biological Chemistry and Drug Discovery, College of Life Sciences, University of Dundee, Dundee, Scotland, UK*

MICHIKO N. FUKUDA • *Tumor Microenvironment Program, Cancer Center, Sanford-Burnham Medical Research Institute, La Jolla, CA, USA*

MINORU FUKUDA • *Scripps Korea Antibody Institute, Chuncheon, Gangwon, Korea; Tumor Microenvironment Program, Cancer Center, Sanford Burnham Medical Research Institute, La Jolla, CA, USA*

NINGGUO GAO • *Department of Pharmacology, University of Texas Southwestern Medical Center, Dallas, Texas, USA*

YIN GAO • *Department of Medicine, Division of Rheumatology, and Department of Biomedical and Molecular Sciences, Queen's University, Kingston, Ontario, Canada*

JOANA GOMES • *IPATIMUP-Institute of Molecular Pathology and Immunology of the University of Porto, Faculdade de Medicina, Instituto de Ciências Biomédicas Abel Salazar, Universidade do Porto, Porto, Portugal*

M. Lucia S. Güther • *Division of Biological Chemistry and Drug Discovery, College of Life Sciences, University of Dundee, Dundee, Scotland, UK*
Kelly G. Ten Hagen • *Developmental Glycobiology Unit, NIDCR, National Institutes of Health, Bethesda, MD, USA*
Robert S. Haltiwanger • *Department of Biochemistry and Cell Biology, Institute of Cell and Developmental Biology, Stony Brook University, Stony Brook, NY, USA*
John A. Hanover • *Laboratory Cell Biochemistry and Biology, NIDDK, National Institutes of Health, Bethesda, MD, USA*
Anne Harduin-Lepers • *Unité de Glycobiologie Structurale et Fonctionnelle, Université Lille Nord de France, Lille 1, Villeneuve d'Ascq, France*
Shingo Hatakeyama • *Department of Urology, Hirosaki University School of Medicine, Hirosaki, Aomori, Japan*
Maria V. Ielmini • *Alberta Glycomics Centre, Department of Biological Sciences, University of Alberta, Edmonton, Alberta, Canada*
Hideyuki Ihara • *Division of Molecular Cell Biology, Department of Biomolecular Sciences, Faculty of Medicine, Saga University, Saga, Japan*
Yoshitaka Ikeda • *Division of Molecular Cell Biology, Department of Biomolecular Sciences, Faculty of Medicine, Saga University, Saga, Japan*
Luis Izquierdo • *Barcelona Centre for International Health Research (CRESIB, Hospital Clinic-Universitat de Barcelona), Barcelona, Spain*
Tongzhong Ju • *Department of Biochemistry, Emory University School of Medicine, Atlanta, GA, USA*
Hiroto Kawashima • *Laboratory of Microbiology and Immunology, School of Pharmaceutical Sciences, University of Shizuoka, Shizuoka, Japan*
Eun Ju Kim • *Laboratory Cell Biochemistry and Biology, NIDDK, National Institutes of Health, Bethesda, MD, USA*
Hiroaki Korekane • *Division of Systems Glycobiology Research Group, RIKEN Global Research Cluster, Wako, Saitama, Japan*
Seung Ho Lee • *Scripps Korea Antibody Institute, Chuncheon, Gangwon, Korea; Tumor Microenvironment Program, Cancer Center, Sanford Burnham Medical Research Institute, La Jolla, CA, USA*
Tony Lefebvre • *Unit of Structural and Functional Glycobiology, University of Lille 1, Villeneuve d'Ascq, France*
Mark A. Lehrman • *Department of Pharmacology, University of Texas Southwestern Medical Center, Dallas, TX, USA*
Martin Loibl • *Cell Chemistry, Centre for Organismal Studies Heidelberg, University of Heidelberg, Heidelberg, Germany*
Vladimir V. Lupashin • *Department of Physiology and Biophysics, University of Arkansas for Medical Sciences, Little Rock, AR, USA*
Ulla Mandel • *Department of Odontology, Copenhagen Center for Glycomics, University of Copenhagen, Copenhagen N, Denmark*
Akio Matsumoto • *Department of Pharmacology, Chiba University Graduate School of Medicine, Chiba, Japan*
Jean-Claude Michalski • *Unit of Structural and Functional Glycobiology, University of Lille 1, Villeneuve d'Ascq, France*

MATIAS A. MUSUMECI • *Department of Biological Sciences, Alberta Glycomics Centre, University of Alberta, Edmonton, AB, Canada*
CORWIN M. NYCHOLAT • *Glycan Microarray Synthesis Core, Consortium for Functional Glycomics, Department of Chemical Physiology, The Scripps Research Institute, La Jolla, CA, USA*
SUGURU OGURI • *Department of Bio-Production, Faculty of Bio-Industry, Tokyo University of Agriculture, Abashiri, Hokkaido, Japan*
KAZUAKI OHTSUBO • *Division of Systems Glycobiology Research Group, RIKEN Global Research Cluster, Wako, Saitama, Japan*
CHIKARA OHYAMA • *Department of Urology, Hirosaki University School of Medicine, Hirosaki, Aomori, Japan*
STEPHANIE OLIVIER-VAN STICHELEN • *Unit of Structural and Functional Glycobiology, University of Lille 1, Villeneuve d'Ascq, France*
AMY J. PADGETT-MCCUE • *Department of Biochemistry and Molecular Biology, The Oklahoma Center for Medical Glycobiology, and The University of Oklahoma Health Sciences Center, Oklahoma City, OK, USA*
JONG YI PARK • *Gyeongbuk Institute for Bio-industry, Andong-si, Gyeongbuk, Korea*
KINNARI B. PATEL • *Centre for Infection and Immunity, Queen's University Belfast, Belfast, UK*
WENJIE PENG • *Glycan Microarray Synthesis Core, Consortium for Functional Glycomics, Department of Chemical Physiology, The Scripps Research Institute, La Jolla, CA, USA*
DANIEL PETIT • *Unité de Génétique Moléculaire Animale, INRA/Université de Limoges, Limoges, France*
JEAN-MICHEL PETIT • *Unité de Génétique Moléculaire Animale, INRA/Université de Limoges, Limoges, France*
IRINA D. POKROVSKAYA • *Department of Physiology and Biophysics, University of Arkansas for Medical Sciences, Little Rock, AR, USA*
PRADMAN K. QASBA • *Structural Glycobiology Section and Basic Science Program, SAIC-Frederick, Inc., Center for Cancer Research Nanobiology Program, Center for Cancer Research, Frederick National Laboratory for Cancer Research, Frederick, MD, USA*
BOOPATHY RAMAKRISHNAN • *Structural Glycobiology Section and Basic Science Program, SAIC-Frederick, Inc., Center for Cancer Research Nanobiology Program, Center for Cancer Research, Frederick National Laboratory for Cancer Research, Frederick, MD, USA*
NAHID RAZI • *AccuDava Inc., La Jolla, CA, USA*
CELSO A. REIS • *IPATIMUP-Institute of Molecular Pathology and Immunology of the University of Porto, Faculdade de Medicina, Instituto de Ciências Biomédicas Abel Salazar, Universidade do Porto, Porto, Portugal*
XIANG RUAN • *Department of Microbiology and Immunology, Centre for Human Immunology, University of Western Ontario, London, ON, Canada*
MAYA K. SETHI • *Department of Cellular Chemistry, Hannover Medical School, Hannover, Germany*
TOSHIAKI K. SHIBATA • *Department of Obstetrics and Gynecology, Hamamatsu University School of Medicine, Hamamatsu, Shizuoka, Japan*

CATHARINA STEENTOFT • *Department of Cellular and Molecular Medicine, Copenhagen Center for Glycomics, University of Copenhagen, Copenhagen N, Denmark*
SABINE STRAHL • *Centre for Organismal Studies Heidelberg, Cell Chemistry, University of Heidelberg, Heidelberg, Germany*
KAZUHIRO SUGIHARA • *Department of Obstetrics and Gynecology, Hamamatsu University School of Medicine, Hamamatsu, Shizuoka, Japan*
TOMOHIKO TAGUCHI • *Department of Health Chemistry, Graduate School of Pharmaceutical Sciences, University of Tokyo, Tokyo, Japan*
SHINJI TAKAMATSU • *Division of Systems Glycobiology Research Group, RIKEN Global Research Cluster, Wako, Saitama, Japan*
HIDEYUKI TAKEUCHI • *Department of Biochemistry and Cell Biology, Institute of Cell and Developmental Biology, Stony Brook University, Stony Brook, NY, USA*
NAOYUKI TANIGUCHI • *Division of Systems Glycobiology Research Group, RIKEN Global Research Cluster, Wako, Saitama, Japan*
BORIS TEFSEN • *Department of Molecular Cell Biology and Immunology, VU University Medical Center, Amsterdam, The Netherlands*
ROXANA ELIN TEPPA • *Unité de Glycobiologie Structurale et Fonctionnelle, Université Lille Nord de France, Lille 1, Villeneuve d'Ascq, France*
E TIAN • *Developmental Glycobiology Unit, NIDCR, National Institutes of Health, Bethesda, MD, USA*
YUKI TOBISAWA • *Department of Urology, Hirosaki University School of Medicine, Hirosaki, Aomori, Japan*
HIROKI TSUKAMOTO • *Division of Molecular Cell Biology, Department of Biomolecular Sciences, Faculty of Medicine, Saga University, Saga, Japan*
MIGUEL A. VALVANO • *Centre for Infection and Immunity, Queen's University Belfast, Belfast, UK*
IRMA VAN DIE • *Department of Molecular Cell Biology and Immunology, VU University Medical Center, Amsterdam, The Netherlands*
ANNE-SOPHIE VERCOUTTER-EDOUART • *Unit of Structural and Functional Glycobiology, University of Lille 1, Villeneuve d'Ascq, France*
MALENE BECH VESTER-CHRISTENSEN • *Department of Cellular and Molecular Medicine, Copenhagen Center for Glycomics, University of Copenhagen, Copenhagen N, Denmark*
ANNA VINNIKOVA • *Department of Medicine, Division of Rheumatology, and Department of Biomedical and Molecular Sciences, Queen's University, Kingston, Ontario, Canada*
PAUL H. WEIGEL • *Department of Biochemistry & Molecular Biology, The Oklahoma Center for Medical Glycobiology, and The University of Oklahoma Health Sciences Center, Oklahoma City, OK, USA*
ROSE A. WILLETT • *Department of Physiology and Biophysics, University of Arkansas for Medical Sciences, Little Rock, AR, USA*
HUI WU • *Department of Pediatric Dentistry, Department of Microbiology, University of Alabama at Birmingham, Schools of Dentistry and Medicine, Birmingham, AL, USA*

REN WU • *Department of Pediatric Dentistry, University of Alabama at Birmingham, Schools of Dentistry and Medicine, Birmingham, AL, USA*
HUA ZHANG • *Department of Pediatric Dentistry, University of Alabama at Birmingham, Schools of Dentistry and Medicine, Birmingham, AL, USA*
LIPING ZHANG • *Developmental Glycobiology Unit, NIDCR, National Institutes of Health, Bethesda, MD, USA*
FAN ZHU • *Department of Pediatric Dentistry, Department of Microbiology, University of Alabama at Birmingham, Schools of Dentistry and Medicine, Birmingham, AL, USA*

Chapter 1

Glycan Microarray Screening Assay for Glycosyltransferase Specificities

Wenjie Peng, Corwin M. Nycholat, and Nahid Razi

Abstract

Glycan microarrays represent a high-throughput approach to determining the specificity of glycan-binding proteins against a large set of glycans in a single format. This chapter describes the use of a glycan microarray platform for evaluating the activity and substrate specificity of glycosyltransferases (GTs). The methodology allows simultaneous screening of hundreds of immobilized glycan acceptor substrates by in situ incubation of a GT and its appropriate donor substrate on the microarray surface. Using biotin-conjugated donor substrate enables direct detection of the incorporated sugar residues on acceptor substrates on the array. In addition, the feasibility of the method has been validated using label-free donor substrate combined with lectin-based detection of product to assess enzyme activity. Here, we describe the application of both procedures to assess the specificity of a recombinant human α2-6 sialyltransferase. This technique is readily adaptable to studying other glycosyltransferases.

Key words Glycosyltransferase, Sialyltransferase, High-throughput, Glycan microarray, Specificity assay

1 Introduction

Glycosyltransferases (GTs) are a large family of enzymes that catalyze the transfer of a sugar residue from a glycosyl donor to a variety of acceptor substrates [1]. Glycans biosynthesized by GTs are essential to many fundamental biological processes in prokaryotes and eukaryotes, as well as in human health and diseases [2]. Unlike other biomolecules (i.e., DNA, RNA, and proteins), the biosynthesis of glycans is non-template driven. However, GTs are generally stringent in their substrate specificities and demonstrate rigid stereoselectivity and regioselectivity in glycosidic linkage formation to secure the assembly of defined structures. As such, GTs have received increased attention as potential molecular targets in chemical biology and drug discovery [3–8]. Recombinant GTs are valuable resources for in vitro studies of glycans and are, particularly, used as synthetic tools for production of complex carbohydrates and glycoconjugates [4, 9–13].

Inka Brockhausen (ed.), *Glycosyltransferases: Methods and Protocols*, Methods in Molecular Biology, vol. 1022, DOI 10.1007/978-1-62703-465-4_1, © Springer Science+Business Media New York 2013

To investigate the potential applications of GTs it is necessary to characterize their substrate specificities and activities. A variety of methods have been employed to evaluate GT activities including assays based on radiochemistry, immunology, chromatography, spectrophotometry, fluorometry, and mass spectrometry [14, 15]. In recent years, glycan microarray technology has offered a high-throughput approach to evaluate the activities and substrate specificities of GTs [16–23]. Incubation of a GT with its appropriate donor substrate on the microarray surface allows simultaneous screening of hundreds of immobilized glycan acceptor substrates.

We previously described a glycan microarray based assay, which was used to evaluate the specificities of several recombinant sialyltransferases and explore their utility for in vitro glycan synthesis [24]. This array was manufactured by the Glycan microrray synthesis Core of the Consortium for Functional Glycomics (CFG, www.functionalglycomics.org) and contained over 200 glycan structures (*see* **Note 1**) covalently bound on glass slides [20]. We performed an in situ incubation reaction on the microarray surface, using a sialyltransferase and a biotin-conjugated cytidine-5′-monophospho-N-acetylneuraminic acid (CMP-9-biotin-Neu5Ac) as donor substrate. In a second step, the array was incubated with fluorescein-conjugated streptavidin to tag the enzyme reaction products and generate a specificity profile for each of the tested enzymes. The resulting detection of fluorescence on the slide surface demonstrated solid-phase sialylation of multiple acceptor substrates. In addition to the known substrates of the tested sialyltransferases the method also revealed additional previously unknown suitable substrates. While the application of the method was demonstrated with sialyltransferases, the assay can be readily adapted to study the activities of other GTs [25–27]. This rapid high-throughput technique is highly sensitive and requires minimal materials for simultaneous screening of hundreds of immobilized glycans. The biotin-conjugated donor substrate can facilitate the direct detection of the incorporated sugar residue. Although, the activities of sialyltransferases are not significantly affected by modifications at C-9 of sialic acid, it should be noted that such substitutions of the donor substrate might not be tolerated by other glycosyltransferases.

Alternatively, label-free CMP-Neu5Ac donor substrate can be used combined with lectin-based detection of enzyme products as validation of the above method [24, 27]. In the case described here we use biotinylated *Sambucus nigra* lectin (SNA-I) coupled with a streptavidin–Alexa488 conjugate to detect α2-6 sialylated products. This lectin-based approach may increase the application of the method especially in cases where the labeled donor is unavailable or not a suitable substrate for the enzyme of interest. However, it is important to note that the lectin-binding approach is an indirect method as detection of the enzyme product is dependent on

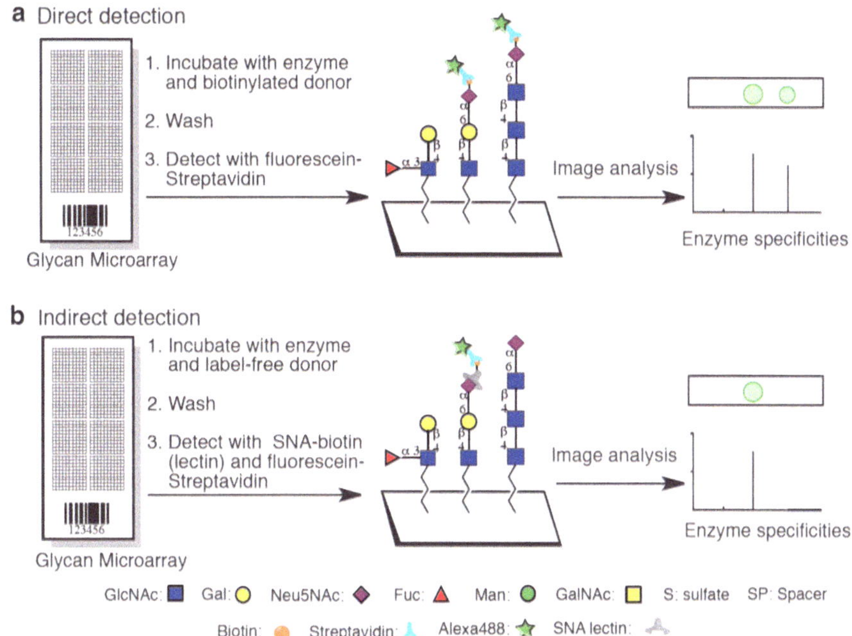

Fig. 1 Principle of evaluating sialyltransferase acceptor specificities on the glycan microarray. (**a**) Direct detection: enzymatic transfer of biotinylated nucleotide sugar donor followed by detection with fluorescently labeled streptavidin–Alexa Fluor488 conjugate; (**b**) Indirect detection: enzymatic transfer of label-free nucleotide sugar donor followed by detection of terminal Neu5Aca2-6 N acetyllactosamine motif with biotinylated SNA-I and fluorescein-streptavidin. It is important to note that the indirect detection of sialylated product is dependent on the lectin-binding specificity. *Symbol structures are based on the Symbol nomenclature proposed in "Essentials of Glycobiology"*

the binding specificity of the lectin. Here, we describe the protocols for both approaches using biotin-conjugated (Fig. 1a) or label-free (Fig. 1b) donor substrates for human α2-6 sialyltransferase activity assessment.

2 Materials

Prepare all solutions using deionized distilled water (ddH_2O, NanoPure Infinity Ultrapure water system Barnstead/Thermolyne) unless indicated otherwise.

2.1 Synthesis of CMP-9-Biotin- Neu5Ac

1. Tris–HCl stock buffer: 100 mM Tris–HCl, 20 mM $MgCl_2$, pH 9.0. Using a graduated cylinder add about 10 mL water to a 200-mL glass bottle. Weigh out 1.58 g of Tris–HCl and 407 mg of $MgCl_2{\cdot}6H_2O$ and transfer to the bottle. Swirl the bottle to help dissolve the solids. Add 80 mL of water to the bottle, mix, and then adjust the pH with NaOH (1 M). Make up to 100 mL with water. Store at 4 °C.
2. 6-Azido-6-deoxy-ManNAc (*see* **Note 2**).

3. Sodium pyruvate (Fisher Scientific, Dallas, TX, USA).
4. Cytidine-5 -triphosphate disodium salt (Sigma-Aldrich Chemical Company, St. Louis, MO, USA).
5. 15 mL Conical Centrifuge tubes (BD Falcon tubes).
6. N-Acetylneuraminic acid aldolase (Toyoba Co. Ltd. Biochemical Department, Tokyo, Japan) (enzyme specific activity: 15 U/mg protein).
7. Recombinant CMP-sialic acid synthetase from *Neisseria meningitidis* (NmCSS) (enzyme specific activity: 290 U/mg protein).
8. Thin Layer Chromatography silica gel plates (EMD Chemicals pre-coated TLC glass plates. Fisher Scientific).
9. TLC plate cutter (Grace Discovery Science, Deerfield, IL).
10. Thin Layer Chromatography glass chamber. Wide-mouth screw-cap jar.
11. Glass micropipettes for spotting sample on TLC plates.
12. Thin Layer Chromatography developing eluent: iPrOH–H_2O–NH_4OH 6:3:2.
13. UV-lamp (Spectroline model ENF-260C, Fisher Scientific).
14. TLC staining solution: 10 % H_2SO_4 in ethanol.
15. Amicon® Ultra-15 Centrifugal Filter Device, 10 kDa Molecular weight cutoff (Millipore Corporation. Fisher Scientific).
16. Centrifuge with swinging bucket or fixed angle rotor with wells/carrier that can accommodate 50 mL tubes.
17. Bio-Gel P-2 gel extra fine (Bio-Rad Laboratories, Hercules, CA).
18. Glass econo-column chromatography column (2.5 × 100 cm) (Bio-Rad).
19. Triphenylphosphine (Sigma-Aldrich).
20. Sulfo-NHS-LC-LC-Biotin (Pierce, Thermo Scientific, Rockford, IL).
21. Tetrahydrofuran (THF, Sigma-Aldrich).

2.2 Sialyltransferase Acceptor Specificity Assays

1. Glycan microarray printed on an NHS-activated glass slides (*see* **Note 3**).
2. VWR black lab marker.
3. Enzymatic reaction buffer: 100 mM Tris–HCl (pH 7.0) stored at 4 °C. Before using, it should be warmed up to room temperature.
4. Sialyltransferase (hST6Gal-I) was over-expressed in insect cells and purified on ion-exchange column chromatography. Stored at 4 °C (*see* **Note 4**).
5. Bovine Serum Albumin standard grade powder was stored at 4 °C (Sigma-Aldrich).

6. Pyrex brand glass tray. Size depends on the number of microarray slides being incubated (Fisher Scientific).
7. Plastic Saran™ wrap.
8. Platform Shaker.
9. Kimwipes.
10. Streptavidin–Alexa488. 1 mg/mL in Tris–HCl (100 mM, pH 7.0) (Invitrogen, Grand Island, NY).
11. C1303-T Slide spinner centrifuge (Labnet International, Inc., Woodbridge, NJ).
12. ScanArray 5000 confocal slide scanner (Perkin Elmer, Boston, MA).
13. ImaGene Image analysis software (BioDiscovery, Inc., El Segundo, CA).
14. Cytidine-5′-monophospho-*N*-acetylneuraminic acid sodium salt (Sigma-Aldrich).
15. PBS 1× (Gibco, Grand Island, NY).
16. Biotinylated *Sambucus nigra agglutinin* (SNA, 10 μg/mL) (Vector Laboratories, Inc., Burlingame, CA).
17. Tween®-20 (Sigma-Aldrich).

3 Methods

3.1 Synthesis of CMP-9-Biotin- Neu5Ac

1. The three-step synthesis described here is outlined in Fig. 2.
2. Weigh out 6-azido-6-deoxy-ManNAc **1** (50 mg, 0.203 mmol, 1 equiv.), sodium pyruvate (67.0 mg, 0.61 mmol) and cytidine-5′-triphosphate disodium salt (161 mg, 0.30 mmol) and transfer the solids to a 15 mL conical centrifuge tube. Dissolve the solids in Tris–HCl (100 mM, 20 mM $MgCl_2$; 9 mL) stock buffer. Adjust to pH 8.6 with NaOH (1 M).

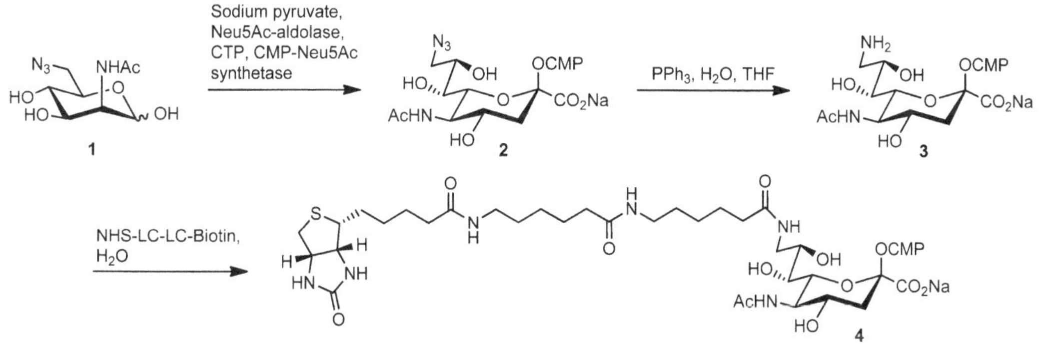

Fig. 2 Synthesis of biotinylated nucleotide sugar donor, CMP-9-biotin-Neu5Ac **4**. 9-Azido CMP-Neu5Ac was enzymatically produced from 6-azido ManNAc with two enzymes, followed by converting azide to free amine with PPh_3. Subsequently, the free amine was coupled with NHS activated biotin to produce the product

3. Add *N*-acetylneuraminic acid aldolase (500 U/mmol ManNAc) and *N. meningitidis* CMP-Neu5Ac synthetase (20 U/mmol CTP) and then gently mix the reaction at 37 °C (*see* **Note 5**).
4. Monitor the pH and adjust with 1 M NaOH to pH 8.3–9.0 as needed.
5. Monitor the reaction by TLC using (iPrOH–H_2O–NH_4OH, 6:3:2) as eluent (*see* **Note 6**).
6. Terminate the completed reaction by passing the mixture through an Amicon® Ultra-15 Centrifugal Filter Device MWCO 10 kDa. Using a pipette transfer the reaction mixture into the top filtering device and centrifuge until all of the liquid has passed through the filter. Freeze the filtrate and then lyophilize the sample (*see* **Note 7**).
7. Dissolve the resulting solid in 20 mM NH_4OH (1 mL) and then load the solution directly onto a column of Bio-Gel P-2 extra fine (2.5 × 100 cm) for purification (*see* **Note 8**). Elute with the same buffer and collect 1 mL fractions.
8. Monitor the fractions by TLC (iPrOH–H_2O–NH_4OH, 6:3:2) (*see* **Note 9**). Combine fractions which contain pure product in a 15 mL centrifuge tube, adjust to pH 9.0 and lyophilize. The product, CMP-9-azido-9-deoxy-Neu5Ac **2**, can be used directly for the next step or stored at −20 °C until needed.
9. Weigh out CMP-9-azido-9-deoxy-Neu5Ac **2** (10.0 mg, 0.0146 mmol), transfer to a glass vial, and add water (1 mL). Swirl the vial to ensure the product is dissolved.
10. Separately, weigh out triphenylphosphine (19.2 mg, 0.0732 mmol, 5 equiv.) in a glass vial and dissolve in THF (1 mL). Swirl the vial to ensure the material is dissolved.
11. Using a glass pipette transfer dropwise the THF solution containing triphenylphosphine to the aqueous solution containing **2**. The solution will turn cloudy as the THF solution is added. Gently swirl the vial to mix the solutions. Add a stir bar to the vial and gently mix the reaction at room temperature overnight.
12. Monitor the reaction by TLC using (iPrOH–H_2O–NH_4OH, 6:3:2) as eluent.
13. When the reaction is completed concentrate the solution under reduced pressure to remove the THF. Dilute the reaction mixture by adding 9 mL of deionized H_2O to the glass vial. Transfer the aqueous solution to a separatory funnel (50 mL). Rinse the vial with additional 10 mL water and then combine the aqueous solutions in the separatory funnel. Total volume will be approx. 20 mL water.
14. Extract the aqueous solution with ethyl acetate (3 × 10 mL) to remove excess triphenylphosphine and the reaction side product,

triphenylphosphine oxide. Add ethyl acetate (10 mL) to the separatory funnel, place the stopper on, and then shake the funnel to mix the solutions (*see* **Note 10**). Following each extraction allow the two layers to separate. The top layer is the organic phase (ethyl acetate) and the bottom is the aqueous phase, which contains the product. Remove the stopper and open the stopcock to drain the aqueous phase into a clean glass vial. Drain the organic layer into a second clean vial. Check the aqueous layer and organic layer by TLC for the presence of triphenylphosphine. Transfer the aqueous solution back to the separatory funnel and repeat extraction with ethyl acetate (10 mL) as needed to remove residual triphenylphosphine. Check the ethyl acetate extracts by TLC for any product. If product is observed back extract the organic phase with H_2O.

15. Combine the aqueous phases and then lyophilize. The resulting product, CMP-9-amino-9-deoxy-Neu5Ac **3**, can be used as it is for the next step or stored at −20 °C until required.
16. Weigh out CMP-9-amino-9-deoxy-Neu5Ac **3** (5 mg, 0.0076 mmol, 1 equiv.) and transfer to a glass vial. Add water (1 mL) to the vial and gently mix to ensure the compound completely dissolves.
17. Weigh out sulfo-NHS-LC-LC-Biotin (7.8 mg, 0.011 mmol, 1.5 equiv.) and add to the reaction vial. Swirl the vial to completely dissolve the solid. Cap the vial and gently mix the reaction at room temperature.
18. Monitor the reaction by TLC (iPrOH–H_2O–NH_4OH, 6:3:2) at 30 min intervals. When the starting material has been consumed, freeze the reaction and lyophilize.
19. Dissolve the resulting white solid in 20 mM NH_4OH (1 mL) and then load the solution directly onto a column of Bio-Gel P-2 (2.5 × 100 cm) and elute with the same buffer. Collect 1 mL fractions.
20. Monitor the fractions by TLC (iPrOH–H_2O–NH_4OH, 6:3:2). Combine the fractions containing pure product into a 15 mL conical centrifuge tube, freeze the aqueous solution, and then lyophilize. The resulting product CMP-9-biotin-Neu5Ac **4** can be stored at −20 °C or used directly for sialylations.

3.2 Sialyltransferase Acceptor Specificity Assays

1. Label the slide with the sample name near the barcode. Mark the glycan-printed side with hydrophobic marker around the print area to form a barrier, and let it dry for several minutes (*see* **Notes 11** and **12**).
2. Pipette 0.5 mL Tris–HCl buffer (100 mM, pH 7.0) containing CMP-9-biotin-Neu5Ac **4** (1 mM), sialyltransferase (50 mU hST6Gal-I) and BSA (3 %) within the hydrophobic marker barrier. Make sure there are no air bubbles trapped (*see* **Notes 13** and **14**).

3. Incubate the mixture on the slide in a sealed humidification chamber while gently shaking on a rotating shaker at room temperature for 90 min. Take care that the speed is not too fast and the liquid on the slide surface does not spill (*see* **Note 15**).
4. Following the incubation period discard the sample by simply pouring it off the slide into the appropriate waste container (*see* **Note 16**).
5. Wash the slide by holding the edges and gently dipping it four times in a Tris–HCl buffer (100 mM, pH 7.0), then four times in ddH_2O (*see* **Note 17**).
6. Clean the backside of the slide with a Kimwipe paper and place back in the humidification chamber (*see* **Note 18**).
7. Pipette 1 mL of Tris–HCl buffer (100 mM, pH 7.0) containing streptavidin–Alexa488 conjugate (1:100 dilution, 1 mg/mL) onto the print surface and incubate in the sealed humidification chamber while gently rotating for 30 min at room temperature.
8. Following incubation repeat **step 4** and **5**.
9. Clean the backside of the slide with a Kimwipe paper.
10. Spin dry the slide in a slide centrifuge at 20 °C for 5 min at 200 rcf (1,024 rpm) (*see* **Note 19**).
11. Scan the slide with a confocal slide array laser scanner at 488 nm wavelength (*see* **Notes 20** and **21**).
12. Process the slide with the image processing program (*see* Fig. 3).

3.3 Detection of Sialylation with SNA-I Lectin

1. Label five slides indicating the different enzyme amounts (0, 0.5, 5, 50, 500 mU) near the barcode. Mark the glycan-printed slide with hydrophobic marker.
2. Pipette onto the slides 0.5 mL Tris–HCl buffer (100 mM, pH 7.0) solutions containing CMP-Neu5Ac (1 mM), hST-6Gal-I (0 mU, 0.5 mU, 5 mU, 50 mU, 500 mU, respectively) and BSA (3 %) (*see* **Note 22**).
3. Repeat **steps 3–6** from Subheading 3.2.
4. Pipette onto the slide, a solution of biotinylated SNA-I (10 μg/mL) in 1 mL phosphate saline buffer (PBS) containing Tween®-20 (0.05 %). Incubate in a humidified chamber at room temperature for 1 h with gentle shaking (*see* **Note 23**).
5. Discard sample by simply pouring it off the slide into the appropriate waste container.
6. Wash the slides one at a time by holding the edges and dipping them ten times in a mixture of 100 mL PBS and 0.01 % Tween®-20, then ten times in 100 mL PBS.
7. Clean the backside of the slide with a Kimwipe and place back in the humidification chamber.

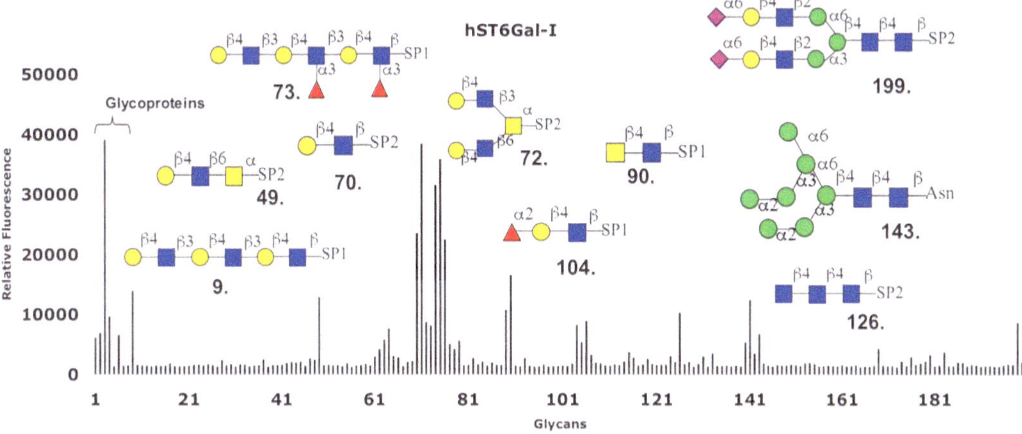

Fig. 3 Evaluation of hST6Gal-I acceptor specificities on the glycan microarray. Biotinylated sugar from sugar nucleotide (1 mM) was transferred to acceptor substrates on the array surface with hST6Gal-I (50 mU), and subsequently overlaid with fluorescently labeled Streptavidin–Alexa Fluor488. Image analysis of each slide using a confocal scanner generated a bar chart describing relative enzyme acceptor substrate affinities. The assay detected a new potential substrate, chitobiose (**126** and **143**), for hST6Gal-I. In addition to the preferential acceptor substrate, type-2 *N*-acetyllactosamine, the enzyme also sialylated LacDiNAc (**90**) and fucosyl-lacNAc (**104**). The numbers along the chart *x*-axis correspond to a particular glycan in the microarray library. *Symbols are described in Fig. 1.* (*Sp1* = $-(CH_2)_2-NH-$, *Sp2* = $-(CH_2)_3-NH-$)

8. Pipette onto the print surface 1 mL of PBS buffer containing streptavidin–Alexa488 conjugate (1:100 dilution, 1 mg/mL), BSA (3 %), and Tween®-20 (0.05 %). Incubate in the sealed humidification chamber while rotating gently for 30 min at room temperature.
9. Discard sample by simply pouring it off the slide into the appropriate waste container.
10. Wash the slides by dipping them ten times in a mixture of 100 mL PBS/0.01 % Tween®-20, ten times in 100 mL PBS, and then in tandem three times in two separate containers with 100 mL water.
11. Repeat steps **9–12** from Subheading 3.2 (*see* Fig. 4).

4 Notes

1. The list of glycans on the microarray has been published [24].
2. The synthesis of 6-azido-6-deoxy-ManNAc is outside the scope of this chapter and the preparation is described elsewhere [28].
3. The CFG glycan microarray consists of a diverse series of defined glycan structures covalently bound to an NHS-activated glass slide (Slide-H) from Schott Nexterion. The glycans were

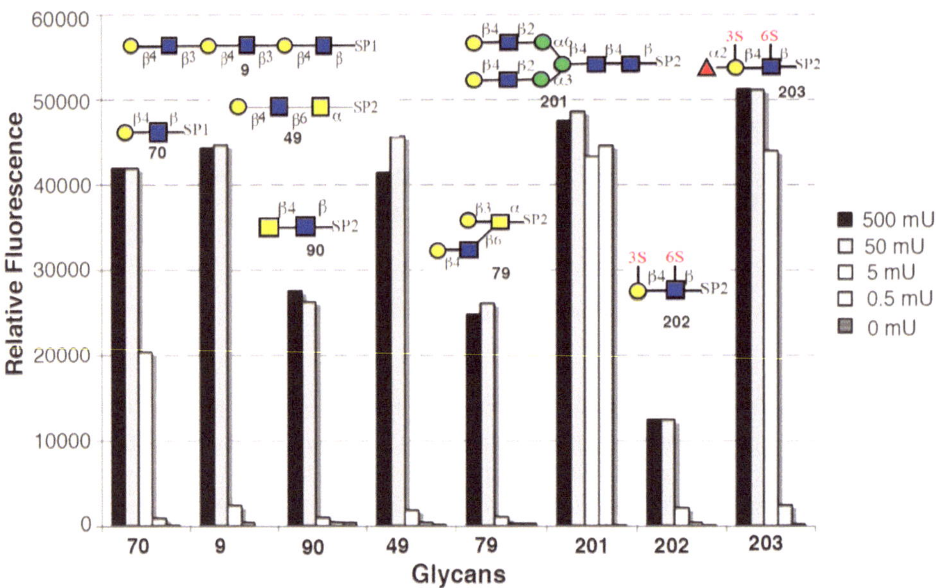

Fig. 4 Detection of unlabeled sialylated acceptors with SNA-I lectin, after incubations with hST6Gal-I (0–500 mU). This graph illustrates the hST6Gal-I activity by comparing the SNA-binding profiles, derived from the in situ incubation of glycan array with various enzyme concentrations and unlabeled CMP-Neu5Ac. The results of this indirect enzyme activity assessment, lectin-based, indicate that hST6Gal-I sialylates type-2 LacNAc very well (**70**, **9**). It also demonstrates a higher rate of sialic acid transferring to biantennary *N*-glycans with terminal lacNAc at lower enzyme concentration (**201**), confirming the higher relative activity for the branched lacNAc acceptor observed in direct detection. The enzyme can also sialylate terminal lacNAc on O-glycans (**49**, **79**), LacDiNAc (**90**), sulfated- and H-type 2-lactose (**202**, **203**) to enable the SNA interactions with these motifs. The results of this indirect detection were consistent with the direct enzyme specificity evaluation shown in Fig. 3. *Symbols are described in Fig. 1.* (*Sp1* = $-(CH_2)_2-NH-$, *Sp2* = $-(CH_2)_3-NH-$)

synthesized by chemoenzymatic methods with over thirty recombinant GTs, expressed in our laboratory at the Consortium for Functional Glycomics. The CFG is a program initially funded by the National Institute of General Medical Sciences (NIGMS). The CFG glycan array was first released with over 200 compounds in 2004 [20]. Since then, this glycan array library has increased in number, size, and complexity of glycans. By the end of the program in 2011, we successfully released the final version of the CFG glycan microarray with 611 structures. This technology is currently available to the public as an investigator-driven resource (for more information visit: www.functionalglycomics.org).

4. The recombinant human α2-6 sialyltransferase (hST6Gal-I) was over-expressed in baculovirus expressing insect cells and purified by affinity column chromatography on

glycosyltransferase-CDP-gel (CalBiochem, San Diego, CA) as reported on the CFG Web page (http://www.functionalglycomics.org).

5. Any sialic acid aldolases, CMP-sialic acid synthetases can be used in this approach.

6. Using the TLC plate cutter to prepare 2 × 5 cm TLC plates, or a size as desired. In pencil, carefully so as not to disturb the silica surface, draw a line 0.75 cm from the bottom along the short edge of the plate. Along this line make three short marks to indicate where to spot the samples. Separately, prepare a control of the starting material by dissolving a few mg of 6-azido-6-deoxy-ManNAc in water. Using a micropipettor spot on the TLC plate ~0.5 μL of the solution containing starting material on the left and center mark. Using a micropipettor spot on the TLC plate ~0.5 μL of the reaction mixture on the center and right marks. The center spot is the co-spot. Place the TLC plate in a desiccator under vacuum for ~10 min to remove the H_2O. Setup the TLC chamber by pouring ~10 mL of eluent into the glass jar. Place the dried TLC plate in the TLC chamber ensuring that the solvent level is not above the pencil line. Remove the plate from the chamber when the solvent front is 0.5 cm from the top of the plate. Mark the position of the solvent front in pencil. Allow the solvent to evaporate. To observe the spots either view the TLC plate under a UV-lamp or dip the plate in a 10 % H_2SO_4 ethanol staining solution, remove the excess solution using a paper towel, and then char the plate on a hot plate. Note: CTP and CMP-NeuAc intermediates will be observable under UV.

7. Refer to the filter device manual for the centrifugal force depending on the type of centrifuge that is used. Centrifuge for 30 min time intervals until the entire sample has passed through the membrane. This step will separate the enzyme from the reaction mixture.

8. Pack the gel filtration column with Bio-Gel P-2 gel to give about a 420 mL hydrated bed volume. Slowly add 145 g dry Bio-Gel P-2 extra fine to 900 mL deionized water in a 2 L beaker. The typical hydrated bed volume is 3 mL/g dry gel. Allow the gel to hydrate for 4 h at room temperature. After hydration, decant half of the supernatant. Transfer the remaining solution and gel to a filter flask and attach to a water aspirator. Degas the solution for 5–10 min with occasional swirling of the flask. Add two bed volumes of degassed buffer to the flask and swirl. Allow gel to settle until 90–95 % of the particles have settled. Decant the supernatant to remove the fine particles. This step can be repeated up to four times to remove >90 % of the fine particles. Fill the column with sufficient buffer to fill 20 % of the column. Pour the gel slurry into the column in a single, smooth movement. Be careful to avoid the introduction of bubbles. When a 5 cm bed has settled, open the

column outlet and allow the column to flow until packed. Pass two bed volumes of eluent through the column prior to using.

9. To monitor the fractions from the gel filtration prepare a 2 × 5 cm TLC plate. Using pencil make a grid mark along the plate with lines ~0.5 cm apart. Number each square with the fraction number and then spot 0.5 μL of each fraction into a separate square. Place the TLC plate on a hot plate to evaporate the H_2O and then observe the plate under UV and/or char with a 10 % H_2SO_4 ethanol staining solution making note which fractions contain UV-active material and/or char. Further check each of these "active" fractions by TLC. Prepare a second TLC plate (3 × 5 cm) as described above (*see* **Note 6**). Using pencil make several evenly spaced marks along the baseline (for 8–10 spots). Using a micropipettor spot 0.5 μL of each fraction on separate marks. Desiccate the plate and then develop using the eluent in the TLC chamber.
10. Caution. Shaking the separatory funnel will result in a buildup of pressure. With the funnel inverted the stopcock should be opened several times during shaking to release any built up pressure. The extraction should be performed in a fume hood while wearing protective eyewear, gloves and lab coat.
11. The hydrophobic barrier can preserve sample volume during incubation and also limit the area the liquid will cover on the array.
12. Glycan printed slides are available from CFG by request (see http://www.functionalglycomics.org).
13. Glycosylation of the maximum glycan surface density (saturation at >50 μM printing concentration) can be saturated with >0.1 mM of the CMP-9-biotin-Neu5Ac using hST6Gal-I for the N-acetyllactosamine (LacNAc) acceptor.
14. The enzyme activity can be determined by radiometric assay with ^{14}C-CMP-Neu5Ac against LacNAc at 37 °C.
15. The chamber can be simply constructed by placing a couple of wet paper towels in the bottom of a Pyrex glass dish, placing the slides on some sort of rack or lid inside the dish and sealing with plastic wrap.
16. If necessary, you should bleach it based on the biohazard policy.
17. This final wash step includes four separate dishes with ddH_2O, and the slides are dipped four times in each dish.
18. Avoid touching the print area.
19. The slide also can be dried under a gentle stream of ultra high purity nitrogen gas for 5 min.
20. Make sure the print surface is facing the laser source.
21. Each scan may lower the fluorescence of the signals.

22. Ensure that the enzyme amount matches with the slide label. Different concentrations of enzyme were used to confirm the indirect lectin-based enzyme activity assessment.
23. *Sambucus nigra lectin* (SNA) binds exclusively to α2-6 sialosides, preferentially on type-2 *N*-acetyllactosamine motifs. *Maackia amurensis* lectin (MAL) binds to α2-3 sialosides.

Acknowledgments

This work was supported by NIGMS grant GM62116 and the Consortium for Functional Glycomics.

References

1. Lairson LL, Henrissat B, Davies GJ, Withers SG (2008) Glycosyltransferases: structures, functions, and mechanisms. Annu Rev Biochem 77:521–555
2. Rademacher TW, Parekh RB, Dwek RA (1988) Glycobiology. Annu Rev Biochem 57:785–838
3. Brossmer R, Gross HJ (1994) Sialic acid analogs and application for preparation of neoglycoconjugates. Methods Enzymol 247:153–176
4. Beyer TA, Sadler JE, Rearick JI, Paulson JC, Hill RL (1981) Glycosyltransferases and their use in assessing oligosaccharide structure and structure-function relationships. Adv Enzymol Relat Areas Mol Biol 52:23–175
5. Pesnot T, Jorgensen R, Palcic MM, Wagner GK (2010) Structural and mechanistic basis for a new mode of glycosyltransferase inhibition. Nat Chem Biol 6:321–323
6. Sindhuwinata N, Munoz E, Munoz FJ, Palcic MM, Peters H, Peters T (2010) Binding of an acceptor substrate analog enhances the enzymatic activity of human blood group B galactosyltransferase. Glycobiology 20:718–723
7. Shih HW, Chen KT, Chen SK, Huang CY, Cheng TJ, Ma C, Wong CH, Cheng WC (2010) Combinatorial approach toward synthesis of small molecule libraries as bacterial transglycosylase inhibitors. Org Biomol Chem 8:2586–2593
8. Brik A, Wu CY, Wong CH (2006) Microtiter plate based chemistry and in situ screening: a useful approach for rapid inhibitor discovery. Org Biomol Chem 4:1446–1457
9. Johnson KF (1999) Synthesis of oligosaccharides by bacterial enzymes. Glycoconj J 16:141–146
10. Wymer N, Toone EJ (2000) Enzyme-catalyzed synthesis of carbohydrates. Curr Opin Chem Biol 4:110–119
11. Blixt O, Razi N (2004) Strategies for synthesis of an oligosaccharide library using a chemoenzymatic approach. In: Wang PG, Ichikawa Y (eds) Synthesis of carbohydrates through biotechnology, vol 873. American Chemical Society, Washington DC, pH 93–112
12. Bowles D, Isayenkova J, Lim EK, Poppenberger B (2005) Glycosyltransferases: managers of small molecules. Curr Opin Plant Biol 8: 254–263
13. Blixt O, Razi N (2006) Chemoenzymatic synthesis of glycan libraries. Methods Enzymol 415:137–153
14. Palcic MM, Sujino K (2001) Assays for glycosyltransferases. Trends Glycosci Glyc 13: 361–370
15. Wagner GK, Pesnot T (2010) Glycosyltransferases and their assays. Chembiochem 11:1939–1949
16. Zhou X, Zhou J (2006) Oligosaccharide microarrays fabricated on aminooxyacetyl functionalized glass surface for characterization of carbohydrate-protein interaction. Biosens Bioelectron 21:1451–1458
17. Park S, Lee MR, Pyo SJ, Shin I (2004) Carbohydrate chips for studying high-throughput carbohydrate-protein interactions. J Am Chem Soc 126:4812–4819
18. Houseman BT, Mrksich M (2002) Carbohydrate arrays for the evaluation of protein binding and enzymatic modification. Chem Biol 9:443–454
19. Seo JH, Kim CS, Hwang BH, Cha HJ (2010) A functional carbohydrate chip platform for

analysis of carbohydrate-protein interaction. Nanotechnology 21:215101

20. Blixt O, Head S, Mondala T, Scanlan C, Huflejt ME, Alvarez R, Bryan MC, Fazio F, Calarese D, Stevens J et al (2004) Printed covalent glycan array for ligand profiling of diverse glycan binding proteins. Proc Natl Acad Sci USA 101:17033–17038
21. Park S, Shin I (2002) Fabrication of carbohydrate chips for studying protein-carbohydrate interactions. Angew Chem Int Ed Engl 41: 3180–3182
22. de Paz JL, Horlacher T, Seeberger PH (2006) Oligosaccharide microarrays to map interactions of carbohydrates in biological systems. Methods Enzymol 415:269–292
23. Zhi ZL, Powell AK, Turnbull JE (2006) Fabrication of carbohydrate microarrays on gold surfaces: direct attachment of nonderivatized oligosaccharides to hydrazide-modified self-assembled monolayers. Anal Chem 78: 4786–4793
24. Blixt O, Allin K, Bohorov O, Liu X, Andersson-Sand H, Hoffmann J, Razi N (2008) Glycan microarrays for screening sialyltransferase specificities. Glycoconj J 25:59–68
25. Park S, Shin I (2007) Carbohydrate microarrays for assaying galactosyltransferase activity. Org Lett 9:1675–1678
26. Serna S, Etxebarria J, Ruiz N, Martin-Lomas M, Reichardt NC (2010) Construction of N-glycan microarrays by using modular synthesis and on-chip nanoscale enzymatic glycosylation. Chemistry 16:13163–13175
27. Serna S, Yan S, Martin-Lomas M, Wilson IB, Reichardt NC (2011) Fucosyltransferases as synthetic tools: glycan array based substrate selection and core fucosylation of synthetic N-glycans. J Am Chem Soc 133: 16495–16502
28. Kong DCM, von Itzstein M (1998) The chemoenzymatic synthesis of 9-substituted 3,9-dideoxy-D-glycero-D-galacto-2-nonulosonic acids. Carbohydr Res 305:323–329

Chapter 2

A Fluorescence-Based Assay for Core 1 β3Galactosyltransferase (T-Synthase) Activity

Tongzhong Ju and Richard D. Cummings

Abstract

Mucin-type O-glycans on glycoproteins in animal cells play important roles in many biological processes. Core 1 β3galactosyltransferase (Core 1 β3GalT, T-synthase) is a key enzyme in the O-glycan biosynthetic pathway. Emerging evidence has shown the importance of O-glycans and the absolute requirement of T-synthase in this pathway. The assessment of the T-synthase activity has historically been conducted using a radioactive method. Here we describe a fluorescence-based assay procedure for T-synthase activity. T-synthase utilizes the acceptor substrate 4-methylumbelliferone-α-GalNAc (GalNAcα-(4-MU)) and the donor substrate UDP-Gal to synthesize the disaccharide product Galβ1,3GalNAcα-(4-MU) structure. This product is specifically hydrolyzed by endo-α-*N*-acetylgalactosaminidase (O-glycosidase) releasing free 4-MU. Free 4-MU is highly fluorescent at pH 9.6–10 and can be easily measured by a fluorescent detector (Ex: 355 nm; Em: 460 nm). This fluorescence-based T-synthase assay is simple, sensitive, reproducible, not affected by enzyme source, and adaptable for high-throughput assays.

Key words O-glycans, Core 1 b3galactosyltransferase, T-synthase, Fluorescent assay, 4-MU

Abbreviations

4-MU	4-Methylumbelliferone
T-synthase	UDP-Gal:*N*-acetylgalactosaminyl-α1-Ser/Thr β3galactosyltransferase
T antigen	Galβ1-3GalNAcα1-Ser/Thr
Tn antigen	GalNAcα1-Ser/Thr
O-glycosidase	Endo-α-*N*-acetylgalactosaminidase
UDP-Gal	Uridine diphosphate galactose
GalNAc	*N*-Acetylgalactosamine (2-acetamido-2-deoxy-D-galactose)
Gal	D-Galactose
Ser/Thr	Serine/Threonine

Inka Brockhausen (ed.), *Glycosyltransferases: Methods and Protocols*, Methods in Molecular Biology, vol. 1022, DOI 10.1007/978-1-62703-465-4_2,

1 Introduction

Mucin type O-glycans on glycoproteins in animal cells play important roles in many biological processes, such as leukocyte recruiting [1–4], lymphocyte homing [5, 6], T cell differentiation [7], angiogenesis [8], and lymphatic vessel development [9]. These O-glycans contain modifications of the common structure GalNAcα-Ser/Thr (Tn antigen); they are synthesized by a set of glycosyltransferases through a series of reactions in the Golgi apparatus. Among them, the Core 1 structure, Galβ1,3GalNAcα1-, also known as the T antigen, is the common precursor of most complex O-glycans. This Core 1 structure is synthesized by the Core 1 β3galactosyltransferase (T-synthase), which transfers Gal from the donor substrate UDP-Gal to the acceptor substrate GalNAcα1-Ser/Thr (Tn antigen) [8, 10]. The T-synthase is a required enzyme in the biosynthesis of mucin-type O-glycans and deletion of the gene for T-synthase in mice eliminates T-synthase activity and results in embryonic lethality [8]. Thus, it is not surprising that deficiencies of T-synthase activity are directly associated with some human diseases, such as Tn-syndrome [11, 12], IgA nephropathy [13], and cancer [14–16]. Furthermore, vertebrate T-synthase has its unique feature in that it requires a specific molecular chaperone, termed Cosmc, to promote correct folding in vivo [17–19] and in vitro [20]. Interestingly, deficiencies of the T-synthase activity have been linked to acquired, somatic mutations in the X-linked *Cosmc* [21–24].

Like other glycosyltransferase activity assays, T-synthase activity is commonly assayed using a radioactive donor UDP-[6-^{3}H]Gal or UDP-[^{14}C]-Gal by measuring the incorporation of ^{3}H- or 14C-Gal into acceptor substrates, such as glycopeptides or other glycosides of only α-GalNAc, then followed by separation procedures to remove residual nucleotide sugars [25–28]. Such methods are time-consuming and impractical, and not useful in high-throughput screening. Therefore, we have developed a simple, sensitive, and high-throughput fluorescence-based method for assaying T-synthase activity in biological samples [29]. This assay is based on the principle that T-synthase utilizes GalNAcα-(4-methylumbelliferone) (GalNAcα-(4-MU)) as its acceptor substrate and UDP-Gal as a donor substrate to form the disaccharide Galβ1-3GalNAcα-(4-MU) (Fig. 1a). The reaction product, but not the acceptor substrate GalNAcα-(4-MU), is efficiently cleaved by endo-α-*N*-acetylgalactosaminidase (O-glycosidase) to release free 4-MU, which is highly fluorescent at high pH (Fig. 1a). Here we describe the detailed procedure of this novel assay. This method will allow for screening of inhibitors and activators of T-synthase, which will aid in studies of O-glycan structures and functions. Additionally, this method makes it possible to screen for a chemical chaperone that might promote T-synthase folding. Discovery of chemical

a The principle of the assay:

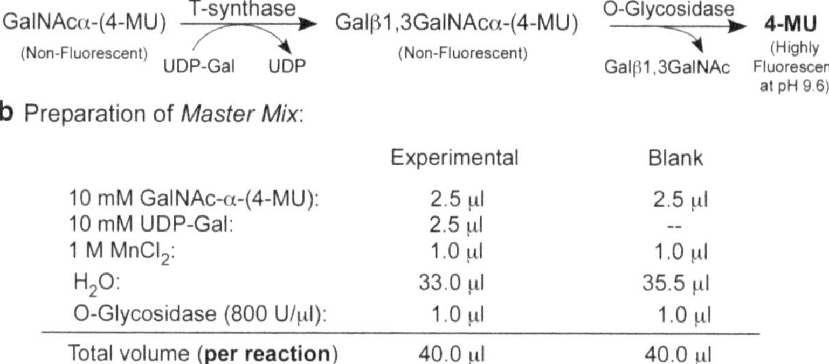

b Preparation of *Master Mix*:

	Experimental	Blank
10 mM GalNAc-α-(4-MU):	2.5 μl	2.5 μl
10 mM UDP-Gal:	2.5 μl	--
1 M $MnCl_2$:	1.0 μl	1.0 μl
H_2O:	33.0 μl	35.5 μl
O-Glycosidase (800 U/μl):	1.0 μl	1.0 μl
Total volume (**per reaction**)	40.0 μl	40.0 μl

c The assay procedure (in *triplicate*)

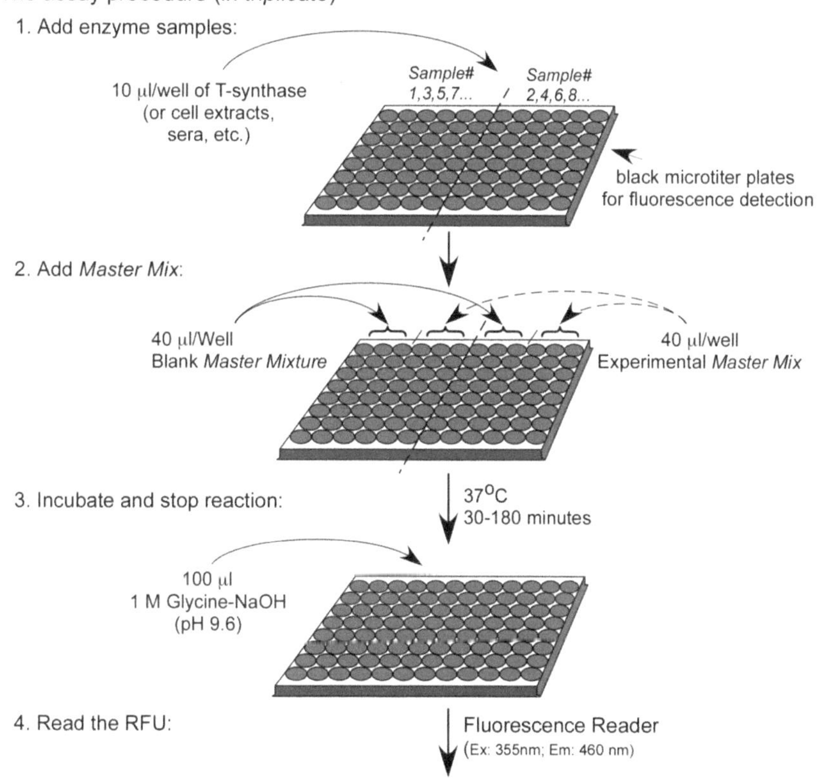

Fig. 1 Illustration of the fluorescent assay for T-synthase activity. (**a**) The principle of the assay: T-synthase utilizes GalNAcα-(4-MU) as its acceptor substrate and UDP-Gal as a donor to form Galβ1-3GalNAcα-(4-MU), which is subsequently quantitatively cleaved by endo-α-*N*-acetylgalactosaminidase (O-glycosidase) to release free 4-MU, which is highly fluorescent at high pH. The fluorescence intensity of 4-MU represents the amount of T-synthase product. (**b**) Preparation of the master mixes: the reaction mixtures including the acceptor GalNAcα-(4-MU), donor UDP-Gal or H_2O, divalent cation Mn^{2+}, detergent, buffer and O-glycosidase are prepared. (**c**) The procedure of the fluorescent T-synthase assay: after adding 10 μl of cell or tissue extracts or sera into each well for six wells, 40 μL of experimental and blank master mixtures is added into corresponding wells. The reaction is incubated at 37 °C for a certain period of time and then 100 μL of 1 M Glycine–NaOH pH 9.6 is added into each well to stop the reaction. The relative fluorescence units (RFU) of free 4-MU is then measured on a fluorescence reader. The activity of T-synthase is directly calculated in terms of pmol product formed over time and per protein concentration. Reproduced and adapted from [29] with permission from Oxford University Press

chaperones that mimic Cosmc function for T-synthase for novel therapeutics is especially important, since several significant human diseases are caused by somatic mutations in *Cosmc* or alterations in *Cosmc* expression that result in a loss of or compromised T-synthase activity [18, 21, 23, 24, 30–33].

2 Materials

Prepare all solutions using deionized water and analytical grade reagents. Prepare and store all reagents at −20 or 4 °C as indicated below. Diligently follow all waste disposal regulations when disposing waste materials. We recommend not adding sodium azide to any reagent.

2.1 Cell or Tissue Extraction Buffer

1. Buffer: 1 M Tris–HCl pH 7.4 (*see* **Note 1**), stored at room temperature.
2. Cell homogenizing buffer: 25 mM Tris–HCl (pH 7.4) containing 150 mM NaCl, and proteinase inhibitor cocktail.
3. Cell extraction buffer: Cell homogenizing buffer plus 0.5 % Triton X-100 (*see* **Note 2**).
4. BCA (bicinchoninic acid) protein assay (Pierce, Thermo Scientific, Rockford, IL, USA).

2.2 Components for T-Synthase Activity Assay

1. Buffer: 1 M MES–NaOH Buffer (pH 6.8), dissolve 21.3 g MES (2-[*N*-morpholino]ethanesulfonic acid) (Sigma-Aldrich, St. Louis, MO, USA) in 60 mL H_2O. Adjust the pH to 6.8 with 5 and 1 N NaOH sequentially and make up to 100 mL in total volume with water (*see* **Note 1**). Store this buffer at 4 °C.
2. Acceptor substrate: 10 mM GalNAcα-(4-MU) (*see* **Note 3**) in 1 M MES–NaOH buffer. Carefully weigh 5 mg of GalNAcα-(4-MU) (M_r 379.39) (Carbosynth Limited, Berkshire, UK) in a 1.5-mL microtube and add 100 μL of DMSO to completely dissolve the reagent. Then add 1,215 μL of 1 M MES–NaOH (pH 6.8) and mix well to generate a 10 mM GalNAcα-(4-MU) stock solution. Aliquot equal portions of this stock solution and store at −20 °C. The solution can also be stored at 4 °C for 2–3 months.
3. Donor Substrate: 10 mM UDP-Gal in water. Carefully weigh 5 mg of UDP-Gal disodium salt (M_r 610.3) (Sigma-Aldrich) in a 1.5-mL microtube and add 810 μL of water to completely dissolve the reagent to generate a 10 mM stock solution. Aliquot 100 μL of this stock solution into 0.5-mL tubes and store at −20 °C. We do not recommend storage at 4 °C for more than 2–3 days.
4. Cation: 1 M $MnCl_2$ in water, stored at 4 °C.

5. Endo-α-*N*-acetylgalactosaminidase (O-glycosidase): 800 units/μL in 25 mM MES–NaOH (pH 6.8). O-glycosidase (40,000,000 units/mL) (New England Biolabs, Ipswich, MA, USA) is stored at −20 °C. Add 1 μL of O-glycosidase to 49 μL of 25 mM MES–NaOH (pH 6.8), mix well to get 800 units/μL solution, and store at 4 °C and use within 1 week.
6. Stop Solution: 1 M Glycine–NaOH (pH 9.6) (*see* **Note 1**), stored at 4 °C.
7. Equipment:
   (a) Black microtiter plates: Costar non-treated 96-well black microtiter plates with flat bottom (Corning Inc., Pittsburgh, PA, USA) were used for the assay. Other brands of black plates suitable for fluorescence-based assays should also be applicable.
   (b) Multichannel pipet: 8-channel or 12-channel P-200.
   (c) A fluorescence plate reader (Victor Multiple-Label Counter) with filter of Ex: 355 nm and Em 460 nm (Perkin Elmer, Waltham, MA, USA). A compatible reader with the same filters will also work.

2.3 Standards

4-MU (*see* **Note 3**): 10, 50, 100, 200, 500, 1,000, 2,000, 5,000, 10,000, and 20,000 nM. 4-MU (M_r 176.17) (Sigma-Aldrich) is dissolved in DMSO at 1.0 mM; generate the solutions at desired concentrations above by dilution with 50 mM MES–NaOH (pH 6.8) (*see* **Note 4**).

3 Methods

Carry out all assay procedures at room temperature unless otherwise specified. However, keep all enzyme sources at 4 °C until aliquotted into plates for the assays.

3.1 Preparation of Cell Extracts and Measurement of Protein Concentration

If the samples are cells or animal tissues, the cell extracts or tissue extracts should be prepared.

1. Add an appropriate amount of homogenizing buffer (1:8, v:v) (*see* **Note 5**) to suspend the cells or tissue samples.
2. Sonicate the cell suspension on ice using the micro-tip for 3 s five times at 15–20 s intervals.
3. Centrifuge sonicated material at low speed (1,000 × *g*) for 5 min at 4 °C to obtain a supernatant, which is designated the post-nuclear supernatant (PNS).
4. Transfer the PNS to a new tube, and add Triton X-100 to 0.5 % (final concentration) (*see* **Note 2**), vortex well, and solubilize for 20 min on ice.

5. Centrifuge the solubilized material at high speed (5,000 × *g*) for 5 min at 4 °C and carefully transfer the supernatant to a new tube, this is termed the cell extract.
6. Make a 1:10 dilution of the cell or tissue extracts with cell extraction buffer, and measure the protein concentration using the micro BCA (bicinchoninic acid) protein assay in duplicate following the manufacturer's instructions with bovine serum albumin as a standard. Set the blanks using cell extraction buffer of the enzyme assay (*see* **Note 6**). The optimum range of protein concentration for T-synthase activity assay is 3–5 mg/mL, but typically dilutions of this are used to validate linearity of the reaction with protein concentration.
7. Keep all cell extracts on ice.

3.2 Procedures for T-Synthase Activity Assay

3.2.1 Preparation of the Master Mix

The reaction is carried out in a total volume of 50 μL containing 500 μM GalNAcα-(4-MU), 500 μM UDP-Gal, 20 mM $MnCl_2$, 0.1 % Triton X-100 (supplied by the cell extracts containing Triton X-100), 800 units of O-glycosidase, in 50 mM MES–NaOH buffer (pH 6.8), and an appropriate amount of enzyme. A 96-well black microtiter plate suitable for fluorescence assays is used. The blank reaction is set up by replacing the donor UDP-Gal with H_2O in the reaction (*see* **Note 7**).

Prepare the Experimental and Blank Master Mix, which include everything except for the enzyme sample (Fig. 1b) (*see* **Note 8**). The volume of the Master Mix is 40 μL per reaction (*see* **Note 9**), and the volume of the enzyme source is 10 μL per reaction to give a final volume of 50 μL per reaction. The total amount of the Master Mix needed depends on how many samples are being assayed. Before preparing the Master Mix, these issues are taken into consideration: (a) assay T-synthase activity in triplicate, (b) the number of enzyme samples being assayed, (c) adding one extra to the sample number (*see* **Note 10**).

The following procedure is for assaying ten enzyme samples in triplicate.

1. Label two 1.7-mL microtubes as T-syn-(+)-Tube, the experimental sample containing T-synthase activity to be measured and T-syn-(−)-Tube, the blank devoid of T-synthase activity, respectively. These are the two Master Mix tubes.
2. Prepare the Master Mix for 11 enzyme samples (10 enzyme samples + 1 extra) in triplicate, therefore a total of 33 reactions (3 × 11):
   (a) Add 1,089.0 μL water (33.0 μL × 33, other components are calculated in the same way) to the T-syn-(+)-Tube and 1,171.5 μL water to the T-syn-(−)-Tube.
   (b) Add 82.5 μL of 10 mM GalNAcα-(4-MU) to both the T-syn-(+)-Tube and T-syn-(−)-Tube, respectively; mix well.
   (c) Add 82.5 μl of 10 mM UDP-Gal to T-syn-(+)-Tube only; mix well.

(d) Add 33 μl of 1 M $MnCl_2$ to both the T-syn-(+)-Tube and T-syn-(−)-Tube, respectively; mix well.

(e) Add 33 μl of O-glycosidase (800 units/μl) to both the T-syn-(+)-Tube and T-syn-(−)-Tube, respectively (*see* **Note 11**); mix well.

(f) Briefly spin the two Master Mix tubes designated T-syn-(+) and T-syn-(−) in a microfuge and place the mixtures on ice.

3.2.2 Setup Reactions and Measure the Fluorescence Intensity

1. Add 10 μL of each cell or tissue extracts to six wells (three for experimental and another three for the blank) (Fig. 1c) (*see* **Note 9**); finish adding all of the samples into the wells.
2. Add 40 μL of T-syn-(−)-Mix to No. 1–3, and 7–9 wells of all samples first and then T-syn-(+)-Mix to No. 4–6, and 10–12 wells of all samples.
3. Add 40 μL of T-syn-(−)-Mix and T-syn-(+)-Mix to two more wells, respectively, and then add 10 μL of cell extraction buffer for the background of the assay (*see* **Note 12**).
4. Seal the plate with a plastic seal and gently shake the plate by hand for a few seconds to mix contents.
5. Place the plate in a 37 °C incubator for 1 or 2 h (*see* **Note 13**).
6. Add 100 μL of stop solution (1 M Glycine–NaOH pH 9.6) (*see* **Note 14**) to each well using an 8- or 12-channel pipet, gently mix well and leave the plate at room temperature for 5 min.
7. Read the relative fluorescence units (RFUs) on a Victor Multiple-Label Counter at a rate of 0.1 s using umbelliferone mode, e.g., Ex: 355 nm and Em 460 nm. If the plate cannot be read immediately, it can be stored at room temperature for a few hour then read since the fluorescence of 4-MU is very stable for at least 12 h (*see* **Note 3**).

3.3 Standard Curve and Determination of the Specificity of 4-MU (See Note 15)

1. Dilute the 1.0 mM solution of 4-MU to the concentrations of 10, 20, 50, 100, 200, 500, 1,000, 2,000, 5,000, and 10,000, and 20,000 nM with 50 mM MES–NaOH (pH 6.8).
2. Add 50 μL of each concentration of 0, 10, 20, 50, 100, 200, 500, 1,000, 2,000, 5,000, 10,000 and 20,000 nM 4-MU to three wells (in triplicate).
3. Seal the plate with a plastic seal.
4. Place the plate in a 37 °C incubator and incubate for 1 h (*see* **Note 16**).
5. Add 100 μL of stop solution (1 M Glycine–NaOH pH 9.6) to each well using a multiple-channel pipet, gently mix well and leave the plate at room temperature for 5 min.
6. Read the relative fluorescence units (RFUs) on a Victor Multiple-Label Counter using umbelliferone mode, e.g., Ex: 355 nm and Em 460 nm.

7. Average the RFU of each concentration of 4-MU.
8. Subtract the background RFU (the 0 nM 4-MU) from RFU of each concentration of 4-MU.
9. Make a curve of RFU on the *Y*-axis versus concentration of 4-MU on the *X*-axis.
10. Calculate the intensity of 4-MU in RFU/pmol, which represents the specific RFU of 4-MU under the experimental conditions. In the conditions of our laboratory, the specific RFU of 4-MU is about 600 RFU/pmol.

3.4 Calculating T-Synthase Activity

1. Average the RFU of blank reactions and experimental reactions from each sample; obtain the net RFU by subtraction of average RFU in blank reaction from the experimental (*see* **Notes 17** and **18**).
2. Calculate the T-synthase Activity (pmol/h-mL):

$$\frac{\text{Net RFU}}{\text{Specific activity}} \times \frac{1\,\text{mL}}{\text{EnzymeVol.(mL)}} \times \frac{1\,\text{h}}{\text{Reaction time (h)}} = \text{pmol / h-mL}$$

3. Calculate the specific activity of T-synthase (pmol/h-mg):

$$\frac{\text{Activity (pmol / h-mL)}}{\text{Protein concentration (mg / mL)}} = \text{(pmol / h-mg)}$$

4 Notes

In comparison to other assay methods for T-synthase activity, this fluorescence-based method has many advantages: (1) the procedure is much simpler and there is no need to separate the product from others reagents; (2) the reagents are much less expensive; 4-MU is a relatively cheap chemical compound, and O-glycosidase is a recombinant bacterial enzyme; by contrast, for the radioactive method, UDP-[^{3}H]-Gal is expensive, and its use requires a C18 column, scintillation vials, scintillation cocktail, and disposal of radioactive waste; (3) the assay generates less waste, and thus creates less harm to the environment, especially by lack of use of long-lived radioisotopes; (4) the assay is sensitive, and is comparable to the radioisotope-based method and shows higher sensitivity than other methods, e.g., HPLC separations of products and reactants; (5) the assay is suitable for a high-throughput format, since it is performed in microtiter plates; and (6) the assay is easy to perform in common biochemistry and clinical laboratories.

1. High concentrations of HCl (12 N) can be used to lower the high starting pH to near the required pH, and thereafter use

1 N HCl to adjust the pH to the desired level. This principle is generally true for making other buffers, including the MES–NaOH and Glycine–NaOH buffers in all experiments.

2. The detergent Triton X-100 is viscous and it is difficult to pipet and accurately measure its volume. It is recommended to prepare 10 % stock Triton solution (w/v) in water first, then to make the working solution. To prepare the 10 % stock Triton solution, weigh 10.0 g of pure Triton X-100 (density 1.07 g/mL) into a 50-mL glass baker and then add 25 mL of water. Stir the solution with a small magnetic stir bar for 30 min, and adjust the total volume to 100 mL with water in a 100-mL glass cylinder. This 10 % Triton X-100 is stored in a 100-mL glass bottle at room temperature or 4 °C.
3. 4-MU is colorless at neutral pH and exhibits a blue fluorescence at pH 7.5. The fluorescence intensity of free 4-MU is pH dependent and increases to a maximum at pH 10. Furthermore, fluorescence of 4-MU is very stable for at least 12 h at pH 10 [34]. Interestingly, its fluorimetric property relies on its 7-hydroxyl group; any modification of this position eliminates fluorescence, including derivatives having sugars linked via a glycosidic bond such as GalNAcα-(4-MU).
4. To accurately assay enzyme activity, the RFU of 4-MU must be measured in the sample. The standard curve of 4-MU is made at the conditions close to the experimental since the assaying buffer for T-synthase is 50 mM MES pH 6.8.
5. Making cell or tissue extracts: the cells or tissues are suspended in an appropriate volume of homogenization buffer. Usually based on the volume or size of the tissues, eight times of homogenizing buffer (1:8 in v:v) is needed. For example, if the volume of cell pellet is 20 μL, then 160 μL of homogenizing buffer is needed.
6. Incubation of BCA with cell extraction buffer alone can generate color or background. Therefore, a blank reaction for BCA protein assay is necessary to obtain an accurate protein concentration.
7. The reaction Buffer: The acceptor substrate GalNAcα-(4-MU) stock is in ~1.0 M MES–NaOH pH 6.8; thus, there is no need to add an extra buffer to the reaction system.
8. Using Master Mixes can greatly minimize pipetting error.
9. If the protein concentration is low or the activity of T-synthase is low, the amount of enzyme samples such as extracts can be increased to 20 μL instead of 10 μL; then the volume of the Master Mix for each reaction can be made at 30 μL per reaction by reducing the water by 10 μL. Alternatively, the reaction time can be extended (also *see* **Note 13**).
10. Make an extra amount of the Master Mixes, since there is always a variation in pipetting, and a little extra ensures having enough to complete all of the required assays.

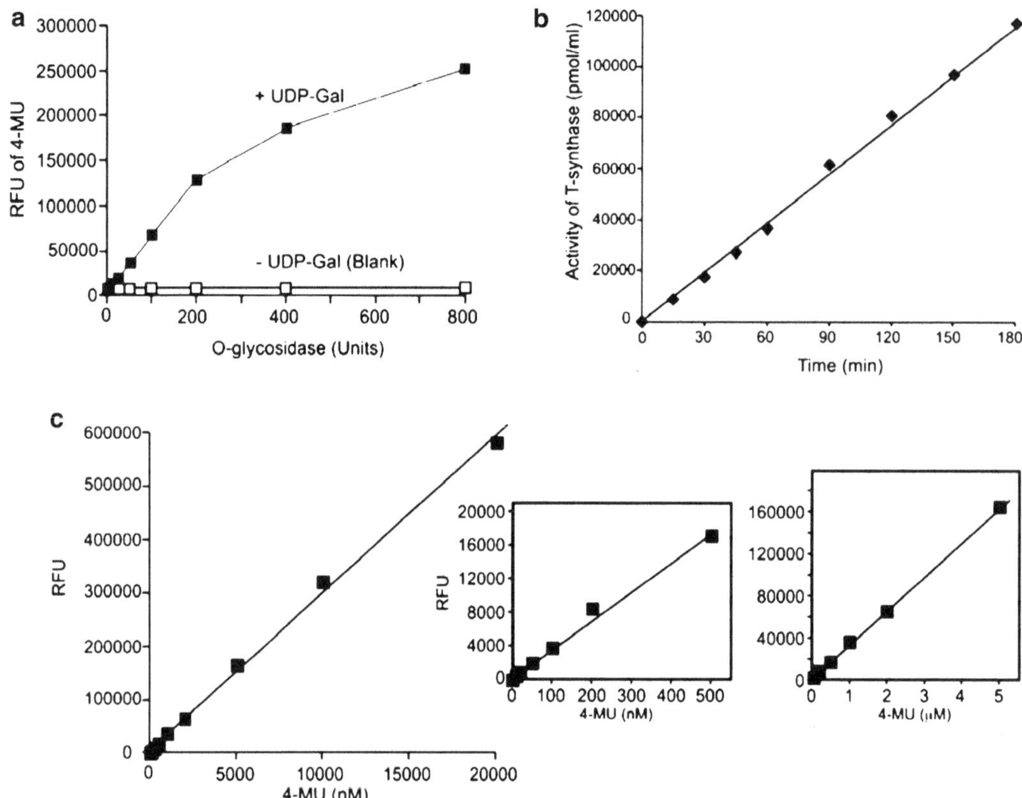

Fig. 2 Characterization of the method: (**a**) Concentration-dependency of O-glycosidase: the T-synthase reactions are set up in the presence of different amounts (0–800 units) of O-glycosidase. The reactions are incubated at 37 °C for 60 min, and the stop solution is added and the fluorescence units are measured. RFUs of 4-MU are plotted versus the O-glycosidase concentration. (**b**) Time course: the reaction with Hi-5 cell extracts containing human recombinant T-synthase is was set up in the same tube and 50 μL reaction mixtures are aliquotted into the black microtiter plate. Reactions are incubated at 37 °C from 0 to 180 min until addition of stop solution as indicated. The fluorescence of 4-MU is measured at the same time and plotted with the incubation time (min). All of the experiments are typically performed in triplicate. (**c**) Standard curve of 4-MU: 50 μL of serial dilutions of standard 4-MU solutions ranging from 0 to 20,000 nM in triplicate are pipetted into plates and incubated at 37 °C for 60 min, and then 100 μL of stop solution are added. The RFUs are measured and plotted against the concentration of 4-MU from 0 to 20,000 nm. The two *insets* show the standard curves representing RFU of 4-MU from concentrations of 0–500 nM, and 0–5,000 nM, respectively. Reproduced from ref. [29] with permission from Oxford University Press

11. O-glycosidase: a sufficient amount of active O-glycosidase is important for this assay. We compared the enzyme from two vendors, New England Biolabs (NEB) and ProZyme. The O-glycosidase from NEB is better and more reliable for this assay. In addition, 800 units of O-glycosidase in each measuring well is sufficient for most assays (*see* Fig. 2a). In other words, the enzyme product is quantitatively cleaved by 800 units of O-glycosidase [29].

12. The differences between the Blank and Background are that "Blank reaction" contains every component of the assay including enzyme except for the donor UDP-Gal, while "Background reaction" has no enzyme. This Background reaction is useful when you want to know the quality of the reagents. For example, if the acceptor GalNAcα-(4-MU) is contaminated by free 4-MU, or O-glycosidase is contaminated by exo-α-*N-acetylgalactosaminidase*, or the sample itself may contain some exo-α-*N*-acetylgalactosaminidase activity, the reading of RFU from the Background reaction will be higher than expected. From our experience, O-glycosidase from NEB is a high-quality enzyme without detectable exo-α-*N*-acetylgalactosaminidase activity [29]. As for the blank reaction, the endogenous α-N-acetylgalactosaminidase probably arises from lysosomal contamination; its activity is low but variable among cell types [29]. In comparison to the background reaction, the RFUs in blank reactions are usually higher than in the Background reactions [29].
13. Incubation time: the reaction time of the T-synthase activity assay depends on the activity of samples, usually from 30 min to 2 h. If the activity is high, the incubation time can be shortened (30–60 min); if activity is low, the reaction time can be extended (2 or 3 h). The reaction rate of T-synthase activity in this method is typically linear up to 3 h (*see* Fig. 2b).
14. The pH of the stop solution: the fluorescence of 4-MU is pH-sensitive. The optimum pH for the fluorescence of 4-MU is 10, based on the literature. From our experience, the highest fluorescence of 4-MU is around pH 9.6 [29].
15. The specific fluorescence intensity or RFU of 4-MU: it is important to calibrate the fluorescence reader, since the sensitivity of 4-MU from different fluorescence readers may vary. Therefore, it is necessary to make a "Calibration Curve" of 4-MU to calculate the specific fluorescence intensity of 4-MU for each reader (*see* Fig. 2c). Once the specific RFU of 4-MU from the fluorescence reader is known and found to be reliable over time, it is not necessary to measure it at each time of assay.
16. This incubation does not involve any reaction, but in such controlled studies it is important to conduct this incubation as for all other samples.
17. Repeat each assay with diluted samples: if the activity of T-synthase is very high, for example, the RFU is over 1,000,000 within 1 h reaction time, the sample needs to be diluted 1:10 and repeat the assay. This will provide more accurate information about the activity of T-synthase.
18. From the triplicate determinations, a standard deviation in activity of each sample can be obtained. First, average the RFU

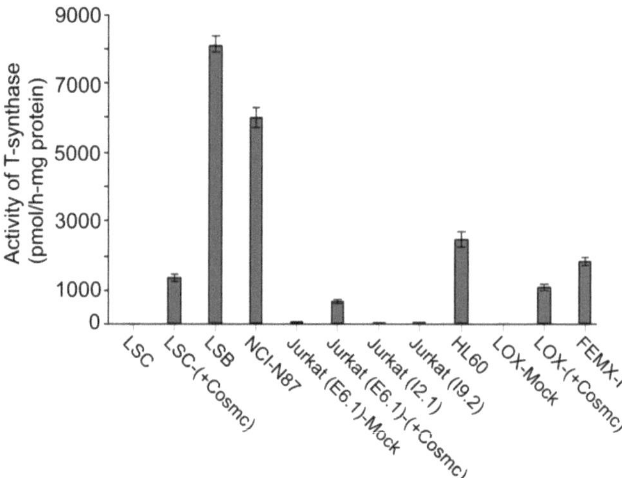

Fig. 3 Application of the fluorescent method for assaying T-synthase activity in the cell extracts from different cell lines: The cell extracts are made from the cell lines indicated and assayed for T-synthase activity in triplicate following the procedure described in Fig. 1b, c. The specific activity (Mean ± SD) of T-synthase from the cell lines are calculated and compared in the bar graph. Human colorectal carcinoma cell lines, LSB and NCI-87 have relatively high T-synthase activity; human leukemia cells HL60 and melanoma cells FEMX-I contain moderate T-synthase activity; while mock-transfected human T-lymphoid Jurkat (clone E6.1), human colorectal carcinoma LSC, and human melanoma LOX cells have little to no T-synthase activity due to mutations in *Cosmc* resulting in an inactive T-synthase; Cosmc-transfectants LSC-(+Cosmc), Jurkat-(+Cosmc) and LOX-(+Cosmc) cells regain moderate T-synthase activity; Human Jurkat cells (clones I2.1 and I9.2) have little T-synthase activity due to having the same mutated *Cosmc* as clone E6.1. Reproduced and adapted from ref. [29] with permission from Oxford University Press

of blank reactions and subtract this averaged RFU of blank from RFU of each of three experimental reactions from each sample to get three net RFUs; then calculate using these net RFUs to get three individual T-synthase activity (pmol/h-mL) or specific activity (pmol/h-mg) from each sample. The mean and standard derivation can be calculated and the activity can be presented as Mean ± SD. The activity of T-synthase from different samples can be compared (*see* Fig. 3).

Acknowledgments

We thank Drs. Jamie Heimburg-Molinaro and Rajindra P. Aryal for helpful suggestions on this manuscript. This work was supported by NIH Grant R01DK80876 (to T.J.).

References

1. Leppanen A, Mehta P, Ouyang YB, Ju T, Helin J, Moore KL, van Die I, Canfield WM, McEver RP, Cummings RD (1999) A novel glycosulfopeptide binds to P-selectin and inhibits leukocyte adhesion to P-selectin. J Biol Chem 274(35):24838–24848
2. Sperandio M (2006) Selectins and glycosyltransferases in leukocyte rolling in vivo. FEBS J 273(19):4377–4389
3. McEver RP, Cummings RD (1997) Role of PSGL-1 binding to selectins in leukocyte recruitment. J Clin Invest 100(11 Suppl): S97–S103
4. McEver RP, Cummings RD (1997) Perspectives series: cell adhesion in vascular biology. Role of PSGL-1 binding to selectins in leukocyte recruitment. J Clin Invest 100(3):485–491
5. Leppanen A, Yago T, Otto VI, McEver RP, Cummings RD (2003) Model glycosulfopeptides from P-selectin glycoprotein ligand-1 require tyrosine sulfation and a core 2-branched O-glycan to bind to L-selectin. J Biol Chem 278(29):26391–26400
6. Yeh JC, Hiraoka N, Petryniak B, Nakayama J, Ellies LG, Rabuka D, Hindsgaul O, Marth JD, Lowe JB, Fukuda M (2001) Novel sulfated lymphocyte homing receptors and their control by a Core1 extension beta 1,3-N-acetylglucosaminyltransferase. Cell 105(7):957–969
7. Fukuda M, Carlsson SR (1986) Leukosialin, a major sialoglycoprotein on human leukocytes as differentiation antigens. Med Biol 64(6):335–343
8. Xia L, Ju T, Westmuckett A, An G, Ivanciu L, McDaniel JM, Lupu F, Cummings RD, McEver RP (2004) Defective angiogenesis and fatal embryonic hemorrhage in mice lacking core 1-derived O-glycans. J Cell Biol 164(3): 451–459
9. Fu J, Gerhardt H, McDaniel JM, Xia B, Liu X, Ivanciu L, Ny A, Hermans K, Silasi-Mansat R, McGee S, Nye E, Ju T, Ramirez MI, Carmeliet P, Cummings RD, Lupu F, Xia L (2008) Endothelial cell O-glycan deficiency causes blood/lymphatic misconnections and consequent fatty liver disease in mice. J Clin Invest 118(11):3725–3737
10. Ju T, Brewer K, D'Souza A, Cummings RD, Canfield WM (2002) Cloning and expression of human core 1 beta1,3-galactosyltransferase. J Biol Chem 277(1):178–186
11. Berger EG (1999) Tn-syndrome. Biochim Biophys Acta 1455(2–3):255–268
12. Cartron JP, Nurden AT (1979) Galactosyltransferase and membrane glycoprotein abnormality in human platelets from Tn-syndrome donors. Nature 282(5739): 621–623
13. Allen AC, Topham PS, Harper SJ, Feehally J (1997) Leucocyte beta 1,3 galactosyltransferase activity in IgA nephropathy. Nephrol Dial Transplant 12(4):701–706
14. Blanchard B, Nurisso A, Hollville E, Tetaud C, Wiels J, Pokorna M, Wimmerova M, Varrot A, Imberty A (2008) Structural basis of the preferential binding for globo-series glycosphingolipids displayed by Pseudomonas aeruginosa lectin I. J Mol Biol 383(4):837–853
15. Springer GF (1984) T and Tn, general carcinoma autoantigens. Science 224(4654): 1198–1206
16. Springer GF, Taylor CR, Howard DR, Tegtmeyer H, Desai PR, Murthy SM, Felder B, Scanlon EF (1985) Tn, a carcinoma-associated antigen, reacts with anti-Tn of normal human sera. Cancer 55(3):561–569
17. Ju T, Aryal RP, Stowell CJ, Cummings RD (2008) Regulation of protein O-glycosylation by the endoplasmic reticulum-localized molecular chaperone Cosmc. J Cell Biol 182(3): 531–542
18. Ju T, Cummings RD (2002) A unique molecular chaperone Cosmc required for activity of the mammalian core 1 beta 3-galactosyltransferase. Proc Natl Acad Sci U S A 99(26):16613–16618
19. Wang Y, Ju T, Ding X, Xia B, Wang W, Xia L, He M, Cummings RD (2010) Cosmc is an essential chaperone for correct protein O-glycosylation. Proc Natl Acad Sci USA 107(20):9228–9233
20. Aryal RP, Ju T, Cummings RD (2010) The endoplasmic reticulum chaperone Cosmc directly promotes in vitro folding of T-synthase. J Biol Chem 285(4):2456–2462
21. Ju T, Cummings RD (2005) Protein glycosylation: chaperone mutation in Tn syndrome. Nature 437(7063):1252
22. Ju T, Lanneau GS, Gautam T, Wang Y, Xia B, Stowell SR, Willard MT, Wang W, Xia JY, Zuna RE, Laszik Z, Benbrook DM, Hanigan MH, Cummings RD (2008) Human tumor antigens Tn and sialyl Tn arise from mutations in Cosmc. Cancer Res 68(6):1636–1646
23. Crew VK, Singleton BK, Green C, Parsons SF, Daniels G, Anstee DJ (2008) New mutations in C1GALT1C1 in individuals with Tn positive phenotype. Br J Haematol 142(4): 657–667
24. Schietinger A, Philip M, Yoshida BA, Azadi P, Liu H, Meredith SC, Schreiber H (2006) A mutant chaperone converts a wild-type protein

into a tumor-specific antigen. Science 314 (5797):304–308
25. Furukawa K, Roth S (1985) Co-purification of galactosyltransferases from chick-embryo liver. Biochem J 227(2):573–582
26. Granovsky M, Bielfeldt T, Peters S, Paulsen H, Meldal M, Brockhausen J, Brockhausen I (1994) UDPgalactose:glycoprotein-N-acetyl-D-galactosamine 3-beta-D-galactosyltransferase activity synthesizing O-glycan core 1 is controlled by the amino acid sequence and glycosylation of glycopeptide substrates. Eur J Biochem 221(3):1039–1046
27. Ju T, Cummings RD, Canfield WM (2002) Purification, characterization, and subunit structure of rat core 1 Beta1,3-galactosyltransferase. J Biol Chem 277(1):169–177
28. Mendicino J, Sivakami S, Davila M, Chandrasekaran EV (1982) Purification and properties of UDP-gal:N-acetylgalactosaminide mucin: beta 1,3-galactosyltransferase from swine trachea mucosa. J Biol Chem 257(7):3987–3994
29. Ju T, Xia B, Aryal RP, Wang W, Wang Y, Ding X, Mi R, He M, Cummings RD (2011) A novel fluorescent assay for T-synthase activity. Glycobiology 21(3):352–362
30. Inoue T, Sugiyama H, Hiki Y, Takiue K, Morinaga H, Kitagawa M, Maeshima Y, Fukushima K, Nishizaki K, Akagi H, Narimatsu Y, Narimatsu H, Makino H (2010) Differential expression of glycogenes in tonsillar B lymphocytes in association with proteinuria and renal dysfunction in IgA nephropathy. Clin Immunol 136(3):447–455
31. Yamada K, Kobayashi N, Ikeda T, Suzuki Y, Tsuge T, Horikoshi S, Emancipator SN, Tomino Y (2010) Down-regulation of core 1 beta1,3-galactosyltransferase and Cosmc by Th2 cytokine alters O-glycosylation of IgA1. Nephrol Dial Transplant 25(12): 3890–3897
32. Qin W, Zhong X, Fan JM, Zhang YJ, Liu XR, Ma XY (2008) External suppression causes the low expression of the Cosmc gene in IgA nephropathy. Nephrol Dial Transplant 23(5):1608–1614
33. Qin W, Zhou Q, Yang LC, Li Z, Su BH, Luo H, Fan JM (2005) Peripheral B lymphocyte beta1,3-galactosyltransferase and chaperone expression in immunoglobulin A nephropathy. J Intern Med 258(5):467–477
34. Mead JA, Smith JN, Williams RT (1955) Studies in detoxication. 67. The biosynthesis of the glucuronides of umbelliferone and 4-methylumbelliferone and their use in fluorimetric determination of beta-glucuronidase. Biochem J 61(4):569–574

Chapter 3

Structural and Biochemical Analysis of a Bacterial Glycosyltransferase

Fan Zhu, Ren Wu, Hua Zhang, and Hui Wu

Abstract

Glycosyltransferases (GTs) are a large family of enzymes that specifically transfer sugar moieties to a diverse range of substrates. The process of bacterial glycosylation (such as biosynthesis of glycolipids, glycoproteins, and polysaccharides) has been studied extensively, yet the majority of GTs involved remains poorly characterized. Besides predicting enzymatic parameters of GTs, the resolution of three-dimensional structures of GTs can help to determine activity, donor sugar binding, and acceptor substrate binding sites. It also facilitates amino acid sequence-based structural modeling and biochemical characterization of their homologues. Here we describe a general procedure to accomplish expression and purification of soluble and active recombinant GTs. Enzymatic characterization, and crystallization of GTs, and data refinement for structural analysis are also covered in this protocol.

Key words Glycosyltransferases, Protein purification, Glycosyltransferase assays, Crystallization and data refinement

1 Introduction

Glycosyltransferases (GTs) are a large family of enzymes that catalyze the transfer of activated sugars to a variety of acceptor molecules; they are important in all domains of life for the biosynthesis of complex carbohydrates and glycoconjugates. Such glycosylation reactions in bacteria are crucial for many fundamental biological processes, including adhesion, signaling, and cell wall biosynthesis [1]. Most characterized bacterial glycoproteins are virulence factors of medically important pathogens [2]. As key players of the glycosylation process, GTs are potential molecular targets in chemical biology and drug discovery. Thus, study of GTs will provide useful information to identify new targets for potential therapeutic and prophylactic measures.

Primary amino acid sequences have been used to predict and classify GTs [3]. Some operationally simple bioassays have been

Inka Brockhausen (ed.), *Glycosyltransferases: Methods and Protocols*, Methods in Molecular Biology, vol. 1022, DOI 10.1007/978-1-62703-465-4_3, © Springer Science+Business Media New York 2013

developed to determine enzymatic activity in the past few years [4]; however, many putative GTs remain uncharacterized since the number of predicated GTs is enormous. Few structural studies have been reported for this large family of enzymes. Only limited numbers of 3D structures from different GT families have been documented. Among these GTs, some catalyze the synthesis of secondary metabolites [5] and others mediate biogenesis of bacterial cell walls or polysaccharides [6, 7]. Several studies reported structures of bacterial GTs that are involved in protein glycosylation. Other than predicting the enzymatic parameters, the resolved high-resolution 3D structure can be used to determine activity of GTs, donor sugar binding, and substrate binding sites and facilitate amino acid sequence-based structure and functional analysis.

Serine-rich repeat glycoproteins (SRRPs) are a growing family of bacterial adhesins found in many streptococci, staphylococci, and other gram-positive bacteria [2]. They have been shown to play important roles in bacterial biofilm formation and pathogenesis [8]. Glycosylation of this family of adhesins is essential for their biogenesis. A number of glycosyltransferases has been implicated in glycosylation of Fap1, a SRRP from an oral streptococcus, *Streptococcus parasanguinis*. Glycosylation of Fap1 is initiated by transferring GlcNAc residues to the Fap1 polypeptide by a two enzyme complex Gtf1 and Gtf2 [9, 10]. A glucosyltransferase (Gtf3) catalyzes the second step of glycosylation of Fap1. Here we use Gtf3 [8, 10] as an example to describe a general procedure to express and purify large quantities of active recombinant GTs for protein crystallization, and structural data refinement.

2 Materials

Prepare all solutions using ultrapure water (prepared by purifying deionized water to attain a sensitivity of 18.6 MΩ cm at 25 °C) and analytical grade reagents. Prepare and store all reagents at 4 °C (unless indicated otherwise). Diligently follow all waste disposal regulations when disposing waste materials. All the procedures for protein purification need to be carried out at 4 °C unless indicated.

2.1 Recombinant Protein Expression Components

1. Plasmid vectors: pET28a-SUMO [8].
2. KOD DNA polymerase kit (EMD Chemicals, Westbury, NY, USA).
3. Restriction enzymes (Promega, Madison, WI, USA).
4. T4 DNA ligase kit (Promega).
5. *E. coli* BL21-Gold competent cell (Invitrogen, Grand Island, NY, USA) (*see* **Note 1**).

6. LB (Luria–Bertani) broth: add 20 g of LB Broth (Fisher Scientific, Rockford, Il, USA) to 1 L of water and autoclave for 30 min.
7. LB agar plates: add 20 g of LB Broth and 15 g agar to 1 L of water. Autoclave for 30 min. Allow LB agar to cool to 55 °C and then add appropriate antibiotics at designated concentrations. Dispense the mixture into sterile petri dishes at room temperature. Store the plates at 4 °C after they are cooled and solidified.
8. Kanamycin sulfate (Fisher Scientific).
9. IPTG (Isopropyl β-D-1-thiogalactopyranoside) (Fisher Scientific).
10. SDS-PAGE: 4–20 % (wt/vol) Tris–glycine gel, 1 mm × 15 well (Invitrogen).

2.2 Protein Purification Components

1. Tris–HCl (Fisher Scientific).
2. NaCl.
3. Imidazole (Fisher Scientific).
4. DTT (Dithiothreitol) (Fisher Scientific).
5. TCEP (Tris (2-carboxyethyl) phosphine) (Fisher Scientific).
6. Binding buffer: 20 mM Tris–HCl pH 8.0, 500 mM NaCl, and 25 mM imidazole.
7. Elution buffer: 20 mM Tris–HCl pH 8.0, 500 mM NaCl, and 500 mM imidazole.
8. Buffer A: 20 mM Tris–HCl, pH 8.0, 100 mM NaCl, 0.3 mM TCEP.
9. Buffer B: 20 mM Tris–HCl, pH 8.0, 1 M NaCl, 0.3 mM TCEP.
10. Buffer G:10 mM Tris–HCl, pH 8.0, 100 mM NaCl, 0.2 mM DTT.
11. SUMO Protease (Life Sensors, MA, USA).
12. Emulsiflex C3 high-pressure homogenizer (Avestin, Gilead, CA, USA).
13. Misonix sonicator 3000 (Fisher Scientific).
14. Spectra molecular porous membrane tubing (Spectrum).
15. AKTA purifier FPLC (GE Healthcare, Piscataway, NJ, USA).
16. HisTrap™ HP column (Ni affinity) (GE Healthcare).
17. HiTrap™ Q HP column (GE Healthcare).
18. HiLoad 16/600 Superdex™ 75 pg (*see* **Note 2**) (GE Healthcare).

2.3 Enzymatic Assay Components

1. Vector to produce recombinant Fap1 substrate, fap1-pGEX-6p1 [10].
2. Vector to modify initial Fap1 glycosylation, *gtf1*/*gtf2*-pVPT [11].

3. Bacterial host, *E. coli* Top 10 (Invitrogen).
4. Activated sugar donor, UDP-[^{3}H]-glucose (Amersham Biosciences, Piscataway, NJ, USA).
5. Glutathione sepharose beads (GE Healthcare).
6. In vitro glycosylation buffer: 50 mM HEPES, pH 7.0, 10 mM $MnCl_2$, 0.01 % bovine serum albumin.
7. Wash buffer: 50 mM HEPES, pH 7.0, 10 mM $MnCl_2$, 0.1 % NP40.
8. NETN buffer: 20 mM Tris–HCl, pH 7.2, 0.1 M NaCl, 1 mM EDTA, 0.5 % NP-40, and protease inhibitor cocktail (1:20, vol/vol).
9. Beckman LS6500 liquid scintillation counter.

2.4 Protein Crystal Screening Components

1. Screening instrumentation, Crystal Phoenix (Art Robbins Instruments, Sunnyvale, CA, USA).
2. Intelli-Plate 96-well flat-bottomed clear polypropylene plates (Hampton Research, Aliso Viejo, CA, USA).
3. Protein crystallization film for 96-well plates (Hampton Research).
4. 24-well crystallization plate (Hampton Research).
5. Natrix, Crystal Screen, Crystal Screen II (Hampton Research).
6. Wizard I, Wizard II, Wizard III (Emerald Biosystem, Bedford, MA, USA).
7. Pre-crystallization kit (PCT) (Hampton Research).

3 Methods

3.1 Plasmid Construction and Protein Expression

1. Amplify the full-length gene *gtf3* from genomic DNA of *S. parasanguinis* FW213 using primer set Gtf3-*Hin*dIII-1F and Gtf3-*Xho*I-987R [10].
2. Purify the PCR product and clone it into pET28a-SUMO [8].
3. Transform the resulting plasmid pET-SUMO-*gtf3* into *E. coli* BL21 Gold (DE3) competent cells using standard transformation protocol.
4. Select the transformants on LB plate with kanamycin (50 μg/mL) and further verify the transformants by PCR and sequencing analysis.
5. Inoculate a single colony into the 10 mL LB medium with kanamycin (50 μg/mL) and grow the culture overnight with shaking (250 rpm) at 37 °C.
6. Inoculate the 10 mL overnight culture into the large flask with 1 L LB medium with kanamycin (50 μg/mL). Grow the bacteria

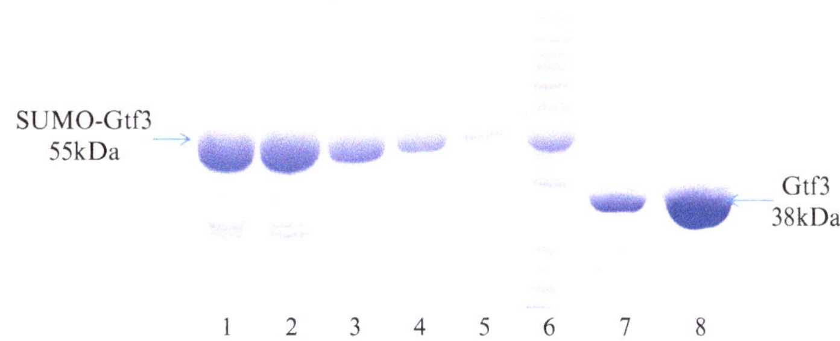

Fig. 1 SDS-PAGE analysis of recombinant Gtf3 protein after purification by HisTrap™ Column and HiTrap™ Q column. 6XHis tagged recombinant Gtf3 was purified on HisTrap™ Column and eluted (*Lanes* 1–5) as described in the Subheading 3.2. Cleavage and dialysis were performed overnight at 4 °C with SUMO protease. Following the dialysis, 6xHis-SUMO and 6xHis tagged SUMO proteases were removed from the flow through by passing through a HisTrap™ Column. The flow through (*Lane* 7) was further purified using HiTrap™ Q column and eluted (*Lane* 8). Protein samples were analyzed on SDS-PAGE after each step (small aliquots were used). *Lane* 6, BenchMark™ Protein Ladder

at 37 °C with shaking at 250 rpm, to an OD_{600} of 0.7–0.8 (*see* **Note 3**).

7. Add IPTG to a final concentration of 0.1 mM and allow the culture to grow at 18 °C overnight to induce protein expression (*see* **Note 4**).

3.2 Protein Purification

1. Harvest bacterial cells by centrifugation (15 min at 6,519 × *g* at 4 °C). Discard supernatant and resuspend cell pellet in binding buffer, and lyse the cells under high pressure, or sonicate the cells 360 rounds of 1 s pulse using automatic sonicator with power output 40 W (*see* **Note 5**).
2. Centrifuge the lysates at 9,740 × *g* for 1 h and filter the supernatant with 0.45 nm filter (*see* **Note 6**).
3. Apply the filtered supernatant to HisTrap™ Column that is pre-equilibrated with 5 bed volumes of binding buffer.
4. After washing with binding buffer (5 bed volumes), elute proteins from the affinity resin by elution buffer using a linear gradient.
5. Check the elution by SDS-PAGE and pool the fractions that contain most of the target proteins (Fig. 1). Cleave the N-terminal His-SUMO tag by incubating the pooled fractions with SUMO Protease during overnight dialysis at 4 °C against 20 mM Tris–HCl pH 8.0, 500 mM NaCl (*see* **Note 7**).
6. Reapply dialyzed protein samples to HisTrap™ Column to remove SUMO Protease, the cleaved SUMO tag and uncleaved proteins (*see* **Note 8**).

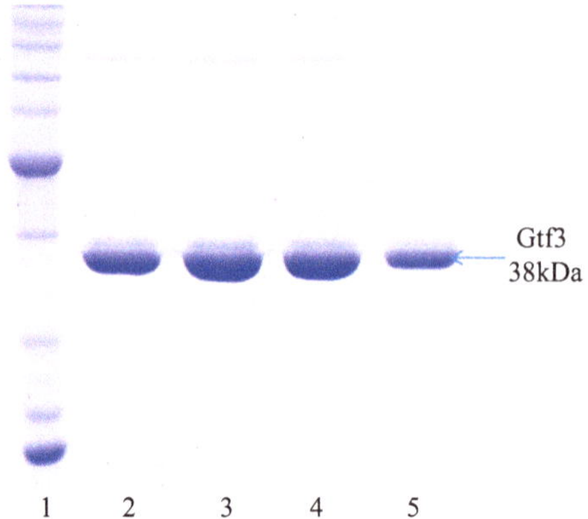

Fig. 2 SDS-PAGE analysis of final Gtf3 protein samples used for crystallization. Peak fractions (*Lanes* 2–5) collected from gel filtration were examined by SDS-PAGE and stained with Commassie Blue. *Lane* 1, BenchMark™ Protein Ladder

7. Collect flow through and apply the harvested protein to anion exchange chromatography on a HiTrap™ Q column equilibrated with Q buffer A.
8. Elute protein samples with Q buffer B using a linear gradient.
9. Collect the fractions containing the target proteins, concentrate them and apply to HiLoad 16/60 Superdex™ 75 preparation grade column which is connected to an AKTA-purifier FPLC system. Set the flow rate at 1 mL/min. Equilibrate the column with buffer G prior to protein loading. Elute fractions using buffer G and collect the fraction every 2 mL using a fraction collector.
10. Analyze protein purity by running fraction samples on SDS-PAGE (Fig. 2). Pool peak fractions and concentrate them to 30 mg/mL for crystallization screen subsequently (*see* **Note 9**).

3.3 Protein Enzymatic Activity Assay

The substrates, metal ions (for metal-dependent GTs), and specific sugar donors are important elements for successful glycosyltransferase enzymatic assays.

1. The substrate of Gtf3 is GlcNAc modified Fap1 [10, 11]. Purify the substrate GlcNAc-modified Fap1 using glutathione sepharose beads from *E. coli* Top 10 (*see* **Note 10**). Specific sugar donor for Gtf3 is UDP-[^{3}H]-glucose.
2. Wash 20 μg substrate Fap1-GlcNAc bound to glutathione sepharose beads five times with the glycosylation buffer.
3. After the last wash, discard the supernatant and suspend the beads in 200 μL glycosylation buffer.
4. Add 2 μg of purified Gtf3 protein and 50 nM of UDP-[^{3}H]-glucose to the glycosylation reaction sequentially and incubate the mixture in an orbital shaker at 60 rpm for 1 h at 37 °C.

5. Wash the beads three times with wash buffer and then transfer the beads to scintillation vials and mix with 5 mL scintillation cocktail.
6. Measure radioactivity transferred to the GlcNAc modified Fap1 substrate from the radiolabeled activated glucose [10] using a scintillation counter (*see* **Note 11**).

3.4 Crystallization and Data Collection

1. Set up initial crystallization screening at room temperature utilizing a Phoenix Crystallization robot and commercially available screening kits such as Natrix, Crystal Screen, Crystal Screen II, Wizard I, Wizard II, and Wizard III by the sitting-drop vapor-diffusion method (*see* **Note 12**).
2. Optimize the crystallization conditions that produce single crystal hits during initial screening, by modifying the pH and the concentrations of metal ions and precipitants (*see* **Note 13**).
3. Mount a single crystal by first soaking for about 10 s in an empirically determined cryoprotectant, and then flash-freeze the crystal by plunging it into liquid nitrogen (*see* **Note 14**).
4. Collect the data using a MAR 300 CCD detector at the Argonne National Laboratory beam line SER-CAT ID-22 (*see* **Note 15**).

3.5 Model Building and Structure Refinement

1. Utilize Phenix software package for model building and structure refinement. Use molecular replacement [12] to solve structures when a known homologous structure is available.
2. Adjust the autobuilt model manually with Coot software. Then refine the structure with Phenix. Repeat that process until the R-work reaches about 0.2 and R-free (multiplied by 10) is the same as or lower than the resolution (*see* **Note 16**).

4 Notes

1. BL21-Gold competent cell is not only good for pET28-SUMO vector expression in LB and also good for protein expression in the minimal medium, which is needed for Seleno-methionine-substituted method (*see* **Note 16**).
2. These columns listed were used in the Gtf3 example; other proteins may need different columns based on the properties of target proteins.
3. It will take approximately 2.5 h to reach the OD value. OD between 0.7 and 0.9 is acceptable. OD higher than 1.0 or lower than 0.6 is not good for protein expression.
4. Cool down the culture first and then add IPTG, which will prevent the production of insoluble proteins in inclusion bodies.

5. It is optional to wash the pellet with PBS once before cell lysis.
6. Filtering the supernatant will eliminate most debris that can clog the affinity column.
7. Check the efficiency of SUMO protease by SDS-PAGE analysis and use an appropriate amount of SUMO protease accordingly.
8. **Steps** 7 and **8** are optional. Continue to **step 9** if the purity of the protein from the flow through is good. And for each step, select appropriate column based on biochemical properties of the target protein.
9. A Pre-Crystallization Test (PCT) kit can be used to select the appropriate protein concentration for crystallization screening.

   The best concentration should be determined empirically for each protein crystal screening.
10. Purify GST fusion proteins using glutathione Sepharose 4B Beads following the manufacturer's instructions. Induce *E. coli* carrying vectors *fap1*-pGEX-6p1 and *gtf1/gtf2*-pVPT to express GlcNAc modified recombinant Fap1 and then lyse the induced cells as described above (*see* Subheading 3.2, **step 1**). Wash 400 μL of glutathione sepharose bead slurry with 1 mL NETN buffer and then mix the lysed supernatant with washed beads and incubate at 4 °C for 3 h. Wash the beads bound with GST fusion Fap1-GlcNAc with 10 mL NETN buffer four times.
11. Optimal reaction pH, temperature, and buffer can be tested empirically to establish a better enzymatic reaction system.
12. Any crystallization system can be used based on availability. Crystal Phoenix from Art Robbins instruments is used for our initial screening. There are also other screening kits from Qiagen, which can be used for screening (http://www.qiagen.com/products/protein/crystallization/default.aspx#ScreeningSuites). We usually place the screening plates at 20 °C at the very beginning but different temperatures can be tested if there is no indication of crystal growth at 20 °C.
13. It is hard to keep every step identical manually during each optimization. In addition, the reservoir from the screening plates will evaporate during the protein crystallization process that will alter the condition; therefore it is important to grow crystals at diverse ranges of pH values or precipitant as you may not be able to obtain the crystals of the same quality at the same condition. For instance, at the initial screening of Gtf3, one condition (0.1 M Succinic Acid pH 7.0, 15 % Polyethylene glycol 3350) gave rise to single crystals (Fig. 3). After optimization, we obtained better crystals from the condition containing 0.1 M Succinic Acid, pH 7.0, 13 % Polyethylene glycol 3350, and 10 % glycerol (Fig. 4).

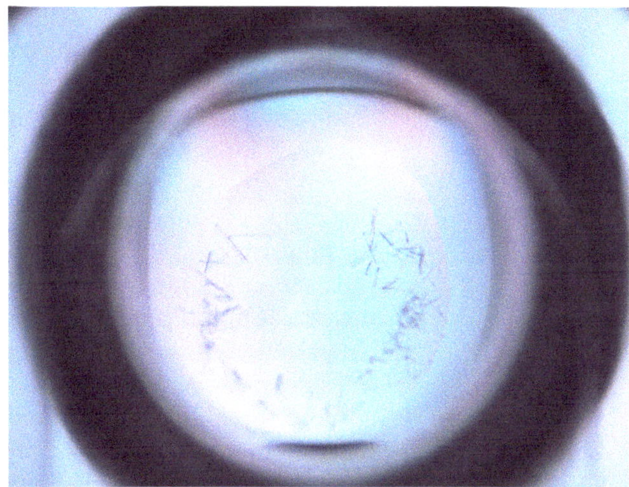

Fig. 3 Crystals obtained from the initial screening of Gtf3. Crystals appeared at one condition from Index screening kit (Hampton Research). The condition contained 0.1 M Succinic Acid pH 7.0, 15 % Polyethylene glycol 3350, and 15 mg/mL Gtf3. The crystals were obtained after overnight incubation at 20 °C

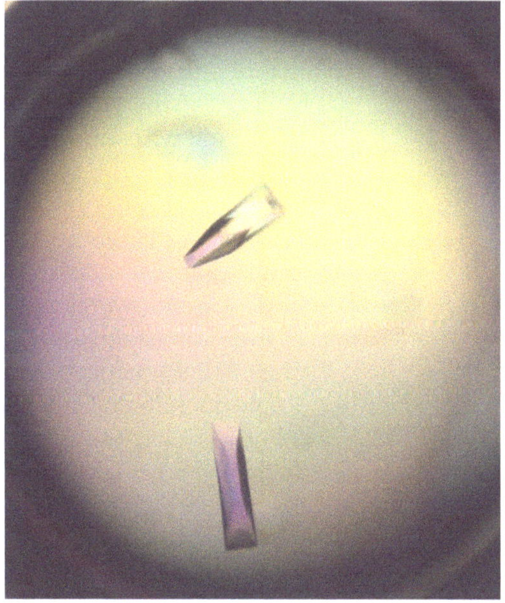

Fig. 4 Crystals obtained after optimization. Single and large crystals were obtained from the condition containing 0.1 M Succinic Acid, pH 7.0, 13 % Polyethylene glycol 3350, and 10 % glycerol

14. We usually use glycerol as cryoprotectant. The cryoprotectant used for Gtf3 was 25 % of glycerol added into 0.1 M Succinic Acid, pH 7.0, 13 % Polyethylene glycol 3350. More concentrated cryoprotectant is needed if the concentration of precipitant is low.

15. Depending on availability, other synchrotron sources, beamlines, and methods can be used to collect data. It is empirical to set up data collection parameters since every crystal is different. Comparing to higher resolution crystals, lower resolution crystals need longer distance away from the detector when the detector size and X-ray wavelength are set. We usually collect one image per angle until a Phi range of 360° is collected. But for bigger unit cell we use smaller angle range. Preprocessing the data during the data collection using HKL 2000 is recommended. One advantage of doing so is to ensure the collected data are useful for structural analysis. The quality of the data can be determined by whether the majority of reflections are covered when integrating the images, and the value of Rsym. Rsym is an internal measure of the errors within a data set. It will be generated after the data are scaled. Rsym $\leq$ 0.05 indicates that the data are good. Rsym around 0.1 means that the data are acceptable. The data are not acceptable if Rsym $\geq$ 0.15. Another advantage of preprocessing the data is that the space group obtained from the scale will provide some hints about how wide the angle should be to solve a structure. Thus, sufficient data can be collected prior to crystal decay because of long period of exposure to X-Ray.
16. Before attempting to solve a protein structure, it is necessary to check if the target protein has any homologous structure solved and what the identity is. If there is a homolog with 50 % or higher identity, the homologous structure can be used to solve target protein structure by molecular replacement. However if there is no homologous structure or the identity is lower than 30 %, single-wavelength anomalous diffraction (SAD) or multiwavelength anomalous diffraction (MAD) [13–15] with metal labeled-residues should be used to solve structure. Seleno-methionine-substituted method is often used. Seleno-methionine-substituted (Se-Met) Gtf3 was produced using a similar protocol, except that a complete amino acid medium with Se-Met substituted the LB medium [16], and that the induction was carried out at 25 °C overnight. Other heavy metals can be used to label crystals if the percentage of methionine is lower than 1 %. Se-Met data are analyzed by Autosol first and then used for autobuilding the protein structure [17]. In the process of solving structures, some important parameters should be taken into consideration such as R-factor and FOM (Figure of merit). The R-factor will be obtained after molecular replacement or Autosol. Normally R-factor should be lower than 0.4. FOM should be around 0.5 (FOM lower than 0.2 indicates that it is impossible to solve the structure). PHENIX and COOT are the common software used to solve structure, and there are other software such as MIRAS, SHARP [18], and CNSsolve [19] that can be used.

Acknowledgments

We thank Dr. Heidi Erlandsen for critical reading of the manuscript. The work was supported by NIH/NIDCR grant R01DE017954 (HW).

References

1. Drickamer K, Taylor ME (1998) Evolving views of protein glycosylation. Trends Biochem Sci 23:321–324
2. Zhou M, Wu H (2009) Glycosylation and biogenesis of a family of serine-rich bacterial adhesins. Microbiology 155:317–327
3. Campbell JA, Davies GJ, Bulone V, Henrissat B (1997) A classification of nucleotide-diphospho-sugar glycosyltransferases based on amino acid sequence similarities. Biochem J 326(Pt 3):929–939
4. Wagner GK, Pesnot T (2010) Glycosyltransferases and their assays. Chembiochem 11:1939–1949
5. Mulichak AM, Losey HC, Walsh CT, Garavito RM (2001) Structure of the UDP-glucosyltransferase GtfB that modifies the heptapeptide aglycone in the biosynthesis of vancomycin group antibiotics. Structure 9:547–557
6. Hu Y, Chen L, Ha S, Gross B, Falcone B, Walker D, Mokhtarzadeh M, Walker S (2003) Crystal structure of the MurG:UDP-GlcNAc complex reveals common structural principles of a superfamily of glycosyltransferases. Proc Natl Acad Sci USA 100:845–849
7. Yuan Y, Barrett D, Zhang Y, Kahne D, Sliz P, Walker S (2007) Crystal structure of a peptidoglycan glycosyltransferase suggests a model for processive glycan chain synthesis. Proc Natl Acad Sci USA 104:5348–5353
8. Zhu F, Erlandsen H, Ding L, Li J, Huang Y, Zhou M, Liang X, Ma J, Wu H (2011) Structural and functional analysis of a new subfamily of glycosyltransferases required for glycosylation of serine-rich streptococcal adhesins. J Biol Chem 286:27048–27057
9. Wu R, Wu H (2011) A molecular chaperone mediates a two-protein enzyme complex and glycosylation of serine-rich streptococcal adhesins. J Biol Chem 286:34923–34931
10. Zhou M, Zhu F, Dong S, Pritchard DG, Wu H (2010) A novel glucosyltransferase is required for glycosylation of a serine-rich adhesin and biofilm formation by *Streptococcus parasanguinis*. J Biol Chem 285:12140–12148
11. Wu R, Zhou M, Wu H (2010) Purification and characterization of an active N-acetylglucosaminyltransferase enzyme complex from Streptococci. Appl Environ Microbiol 76:7966–7971
12. Dodson E (2008) The befores and afters of molecular replacement. Acta Crystallogr D Biol Crystallogr 64:17–24
13. McRee DE (ed) (1999) Practical protein crystallography, 2nd edn. Academic, St. Louis, MO
14. Hendrickson WA, Horton JR, LeMaster DM (1990) Selenomethionyl proteins produced for analysis by multiwavelength anomalous diffraction (MAD): a vehicle for direct determination of three-dimensional structure. EMBO J 9:1665–1672
15. Hendrickson WA, Horton JR, Murthy HM, Pahler A, Smith JL (1989) Multiwavelength anomalous diffraction as a direct phasing vehicle in macromolecular crystallography. Basic Life Sci 51:317–324
16. Doublie S (1997) Preparation of selenomethionyl proteins for phase determination. Methods Enzymol 276:523–530
17. Terwilliger TC, Grosse-Kunstleve RW, Afonine PV, Moriarty NW, Zwart PH, Hung LW, Read RJ, Adams PD (2008) Iterative model building, structure refinement and density modification with the PHENIX AutoBuild wizard. Acta Crystallogr D Biol Crystallogr 64:61–69
18. Messerschmidt A (2007) X-ray crystallography of biomacromolecules: a practical guide, 1st edn. Wiley-VCH, Weinheim, Germany, p 318
19. Brunger AT, Adams PD, Clore GM, DeLano WL, Gros P, Grosse-Kunstleve RW, Jiang JS, Kuszewski J, Nilges M, Pannu NS, Read RJ, Rice LM, Simonson T, Warren GL (1998) Crystallography & NMR system: a new software suite for macromolecular structure determination. Acta Crystallogr D Biol Crystallogr 54:905–921

Chapter 4

Study of the Biological Functions of Mucin Type Core 3 *O*-glycans

Seung Ho Lee and Minoru Fukuda

Abstract

Core 3 *O*-glycan is very short glycan structure which is composed of one *N*-acetylglucosamine and one *N*-acetylgalactosamine. The core 3 *O*-glycan structure is synthesized by core3 synthase (beta 1, 3-*N*-acetylglucosaminyltransferase 6) using UDP-*N*-acetylglucosamine as substrate. We revealed that the core 3 *O*-glycan structure modulates prostate cancer formation and gastrointestinal cell differentiation through regulating the heterodimerization of α2β1 integrin [1, 2] and cell surface expression of differentiation marker proteins [3] respectively. This chapter describes the way to determine the functions of core 3 *O*-glycan in tumor formation and gastrointestinal cell differentiation.

Key words Core 3 *O*-glycan, Core 3 synthase, Integrin α2β1, Gastrointestinal cell differentiation sucrase isomaltase, Dipeptidylpeptidase IV

1 Introduction

Cell surface carbohydrates are closely associated with cancer malignancy [4, 5] and cell differentiation [6, 7]. Although several specific glycan structures such as the β1,6 GlcNAc structure [8, 9] and polylactosamine structures [10] have been defined to be functional glycan structures, the biological functions of numerous glycan structures still remain to be determined. Mucin type *O*-glycans are mainly classified by four different structures; core 1, core 2, core 3, and core 4 [1]. Among them core 1 and core 2 *O*-glycans are well-known tumor associated structures [11–14]. However, functions of core 3 and core 4 *O*-glycan structures are not well elucidated. Core 3 *O*-glycans are normally detected and synthesized (Fig. 1) in the gastrointestinal tract; however reduced expression levels were detected in various cancer tissues [1, 15]. In addition, the function of core 3 *O*-glycans during gastrointestinal cell differentiation has not been studied. To reveal the biological functions of core 3 *O*-glycans, we generated core 3 *O*-glycan

Inka Brockhausen (ed.), *Glycosyltransferases: Methods and Protocols*, Methods in Molecular Biology, vol. 1022, DOI 10.1007/978-1-62703-465-4_4,

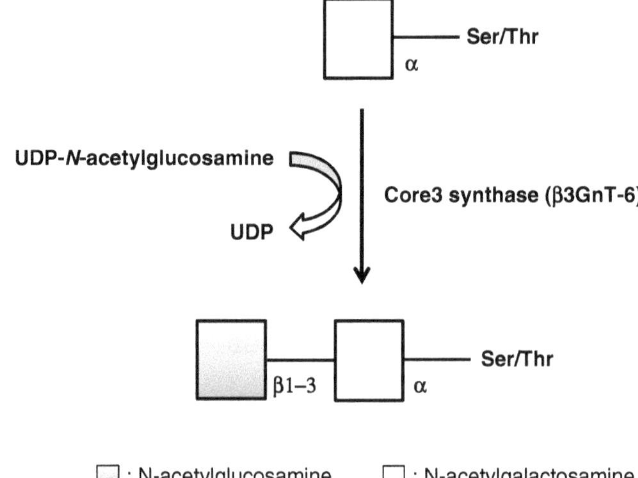

Fig. 1 Biosynthetic pathway of core 3 *O*-glycan. GalNAcα-Ser/Thr is converted by core 3 synthase (β3GnT-6) to GlcNAcβ1-3GalNAcα-Ser/Thr (core 3 structure)

overexpressing prostate cancer cells and gastrointestinal cancer cells. In this chapter we describe how functional studies are carried out. Interestingly, increased core 3 *O*-glycan structures attenuated the prostate tumorigenesis [1] as well as cell surface expression of gastrointestinal differentiation markers such as sucrase isomaltase (SI) and dipeptidylpeptidase IV (DPP-IV) [3]. These results suggest that the core 3 *O*-glycan is a functional cell surface glycan structure with important roles in tumorigenesis and gastrointestinal cell differentiation.

2 Materials

2.1 Cell Culture and Transfection

1. Prostate cancer cell lines PC3, LnCaP, and colonic carcinoma cell line HT-29 cells (American Type Culture Collection, USA).
2. PC3 and LnCaP cells were maintained with RPMI-1640 (Gibco, Carlsbad, CA, USA) with 10 % fetal bovine serum.
3. Caco-2 cells were maintained with DMEM high glucose (Gibco) with 10 % fetal bovine serum.
4. Lipofectamine 2000 reagent (Invitrogen, Grand Island, NY, USA).
5. pcDNA3.1/core 3: core 3 synthase expression plasmid [15].

2.2 Tumor Formation Assay

1. Mice: BALBc nude (nude/nude), 6–8 week old males obtained from Taconic (Oxnard, CA, USA).
2. Scissors (FineSicence, Foster City, CA, USA), 50 μL syringe (Hamilton #80530), 30 gauge custom needle, Autoclip 9 mm (Clay Adams #7631 Becton Dickinson), 1.25 % avertin.

2.3 Semiquantitative RT-PCR

1. Trizol (Invitrogen).
2. SuperScript II (Invitrogen).
3. Oligo(dT) primer (Invitrogen).
4. RNase Inhibitor (Promega, Madison, WI, USA).

2.4 Heterodimerization Assay

1. Polyclonal beta1 integrin antibody (Ab1952, Chemicon, Billerica, MA, USA).
2. Rabbit anti-alpha2 integrin antibody (Ab1936, Chemicon).
3. HRP conjugated anti-rabbit IgG.
4. Cell lysis buffer: 20 mM Tris–HCl, pH 7.4, 150 mM NaCl, 5 mM EDTA, 1 %(w/v) Nonidet P-40, 5 mM sodium pyrophosphate, 10 mM NaF, 1 mM sodium orthovanadate, 10 mM beta-glycerophosphate, 1 mM phenylmethylsulfonyl fluoride, and protease inhibitor mixture.
5. ECL reagent (Amersham Biosciences, Piscataway, NJ, USA).
6. Strip buffer: 1 N NaOH.
7. Washing buffer (TBST): 20 mM Tris–HCl, pH 7.4, 0.15 M NaCl, 0.05 % Tween 20.
8. Blocking buffer: Washing buffer including 5 % Skim milk.
9. Vectastain ABC kit (Vector Laboratories, Inc., Burlingame, CA, USA).
10. Image J program.

2.5 Cell Differentiation and Immunocytochemistry

1. Goat anti-SI antibody and rabbit anti-CD26(DPP-IV) antibody (Santa Cruz Biotechnology, Santa Cruz, CA, USA).
2. Alexa Fluor 488-labeled anti-goat IgG (Invitrogen).
3. Alexa Fluor 488-labeled anti-rabbit IgG (Invitrogen).

2.6 Cell Surface Biotinylation

1. Sulfo-succinimidobiotin (Pierce, Rockford, IL, USA).
2. Washing buffer: 50 mM NH4Cl and PBS containing 1 mM $MgCl_2$ and 0.1 mM $CaCl_2$.
3. Vectastain ABC Kit (Vector Laboratories).

3 Methods

3.1 Construction of Core 3 Overexpressing Cell Lines

1. 1 μg pcDNA3.1/core 3 plasmid will be transfected to PC3, LnCaP, and Caco-2 cells using Lipofectamine 2000 with serum-free media (6-well plate, 70 % cell confluence).
2. After 6 h incubation, media will be exchanged with regular media (including 10 % serum) followed by further incubation for 12–16 h.

3. 1/100 volume of cells will be placed into a 10 cm dish with G418 (200 μg/mL for LnCaP, PC3, and 1 mg/mL for HT-29 cells) (*see* **Note 1**).
4. Incubate dish for 2–3 weeks.
5. Pick single colonies and transfer each to a single well.
6. Incubate each colony for 2–3 weeks.
7. Each clone will be split into two different wells (one for making stock and the other for checking the expression of core 3 synthase mRNA).
8. Isolate the total RNA and check the expression of core 3 synthase mRNA by RT-PCR.
   - PCR conditions: 35 cycles of 94 °C for 30 s, 56 °C for 30 s, and 72 °C for 30 s and one cycle at 72 °C for 5 min.
   - PCR primers for core 3 synthase: 5′-agcactgcagcagtggttc-3′, 5′-gaggaaggtgtccgcgaag-3′
   - PCR primers for GAPDH: 5′-cctggccaaggtcatccatgaca-3′, 5′-atgaggtccaccaccctgttgct-3′
9. Separate PCR products by electrophoresis on 1 % agarose gels.

3.2 In Vivo Tumor Formation Assay

1. Wash the PC3 and LnCaP cells three times with PBS.
2. Add enzyme free EDTA solution to those cells and incubate.
3. Harvest cells and centrifuge (3,000 × *g*, 5 min).
4. Remove the supernatants and then wash with PBS.
5. Repeat two more times (**steps. 3–4**).
6. Count the cell numbers and concentrate volumes.
7. 2×10^6 cells of PC3 and 5×10^7 cells of LnCaP cells were used for tumor formation respectively (*see* **Note 2**).

3.2.1 Orthotopic Tumor Cell Injection

1. Anesthetize mice with avertin (1.25 %, 500 μL/mouse).
2. 1–2 min after anesthetizing, perform laparotomy.
3. Suspend 2×10^6 cells of PC3 and mock transfected PC3 cells in 20 μL of serum free RPMI-1640 medium and inoculate into the posterior lobe (*see* **Note 3**).
4. 9 mm autoclips will be used for closing the wound.
5. 8 weeks later, tumor in prostate will be taken out and weighed (Fig. 2).

3.3 Heterodimerization Assay

1. Wash cells with ice-cold PBS.
2. Add the lysis buffer and harvest the cells into a 1.5 mL tube.
3. Incubate on ice for 2–3 h.
4. Centrifuge (15,000 rpm (25,000 × *g*) for 15 min) and transfer supernatant to a new tube.

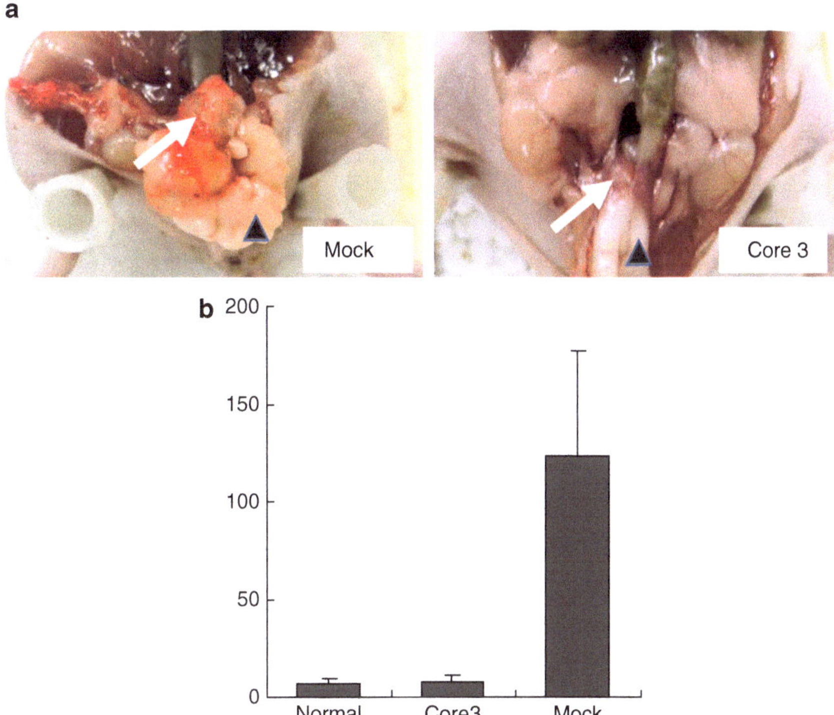

Fig. 2 Core 3 *O*-glycan suppresses tumor formation. Mock-transfected and core 3 expressing PC3 cells were orthotopically inoculated into the prostate of nude mice, and animals were sacrificed 8 weeks later. (**a**) Mock-transfected PC3 cells produced a large prostate tumor (*arrow*) whereas tumors produced by core 3-expressing PC3 were not observed in the prostate (*arrow*). Seminal Vesicles are also seen (*arrowhead*). (**b**) Wet weights of eight prostates are shown. Reproduced from ref. [1]

5. After measuring protein concentration, incubate the anti-alpha2 and anti-beta1 integrin antibody with 500 μg of total protein, respectively.
6. Add 10 μL of protein A sepharose to each tube and incubate overnight.
7. Centrifuge (10,000 rpm (11,000 × *g*) for 30 s), discard supernatant, and then wash with PBS.
8. Repeat **step 7** two times.
9. Add SDS-loading buffer and then boil each tube.
10. Centrifuge (10,000 rpm (11,000 × *g*) for 1 min) and take out supernatant.
11. Load equal volumes of supernatant to 8 % SDS PAGE gel.
12. After finishing the running, transfer to PVDF membrane.
13. After blotting with anti-alpha2 antibody, reprobe the membrane (1 N NaOH buffer, 1 min) and then wash with TBST three times (*see* **Note 4**).

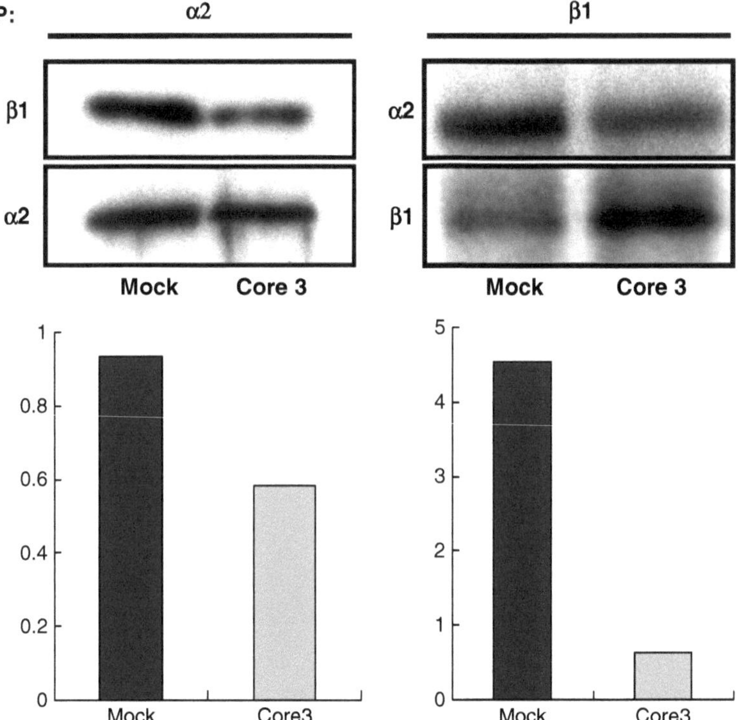

Fig. 3 Impaired α2β1 integrin association in core 3 *O*-glycan expressing cells. α2 Integrin was immunoprecipitated (IP) with rabbit anti-α2 integrin antibody, and the blot was incubated with mouse monoclonal anti-β1 integrin antibody. The membrane was then stripped by incubation with 1 N NaOH for 1–2 min and then incubated with anti-α2 integrin antibody. In parallel, β1 integrin was first immunoprecipitated by polyclonal anti-β1 integrin antibody, and the immunoprecipitates were sequentially incubated with anti-α2 antibody and anti-β1 antibody. The experiments were repeated three times, and a representative result is shown. Heterodimerization rate (α2/β1 or β1/α2) was estimated by scanning the gel and is tabulated in the *lower panel*. Reproduced from ref. [1]

14. Incubate membrane with blocking buffer (30 min), and then add anti-beta1 antibody and incubate overnight at 4 °C.
15. After washing three times with TBST buffer, add HRP conjugated anti rabbit IgG.
16. Signal is detected with ECL kit.
17. Measure the band intensity using Image J program.
18. Calculate the heterodimerization ratio.

    Heterodimerization ratio = Intensity of alpha2 integrin/ Intensity of beta1 integrin (Fig. 3).

3.4 Cell Differentiation and Cell Surface Biotinylation

1. Change the media from DMEM high glucose to RPMI-1640 for cell differentiation (*see* **Note 5**).
2. Culture 3 days more after confluence of mock and core 3 synthase transfected HT-29 cells.

3. Cells will be biotinylated (1 mg/mL, 1 h on ice), washed one time withNH4Cl and three times with PBS containing 0.1 mM $CaCl_2$, 1 mM $MgCl_2$ (*see* **Note 6**).
4. Biotinylated cells will be subjected to lysis and immunoprecipitated with anti-DPP-IV antibody.
5. Immunoprecipitates will be separated in 6 % SDS PAGE and transferred to PVDF membrane.
6. Signals will be detected with ECL kit.

3.5 Immunocytochemistry

1. Differentiated (3 days after confluence) mock and core 3 synthase overexpressing Caco-2 cells will be used for immunocytochemistry.
2. Cells are washed with PBS containing 0.1 mM $CaCl_2$, 1 mM $MgCl_2$ and then fixed with 4 % paraformaldehyde solution for 10 min.
3. Cells are incubated for blocking with PBS containing 5 % BSA for 1 h at room temperature.
4. Anti-SI (1/100) and Anti-DPPIV (1/100) antibody will be added and then incubated for 2 h at room temperature.
5. After washing three times with PBS, Alexa conjugated anti-goat IgG (1/300) and anti-rabbit (1/300) IgG are used to detect the DPP-IV signal (*see* **Note 6**).
6. Determine the apical or basolateral expression of DPP-IV according to Fig. 4 (*see* **Note** 7).

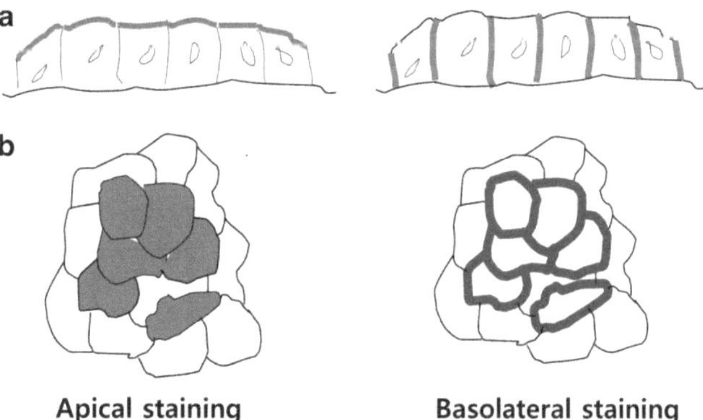

Fig. 4 Schematic apical and basolateral expression. Apical and basolateral expression patterns of confluent cells are shown. When the cells are vertically examined for the expression of target protein, apical signals are found on the upper side of cells (**a**, *left figure*) and basolateral signals between cells (**a**, *right* figure). Apical signals are found on the entire area of cells when cells are examined horizontally (**b**, *left* figure), whereas basolateral signals are found at the outer border of cells (**b**, *right* figure)

4 Notes

1. To make core 3 synthase overexpressing cells, it is important to determine the antibiotic concentration considering the cell types. Since the core 3 *O*-glycan structure could attenuate integrin function in prostate cancer cells [1], it is very hard to make one single clone which has highly expressing core 3 synthase. To have a higher chance of isolating a single positive colony, the use of low concentrations of antibiotic is helpful, and it is better to split the low number of transfected cells into a 10 cm dish (1/100 dilution) for selection with a low concentration of antibiotic.
2. Doubling time of LnCaP is less than that of PC3 cells. Therefore, it is recommended to use large numbers of LnCaP cells for in vivo tumor formation assay.
3. Since the space of prostate has limitation, it is better to concentrate the injection volume up to 20 μL. After injecting tumor cells, it is hard to see the size of the tumor in the prostate. It is recommended to check the size of tumor by touching with hands before sacrificing the mice [16].
4. Since integrin is associated with two different alpha and beta chains, the heterodimerization ratio could be calculated by measuring co-precipitated integrin chain. For example, when the alpha 2 antibody was used for first immunoprecipitation, beta 1 integrin should be co-precipitated and the amount of beta 1 integrin could be detected in the same membrane. Therefore, the membrane used for detection of alpha 2 integrin should be handled carefully. After finishing the first detection, wash the membrane briefly and then reprobe the membrane with 1 N NaOH solution. More than 5 min incubation is not recommended. It is better to keep the membrane in washing buffer at 4 °C for further use. Increased core 3 *O*-glycans attenuated the heterodimerization of α2β1 integrin which results in decreased prostate tumor formation in vivo (Figs. 2 and 3). With these results, we concluded that core 3 *O*-glycan is a functional glycan structure in prostate tumorigenesis.
5. Among the gastrointestinal cells, Caco-2 cells are able to differentiate spontaneously, whereas a reduced glucose concentration in the medium is required to differentiate HT-29 cells. Therefore, when HT-29 cells reach confluence, replace the media from DMEM-high glucose to RPMI-1640 for differentiating the HT-29 cells.

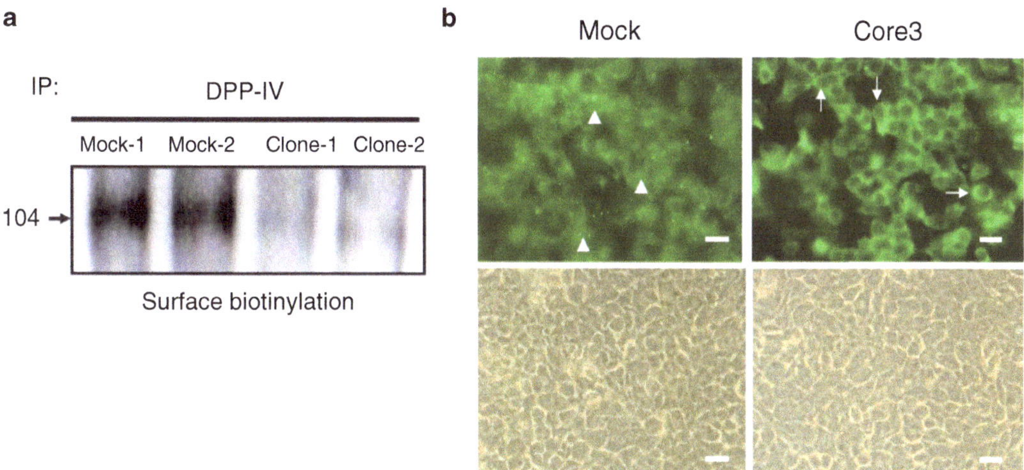

Fig. 5 (**a**) Cell surface expression of DPP-IV is decreased in core 3 synthase expressing HT-29 cells compared with mock-transfected cells. Differentiated mock- and core 3 synthase-transfected HT-29 cells were biotinylated, and cell surface expression of DPP-IV was compared after immunoprecipitation (IP) with DPP-IV antibody. (**b**) DPP-IV cell surface expression was examined by immunofluorescence. Both mock- and core 3 synthase-transfected HT-29 cells were stained with rabbit anti-CD26 (DPP-IV) antibodies in non-permeable conditions. Cell density of the depicted areas is shown by phase-contrast microscopy (*lower panels*). *Arrowhead* and *arrows* indicate the apical and basolateral area, respectively. Scale bar, 20 μm. Reproduced from ref. [3]

6. For labeling the proteins which are located in the apical area, cell surface biotinylation is a good way to detect proteins. When cells reach confluence, plates should be placed on ice to inhibit the internalization of cell surface proteins. Ice cold biotin solution must be used and washing step with NH_4Cl for quenching the extra biotin on cell surfaces is important for comparison of the expression levels. The biotin signal is sensitive if the avidin-HRP is used for developing the signal; this is good for a comparison of cell surface proteins which are weakly expressed.

7. DPP-IV is one of the important gastrointestinal cell differentiation marker protein. Mock transfected HT-29 cell showed apical expression of DPP-IV; however the signal of DPP-IV was located in the basolateral area in core 3 *O*-glycan overexpressing HT-29 cells (Fig. 5). A different expression pattern of the green fluorescence signal was mainly detected in the outer border of core 3 *O*-glycan transfected cells. However, most of the signal was shown on the entire area of cells of mock transfected cells. These results indicated that DPP-IV positive green signals were affected by cell surface core 3 *O*-glycans. With these results, we concluded that the core 3 *O*-glycans have a function in proper sorting of DPP-IV which is important in gastrointestinal cell differentiation.

Acknowledgments

This study was supported by Leading Foreign Research Institute Recruitment Program through the National Research Foundation of Korea (NRF) funded by the Ministry of Education, Science and Technology (MEST) (K2100311815-11E0111-02010), and NIH grants R01 CA033000 and P01 CA071932.

References

1. Lee SH, Hatakeyama S, Yu SY, Bao X, Ohyama C, Khoo KH, Fukuda MN, Fukuda M (2009) Core3 O-glycan synthase suppresses tumor formation and metastasis of prostate carcinoma PC3 and LNCaP cells through down-regulation of alpha2beta1 integrin complex. J Biol Chem 284:17157–17169
2. Lee SH, Fukuda M (2010) Core3 glycan as tumor suppressor. Methods Enzymol 479: 143–154
3. Lee SH, Yu SY, Nakayama J, Khoo KH, Stone EL, Fukuda MN, Marth JD, Fukuda M (2010) Core2 O-glycan structure is essential for the cell surface expression of sucrase isomaltase and dipeptidyl peptidase-IV during intestinal cell differentiation. J Biol Chem 285: 37683–37692
4. Hakomori S (2002) Glycosylation defining cancer malignancy: new wine in an old bottle. Proc Natl Acad Sci USA 99:10231–10233
5. Nakamori S, Kameyama M, Imaoka S, Furukawa H, Ishikawa O, Sasaki Y, Kabuto T, Iwanaga T, Matsushita Y, Irimura T (1993) Increased expression of sialyl Lewisx antigen correlates with poor survival in patients with colorectal carcinoma: clinicopathological and immunohistochemical study. Cancer Res 53: 3632–3637
6. Alfalah M, Jacob R, Preuss U, Zimmer KP, Naim H, Naim HY (1999) O-linked glycans mediate apical sorting of human intestinal sucrase-isomaltase through association with lipid rafts. Curr Biol 9:593–596
7. Naim HY, Sterchi EE, Lentze MJ (1988) Biosynthesis of the human sucrase-isomaltase complex. Differential O-glycosylation of the sucrase subunit correlates with its position within the enzyme complex. J Biol Chem 263:7242–7253
8. Kang R, Saito H, Ihara Y, Miyoshi E, Koyama N, Sheng Y, Taniguchi N (1996) Transcriptional regulation of the N-acetylglucosaminyltransferase V gene in human bile duct carcinoma cells (HuCC-T1) is mediated by Ets-1. J Biol Chem 271:26706–26712
9. Nakahara S, Saito T, Kondo N, Moriwaki K, Noda K, Ihara S, Takahashi M, Ide Y, Gu J, Inohara H, Katayama T, Tohyama M, Kubo T, Taniguchi N, Miyoshi E (2006) A secreted type of beta1,6 N-acetylglucosaminyltransferase V (GnT-V), a novel angiogenesis inducer, is regulated by gamma-secretase. FASEB J 20:2451–2459
10. Zenita K, Kirihata Y, Kitahara A, Shigeta K, Higuchi K, Hirashima K, Murachi T, Miyake M, Takeda T, Kannagi R (1988) Fucosylated type-2 chain polylactosamine antigens in human lung cancer. Int J Cancer 41: 344–349
11. Dalziel M, Whitehouse C, McFarlane I, Brockhausen I, Gschmeissner S, Schwientek T, Clausen H, Burchell JM, Taylor-Papadimitriou J (2001) The relative activities of the C2GnT1 and ST3Gal-I glycosyltransferases determine O-glycan structure and expression of a tumor-associated epitope on MUC1. J Biol Chem 276:11007–11015
12. Brockhausen I (2006) Mucin-type O-glycans in human colon and breast cancer: glycodynamics and functions. EMBO Rep 7:599–604
13. Julien S, Krzewinski-Recchi MA, Harduin-Lepers A, Gouyer V, Huet G, Le Bourhis X, Delannoy P (2001) Expression of sialyl-Tn antigen in breast cancer cells transfected with the human CMP-Neu5Ac: GalNAc alpha2,6-sialyltransferase (ST6GalNac I) cDNA. Glycoconj J 18:883–893
14. Tsuboi S, Sutoh M, Hatakeyama S, Hiraoka N, Habuchi T, Horikawa Y, Hashimoto Y, Yoneyama T, Mori K, Koie T, Nakamura T, Saitoh H, Yamaya K, Funyu T, Fukuda M, Ohyama C (2011) A novel strategy for evasion of NK cell immunity by tumours expressing core2 O-glycans. EMBO J 30:3173–3185
15. Iwai T, Kudo T, Kawamoto R, Kubota T, Togayachi A, Hiruma T, Okada T, Kawamoto T, Morozumi K, Narimatsu H (2005) Core 3 synthase is down-regulated in colon carcinoma and profoundly suppresses the metastatic potential of carcinoma cells. Proc Natl Acad Sci USA 102:4572–4577
16. Hatakeyama S, Yamamoto H, Ohyama C (2010) Tumor formation assays. Methods Enzymol 479:397–411

Chapter 5

Generation of Anti-sulfated Glycan Antibodies Using Sulfotransferase-Deficient Mice

Hiroto Kawashima

Abstract

Anti-carbohydrate monoclonal antibodies (mAbs) are very useful in the functional analysis of complex carbohydrates in vivo. However, such mAbs are difficult to generate, largely because a wide variety of complex carbohydrates is intrinsically expressed in mice and rats and because the antigenicities of glycans are generally poor. In this chapter, I describe an efficient method for generating anti-carbohydrate mAbs using glycan-synthesizing enzyme-knockout mice in which the glycan structures formed by the missing enzymes should be highly antigenic. As an application of this method, I describe the generation of anti-sulfated glycan mAbs using sulfotransferase-deficient mice and the immunohistochemical detection of sulfated glycans involved in lymphocyte homing in both humans and mice.

Key words Sulfotransferase, Lymphocyte homing, High endothelial venule, L-selectin, Anti-carbohydrate antibody

1 Introduction

Recent studies using gene-targeted mice have revealed various physiological functions of sulfated glycans. Our group [1] and others [2] previously generated mice deficient in two *N-acetylglucosamine-6-O*-sulfotransferases (GlcNAc6STs), GlcNAc6ST-1 and GlcNAc6ST-2 (also called HEC-GlcNAc6ST or L-selectin ligand sulfotransferase), and showed that GlcNAc-6-*O*-sulfation of the L-selectin ligand oligosaccharides expressed in high endothelial venules (HEVs) in peripheral lymph nodes (PLNs) plays a major role in lymphocyte homing to PLNs. In that work, we performed a detailed analysis of the carbohydrate structures of glycosylation-dependent cell adhesion molecule-1 (GlyCAM-1) expressed in HEVs by conventional biochemical methods to provide a link between the structural changes in glycans and their functions. However, in the absence of specific anti-carbohydrate antibodies, carbohydrate structural analysis is very laborious and

Inka Brockhausen (ed.), *Glycosyltransferases: Methods and Protocols*, Methods in Molecular Biology, vol. 1022, DOI 10.1007/978-1-62703-465-4_5,

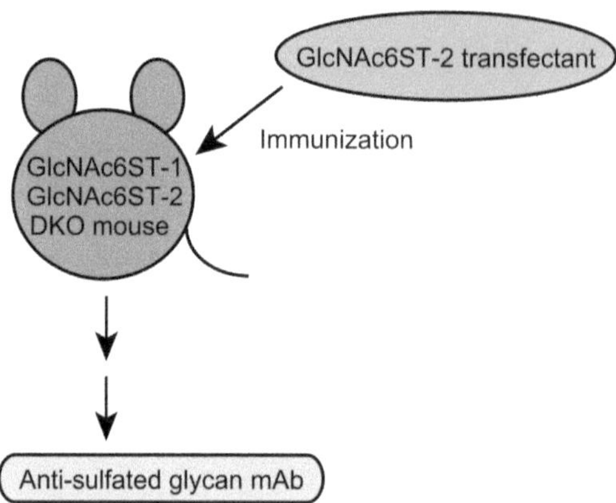

Fig. 1 A strategy for generating anti-sulfated glycan mAbs. GlcNAc6ST-1/GlcNAc6ST-2 DKO mice are immunized with GlcNAc6ST-2-overexpressing cells. The carbohydrate structures formed by the missing sulfotransferase are expected to be highly antigenic in the DKO mice.

time-consuming, especially when analyzing the structures expressed in a minor cell population such as in HEVs. In this chapter, I describe an efficient method for generating anti-carbohydrate monoclonal antibodies (mAbs) and the application of these antibodies to the immunohistochemical determination of glycan structures expressed in HEVs.

Over the past few decades, many of the genes encoding glycan-synthesizing enzymes, such as glycosyltransferases and sulfotransferases, have been identified. Extensive in vivo functional analyses of complex carbohydrates have been performed using knockout mice deficient in those enzymes. Thus, a number of gene-targeted mice deficient in various glycan-synthesizing enzymes are now available. The concept behind the method I describe in this chapter is as follows: Glycan-synthesizing enzyme-deficient mice are immunized with transfected cells overexpressing the missing glycan-synthesizing enzyme (Fig. 1). Because the products of the enzyme should be highly antigenic in the knockout mice, I hypothesized that anti-carbohydrate mAbs could be efficiently generated by this method.

The generation of anti-bisecting-GlcNAc antisera in *N*-acetylglucosaminyltransferase III-deficient mice immunized with cells expressing this glycosyltransferase was previously reported [3]. In addition, the generation of anti-*N*-glycolylneuraminic acid (Neu5Gc) antisera in CMP-*N*-acetylneuraminic acid (Neu5Ac) hydroxylase (Cmah)-deficient mice immunized with Neu5Gc-expressing thymocytes from wild-type (WT) mice was also reported [4]. However, those studies did not try to generate anti-carbohydrate mAbs. In contrast, our group recently generated anti-carbohydrate mAbs based on the idea described above [5]. We generated two anti-sulfated glycan mAbs, named S1 and S2, by immunizing the

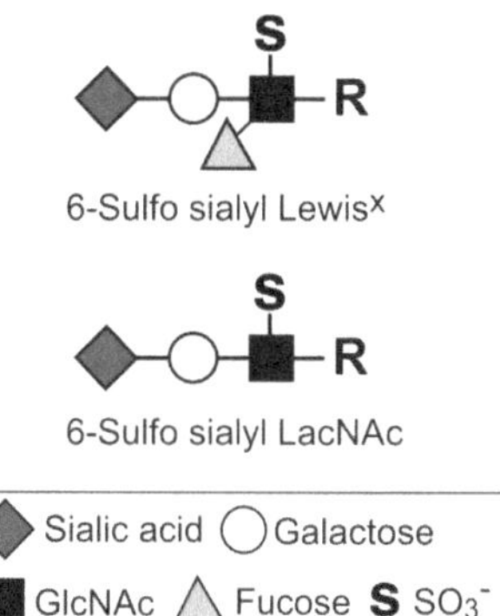

Fig. 2 6-Sulfo sialyl Lewis^x and 6-sulfo sialyl LacNAc structures. Both *N*- and *O*-glycans of glycoproteins can be modified with 6-sulfo sialyl Lewis^x and 6-sulfo sialyl LacNAc structures. The minimum structure recognized by the mAbs S1 and S2 is 6-sulfo sialyl LacNAc [5]

GlcNAc6ST-1 and GlcNAc6ST-2 double-deficient (DKO) mice with transfected cells expressing a sulfotransferase, GlcNAc6ST-2 (Fig. 1). These mAbs specifically recognize 6-sulfo sialyl Lewis^x and 6-sulfo sialyl *N*-acetyllactosamine (LacNAc) (Fig. 2). Using a similar strategy, Arata-Kawai et al. also reported the generation of an anti-sulfated glycan mAb CL40 with similar binding specificities [6].

In the method described here, transfected cells are used as an immunogen because it is easy to modify their carbohydrate structures by introducing cDNAs that encode glycan-modifying enzymes. As described below, by transfecting GlcNAc6ST-2-expressing cells with cDNAs encoding Cmah to modify the terminal sialic acid, mAbs reactive with mouse tissues rich in Neu5Gc can be successfully generated. Because the cDNAs of various glycan-modifying enzymes and a number of mutant mice deficient in glycosyltransferases and sulfotransferases are now available, I believe that the method described herein will be widely used for the generation of various anti-carbohydrate mAbs.

2 Materials

2.1 Mice

1. GlcNAc6ST-1/GlcNAc6ST-2 DKO, fucosyltransferase (FucT)-IV/FucT-VII DKO, and core 1 β1,3-*N*-acetylglucosaminyltransferase (C1β3GnT)/core 2 β1,6-*N*-acetylglucosaminyltransferase (C2GnT)-I DKO mice are backcrossed to C57BL/6 wild-type (WT) mice and maintained as described previously [1, 7, 8].
2. C57BL/6 WT mice (Japan SLC, Hamamatsu, Japan).
3. The mice are treated in accordance with the guidelines of the Animal Research Committee of the University of Shizuoka.

2.2 Cells

1. CHO cells stably expressing human CD34, human FucT-VII, human C1β3GnT, human C2GnT-I, and mouse GlcNAc6ST-2 (CHO/CD34/F7/C1/C2/GlcNAc6ST-2) [9] are cultured in DME/F-12 medium (Sigma-Aldrich, St. Louis, MO, USA) supplemented with 10 % fetal bovine serum (FBS) (HyClone, Logan, UT, USA) and penicillin–streptomycin (Invitrogen, Carlsbad, CA, USA) and are maintained in a humidified incubator (37 °C, 5 % CO_2) (*see* **Note 1**).
2. P3X63Ag8.653 myeloma cells are obtained from the American Type Culture Collection.

2.3 Culture Medium

1. Culture medium A: RPMI-1640 medium (Sigma-Aldrich) supplemented with 15 % FBS, penicillin–streptomycin, 50 μM 2-mercaptoethanol, and OPI media supplement (Sigma-Aldrich).

2.4 Microscopes and Related Equipments

1. Confocal laser scanning microscope (LSM510 META; Carl Zeiss, Inc., Thornwood, NY, USA).
2. AX-80 microscope (Olympus, Center Valley, PA, USA).
3. DP72 CCD camera (Olympus).
4. DP2-BSW software (Olympus).

2.5 Other Materials and Kits

1. FuGENE 6 transfection reagent (Roche, Indianapolis, IN, USA).
2. Imject Alum (Thermo Fisher Scientific, Rockford, IL, USA).
3. 50 % PEG solution (Sigma-Aldrich).
4. HAT supplement (Invitrogen).
5. HT supplement (Invitrogen).
6. Mouse monoclonal antibody isotyping kit (GE Healthcare, Uppsala, Sweden).
7. Pristane (2,6,10,14-tetramethyl pentadecane; Sigma-Aldrich).
8. Sephacryl S-300 (GE Healthcare).
9. BCA protein assay kit (Thermo Fisher Scientific).
10. OCT compound (Sakura Finetek, Tokyo, Japan).
11. Wax pencil (ImmEdge hydrophobic barrier pen, Vector Laboratories, Burlingame, CA, USA)
12. AlexaFluor 594-labeled goat anti-mouse IgM (Invitrogen).
13. Fluoromount (Diagnostic Biosystems, Pleasanton, CA, USA).
14. HRP- and anti-mouse Ig-conjugated EnVision + polymer (Dako, Glostrup, Denmark).

15. 3,3′-Diaminobenzidine (DAB; Dojindo, Kumamoto, Japan).
16. Hematoxylin (Sigma-Aldrich).
17. Protein G-Sepharose column (GE Healthcare).
18. EZ-Link Sulfo-NHS-LC-Biotin (Thermo Fisher Scientific).
19. AlexaFluor 488-labeled goat anti-mouse IgG (Invitrogen).

3 Methods

3.1 Generation of Anti-sulfated Glycan mAbs

1. Transiently transfect CHO/CD34/F7/C1/C2/GlcNAc6ST-2 cells cultured in a 100-mm dish with cDNAs encoding mouse Cmah, which generates CMP-Neu5Gc from CMP-Neu5Ac [10], using the FuGENE 6 transfection reagent (*see* **Note 2**).
2. After 48 h of transfection, wash the cells with PBS and disperse them with 0.5 mM EDTA in PBS.
3. After centrifugation, suspend the cells in PBS and mix the cell suspension with Imject Alum at a ratio of 1:1 in a final volume of 600 μL.
4. Vortex vigorously for 30 min.
5. Immunize GlcNAc6ST-1/GlcNAc6ST-2 DKO mice intraperitoneally three times at 2-week intervals with the cell suspension (250 μL/mouse).
6. 4 days after the final immunization, fuse lymphocytes from the spleens of the DKO mice with P3X63Ag8.653 myeloma cells (lymphocytes:myeloma cells = 3:1) in the presence of 50 % PEG solution.
7. After centrifugation, suspend the cells in culture medium A (1×10^6 lymphocytes/mL).
8. Plate 100 μL/well (1×10^5 lymphocytes/well) of the cell suspension into 96-well tissue culture plates and culture the cells in a humidified incubator (37 °C, 5 % CO_2).
9. Add 100 μL/well of culture medium A containing HAT supplement (2× concentration) 24 h after fusion.
10. After the colonies are formed, select hybridomas that secrete anti-sulfated glycan antibodies into the culture supernatant that are reactive with the HEVs of WT mice but not with those of GlcNAc6ST-1/GlcNAc6ST-2 DKO mice by immunofluorescence as described in Subheading 3.3 (*see* **Note 3**).
11. Clone hybridomas secreting anti-sulfated glycan antibodies by limiting dilution. Culture the cells in culture medium A containing HT supplement.

12. After the colonies are formed, select hybridomas that secrete anti-sulfated glycan antibodies into the culture supernatant as described in **step 10** (*see* **Note 3**).
13. Determine the isotypes of the mAbs using a mouse monoclonal antibody isotyping kit.

3.2 Purification of Anti-sulfated Glycan mAbs

1. Culture the established clones in culture medium A in a humidified incubator (37 °C, 5 % CO_2), wash the cells with PBS, and inject them intraperitoneally into BALB/c Slc-*nu/nu* mice (5.0×10^6 cells in 300 μL PBS/mouse) that have been pre-injected intraperitoneally with 500 μL/mouse of pristane a few weeks before the cell injection.
2. House the nude mice under specific pathogen-free conditions in the animal facility.
3. After confirming that the abdomens of the nude mice swelled, collect the ascitic fluid from the abdominal cavities into a 15-mL centrifuge tube using a glass pipette (*see* **Note 4**).
4. After settling the ascitic fluid at room temperature for 10 min, vortex vigorously for 30 s, and centrifuge the tube at $2{,}600 \times g$ for 5 min. Collect supernatants and store them at −80 °C until use.
5. If the isotype of the mAb is IgM, apply the ascitic fluid to a Sephacryl S-300 gel filtration column (1.5×100 cm) equilibrated with PBS (*see* **Note 5**).
6. Collect fractions (2 mL/fraction) at a flow rate of 12 mL/h.
7. Determine the protein concentration of each fraction using a BCA protein assay kit.
8. Apply the fractions to a 10 % SDS-PAGE gel to determine the purity of the mAb.
9. The purified mAb fractions (>95 % purity) thus obtained are used for the following experiments (*see* **Note 6**).

3.3 Immunofluorescence

1. Embed mouse PLNs from WT and DKO mice in OCT compound, place them on dry ice until they are frozen, and store them at −80 °C until sectioning.
2. Prepare frozen sections (7-μm thick), and fix them in ice-cold acetone for 5 min.
3. Dip the sections in PBS for 5 min.
4. Repeat **step 3** twice.
5. Remove excess fluid by gently tapping against tissue paper and carefully wiping around the tissue sections.
6. Circle the tissue sections with a wax pencil.

7. Apply 300 μL of 3 % BSA in PBS to block nonspecific binding sites for 1 h.
8. Incubate the sections with the culture supernatants of hybridomas or with 5 μg/mL of purified anti-sulfated glycan mAbs in PBS containing 0.1 % BSA for 1 h.
9. Wash the sections three times with PBS containing 0.1 % BSA.
10. If the isotype of the mAb is IgM, incubate the sections with 0.5 μg/mL AlexaFluor 594-labeled goat anti-mouse IgM in PBS containing 0.1 % BSA (*see* **Note 7**).
11. Mount the sections using Fluoromount.
12. Obtain images using a confocal laser scanning microscope using a 40× water immersion objective (*see* **Notes 8**).

3.4 Histological Examination of Human Tissues

1. Prepare 3-μm thick sections of the formalin-fixed human tonsil specimens (*see* **Note 9**).
2. Deparaffinize the sections in xylene and rehydrate them in ethanol.
3. Retrieve the antigens by boiling the sections in 10 mM Tris–HCl buffer (pH 8.0) containing 1 mM EDTA for 20 min in a microwave oven (*see* **Note 10**).
4. Quench the endogenous peroxidase activity of the tissue by soaking in absolute methanol containing 0.3 % hydrogen peroxide for 30 min.
5. Block the nonspecific protein binding sites in the tissue with 1 % BSA in TBS (20 mM Tris–HCl, 0.15 M NaCl, pH 7.4) for 15 min.
6. Incubate the sections overnight with the anti-sulfated glycan mAb such as S1 or S2 (5 μg/mL) in TBS containing 5 % BSA at 4 °C.
7. After washing with TBS, incubate the sections with HRP- and anti-mouse Ig-conjugated EnVision+polymer for 30 min according to the procedure provided by the manufacturer.
8. After washing with TBS, develop the color reaction in TBS containing 0.2 % DAB and 0.02 % hydrogen peroxide for 7 min.
9. Counterstain the sections briefly with hematoxylin.
10. Observe the slides using an AX-80 microscope with a 100× oil-immersion objective lens.
11. Take pictures using a DP72 CCD camera with DP2-BSW software (*see* **Note 11**).

4 Notes

1. FBS should be heat-inactivated for 30 min at 56 °C.
2. To my knowledge, none of the previous anti-sialyl Lewis^x and anti-6-sulfo sialyl Lewis^x mAbs react with the HEVs of C57BL/6 WT mice [11, 12], most likely because a large proportion of the terminal sialic acid in the WT mice is Neu5Gc, whereas those mAbs react with glycans modified with Neu5Ac [11]. Therefore, cells should be transiently transfected with an expression vector encoding Cmah, which generates CMP-Neu5Gc from CMP-Neu5Ac [10] in this step to obtain mouse tissue-reactive anti-sulfated glycan mAbs.
3. Make frozen stocks of the hybridomas that secrete anti-sulfated glycan antibodies at these steps by standard procedures.
4. It usually takes 10–14 days until the abdomens of the nude mice swell sufficiently for the collection of ascitic fluid as described in this step. To collect the remaining ascitic fluid, wash the abdominal cavity with 3 mL of PBS per mouse and collect the fluid using a glass pipette.
5. If the isotype of the mAb is IgG, purify the mAb using a Protein G-Sepharose column according to the manufacturer's protocol.
6. In some applications, purified mAbs should be conjugated with EZ-Link Sulfo-NHS-LC-Biotin according to the manufacturer's protocol.
7. If the isotype of the mAbs is IgG, incubate the sections with 0.5 μg/mL AlexaFluor 488-labeled goat anti-mouse IgG in PBS containing 0.1 % BSA.
8. Typical results of immunofluorescence with the anti-sulfated glycan mAbs S1 and S2 using frozen sections of PLNs from WT and various DKO mice are shown in Fig. 3. Both S1 and S2 bind well to the HEVs of WT and FucT-IV/FucT-VII DKO mice but fail to bind those of GlcNAc6ST-1/GlcNAc6ST-2 DKO mice. The binding of S1 to the HEVs of C1β3GnT/C2GnT-I DKO mice is almost completely abolished, whereas clear binding of S2 to the HEVs of C1β3GnT/C2GnT-I DKO mice can be detected, although the binding is slightly reduced. These results indicate that S1 preferentially binds sulfated *O*-glycans in HEVs, whereas S2 binds both sulfated *N*- and sulfated *O*-glycans in HEVs [5].
9. The use of human tissue sections was approved by the Ethical Committee of Shinshu University School of Medicine.
10. This step is required when the staining with the mAbs is sensitive to formalin-fixation.
11. Typical results of the immunohistochemical staining with S1 and S2 using human tonsil tissue sections are shown in Fig. 4. Both S1 and S2 react well with the HEVs of human tonsils [5].

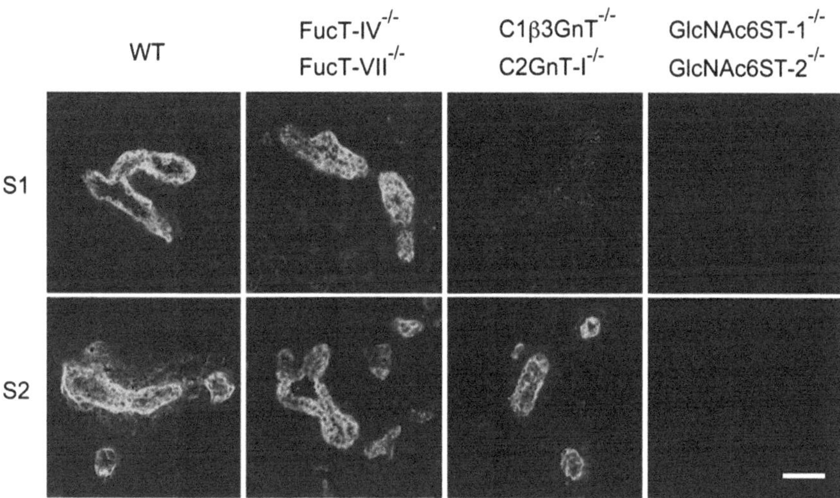

Fig. 3 Immunofluorescence of frozen mouse PLN sections from WT and various gene-targeted mice. Frozen sections (7-μm) of PLNs from WT, FucT-IV/FucT-VII DKO, C1β3GnT/C2GnT-I DKO, and GlcNAc6ST-1/GlcNAc6ST-2 DKO mice were stained with the S1 and S2 mAbs as described in Subheading 3. Bar, 50 μm. Modified from [5]

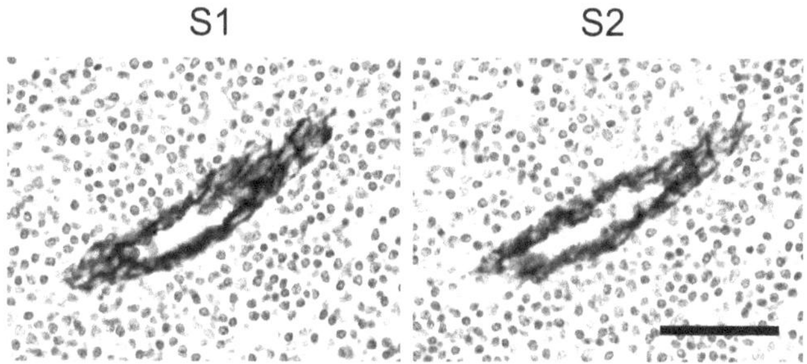

Fig. 4 Immunohistochemical staining of human tonsil sections with the S1 and S2 mAbs. Paraffin-embedded sections of human tonsil tissue were stained with the S1 and S2 mAbs as described in Subheading 3. Bar, 50 μm. Modified from [5]

Acknowledgments

I would like to thank Drs. Jotaro Hirakawa (University of Shizuoka, School of Pharmaceutical Sciences), Minoru Fukuda (Sanford-Burnham Medical Research Institute), and Motohiro Kobayashi (Shinshu University, School of Medicine) for collaboration. This work was supported in part by Grants-in-Aid for Scientific Research, Category (B) and Grants-in-Aid for Scientific Research on Priority Areas, Dynamics of Extracellular Environments, from the Ministry of Education, Culture, Sports, Science and Technology, Japan (21390023 and 20057022, respectively).

References

1. Kawashima H, Petryniak B, Hiraoka N, Mitoma J, Huckaby V, Nakayama J, Uchimura K, Kadomatsu K, Muramatsu T, Lowe JB, Fukuda M (2005) *N*-acetylglucosamine-6-*O*-sulfotransferases 1 and 2 cooperatively control lymphocyte homing through L-selectin ligand biosynthesis in high endothelial venules. Nat Immunol 6:1096–1104
2. Uchimura K, Gauguet JM, Singer MS, Tsay D, Kannagi R, Muramatsu T, von Andrian UH, Rosen SD (2005) A major class of L-selectin ligands is eliminated in mice deficient in two sulfotransferases expressed in high endothelial venules. Nat Immunol 6:1105–1113
3. Lee J, Park SH, Stanley P (2002) Antibodies that recognize bisected complex *N*-glycans on cell surface glycoproteins can be made in mice lacking *N*-acetylglucosaminyltransferase III. Glycoconj J 19:211–219
4. Tahara H, Ide K, Basnet NB, Tanaka Y, Matsuda H, Takematsu H, Kozutsumi Y, Ohdan H (2010) Immunological property of antibodies against *N*-glycolylneuraminic acid epitopes in cytidine monophospho-*N*-acetylneuraminic acid hydroxylase-deficient mice. J Immunol 184:3269–3275
5. Hirakawa J, Tsuboi K, Sato K, Kobayashi M, Watanabe S, Takakura A, Imai Y, Ito Y, Fukuda M, Kawashima H (2010) Novel anti-carbohydrate antibodies reveal the cooperative function of sulfated *N*- and *O*-glycans in lymphocyte homing. J Biol Chem 285:40864–40878
6. Arata-Kawai H, Singer MS, Bistrup A, Zante A, Wang YQ, Ito Y, Bao X, Hemmerich S, Fukuda M, Rosen SD (2011) Functional contributions of *N*- and *O*-glycans to L-selectin ligands in murine and human lymphoid organs. Am J Pathol 178:423–433
7. Mitoma J, Bao X, Petryanik B, Schaerli P, Gauguet JM, Yu SY, Kawashima H, Saito H, Ohtsubo K, Marth JD, Khoo KH, von Andrian UH, Lowe JB, Fukuda M (2007) Critical functions of *N*-glycans in L-selectin-mediated lymphocyte homing and recruitment. Nat Immunol 8:409–418
8. Homeister JW, Thall AD, Petryniak B, Maly P, Rogers CE, Smith PL, Kelly RJ, Gersten KM, Askari SW, Cheng G, Smithson G, Marks RM, Misra AK, Hindsgaul O, von Andrian UH, Lowe JB (2001) The α(1, 3)fucosyltransferases FucT-IV and FucT-VII exert collaborative control over selectin-dependent leukocyte recruitment and lymphocyte homing. Immunity 15:115–126
9. Yeh JC, Hiraoka N, Petryniak B, Nakayama J, Ellies LG, Rabuka D, Hindsgaul O, Marth JD, Lowe JB, Fukuda M (2001) Novel sulfated lymphocyte homing receptors and their control by a Core1 extension β1,3-N-acetylglucosaminyltransferase. Cell 105: 957–969
10. Kawano T, Koyama S, Takematsu H, Kozutsumi Y, Kawasaki H, Kawashima S, Kawasaki T, Suzuki A (1995) Molecular cloning of cytidine monophospho-N-acetylneuraminic acid hydroxylase. Regulation of species- and tissue-specific expression of N-glycolylneuraminic acid. J Biol Chem 270:16458–16463
11. Mitoma J, Miyazaki T, Sutton-Smith M, Suzuki M, Saito H, Yeh JC, Kawano T, Hindsgaul O, Seeberger PH, Panico M, Haslam SM, Morris HR, Cummings RD, Dell A, Fukuda M (2009) The *N*-glycolyl form of mouse sialyl Lewis X is recognized by selectins but not by HECA-452 and FH6 antibodies that were raised against human cells. Glycoconj J 26:511–523
12. Kannagi R, Ohmori K, Kimura N (2009) Anti-oligosaccharide antibodies as tools for studying sulfated sialoglycoconjugate ligands for siglecs and selectins. Glycoconj J 26:923–928

Chapter 6

Fluorescent Microscopy as a Tool to Elucidate Dysfunction and Mislocalization of Golgi Glycosyltransferases in COG Complex Depleted Mammalian Cells

Rose A. Willett, Irina D. Pokrovskaya, and Vladimir V. Lupashin

Abstract

Staining of molecules such as proteins and glycoconjugates allows for an analysis of their localization within the cell and provides insight into their functional status. Glycosyltransferases, a class of enzymes which are responsible for glycosylating host proteins, are mostly localized to the Golgi apparatus, and their localization is maintained in part by a protein vesicular tethering complex, the conserved oligomeric Golgi (COG) complex. Here we detail a combination of fluorescent lectin and immuno-staining in cells depleted of COG complex subunits to examine the status of Golgi glycosyltransferases. The combination of these techniques allows for a detailed characterization of the changes in function and localization of Golgi glycosyltransferases with respect to transient COG subunit depletion.

Key words Conserved oligomeric Golgi (COG) complex, Golgi, Glycosyltransferases, Immunofluorescence, Lectin, siRNA knockdown

1 Introduction

The conserved oligomeric Golgi (COG) complex is a hetero-oligomeric protein complex that functions to tether intra-Golgi vesicles during vesicular trafficking. Vesicular trafficking, which occurs in both an anterograde (forward) and retrograde (reverse) direction [1], is responsible for maintaining the localization of resident Golgi proteins, like glycosyltransferases, to their correct Golgi cisternae. Maintaining the correct localization of glycosyltransferases is crucial for the accurate glycosylation of host proteins [2].

The COG complex consists of eight proteins, named COG1-8 [3–6] which have been grouped into two lobes, COG1-4 in lobe A, and COG5-8 in lobe B [6–8]. Mutations or depletions of COG complex subunits result in the improper glycosylation of the total cellular glycoconjugates [2, 9–11]. In humans, these defects in glycosylation manifest in multiple organ system pathologies

Inka Brockhausen (ed.), *Glycosyltransferases: Methods and Protocols*, Methods in Molecular Biology, vol. 1022, DOI 10.1007/978-1-62703-465-4_6, © Springer Science+Business Media New York 2013

referred to as congenital disorders of glycosylation (CDG) [12]. Currently, patients with CDG's stemming from defects in COG subunits COG1, COG4, COG5, COG6, COG7, and COG8 have been identified [13–21].

To properly assess the abnormalities of steady state conditions for Golgi glycosyltransferases that result from COG complex depletion, we have employed the use of fluorescent microscopy. The glycosyltransferases MAGT1 (α-1,3-mannosyl-glycoprotein 2-beta-*N*-acetylglucosaminyltransferase), MAN2A1 (α-mannosidase II), and ST6GAL1 (β-galactoside α-2, 6-siayltransferase 1) are all localized to the Golgi apparatus under wild type conditions (MAGT1 localizes to *cis*-Golgi membranes, MAN2A1 localizes to medial-Golgi membranes, and ST6GAL1 localizes to *trans*-Golgi membranes). To determine the localization of these glycosyltransferases, we use stable cell lines expressing a tagged version of MAGT1-myc, MAN2A1-VSV, and ST6GAL1-VSV. Without a functional COG complex, the retrograde vesicles used to recycle the glycosyltransferases are not tethered to the Golgi cisternae where they function, and subsequently the protein is not recycled. Upon depletion of the COG complex the enzymes are now partially mislocalized, being found on both small vesicle-like membranes and fragmented Golgi mini stacks (Fig. 1) [2]. The mislocalization of these enzymes results in the incomplete processing of glycoconjugates. In particular, reduced activity of MAGT1 and MAN2A1 will increase a population of glycoconjugates with the immature terminal mannose residues, while the reduced activity of ST6GAL1 will increase a population of glycoconjugates with terminal nonreducing *N*-acetyl-D-glucosaminyl residues. A majority of cellular glycoconjugates are destined for the plasma membrane and therefore are detectable by lectin staining.

In this study we have combined siRNA induced knockdown of individual COG subunits with differential lectin staining techniques to determine the effect of COG subunit depletion on the plasma membrane population of glycoconjugates. Through our studies we found two lectins, *Griffonia simplicifolia* lectin-II (GS-II; binds with high selectivity to terminal, nonreducing α- and β-*N*-acetyl-D-glucosaminyl (GlcNAc) residues of glycoconjugates) and *Galanthus nivalus* lectin (GNL; binds to terminal mannose residues of glycoconjugates) that specifically bind to immature glycoconjugates localized on plasma membrane in COG depleted cells (Fig. 2). HeLa cells treated with a scrambled siRNA were used as a negative control. The ldlB (COG1 KO) and ldlC (COG2 KO) CHO cells were used as positive controls. Control CHO cells were not capable of binding PNA-rhodamine, while both ldlB and ldlC cells were intensively decorated with this lectin, validating our staining procedure (data not shown). The plasma membrane of all tested COG KD cells was specifically stained with GS-II and GNL. Likewise, the plasma membrane of both ldlB and ldlC cells

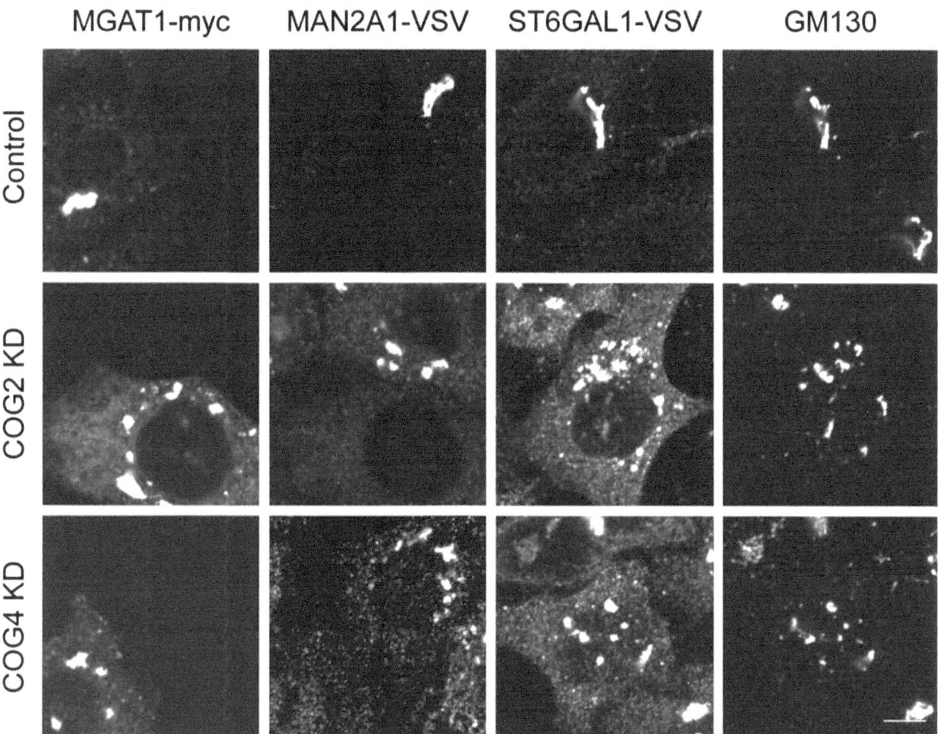

Fig. 1 Localization of Golgi enzymes in COG KD cells. HeLa cells that stably express VSV or myc-tagged Golgi enzymes were mock-transfected or transfected with siRNA to COG2 or COG4. 96 h after transfection cells were fixed, stained with antibodies as indicated and analyzed by laser confocal fluorescent microscopy. In control cells MGAT1-myc (*cis*-Golgi localized glycosyltransferase), MAN2A-VSV (medial-Golgi localized glycosidase), and ST6GAL1-VSV (*trans*-Golgi localized glycosyltransferase) localized to the perinuclear area and co-localized with the Golgi marker GM130. In COG2 and COG4 depleted cells the glycosyltransferases were severely mis-localized, now being found in the periphery on vesicle-like structures as well as fragmented Golgi mini-stacks (Golgi marker GM130 positive membranes). These results indicate a severe mislocalization of glycosyltransferases upon depletion of COG complex subunits

was also distinctly stained with GS-II and GNL. No staining was observed for either control HeLa or CHO cells. This indicated that the COG subunit knockdown cells express immature plasma membrane-localized galactosylated *N*-glycans and glycoconjugates with an increased amount of terminal mannose residues.

2 Materials

2.1 siRNA Induced Knockdown of COG Subunit Components

1. Coverslips: #1.5 12 mm, 0.17 mm thickness, round glass coverslips (Warner Instruments; Hamden, CT, USA) (maximum of five coverslips per well on a 6 well plate).
2. Culture dishes: 6 well tissue culture plates (TPP).
3. HeLa cells, wild type or stably expressing tagged glycosyltransferase (MAGT1-myc (α-1,3-mannosyl-glycoprotein

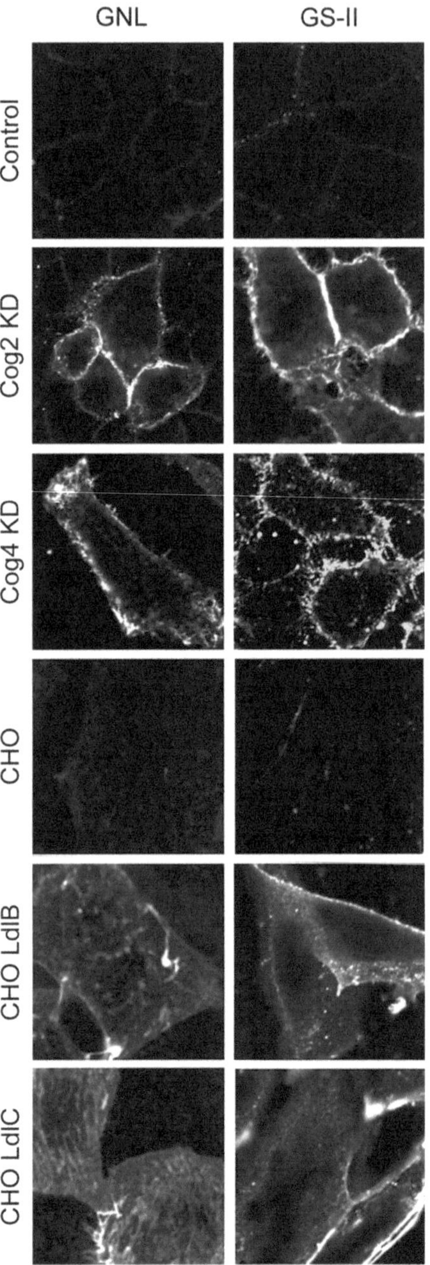

Fig. 2 Lectin staining of COG complex depleted HeLa cells reveals altered glycosylation of plasma membrane glycoconjugates. HeLa cells were mock-transfected or transfected with siRNA to COG2 or COG4. 96 h after transfection, HeLa cells, and CHO cells for control staining, were fixed, stained with GS-II-Alexa 594 or GNL-Alexa 647 lectins for 30 min and analyzed by wide-field fluorescent microscopy. The control HeLa cells and control CHO cells were negative for both lectins, indicating that plasma membrane localized glycoconjugates' polysaccharide chains are completely mature. In cells depleted of either COG2 or COG4, the plasma membrane was extensively labelled with both GNL and GS-II lectins. Likewise, the plasma membranes of both ldlB and ldlC cells were also distinctly stained with GNL and GS-II. This indicates that the COG subunit knockdown cells express plasma membrane-localized glycoconjugates with an increased amount of terminal mannoses and GlcNAc

2-beta-*N*-acetylglucosaminyltransferase)) [22] grown on coverslips to 30 % confluency on a 6 well plate.

4. HeLa cells, wild type or stably expressing tagged glycosidase MAN2A1-VSV (α-mannosidase II) [23] grown on coverslips to 30 % confluency on a 6 well plate.
5. HeLa cells, wild type or stably expressing tagged glycosyltransferase ST6GAL1-VSV (β-galactoside α-2, 6-sialyl-transferase 1) [23] grown on coverslips to 30 % confluency on a 6 well plate.
6. Dulbecco's Phosphate Buffered Saline (dPBS 1×) without calcium and magnesium (Thermo Fisher Scientific Inc; Waltham, MA).
7. Growth Media: dilute 50 mL of heat inactivated Fetal Bovine Serum (FBS) (Atlas Biologicals Inc., Fort Collins, CO, USA) in 450 mL of in DMEM/F-12 50/50 medium supplemented with 15 mM HEPES, 2.5 mM L-glutamine (Invitrogen; Carlsbad, CA, USA). Filter solution in 0.45 μm PES (Corning, Lowell, MA, USA) filtration system.
8. Transfection media: Opti-MEM® I Reduced Serum Media buffered with HEPES and sodium bicarbonate and supplemented with hypoxanthine, thymidine, sodium pyruvate, L-glutamine, trace elements and growth factors (Invitrogen).
9. Gibco® 0.25 % Trypsin-EDTA (1×) phenol red (Invitrogen).
10. Lipofectamine RNAiMAX siRNA Transfection Reagent (Invitrogen).
11. SiRNA: siGENOME siRNA—Human COG2 (target sequence: GGGCAGTTGATGAACGAAT), ON-TARGETplus siRNA—Human COG4 (target sequence: GTGCTGAAATCCACCTTTA), and control scrambled ON-TARGETplus siRNA) (Dharmacon; Chicago, IL, USA).
12. Primers: hCOG2 Forward: GGACACGCTCTGCTTCGACA; hCOG2 Reverse: ACAGAAAGCTGGTTGAGGGC; and hCOG4 Forward: TCTGCAGGTGGAATGTGACAGACA; hCOG4 Reverse: CTGTGCATGATGTTCACGGCACTT (Invitrogen).

2.2 Lectin-Staining of Intact Cell Components

1. HeLa cells transfected with siRNAs grown on coverslips to 70 % confluency on a 6 well plate.
2. CHO cells, CHO ldl (low density lipoprotein) B cells (stable COG1 knockout), and CHO ldlC cells (stable COG2 knockout) [24–26] grown on coverslips to 70 % confluency.
3. Coverslips: #1.5 12 mm, 0.17 mm thickness, round glass coverslips (Warner Instruments).
4. Lectins: *Griffonia simplicifolia* lectin II (GSII)-Alexa 594 (100 μg/mL, Invitrogen), *Galanthus nivalus* lectin

(GNL)-Fluorescein (20 μg/mL, Vector laboratories; Burlingame, CA, USA) (*see* **Note 1**).

5. Dulbecco's Phosphate Buffered Saline (dPBS 1×) without calcium and magnesium (Thermo Fisher Scientific).
6. Cell fixative solution: 1 % solution paraformaldehyde solution in dPBS prepared by diluting 16 % stock solution (Electron Microscopy Sciences; Hatfield, PA, USA) (*see* **Note 2**).
7. Quenching solution: 50 mM NH_4Cl prepared by dissolving 134 mg of NH_4Cl (Sigma-Aldrich, St. Louis, MO, USA) in 50 mL of dPBS. Store at 4 °C.
8. Blocking A solution: 1 % BSA prepared by dissolving 1 g of Bovine serum albumin (BSA, Fraction V) (Research Products International Corporation; Mount Prospect, IL, USA) in 100 mL dPBS. Filter completely dissolved solution in Corning 250 mL 0.22 μm PES filter system. Store at 4 °C.
9. 4′,6-Diamidino-2-phenylindole dihydrochloride (DAPI) (Sigma-Aldrich).
10. Mounting media: Prolong® Gold antifade mounting media from Invitrogen.
11. Glass slides: Fisherbrand frosted microscope slides (precleaned).
12. Parafilm.
13. Vacuum apparatus for collecting waste: Büchner flask, with extended intake tubing, connected to a vacuum source.
14. Zeiss Axiovert 200 M fluorescent microscope.

2.3 Immunofluorescence Staining Components

1. HeLa cells stably expressing tagged glycosyltransferases (MAGT1-myc (α-1,3-mannosyl-glycoprotein 2-beta-*N-acetyl glucosaminyltransferase*), MAN2A1-VSV (α-mannosidase II), ST6GAL1-VSV (β-galactoside α-2, 6-siayltransferase 1) transfected with siRNAs grown on coverslips to 70 % confluency on a 6 well plate.
2. Coverslips: #1.5 12 mm, 0.17 mm thickness, round glass coverslips (Warner Instruments).
3. Dulbecco's Phosphate Buffered Saline (dPBS 1×) without calcium and magnesium.
4. 20 % (w/v) Triton X-100 solution prepared by weighing out 10 g of Triton X-100 (Sigma-Aldrich) and diluting it to a total volume of 50 mL with Milli-Q water (*see* **Note 3**).
5. Cell fixative solution: 4 % paraformaldehyde solution in dPBS prepared by diluting 250 μL of 16 % stock solution (Electron Microscopy Sciences) in 750 μL of dPBS (*see* **Note 2**).
6. Cell permeabilizing solution: 0.1 % Triton solution prepared by diluting 250 μL of 20 % Triton X-100 stock in 50 mL of dPBS. Store at 4 °C.

7. Quenching solution: 50 mM NH_4Cl prepared by dissolving 134 mg of NH_4Cl (Sigma-Aldrich) in 50 mL of dPBS. Store at 4 °C.
8. Blocking B solution: 1 % BSA, 0.1 % saponin prepared by dissolving 1 g of Bovine serum albumin (BSA, Fraction V) (Research Products International Corporation) and 100 mg saponin (Sigma-Aldrich) in 100 mL dPBS. Filter completely dissolved solution in Corning 250 mL 0.22 μm PES filer system. Store at 4 °C.
9. Diluent solution: 1 % fish gelatin, 0.1 % saponin prepared by dissolving 1 g gelatin from cold water fish skin (Sigma-Aldrich) and 100 mg saponin (Sigma-Aldrich) in 100 mL dPBS. Filter completely dissolved solution in Corning 250 mL 0.22 μm PES filter system. Store at 4 °C.
10. Primary antibodies: Anti-myc tag rabbit polyclonal antibodies (Bethyl Laboratories; Montgomery, TX, USA) 1:3,000 dilution in diluent solution. Anti-VSV tag rabbit polyclonal antibodies (Bethyl Laboratories) 1:400 dilution in diluent solution. Anti-GM130 mouse monoclonal antibodies (BD Biosciences; San Jose, CA, USA) 1:400 in diluent solution.
11. Secondary antibodies: Anti-rabbit HiLyte 488 1:400 in diluent solution, anti-mouse HiLyte 555 (for GM130) 1:1,000 in diluent solution (AnaSpec, Inc., San Jose, CA, USA).
12. Mounting media: Prolong® Gold antifade mounting media with Dapi (Invitrogen).
13. Glass slides: Fisherbrand frosted microscope slides (precleaned).
14. Parafilm.
15. Vacuum apparatus for collecting waste: Büchner flask, with extended intake tubing, connected to a vacuum source.
16. Zeiss LSM510 laser inverted microscope outfitted with confocal optics, 63× oil 1.4 numerical aperture (NA) objective. Image acquisition is controlled with LSM510 software (Release Version 4.0 SP1).

3 Methods

3.1 siRNA Induced Knockdown of COG Subunits

All steps are performed under a sterile hood. Gloves are worn at all times to prevent contamination. Two wells for each transfection were used, one well with coverslips, and one well without coverslips for knockdown efficiency analysis.

1. Plate wild type HeLa cells, or HeLa cells stably expressing tagged glycosyltransferases, on 6 well culture dishes with coverslips one day prior to transfection in 10 % FBS DMEM/F-12 media that does not contain any antibiotics so that the day of

the transfection the cells are 30 % confluent and evenly spread (*see* **Note 4**). Grow cells at 37 °C and 5 % CO_2 in a 90 % humidified incubator.

2. Prepare transfection solutions as detailed by manufacturer's protocol. For a 6 well plate: in a 1.5 mL microcentrifuge tube dilute 5 μL of Lipofectamine™ RNAiMAX in 245 μL of Opti-MEM®, set aside and let incubate for 5–10 min. In a separate tube, combine 10 μL of 20 μM hCOG2 or hCOG4 siRNA stock with 240 μL of Opti-MEM®, gently mixing the solution. After 10 min, combine the diluted siRNA with the diluted Lipofectamine™ RNAiMAX and incubate for 20 min (*see* **Note 5**).
3. While solution is incubating, wash cells two times with sterile dPBS. Remove residual PBS and incubate cells in 2 mL of Opti-MEM®.
4. After 20 min of incubation, add in a drop-wise manner the siRNA–Lipofectamine™ RNAiMAX complexes to their corresponding wells. Mix gently by rocking the plate back and forth.
5. Incubate the cells for 12 h at 37 °C, 5 % CO_2, and 90 % humidity, then remove transfection solution and replace with 10 % FBS—DMEM/F-12 growth media and allow cells to recover for an additional 12 h.
6. 24 h after the transfection, repeat the siRNA knockdown as described above [2–5].
7. Incubate cells for a total of 96 h after the first transfection, and then proceed to harvesting and staining.
8. Knockdown efficiency is determined 48 h after the first transfection by qRT-PCR using primers for COG2 and COG4 (*see* **Note 6**).

3.2 Lectin Staining of COG Subunit Depleted Cells

Lectin staining is performed with both fixed and unfixed cells (*see* **Notes 7** and **8**). All steps are performed at room temperature. Solutions are stored at 4 °C until use. Method corresponds to Fig. 1.

1. Untreated (control) and siRNA treated cells are grown on 12 mm glass coverslips at 70 % of confluency. 96 h after the siRNA-induced transfection cells are rinsed with dPBS to remove the broken cell material and growth media.
2. Coverslips are taken out and placed on parafilm sheet at room temperature with the cell covered surface facing up (do not let cells dry at any step!).
3. Incubate coverslips in 100 μL freshly made 1 % paraformaldehyde in dPBS solution for 10 min.
4. Remove the residual 1 % paraformaldehyde by pipette (*see* **Note 9**) and rinse coverslips two times with 200 μL of dPBS.
5. Incubate coverslips in 100 μL of 50 mM NH_4Cl in PBS for 5 min.

6. Remove 50 mM NH_4Cl and wash coverslips with 100 μL of dPBS.
7. Incubate coverslips in 100 μL of 0.1 % BSA in dPBS for 10 min.
8. Stain with lectins (100–150 μL) diluted to proper concentration (*see* Subheading 2.2) for 1 h at room temperature. Place a small box over surface that coverslips are protected from light.
9. Remove lectin solution by vacuum and wash coverslips four times with 200 μL of 1× dPBS, incubating for 2 min for each wash.
10. Incubate coverslips with DAPI (1 μL diluted in 5 mL dPBS) for 30 s (*see* **Note 10**).
11. Pick up coverslip with forceps and immerse them into beaker of dPBS ten times, followed by dipping in beaker with Milli-Q water ten times.
12. Gently remove excess liquid by tapping a Kimwipe to the edge of coverslip.
13. Place coverslip cell side down on glass slide with a small drop (5 μL) of mounting media.
14. Once all coverslips are mounted to glass slide (no more than 6 per slide) remove excess solution with vacuum.
15. Store slides on a flat, dry surface protected from light (slide book), and let cure overnight.
16. Image coverslips with the 63× oil 1.4 numerical aperture (NA) objective of a Zeiss Axiovert 200 M fluorescent microscope.

3.3 Immunofluorescence Staining of Glycosyltransferases in COG Subunit Depleted Cells

All steps are performed at room temperature. Solutions are stored at 4 °C until use (*see* **Note 8**). Method corresponds to Fig. 2.

1. Coverslips with control or siRNA treated cells grown to near confluency (70 %) are removed from culture dishes and placed on parafilm covered surface with the cell covered surface facing up (do not let cells dry at any step!).
2. Wash coverslips rapidly three times by adding 200 μL 1× dPBS to each coverslip, removing the used solution.
3. Remove the residual 1× dPBS and incubate coverslips in 100 μL of 4 % paraformaldehyde for 15 min.
4. Remove 4 % paraformaldehyde solution by pipette and incubate coverslips in 100 μL of 1 % Triton X-100 for 1 min.
5. Remove 1 % Triton X-100 solution and incubate coverslips in 100 μL of 50 mM NH_4Cl for 5 min.
6. Remove 50 mM NH_4Cl solution and incubate coverslips with 100 μL of 1 % BSA, 0.1 % saponin 10 min, remove this solution and then repeat incubation with the same solution for another 10 min.

7. Remove 1 % BSA, 0.1 % saponin solution and add 100 μL of diluted primary antibody solution to coverslips and incubate for 40 min. For Myc tagged cells, the primary antibody solution is 1 mL of diluted anti-myc antibodies (1 μL antibodies in 3 mL 1 % fish gelatin, 0.1 % saponin) and 2.5 μL of anti-GM130 antibodies. For VSV tagged cells, the primary antibody solution is 2.5 μL anti-VSV antibodies and 2.5 μL anti-GM130 antibodies diluted in 1 mL 1 % fish gelatin, 0.1 % saponin (*see* **Note 11**).
8. Remove primary antibody solution and wash coverslips four times with 200 μL of 1× dPBS, incubating for 2 min for each wash.
9. Remove dPBS and add 100 μL of diluted secondary antibodies to coverslips and incubate for 30 min. The secondary antibody solution is 1 μL anti-rabbit HiLyte 488 and 1 μL anti-mouse HiLyte 555 in 1 mL 1 % fish gelatin, 0.1 % saponin (*see* **Note 11**). Place a small box over surface to protect samples from light.
10. Remove secondary antibody solution and wash coverslips five times with 1× dPBS, incubating for 2 min for each wash.
11. Pick up coverslip with forceps and immerse them into beaker of dPBS ten times, followed by dipping in beaker with Milli-Q water ten times.
12. Gently remove excess liquid by tapping a Kimwipe to the edge of coverslip.
13. Place coverslip cell side down on glass slide with a small drop (5 μL) of mounting media.
14. Once all coverslips are mounted to glass slide (no more than 6 per slide) remove excess of mounting media with vacuum.
15. Store slides on a flat, dry surface protected from light (slide book), and let cure overnight.
16. Image coverslips with the 63× oil 1.4 numerical aperture (NA) objective of a LSM510 Zeiss Laser inverted microscope outfitted with confocal optics. Image acquisition is controlled with LSM510 software (Release Version 4.0 SP1).

4 Notes

1. Lectins are stored in the dark at 4 °C. In addition to commercially available fluorescent-labelled lectins, unlabelled lectins can be purchased from Vector laboratories and labelled with Alexa-647 protein-labelling kit (Invitrogen).
2. Fixative solution works best when diluted fresh before start of experiment. Seal unused 16 % paraformaldehyde with parafilm and store away from light.

3. Rock at 4 °C overnight and store at 4 °C until use. Do not store for more than 1 month.
4. The actual number of cells to achieve desired confluency varies depending on cell line. Differences in size and growth rates should be considered when plating cells for transfection.
5. siRNA is extremely fragile and should be handled very delicately. When diluting siRNA only gently mix it with the Opti-MEM®. For best results, do not use stock aliquot more than two times because continually thawing will destroy siRNA.
6. qRT-PCR was used to validate the efficiency knockdowns because reliable commercial antibodies against COG2 and COG4 are not available. Actin qRT-PCR was used as a control.
7. For lectin staining of unfixed cells, all steps are performed in cold room at 4 °C. Protocol was done as follows: Rinse coverslips in dPBS (chilled to 4 °C) and incubate for 15 min to reduce endocytosis of lectin. After cells have been chilled, incubate with lectins for 20 min (lectin solutions should be chilled to 4 °C as well). Wash coverslips 4 times with cold dPBS and then fix with 4 % paraformaldehyde (prepared fresh and chilled to 4 °C). Mount coverslips to slides as indicated above. While both methods will yield the same result, we have found that the aforementioned protocol (staining then fixing) is simpler and more reproducible.
8. It is important to never let the coverslips dry during the staining procedure.
9. With the exception of the paraformaldehyde, the removal of all of the staining solutions can be done either by pipette, or with a vacuum (*see* Subheading 2). We prefer to use the vacuum because it allows for a faster removal of solutions, thereby increasing the accuracy of incubation time.
10. DAPI staining is optional.
11. Diluted antibodies should be thoroughly mixed and then centrifuged at 20,000 × *g* for 1 min. This step removes any large particles that could provide a high background during visualization.

References

1. Shorter J, Warren G (2002) Golgi architecture and inheritance. Annu Rev Cell Dev Biol 18:379–420
2. Pokrovskaya ID, Willett R, Smith RD, Morelle W, Kudlyk T, Lupashin VV (2011) COG complex specifically regulates the maintenance of Golgi glycosylation machinery. Glycobiology 21(12):1554–1569
3. Suvorova ES, Kurten RC, Lupashin VV (2001) Identification of a human orthologue of Sec34p as a component of the cis-Golgi vesicle tethering machinery. J Biol Chem 276: 22810–22818
4. Suvorova ES, Duden R, Lupashin VV (2002) The Sec34/Sec35p complex, a Ypt1p effector required for retrograde intra-Golgi trafficking, interacts with Golgi SNAREs and COPI vesicle coat proteins. J Cell Biol 157:631–643
5. Whyte JR, Munro S (2001) The Sec34/35 Golgi transport complex is related to the

exocyst, defining a family of complexes involved in multiple steps of membrane traffic. Dev Cell 1:527–537
6. Ungar D, Oka T, Brittle EE, Vasile E, Lupashin VV, Chatterton JE, Heuser JE, Krieger M, Waters MG (2002) Characterization of a mammalian Golgi-localized protein complex, COG, that is required for normal Golgi morphology and function. J Cell Biol 157:405–415
7. Ungar D, Oka T, Vasile E, Krieger M, Hughson FM (2005) Subunit architecture of the conserved oligomeric golgi complex. J Biol Chem 280:32729–32735
8. Fotso P, Koryakina Y, Pavliv O, Tsiomenko AB, Lupashin VV (2005) Cog1p plays a central role in the organization of the yeast conserved oligomeric golgi complex. J Biol Chem 280:27613–27623
9. Bruinsma P, Spelbrink RG, Nothwehr SF (2004) Retrograde transport of the mannosyltransferase Och1p to the early Golgi requires a component of the COG transport complex. J Biol Chem 279:39814–39823
10. Shestakova A, Zolov S, Lupashin V (2006) COG complex-mediated recycling of Golgi glycosyltransferases is essential for normal protein glycosylation. Traffic 7:191–204
11. Kingsley DM, Kozarsky KF, Segal M, Krieger M (1986) Three types of low density lipoprotein receptor-deficient mutant have pleiotropic defects in the synthesis of N-linked, O-linked, and lipid- linked carbohydrate chains. J Cell Biol 102:1576–1585
12. Reynders E, Foulquier F, Annaert W, Matthijs G (2011) How Golgi glycosylation meets and needs trafficking: the case of the COG complex. Glycobiology 21:853–863
13. Wu X, Steet RA, Bohorov O, Bakker J, Newell J, Krieger M, Spaapen L, Kornfeld S, Freeze HH (2004) Mutation of the COG complex subunit gene COG7 causes a lethal congenital disorder. Nat Med 10:518–523
14. Foulquier F, Ungar D, Reynders E, Zeevaert R, Mills P, Garcia-Silva MT, Briones P, Winchester B, Morelle W, Krieger M, Annaert W, Matthijs G (2007) A new inborn error of glycosylation due to a Cog8 deficiency reveals a critical role for the Cog1-Cog8 interaction in COG complex formation. Hum Mol Genet 16:717–730
15. Foulquier F, Vasile E, Schollen E, Callewaert N, Raemaekers T, Quelhas D, Jaeken J, Mills P, Winchester B, Krieger M, Annaert W, Matthijs G (2006) Conserved oligomeric Golgi complex subunit 1 deficiency reveals a previously uncharacterized congenital disorder of glycosylation type II. Proc Natl Acad Sci USA 103:3764–3769
16. Kranz C, Ng BG, Sun L, Sharma V, Eklund EA, Miura Y, Ungar D, Lupashin V, Winkel RD, Cipollo JF, Costello CE, Loh E, Hong W, Freeze HH (2007) COG8 deficiency causes new congenital disorder of glycosylation type IIh. Hum Mol Genet 16: 731–741
17. Ng BG, Kranz C, Hagebeuk EE, Duran M, Abeling NG, Wuyts B, Ungar D, Lupashin V, Hartdorff CM, Poll-The BT, Freeze HH (2007) Molecular and clinical characterization of a moroccan Cog7 deficient patient. Mol Genet Metab 91:201–204
18. Zeevaert R, Foulquier F, Jaeken J, Matthijs G (2008) Deficiencies in subunits of the conserved oligomeric Golgi (COG) complex define a novel group of congenital disorders of glycosylation. Mol Genet Metab 93:15–21
19. Reynders E, Foulquier F, Leao Teles E, Quelhas D, Morelle W, Rabouille C, Annaert W, Matthijs G (2009) Golgi function and dysfunction in the first COG4-deficient CDG type II patient. Hum Mol Genet 18:3244–3256
20. Paesold Burda P, Maag C, Troxler H, Foulquier F, Kleinert P, Schnabel S, Baumgartner M, Hennet T (2009) Deficiency in COG5 causes a moderate form of congenital disorders of glycosylation. Hum Mol Genet 18(22): 4350–4356
21. Lubbehusen J, Thiel C, Rind N, Ungar D, Prinsen BH, de Koning TJ, van Hasselt PM, Korner C (2010) Fatal outcome due to deficiency of subunit 6 of the conserved oligomeric Golgi complex leading to a new type of congenital disorders of glycosylation. Hum Mol Genet 19:3623–3633
22. Nilsson T, Hoe MH, Slusarewicz P, Rabouille C, Watson R, Hunte F, Watzele G, Berger EG, Warren G (1994) Kin recognition between medial Golgi enzymes in HeLa cells. EMBO J 13:562–574
23. Miles S, McManus H, Forsten KE, Storrie B (2001) Evidence that the entire Golgi apparatus cycles in interphase HeLa cells: sensitivity of Golgi matrix proteins to an ER exit block. J Cell Biol 155:543–555
24. Krieger M, Brown M, Goldstein J (1981) Isolation of Chinese hamster cell mutants defective in the receptor-mediated endocytosis of low density lipoprotein. J Mol Biol 150:167–184
25. Podos SD, Reddy P, Ashkenas J, Krieger M (1994) LDLC encodes a brefeldin A-sensitive, peripheral Golgi protein required for normal Golgi function. J Cell Biol 127:679–691
26. Chatterton JE, Hirsch D, Schwartz JJ, Bickel PE, Rosenberg RD, Lodish HF, Krieger M (1999) Expression cloning of LDLB, a gene essential for normal Golgi function and assembly of the ldlCp complex. Proc Natl Acad Sci USA 96:915–920

Chapter 7

A Practical Approach to Reconstruct Evolutionary History of Animal Sialyltransferases and Gain Insights into the Sequence–Function Relationships of Golgi-Glycosyltransferases

Daniel Petit, Roxana Elin Teppa, Jean-Michel Petit, and Anne Harduin-Lepers

Abstract

In higher vertebrates, sialyltransferases catalyze the transfer of sialic acid residues, either Neu5Ac or Neu5Gc or KDN from an activated sugar donor, which is mainly CMP-Neu5Ac in human tissues, to the hydroxyl group of another saccharide acceptor. In the human genome, 20 unique genes have been described that encode enzymes with remarkable specificity with regards to their acceptor substrates and the glycosidic linkage formed. A systematic search of sialyltransferase-related sequences in genome and EST databases and the use of bioinformatic tools enabled us to investigate the evolutionary history of animal sialyltransferases and propose original models of divergent evolution of animal sialyltransferases. In this chapter, we extend our phylogenetic studies to the comparative analysis of the environment of sialyltransferase gene loci (synteny and paralogy studies), the variations of tissue expression of these genes and the analysis of amino-acid position evolution after gene duplications, in order to assess their sequence–function relationships and the molecular basis underlying their functional divergence.

Key words Sialyltransferases, Molecular phylogeny, Phylogenomics, Evolution, Golgi, Glycosylation, Cloning

1 Introduction

Glycobiology is one of the most important fields of life science and glycosylation represents one of the most complex and important modification of proteins and lipids. Glycosylation involves the sequential and coordinated action of several Golgi localized enzymes known as glycosyltransferases that catalyze the transfer of a unique monosaccharide residue from an activated sugar donor to specific acceptor molecules which are either glycoproteins or glycolipids. Over 180 carbohydrate active enzymes named

Inka Brockhausen (ed.), *Glycosyltransferases: Methods and Protocols*, Methods in Molecular Biology, vol. 1022, DOI 10.1007/978-1-62703-465-4_7, © Springer Science+Business Media New York 2013

glycogenes have been identified in the human genome [1]. Among the last acting enzymes, 20 human sialyltransferases (ST) have been identified, their substrates and not substrates specificities characterized essentially using biochemical approaches and their corresponding cDNA have been cloned (for reviews *see* refs. 2–4) clearly demonstrating that sequence complexity of glycosyltransferases leads to a variety of glycan functions. Members of the multigene sialyltransferase superfamily are involved in the biosynthesis of sialylated molecules that play important role in a variety of cellular processes. Phenotypic changes have been observed due to mutated or disrupted sialyltransferase genes leading to defective sialic acid metabolism that are the molecular basis of several diseases and microbial infections [5, 6], congenital disorders of glycosylation [7], and cancers [8]. Despite progress made these past years in mammalian sialyltransferase structural glycobiology [9, 10], still very little is known concerning their biological functions and it is difficult to use structure-based approaches for function prediction. In order to gain a global understanding of these animal glycogenes, we document their evolutionary relationships thus providing a framework to elucidate their common mechanisms and establish relationships between glycogenomics and sialomes expressed by cells and animals used as model organisms such as amphioxus or zebrafish [11–13]. The sialyltransferase superfamily in Metazoan is evolutionarily ancient since a unique α2,6-sialyltransferase (ST6Gal) sequence could be identified in the genome of the sponge *Oscarella carmela* [14]. It appears that a limited number of sialyltransferase ancestors have diversified during chordate evolution via successive rounds of gene duplication, diverged through several genetic mechanisms such as nucleotide polymorphism, tandem duplication and translocation (TDT), polyploidy or loss of genes, and acquired new specificity from subtle sequence changes [3, 4, 14, 15]. In this chapter, we provide a comprehensive coverage of the methodological bioinformatic approaches and in silico analysis that have significantly enriched our understanding of sequence–function relationships of these Golgi-glycosyltransferases.

2 Materials

All the programs described in this paper may run on a personal computer with an Internet connection.

1. PHYLIP 3.69 package for Windows 3.1 [16] and PhyML version 2.4 [17] are downloaded from the http://evolution.genetics.washington.edu/phylip/getme.html Web site at the Department of Genetics of the University of Washington and from the http://www.atgc-montpellier.fr/phyml/ Web site, respectively.

2. MEGA 5.0 for Windows is downloaded from the http://www.megasoftware.net/ Web site [18].
3. PAST 2.11 is downloaded from the Web site http://www.nhm.uio.no/norlex/past/download.html [19].
4. T-Coffee package is downloaded from the site http://www.tcoffee.org/ or used via Web http://tcoffee.crg.cat/ [20].
5. The HUGO Gene Nomenclature Committee (HGNC) is found at http://www.genenames.org/ [21].
6. Treeview can be downloaded from the Web site: http://treeview-x.en.softonic.com/ [22].

3 Methods

Phylogenomic inference of animal sialyltransferase function consists in a series of steps depicted in Fig. 1. The general strategy adopted to identify and assess phylogenetic distribution of putative gene and protein sialyltransferase sequences in more distantly related genomes relies on sequence-based screening methods with known human or mouse complete sialyltransferase coding region. Multiple sequence alignment of identified sequences is a critical step for subsequent analyses and phylogenetic tree construction. Molecular phylogeny and phylogenomics enable us to evaluate their duplicative origin and gain insights into their evolutionary history and sequence/function relationships. This strategy cannot uncover nonhomologous enzymes with similar biochemical functions such as bacterial sialyltransferases. Data mining of sialyltransferase homologues, both paralogues and orthologues, is greatly facilitated by the presence of four consensus peptide sequences named sialylmotifs found in the catalytic domain of all animal sialyltransferases [3, 23, 24]. Sialylmotifs are functional signatures of these enzymes since they are essential for their enzymatic activity [24–29]. Four main ST families are distinguished based on their substrate specificity and glycosidic linkage formed. They are named ST8Sia, ST6Gal, ST6GalNAc, and ST3Gal according to the nomenclature proposed by Tsuji et al. [30]. Each of these sialyltransferase families is characterized by family motifs described more recently [3, 31], which are also useful for sialyltransferase identification. Mammalian sialyltransferase genes are named according to the HUGO nomenclature committee and a refined nomenclature is proposed in the notes section to account for the vertebrate sialyltransferase diversity described recently (*see* **Note 1**). Interestingly, all the animal sialyltransferase genes are gathered in the group GT29 of the CAZy classification of glycosyltransferases [32] suggesting their common origin.

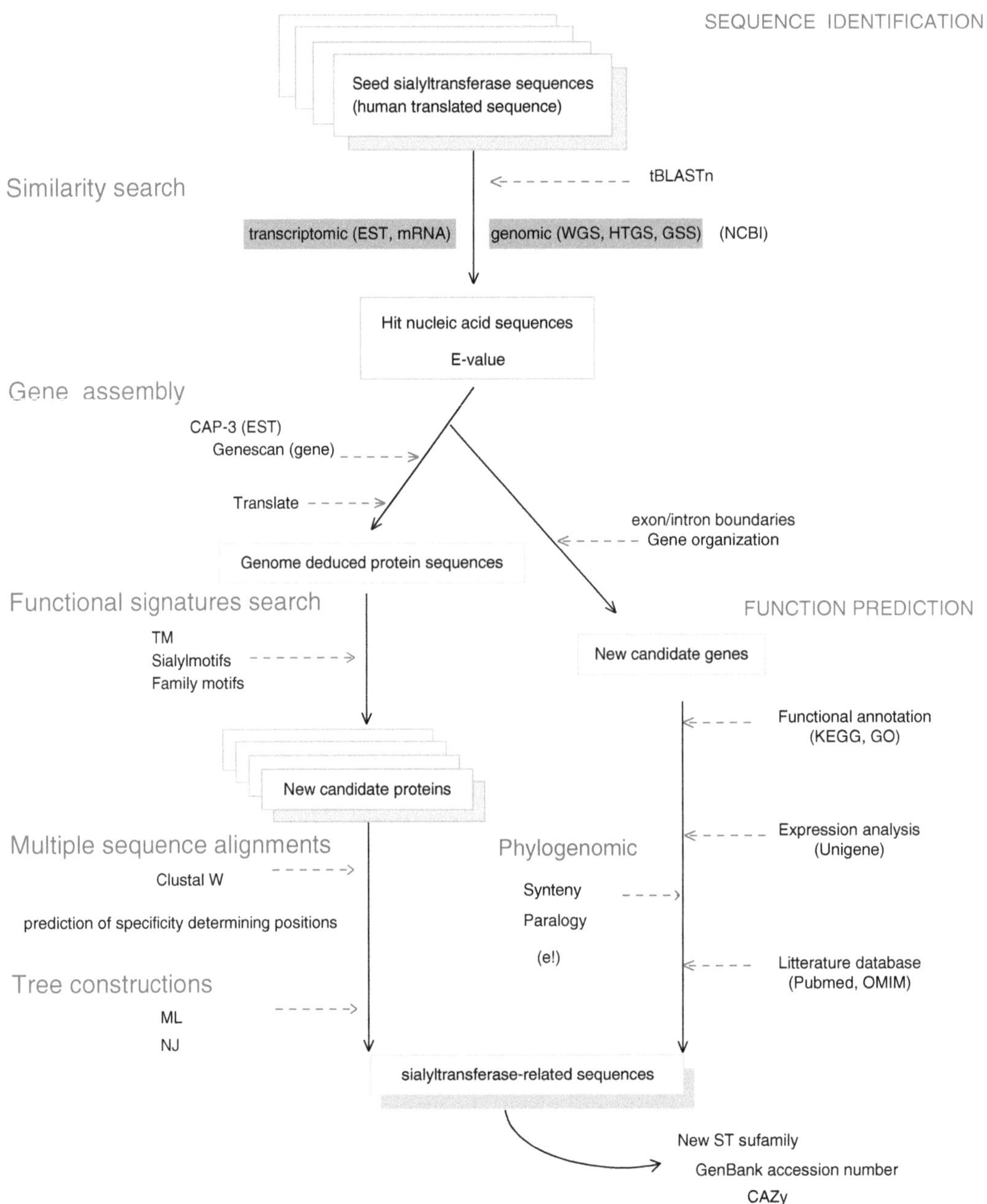

Fig. 1 Flowchart for sialyltransferase-related sequence identification and function prediction. Various sequenced-based approaches are used for the identification of sialyltransferases. This drawing illustrates the various steps of similarity searches with user input (protein or nucleotide known sialyltransferase), of gene assembly, of functional signature and motif search, of multiple sequence alignments, and of tree construction. Information is finally gathered for classification and functional inference

3.1 Mining of Sialyltransferase-Related Sequences

3.1.1 Databases

There are two kinds of major sequence repositories used for the identification of sialyltransferases (*see* **Note 2**) gathered in Table 1. We mostly use genome browser of the University of California Santa Cruz (UCSC) [33] of Ensembl [34] and of the NCBI or general publicly available genomic or transcriptomic databases of the NCBI

Table 1
Summary of available Web-online sites of genomes databases

Databases	URL	References
EMBL	http://www.ebi.ac.uk/embl/	[61]
GenBank	http://www.ncbi.nlm.nih.gov/genbank/GenbankSearch.html	[62]
DDBJ	http://blast.ddbj.nig.ac.jp/top-e.html	[63]
Uniprot	http://www.uniprot.org/	[69]
Joint genome institute (JGI)	http://genome.jgi.doe.gov	[70]
TGI	http://compbio.dfci.harvard.edu/	[65]
Welcome trust Sanger Institute	http://www.sanger.ac.uk/resources/databases/	
Kyoto encyclopedia of gene and genome KEGG genes	http://www.genome.jp/kegg/genes.html	[71, 72]
CAZy	http://www.cazy.org/	[32]
GGDB glycogene database	http://riodb.ibase.aist.go.jp/rcmg/ggdb/	[1]
Consortium for functional glycomics CFG GT database	http://www.functionalglycomics.org/	[73]
ZFIN	http://zfin.org/	[74]
BRENDA	http://www.brenda-enzymes.org/	[75]
Pfam	http://www.sanger.ac.uk/resources/databases/pfam.html	[43]
HUGE	http://www.kazusa.or.jp/huge/	[76]

Information sources of sialyltransferase-related databases

with the BLAST tool and a protein sequence query to extract new protein and nucleotide sialyltransferase-related sequences, as described in Fig. 2. In order to search for new sequences:

1. Open the NCBI BLAST page (Table 2).
2. Choose a Basic BLAST algorithm like tBLASTn, which screens translated nucleotide database using a translated nucleotide query. Paste the human protein sequence in the section “enter query sequence”. The preferred query sequence format for the BLAST program is the FASTA format, which begins with a “greater-than” symbol in the line of description, followed by a line of sequence data. Input data can be introduced either by copy/paste, file loading, database fetch, accession numbers or GI’s or selection of the output from a previous analysis. There is also the option of searching with a segment of the query sequence.

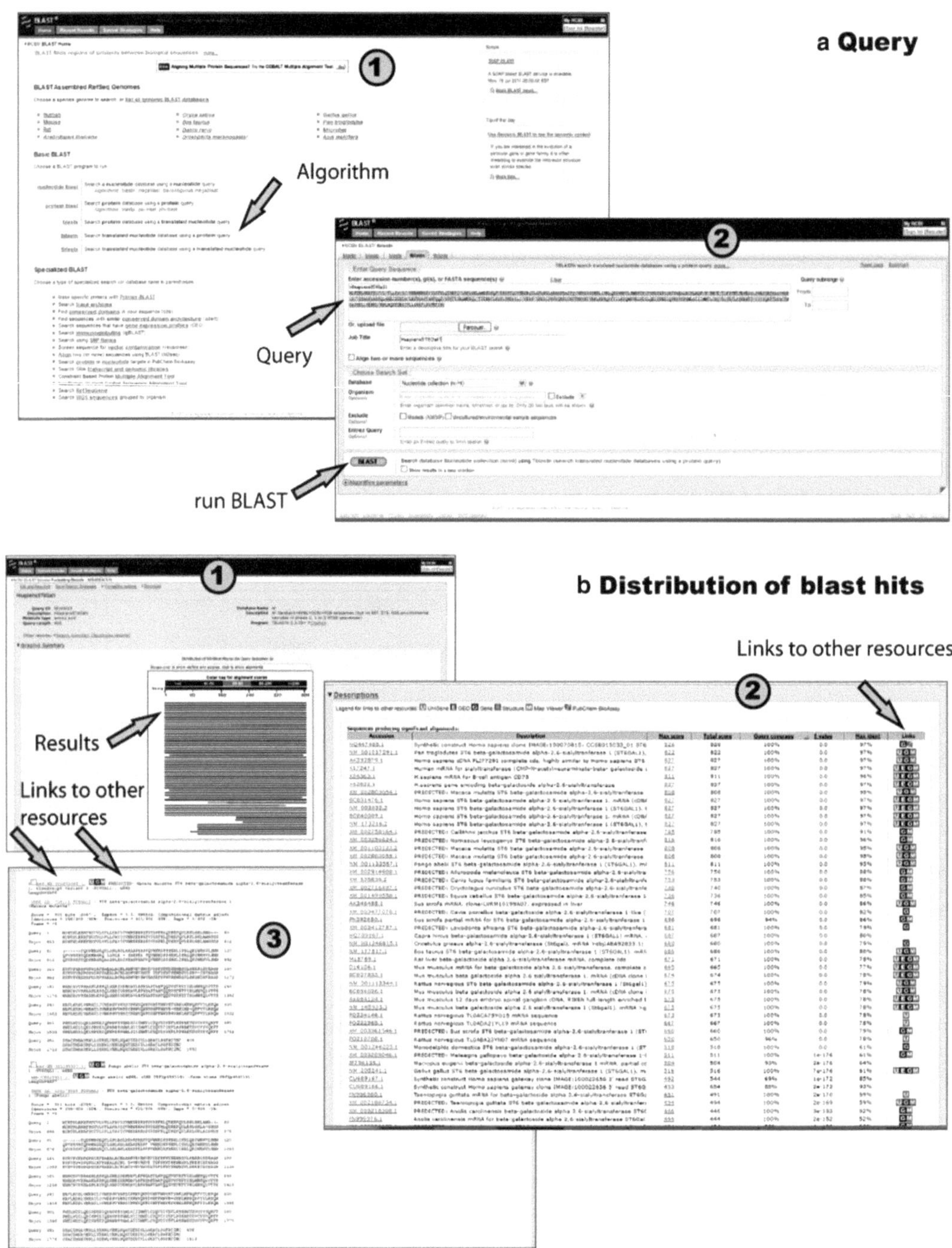

Fig. 2 BLAST-based methods of identification of sialyltransferase sequences. The following screenshots illustrate the process of sequence analysis using BLAST. (**a**) Query sequence algorithm and (**b**) Distribution of Blast hits, links to other resources, local alignments. The *labels* 1, 2, or 3 in each panel indicate the successive steps of analysis

Table 2
Summary of available Web-online sites of tools for protein and nucleotide sequence analyses

Tools	URL	References
NCBI BLAST	http://blast.ncbi.nlm.nih.gov/Blast.cgi	[66]
BLAT search genome	http://genome.ucsc.edu/cgi-bin/hgBlat	[33]
EBI	http://www.ebi.ac.uk/Tools/sequence.html	[77]
EBI transeq	http://www.ebi.ac.uk/Tools/emboss/transeq/	[77]
CAP3	http://pbil.univ-lyon1.fr/cap3.php	[35]
Genscan	http://genes.mit.edu/GENSCAN.html	[36]
Splice site prediction	http://www.fruitfly.org/seq_tools/splice.html	[78]
ClustalW at PBIL	http://npsa-pbil.ibcp.fr/cgi-bin/npsa_automat.pl?page=npsa_clustalw.html	[38, 39]
ClustalX 2.0	http://bips.u-strasbg.fr/fr/Documentation/ClustalX/	[79]
Dialign	http://bibiserv.techfak.uni-bielefeld.de/dialign/submission.html	[40]
Expasy	http://expasy.org/genomics	[42]
WebLogo	http://weblogo.berkeley.edu/logo.cgi	[80]
Gblocks	http://molevol.cmima.csic.es/castresana/Gblocks_server.html	[46, 47]
PHYLIP	http://evolution.genetics.washington.edu/phylip.html	[16]
TimeTree	http://www.timetree.org/	[51]
Paralogon	http://wolfe.gen.tcd.ie/dup/	[52]
Synteny database	http://teleost.cs.uoregon.edu/synteny_db/	[53]
Synteny database	http://www.dyogen.ens.fr/genomicus-64.01/cgi-bin/search.pl	[54]
T-Coffee	http://www.tcoffee.org/	[20]

3. Choose in the search set section either a genomic database (HTGS: high throughput genomic sequence, WGS: whole genome shotgun reads or GSS: genomic survey sequence) or a transcriptomic database (nr: non redundant GenBank, Refseq RNA, EST: Expressed Sequence Tag, TSA: Transcriptome Shotgun Assembly). Organism can be specified.
4. Analyze distribution of BLAST hits on query sequence. Further information can be obtained through the use of links with other databases (Unigene, NCBI map viewer…). Manually analyze the identified sequences (Scores, predicted sequences, mRNA full length sequence) and discard any false-positive hits by investigating the results of multiple sequence alignments.
5. Save the newly identified sequences, both translated protein and nucleotidic sequence into a file on your disk for subsequent sequence analysis.

3.1.2 The Basic Local Alignment Search Tool

Sialyltransferase homologous sequences are extracted via BLAST (Basic Local Alignment Search Tool) searches (*see* **Note 3** and Table 2). Looking for new sialyltransferase-related sequences, it might be desirable to change the number of aligned sequences displayed (100 by default) in the algorithm parameter section and to take into account the percentage of coverage of the alignment, in order to eliminate hits that represent fractions of domains. Homologous sequences with a current cut-off E-value of 0.01 coding for a sialyltransferase domain are retained for further sequence analysis.

We have used also BLAST-like Alignment Tool (Blat) [33], which is a fast alignment tool like BLAST for nucleotide sequences. It is commonly used for gene identification to find sequences with 95 % or greater similarities, for chromosomal walking and to look up the location of a sequence in the genome or determine the exon structure of an mRNA. We carry out location of a nucleotide sequence within a genome using BLAT:

1. Open the BLAT Search genome page (Table 2).
2. Select genome, assembly and query type.
3. Paste the nucleotide sequence of sialyltransferase or FASTA-formatted list into the edit box or enter file name in the upload sequence text box.
4. The detail link in the search result list gives the alignments of the sequences to the genome.
5. Translation of nucleotide sequence is carried out at the EBI Transeq Web site (Table 2).
6. Save the newly identified sequences into a file on your disk.

3.1.3 DNA Analysis and Gene Assembly

It is desirable to reconstitute and analyze the sialyltransferase gene organization and identify the open reading frame. Newly identified sialyltransferase gene organization may then be compared to previously described human gene organization. The shotgun sequence projects have generated short reads that need to be assembled into long sequences. Two different strategies are followed: expressed sequence tag (EST) assembly and genomic reconstruction.

1. Perform an initial clustering of related EST, contigs of the different EST sequences corresponding to each sialyltransferase with CAP-3 (Table 2) [35].
2. As described above, the different general and species specific genomic databases (HTGS, WGS, GSS) are also mined for potential sialyltransferase genes. Putative exons are predicted using the Genscan server at the MIT (Table 2) [36].
3. The best exon/intron boundaries following the most common AG/GT rule are search at the Berkeley drosophila genome project site (Table 2) for splice site prediction [37].

4. The EST contig sequences and the genomic sequences corresponding to each hypothetical sialyltransferase gene of each species are compared to correct eventual splicing errors.
5. Translate the nucleic acid sequence into protein sequence with Transeq at EBI Web site (Table 2).

3.2 Sequence Analysis and Multiple Sequence Alignments

We commonly use ClustalW [38, 39] version 1.7, July 1997 or the graphical version ClustalX 2.0 to perform fast multiple sequence alignments to assign identified sequences to sialyltransferases (*see* **Note 4**). Genome deduced protein sequences are then analyzed both manually and with bioinformatic tools (Table 2). Sialyltransferases are Golgi resident type II glycoproteins with a unique transmembrane domain located towards the N-terminus part of the protein. Their cytoplasmic/transmembrane/stem region (CTS) is usually not well conserved in various animal species. The variations in this non-catalytic domain come from shifts of the reading frame over a limited range of amino acid residues and from small indels in the coding gene [14]. Global alignments, such as performed by Clustal programs are poorly informative in this case and it might be useful to align the CTS domain of sialyltransferases using software allowing local alignments like Dialign [40] (Table 2). Within their catalytic domain, sialyltransferase sequences show three levels of amino acid sequence conservation (Fig. 3): (a) The sialylmotifs L, S, III, and VS found in all the animal sialyltransferases, (b) the Family motifs characteristic of each sialyltransferase family and (c) subfamily motifs found uniquely in each of the vertebrate sialyltransferase subfamily. These various peptidic motifs have proven to be useful hallmark for sialyltransferase identification [3, 31, 41].

3.2.1 Motifs for Sialyltransferase Protein Identification

1. Search for transmembrane domain with TMPred software as implemented at the SIB ExPASy Bioformatics Resources Portal (Table 2) [42].
2. Search for the Pfam domain characteristic of glycosyltransferases at the Pfam database Web site (Table 1) [43]. Sialyltransferases of the CAZy GT29 family show the Pfam domain PF00777.
3. Multiple protein sequence alignments using ClustalW freely accessible at the PBIL Web site (Table 2).
4. Presence of conserved motifs (sialylmotifs L, S, III, and VS) and of the family motifs for ST6Gal, ST3Gal, ST6GalNAc, and ST8Sia is detected manually in the catalytic domain of sialyltransferase sequences. Figure 3 is a graphical representation of these patterns within multiple sequence alignments of vertebrate sialyltransferases analyzed and visualized at the WebLogo site (Table 2) at Berkeley. Briefly, conservation of a particular position is calculated as the difference between the

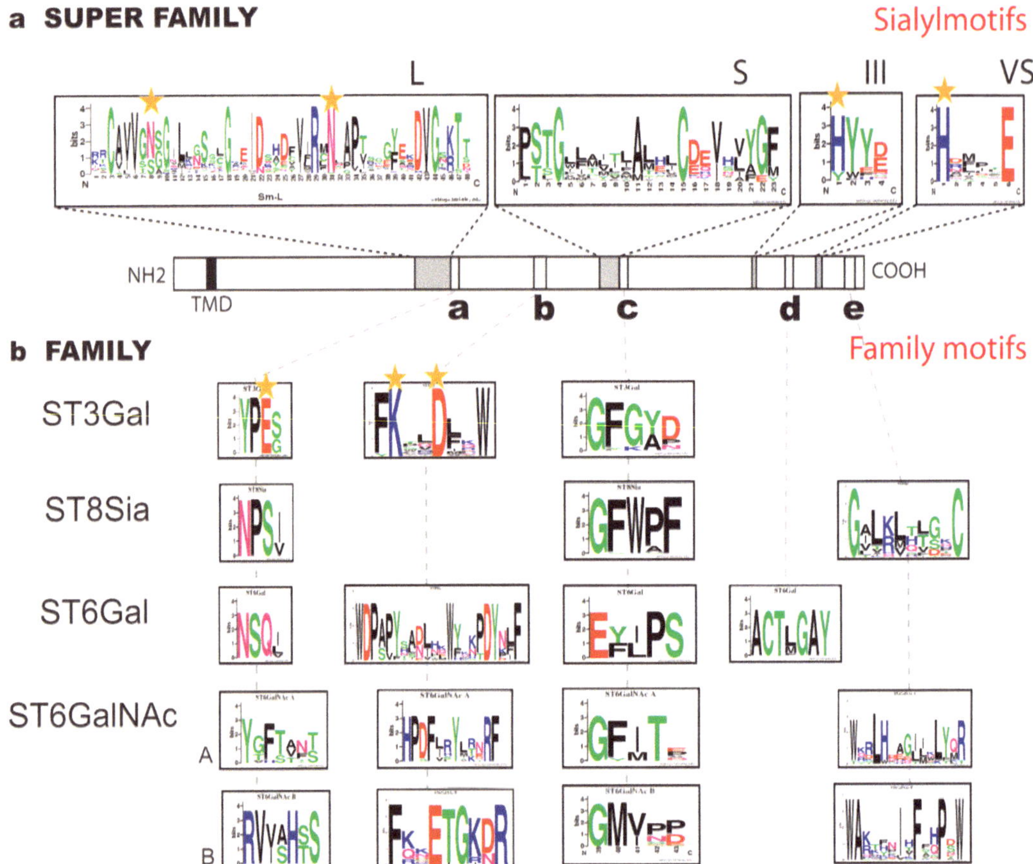

Fig. 3 Motifs for sialyltransferase protein identification. The transmembrane domain (TMD) of the sialyltransferase-related sequence is determined using the TMPRED program available from the ExPASy proteomics server and is represented by a *black rectangle*. The catalytic domain of sialyltransferases is represented by the *white rectangles*, the sialylmotifs L, S, III, and VS by *grey boxes*, and the a, b, c, d, and e family motifs by *light grey boxes*. Multiple sequence alignments of 125 vertebrate sialyltransferase sequences have been performed with ClustalW at PBIL and sequence logos were created using WebLogo (version 2.8.2). Motifs represented in (**a**) are those of the superfamily of animal sialyltransferases which correspond to the first level of amino acid conservation. Similarly, multiple sequence alignments of 42 vertebrate ST3Gal, of 12 vertebrate ST6Gal, of 11 vertebrate ST6GalNAc I and ST6GalNAc II designated as ST6GalNAc (A) and 23 ST6GalNAc III, ST6GalNAc IV, ST6GalNAc V, and ST6GalNAc VI vertebrate sequences, designated as ST6GalNAc (B) were used for WebLogo representation of the various family motifs. *Yellow stars* indicate amino acid residues implicated in substrate binding as demonstrated by X-ray crystallography of porcine ST3Gal I [9]. In the logos, letter amino acid symbols are colored according to their chemical properties: polar amino acids (G, C, S, T, Y) are *green*, basic (K, R, H) are *blue*, acidic (D, E) are *red*, hydrophobic (A, V, L, I, P, W, F, M) are *black*, and neutral polar amino acids (N, Q) are *pink*. The overall height of the stacks indicates the sequence conservation at a given position, while the height of symbols within the stack indicates the relative frequency of each amino acid at that position [80]

maximum possible entropy and the entropy of the conserved symbol distribution. It has to be mentioned that for alignments with less than 20 nucleotide or 40 protein sequences, the entropy is underestimated. This bias can be partially corrected using the small sample correction option.

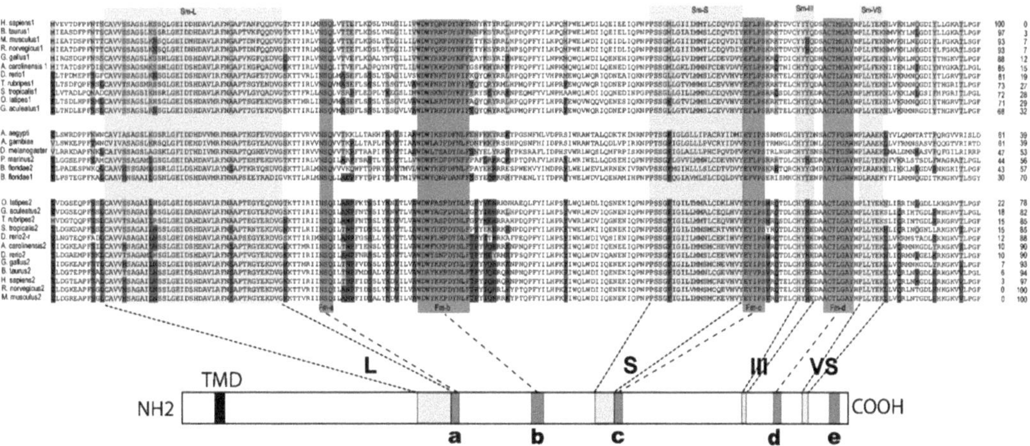

Fig. 4 Amino acid positions conserved in subfamilies. Multiple sequence alignments are carried out at PBIL and the selection of informative positions in the sialyltransferase sequences is carried out using Gblocks. Then, manual specificity-determining position prediction is achieved as described previously [15, 45]

5. We also detect manually specificity-determining amino acid positions in each subfamily by comparing the amino acid position conservation within each subfamily. Briefly, related amino acid residues are grouped according to their chemical properties based on a chemical alphabet comprising five groups which are acidic or amide (E, D, Q, N), hydrophobic (I, L, V, M), aromatic (F, Y, W), basic (R, H, K), hydroxyl (S, T). The remaining four amino acids: A, G, P and C are analyzed separately, as it has been described previously for protein fold recognition [44]. Amino acids of the same group are considered equivalent for the definition of a conserved position. Amino acid positions conserved at more than 50 % in the different sialyltransferase sequences of one subfamily are colored. We then count the positions and percentage of amino acid positions that are conserved in each subfamily to classify sialyltransferase sequences (Fig. 4) [15, 45].

3.2.2 Internet Resources

1. All newly translated protein sequences are gathered in one file with a FASTA format.
2. Protein sequences are aligned with ClustalW and the multiple sequence alignment is examined carefully to identify sialylmotifs and conserved amino acid residues. Importantly, protein sequences producing potential biases in the subsequent phylogenetic reconstruction are eliminated and automated predicted sequences with wrong splicing errors and sequences that are too distantly related from the rest of the set are removed from further analysis.
3. Sequences are realigned and output file is saved in PIR, NBRF/PIR or FASTA format.

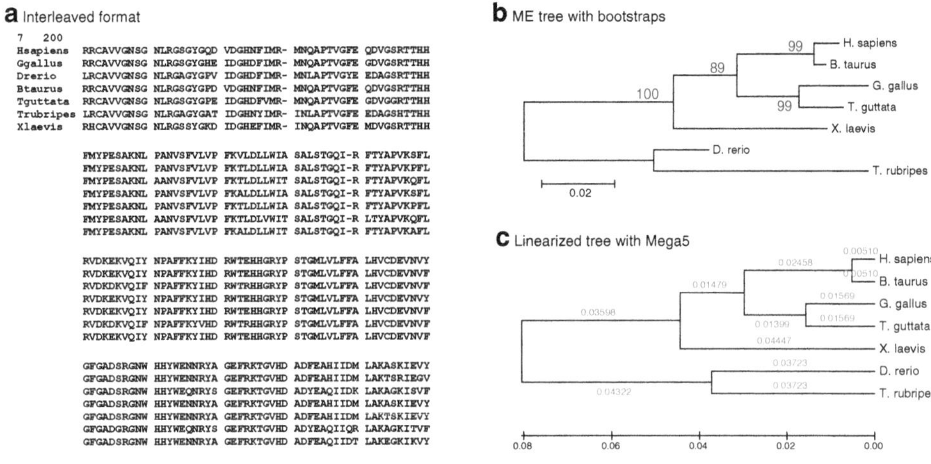

d Linearized tree in newick format

(((Hsapien2:0.00691,Btaurus2:0.00329)0.9970:0.02136,(Ggallus2:0.01907,Tguttat2:0.01230)0.9960:0.01720)0.8930:0.01758,Xlaevis2:0.04263,(Drerio2:0.01520,Trubrip2:0.05926)1.0000:0.07657);

Fig. 5 Phylogenetic tree production. (**a**) Interleaved format. Protein sequence alignment of seven vertebrate ST3Gal II catalytic domains. The access numbers of each sequence in GenBank are as follows: *T. guttata:* XP_002188564; *T. rubripes:* CAF25175; *G. gallus*: NP_989811; *X. laevis*: NP_001084518; *D. rerio*: NP_001004012; *B. Taurus*: NP_001002892; *H. sapiens*: AAH36777. The alignment is according to the PHYLIP interleaved format. (**b**) Distance tree obtained by Minimum Evolution (ME). Distance tree obtained by ME method of the alignment file in interleaved format using MEGA 5.0. The bootstrap percentages are calculated from 2,000 replicates. (**c**) Linearized tree. Linearized tree derived from the previous one, obtained using MEGA 5.0. The branch lengths are used for datation of duplication events. (**d**) Parenthetic form of ME tree corresponding to Fig. 5b in Newick format. The bootstrap values and the branch lengths are indicated

4. The selection of informative positions in the sialyltransferase sequences is carried out using Gblocks (Table 2), which eliminates poorly aligned positions and divergent regions of a DNA or protein alignments so that multiple sequence alignments produced become suitable for phylogenetic analysis [46, 47]. The output is saved in the PHYLIP format for subsequent phylogeny analysis.

3.3 Phylogenetic Tree Constructions

Among the different methods tested to reconstruct the phylogeny of sialyltransferases, we obtained the most satisfactory results using the two distance methods Minimum Evolution and Maximum Likelihood implemented in MEGA 5.0 [18] and in PhyML version 2.4 [17], both using the JTT model of amino-acid substitution. The alignment file is organized according to the PHYLIP "interleaved" format, similar to the output of alignment program indicating the number of sequences and their length on the first line. The first block then includes the names of sequences followed by the beginning of sequences themselves. The following parts of each sequence appear in blocks on successive lines, each sequence starting at the eleventh column (Fig. 5a). Finally, evolutionary tree construction requires high quality and accurate multiple sequence alignments, which is a challenge for large scale phylogeny estimation (*see* **Note 5**).

3.3.1 Robustness of Phylogenetic Trees

The reliability of the branching pattern of a tree is assessed by the bootstrap method [16].

1. In the bootstrapping non parametric strategy, the dataset of aligned sequences is used to produce multiple resampled versions of the original multiple sequence alignment, where each site can be randomly doubled or removed, the total number of sites remaining constant. The SEQBOOT program from the PHYLIP package version 3.69 (Table 2) can generate these new datasets, whereas this operation is automated in MEGA 5.0.
2. A tree is constructed for each new dataset and a consensus tree is calculated based on a majority rule. In PHYLIP package, this can be done with the CONSENSUS program. The score for each branch is calculated from the relative number of bootstrapped trees (>2,000 replicates) supporting this node (Fig. 5b). Bootstrap values superior to 80 % are reported at the left of each divergence point on trees [48].

3.3.2 Visualization of the Trees

The tree saved in parenthetic format known as "newick" format can be drawn using several programs such as PHYLIP package, MEGA 5.0, Treeview and others. The length of each branch can be indicated at the right side of each closing parenthesis or sequence, with two dots on its left (Fig. 5d).

3.4 Datation of Duplication

The duplication event can simply be positioned by its place within the tree, e.g., within vertebrate lineage, after the divergence of Lamprey from Vertebrates with jaws (Gnathostomata) (*see* Fig. 6). If a more precise indication is needed, the simplest way to calculate the date of a duplication event is to hypothesize that there is a linear accumulation of mutations with time of divergence known as molecular clock. Of course, this way suffers many exceptions as the evolution rate may change from one lineage to another [49]. To overcome this problem, we take into account the maximum of known calibrated dates and the corresponding lengths of branches in the linearized tree.

1. Using MEGA 5.0, the best tree has to be linearized so that the branches leading to terminal sequences are plotted on a vertical line (Fig. 5c).
2. The divergence time of the different lineages are recorded from present. Several resources can be taken into account: paleontological evidence or molecular clock calculations (see Table 3) [14, 50]. We also use the TimeTree2 Web resource (Table 2), which contains time trees reported from molecular clock analyses in 910 published studies and 17,341 species [51].
3. Use the datasets corresponding to branch lengths and dates to calculate Pearson's correlation coefficient. If it is significant, calculate the regression equation using for example PAST 2.11 or later version [19]. The date of duplication event has to be

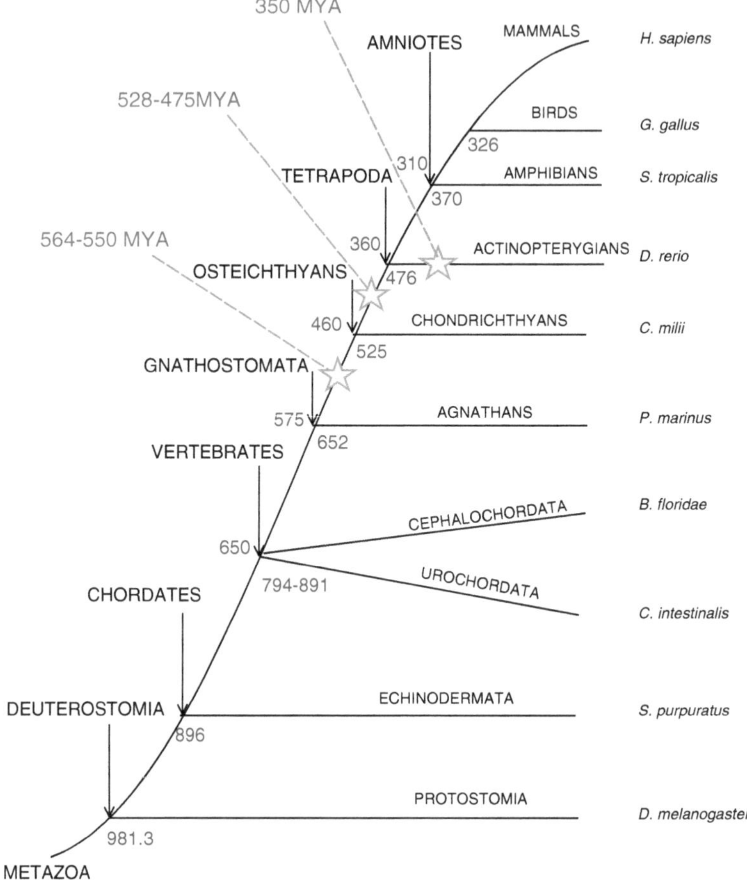

Fig. 6 Tree and Calibration. Illustration of the metazoan tree of life showing calibration date proposed by Blair and Hedges [50] or Otsuka et al. [81] indicated in *grey* below each branch. Numbers indicated on the *left side* of the tree are the ones used for datation of sialyltransferase duplication events. The stars indicate genome duplication events

deduced from the regression equation, taking the corresponding branch length as x. Confidence intervals at 95 % are calculated as 1.96 times the standard deviations of regression equation residues [4].

3.5 Synteny/Paralogy

Evolutionary analysis and comparative genomics remain a pivotal strategy for glycobiologists to understand the function of sialyltransferase sequences. Thus, it is desirable to reconstruct the history of the genes neighboring the sialyltransferase gene locus of interest from deuterostomes to human, in order to estimate the degree of conservation of its genic environment and trace its evolutionary relationships. Shared synteny, known also as conserved synteny found in Vertebrate genomes described a block of gene loci around the gene of interest that retain their relative positions and orders on chromosomes over evolutionary time. Neighboring genes

Table 3
Calibration table

Divergence of organisms	Calibration used for sialyltransferase	Median from TimeTree	Calibration from refs. [50] and [81]
Protostomia/deuterostomia			981.3
Echinodermata/chordates			896
Cephalochordates/vertebrates	650 MYA	720	794–891
Urochordates/vertebrates	650 MYA	720	794–891
Agnathans/gnathostomata	575 MYA	500	652
Chondrichthyans/osteichthyans	460 MYA	462.5	525
Actinopterygians/tetrapoda	360 MYA	436.8	476
Amphibian/amniota	310 MYA	359.1	370
Birds/mammals		323	326
WGDR1	564–550 MYA		
WGDR2	528–475 MYA		
WGDR3 (Teleosts)	350 MYA		

Calibration used for sialyltransferase divergence is indicated in the first column and reported in Fig. 7; divergence median time calculated in TimeTree [51] is indicated in the second column and the last column indicates calibration proposed by Blair and Hedges 2005. Whole genome duplication round (WGDR) 1, 2, and 3 are also indicated

might be different in a given animal species compared to the others following gene translocation as described previously for *st6gal1* genes in Teleosts (Bony fishes with terminal mouth) [14]. This process can be related to a change in regulation and tissue expression level. In the case of a whole genome doubling (WGD) events, which took place in the first times of Vertebrates or Teleosts evolution (Fig. 6), pairs of chromosomes are detected in genomes that are named paralogons and are the historical remnants of WGD.

3.5.1 Manual Analysis

The genomic segment bearing the sialyltransferase gene locus and its neighboring genes can be displayed using Ensembl or Mapviewer (NCBI) sites for various genomes. It is possible to list and compare manually the surrounding genes of the sialyltransferase locus and its paralogues (in the same genome) and/or orthologues (in various animal species). This method is fastidious and not exhaustive since the genes of the paralogons can be positioned away from their ancestral location as a result of frequent inversions, or might have disappeared through deletion events.

3.5.2 Automated Approaches

The paralogon Web site of Wolfe and McLysaght (*see* the Trinity College Dublin Web site in Table 2) [52] gives a friendly way to visualize paralogons within the human genome. The user has to choose the chromosomes bearing the paralogous sialyltransferase genes, with self-defined parameters. The result is given by a window showing the list of genes included in the paralogon, and the mapping of the corresponding segments on the human caryotype. To assess syntenies including sialyltransferase genes between human and other vertebrates, we have used the Synteny database Web site at the University of Oregon (Table 2) [53] and the Genomicus database Web site at Dyogen (Table 2) [54]. The Synteny Database detects conserved synteny by examining datasets from Ensembl version 61 and version 64.01, respectively generated by a reciprocal best hit analysis.

3.6 Substitution Rate

In order to visualize the density of amino acid changes of the sialyltransferase proteins, we use the method developed in Petit et al. [55, 56]. It provides a view of the substitution rate variations in and between the sialylmotifs to estimate the selective pressure on each part of the sequence.

1. The alignment of amino-acid sequences, cleaned from insertions, is treated by PROTPARS, a parsimony program of PHYLIP package (Table 2) [16], using the parenthetic topology obtained in the best tree as user tree, in order to export the number of changes site by site (Fig. 7a, b).
2. For each site, this number is divided by the number of sequences, allowing drawing the profile of site change rates. This profile is easier to interpret when a smoothing over a window of 5–9 amino acids is applied. For a size window of 5, we use a method for pondering in order to get a smoothed curve:

$$\mathrm{Val}(n) = \left(\mathrm{val}(n-2) + 2\quad \mathrm{val}(n-1) + 3\quad \mathrm{val}(n) + 2\quad \mathrm{val}(n+1) + \mathrm{val}(n+2)\right) / 9.$$

Figure 7c illustrates such an analysis carried out with 15 amino-acid sequences from Fish, Birds, Lizard, and Mammals corresponding to the catalytic part of ST6Gal I and ST6Gal II sequences. Both

Fig. 7 (continued) CAI39644.1; *D. rerio*: CAF29495.1; *T. nigroviridis*: CAL44592.1; *G. aculeatus*: CBQ74104.1. Each amino acid position site is at the cross of tens (indicated in line at the *left side* of (**b**)) and units (indicated in column on the top of (**b**)). (**c**) The moving average, in other words mean change numbers (*y* axis) in ST6Gal I (*blue curve*) and in ST6Gal II (*red curve*) are plotted in parallel to the sialylmotifs (indicated in *green*) and to the family motif (indicated in *violet*) on the *x* axis. For each site, this number was divided by the number of sequences, allowing drawing the profile of site change rates. This profile is easier to interpret when a smoothing over a window or 5–9 is applied. For a size window of 5, we use a ponderation method, in order to get a smoothed curve: $\mathrm{Val}(n) = \left(\mathrm{val}(n-2) + 2\quad \mathrm{val}(n-1) + 3\quad \mathrm{val}(n) + 2\quad \mathrm{val}(n+1) + \mathrm{val}(n+2)\right) / 9$ As expected, the mean change numbers are low in the motifs L, S, III, and VS, as well as in the motifs specific to ST6Gal, i.e., motifs a, c, and d [3, 31]. The S motif is not well conserved, indicative of particularities of ST6Gal catalytic activity, relative to other sialyltransferases

a ST6Gal I

	0	1	2	3	4	5	6	7	8	9
0		9	11	5	0	0	0	0	1	0
10	0	0	0	2	1	4	0	4	0	0
20	5	1	0	0	4	0	0	0	0	3
30	0	0	0	1	0	0	2	7	4	2
40	1	6	0	0	0	4	0	0	0	2
50	0	0	5	0	0	0	1	2	1	3
60	2	9	5	6	0	0	7	3	2	2
70	0	2	5	0	4	0	1	4	0	0
80	0	2	3	1	2	4	3	1	5	3
90	0	4	4	4	1	0	0	5	1	1
100	9	6	2	1	6	0	0	6	5	1
110	0	9	0	0	0	1	0	1	4	0
120	3	2	2	0	1	2	0	2	3	4
130	0	2	3	2	4	1	9	0	0	2
140	0	0	0	0	0	0	2	1	0	2
150	4	4	0	0	3	1	0	1	3	0
160	1	1	0	0	1	0	0	0	2	0
170	3	0	2	3	0	1	0	1	0	3
180	2	6	0	6	0	0	0	1	0	0
190	0	0	0	0	0	2	0	0	0	2
200	1	0	3	2	0	5	0	4	1	2
210	0	0	0	3	4	0	1	4	0	0
220	1	0	1	5	8	6	3	0		

b ST6Gal II

	0	1	2	3	4	5	6	7	8	9
0		2	6	3	0	0	0	0	4	0
10	0	0	1	1	0	2	0	1	0	0
20	6	0	0	0	3	0	1	0	0	0
30	0	0	0	2	0	0	0	3	2	0
40	0	2	0	0	0	2	1	0	0	5
50	0	0	0	0	0	0	0	0	3	0
60	0	6	2	3	0	6	5	1	3	2
70	0	0	2	1	3	0	1	0	0	0
80	0	1	0	0	2	5	2	0	4	10
90	1	3	5	3	0	0	0	1	0	0
100	3	0	0	6	2	1	0	6	7	5
110	1	9	0	1	0	0	0	0	0	0
120	3	1	3	0	2	0	0	3	2	0
130	0	3	0	2	4	0	3	0	0	0
140	0	0	0	0	0	0	0	0	0	1
150	1	4	0	0	4	2	0	9	4	2
160	0	0	0	0	0	1	0	0	7	0
170	0	1	2	0	0	0	0	0	0	5
180	0	1	0	1	0	0	0	0	0	0
190	0	0	0	0	1	1	0	0	4	0
200	1	1	0	3	1	7	0	5	4	6
210	2	0	6	4	0	0	1	0	2	0
220	0	0	2	4	6	5	8	0		

c Positions of substituted sites

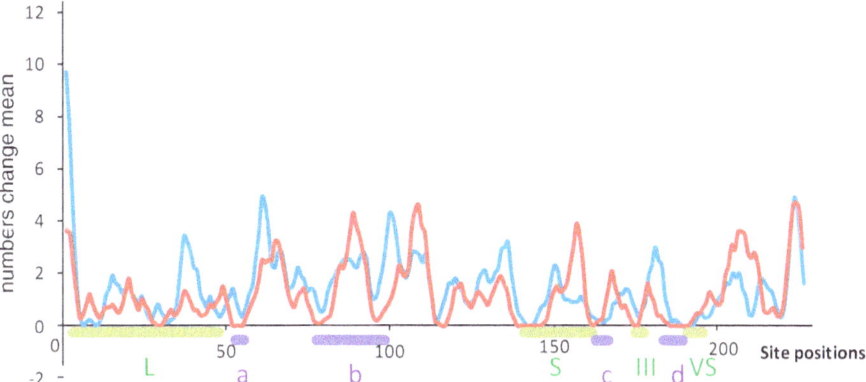

Fig. 7 Position of sites according to substitution rates of amino acids. (**a**) Number of amino acid substitutions for each site during the evolution of ST6Gal I in Vertebrates, resulting from PROTPARS (PHYLIP package) analysis of 15 vertebrate ST6Gal I sequences. The access numbers of each sequence in GenBank (except for *D. novemcinctus* ST6Gal I) are as follows: *H. sapiens*: NP_003023; *M. mulatta*: XP_001103123.1; *C. porcellus*: XP_003477126.1; *M. musculus*: BAA03680; *B. taurus:* CAA75385; *C. familiaris*: XP_535839.2; *D. novemcinctus:* ENSDNOP00000004279 (Ensembl, release 65); *A. carolinensis*: CBQ74101; *G. gallus*: NP_990572; *T. guttata*: CBQ74105; *G. aculeatus*: CBQ74103; *D. rerio*: CAG32837; *O. latipes*: CAI39643.1; *T. rubripes*: CAG32836; *T. nigroviridis*: CAI29183. Each amino acid position site is at the cross of tens (indicated in line at the *left side* of (**a**)) and units (indicated in *column* on the top of (**a**)). For example, the 84th site has 2 substitutions. (**b**) Number of amino acid substitutions for each site during the evolution of ST6Gal II in vertebrates, resulting from PROTPARS (PHYLIP package) analysis of 15 vertebrate ST6Gal II sequences. The access numbers of each sequence in GenBank are as follows: *H. sapiens*: BAC24793.1; *M. mulatta*: XP_001109602; *M. musculus*: NP_766417.1; *C. porcellus*: XP_003471791; *O. cuniculus*: XP_002709692.1; *B. Taurus*: CAI29185.1; *A. carolinensis*: CBQ74102; *G. gallus*: CAF29497.1; *T. guttata*: CBQ74106; *S. tropicalis*: CAF29496.1; *T. rubripes*: CAI29184.1; *O. latipes*:

proteins show parallel variations and as expected, the mean change numbers are low in the motifs L, III, and VS, as well as in the motifs specific to the ST6Gal family, i.e., a, c, and d [3, 31]. A more detailed view can be obtained considering the profiles in the different transitions between ancestors of the different lineages.

3.7 Positive Selection

A long branch in a tree can be the result of a weak selection pressure, following duplication event for example, or a positive selection linked to new interactions of the encoded enzyme in a given evolutionary lineage. The evidence of positive selection or adaptative natural selection is assessed using the branch-site method implemented in PAML version 4 [57], as previously described [58, 59].

1. In order to quantify the impact of natural selection on molecular evolution, we calculate the ratio of the non-synonymous substitution rate (dN) to the synonymous substitution rate (dS) for each sialyltransferase subfamily ($\omega = dN/dS$), in the branch of interest (ω (i) for one sialyltransferase subfamily) and in the background branches (ω (b) for the remaining sialyltransferase subfamilies).
2. Align the total number of sialyltransferase sequences in multiple sequence alignments within the informative sites selected by Gblocks. The best tree topology retrieved among the different tested methods should be chosen.

3.8 Expression Analysis

In order to compare the tissue expression level of target genes in Vertebrates, it is useful to consider the information given in EST Unigene at the NCBI database.

1. Construction of the dataset. For every tissue in a given species, the total numbers of EST for all genes and for the target gene are recorded, so it is possible to calculate the expression rate of the gene in each tissue. For an easy reading of numbers, it is useful to multiply these rates by $10e^6$. The rows where EST expression is limited to one organism are discarded. The database includes the name of a gene in a given organism in columns and the name of tissue/organ in rows, each cell containing the expression rate, or "?" if the data is missing or non-applicable.
2. This table is submitted to multivariate analyses (Principal component analysis and Cluster analysis) using PAST 2.11 [19]. According to the method described by Ermonval et al. [60], PCA allows projecting the dataset onto a two-dimensional plan, each column factor represented by a vector according to pair-wise correlations; the higher the correlation between two factors, the more acute the angle between the vectors. In this plan, the target genes corresponding to each species are projected in the direction of their greatest values (Fig. 8).

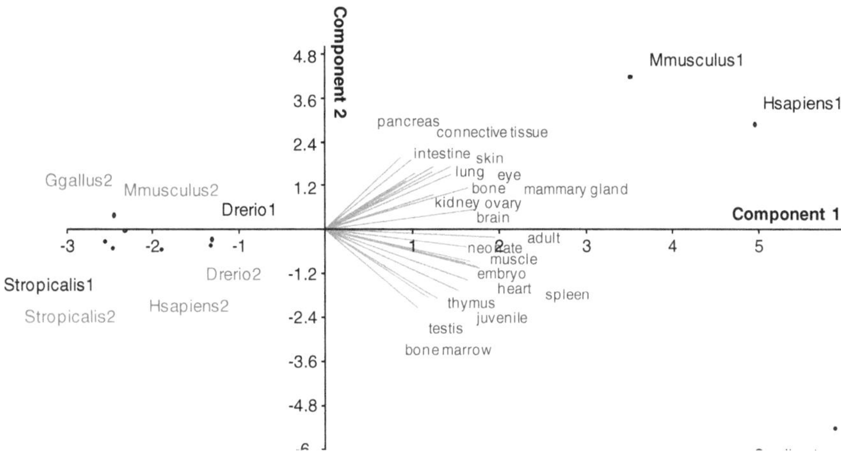

Fig. 8 *St6gal* EST profile of expression. Principal component analysis obtained using PAST 2.11. The vectors corresponding to each tissue indicate high values in the direction of the component 1. The more acute the angle, the more correlated the expression in the different species. Sequences of ST6Gal II have low values for all the tissues and are positioned on the opposite side of the vectors. Sequences of ST6Gal I are distributed in two opposite areas of the plane, indicative of a change in their expression during vertebrate evolution

3. The EST ratios are log-transformed to normalize the distribution and then submitted to a two-way clustering using Euclidean distance as measure of similarity, using PAST 2.11. The coloration intensity of each case in the table is in proportion to the values. This procedure allows showing the tissues sharing similar expression levels and organisms sharing a common gene expression pattern.

4 Notes

1. The HUGO Gene Nomenclature Committee (HGNC) has assigned unique gene symbols and names to over 33,000 human loci. The human sialyltransferase genes are named accordingly and the sialyltransferase nomenclature used is according to Tsuji et al. [30]. However, vertebrate genomes contain numerous duplicated genes that may result from various duplication events that need to be studied:
   (a) A genome-wide duplication in the ray fin fish lineage (known as WGDR3 giving rise to two copies of a gene that is present as a single copy in mammals) as it is the case for *st6gal* genes [14]. In these cases, symbols for the two zebrafish genes is the same as the approved symbol of the mouse sialyltransferase orthologue followed by "-r1" or "-r2" standing for "-related" to indicate that they are duplicate copies. It is also important to provide evidence by mapping that the two copies reside on duplicated chro-

mosome segments (paralogons). These analyses have been carried out for the fish *st8sia* and *st6gal* genes [4, 14], but not yet for the fish st3gal and st6galnac genes.

(b) Duplicated genes may also result from single gene duplication events (tandem duplication) in a genome and in this case symbols used for the two paralogous sialyltransferase genes is the same as the approved symbol of the mouse sialyltransferase orthologue followed by A and B. As an example, the *Onchorhynchus mykiss st8sia2* genes A and B described in ref. [4].

(c) Finally, duplicates that resulted prior to the divergence of ray fin and lobe fin fish are considered as new sub-families; then, a new subfamily number is given and no suffix is attributed. For instance *st8sia7* found in Cyprinidae fish is also found in the lizard *A. carolinensis* attesting of its origin.

2. Two types of databases can be used: the first ones are archival resources, which are general, publicly available databases of nucleotide sequences such as EMBL bank [61], GenBank [62], and DDBJ [63] maintained by the EBI (European Bioinformatic Institute), by the NCBI (National Center for Biotechnology Information) and by the NIG (National Institute of Genetics), respectively. The three databases exchange new and updated data on a daily basis to achieve optimal synchronization and sequences submitted to EMBL/DDBJ/GenBank receive unique identifier. Recently, the three major databases of protein sequences Swiss-Prot, TrEMBL (Translated EMBL nucleotide sequence library) and PIR (protein identification resource) have assembled in one unique resource UniProt (United Protein databases) [64].

   The second ones are curated and specialized resources such as TGI database (maintained at Dana Farber Cancer Institute (DFCI)) [65], JGI, CAZy, KEGG, GGDB, CFG glycoenzyme, BRENDA, Pfam.

3. Sialyltransferase homologous sequences are extracted via BLAST (Basic Local Alignment Search Tool) searches, a family of sequence alignment algorithms developed by Altschul [66], which allows local alignments. The most widely used is tBLASTn (version 2.2.25) and other advanced BLAST algorithms like MegaBLAST, which allows variable parameter setting search as preferred statistical matrices found at the NCBI Web site (Table 2). PSI-BLAST (Position-Specific Iterated BLAST) or PHI-BLAST (Pattern-Hit Initiated BLAST) and BLAST 2 sequences are rather protein BLAST that may be coupled with a motif search.

4. Clustal does not allow editing of sequences and alignments, which might be performed using SeaView [67] to add or

remove gaps in one or several sequences simultaneously in an interactive way. This is a multiplatform program to perform multiple sequence alignment and phylogenetic trees through a graphical interface. SeaView integrated ClustalW and Muscle programs for sequence alignments and PhyML version 3 for maximum-likelihood phylogenetic tree construction. Their utility is comparable to MEGA 5.0; however, it is less versatile than MEGA for pair wise distance computations and lacks several features, its advantage is that it is available for all major computer platforms.

5. It is important to note that evolutionary tree construction requires high quality and accurate multiple sequence alignment and that the performance of different algorithms to create multiple sequence alignment depends on the number of sequences and their diversity.

   (a) It is thus desirable to use tools improving the accuracy of alignment and assessing their quality. As a general rule for aligning few sequences (less than 100) T-Coffee and Clustal algorithms are a good choice, while for a higher number of sequences, programs such as MUSCLE or MAFT show a better accuracy [20, 68]. Clustal programs can incorporate secondary structure information provided in SWISS-PROT, CLUSTAL, or GDE format to enhance the penalty of gaps and obtain more biologically meaningful alignments. In the resulting alignment, the gaps are rather inserted into the areas outside the described secondary structures. T-Coffee programs can incorporate three-dimensional structure information of a known structure with at least 60 % identity to a sequence in the dataset. Since protein structures evolve slower than sequences, this will improve multiple sequence alignments produced.

   (b) Besides the choice of alignment algorithm, it is necessary to carefully select sequences to be included in the dataset through calculation of the average level of identity between each sequence and the others. Indeed, distantly related sequences with less than 20 % sequence identity (outlayers) have a strong negative impact on the whole alignment and sequences with more than 60 % sequence identity (redundant) are not very informative and consequently should be removed.

   (c) Finally, quality of the alignment is assessed either by (a) comparing the residues aligned with the spatial overlap of amino acids in aligned 3D structures (when available), or (b) comparing the alignment with biological reference alignments which is also useful to identify correct blocks within a difficult multiple alignment, or (c) using functional information and assessing alignment of functionally related residues.

All the strategies mentioned are available in the T-Coffee package (Table 2).

Acknowledgments

The project was financially supported by the University of Lille1 (ppf bioinformatique de Lille1), by the French Centre National de la Recherche Scientifique (CNRS), and by the Agence Nationale de la Recherche (ANR) project GalFish (ANR-2010-BLAN-120401). An Erasmus Mundus doctoral fellowship was attributed to R.E. Teppa.

References

1. Kikuchi N, Narimatsu H (2006) Bioinformatics for comprehensive finding and analysis of glycosyltransferases. Biochim Biophys Acta 1760:578–583. doi:10.1016/j.bbagen.2005.12.024
2. Harduin-Lepers A, Stokes DC, Steelant WF, Samyn-Petit B, Krzewinski-Recchi MA, Vallejo-Ruiz V, Zanetta JP, Augé C, Delannoy P (2000) Cloning, expression and gene organization of a human Neu5Ac alpha 2-3Gal beta 1-3GalNAc alpha 2,6-sialyltransferase: hST6GalNAcIV. Biochem J 352(Pt 1):37–48
3. Harduin-Lepers A (2010) Comprehensive analysis of sialyltransferases in vertebrate genomes. Glycobiol Insights 29. doi:10.4137/GBI.S3123
4. Harduin-Lepers A, Petit D, Mollicone R, Delannoy P, Petit J-M, Oriol R (2008) Evolutionary history of the alpha2,8-sialyltransferase (ST8Sia) gene family: tandem duplications in early deuterostomes explain most of the diversity found in the vertebrate ST8Sia genes. BMC Evol Biol 8:258. doi:10.1186/1471-2148-8-258
5. Varki A (2009) Ajit Varki: on the origin of maladies. Interviewed by Amy Maxmen. J Exp Med 206:1836–1837. doi:10.1084/jem.2069pi
6. Schauer R (2009) Sialic acids as regulators of molecular and cellular interactions. Curr Opin Struct Biol 19:507–514. doi:10.1016/j.sbi.2009.06.003
7. Hu H, Eggers K, Chen W, Garshasbi M, Motazacker MM, Wrogemann K, Kahrizi K, Tzschach A, Hosseini M, Bahman I, Hucho T, Mühlenhoff M, Gerardy-Schahn R, Najmabadi H, Ropers HH, Kuss AW (2011) ST3GAL3 mutations impair the development of higher cognitive functions. Am J Hum Genet 89:407–414. doi:10.1016/j.ajhg.2011.08.008
8. Harduin-Lepers A (2012) Sialyltransferases functions in cancers. Front Biosci E4:499. doi:10.2741/396
9. Rao FV, Rich JR, Rakić B, Buddai S, Schwartz MF, Johnson K, Bowe C, Wakarchuk WW, Defrees S, Withers SG, Strynadka NCJ (2009) Structural insight into mammalian sialyltransferases. Nat Struct Mol Biol 16:1186–1188. doi:10.1038/nsmb.1685
10. Audry M, Jeanneau C, Imberty A, Harduin-Lepers A, Delannoy P, Breton C (2011) Current trends in the structure-activity relationships of sialyltransferases. Glycobiology 21:716–726. doi:10.1093/glycob/cwq189
11. Chang L-Y, Harduin-Lepers A, Kitajima K, Sato C, Huang C-J, Khoo K-H, Guérardel Y (2009) Developmental regulation of oligosialylation in zebrafish. Glycoconj J 26:247–261. doi:10.1007/s10719-008-9161-5
12. Chang L-Y, Mir A-M, Thisse C, Guérardel Y, Delannoy P, Thisse B, Harduin-Lepers A (2009) Molecular cloning and characterization of the expression pattern of the zebrafish alpha2, 8-sialyltransferases (ST8Sia) in the developing nervous system. Glycoconj J 26:263–275. doi:10.1007/s10719-008-9165-1
13. Guérardel Y, Chang L-Y, Fujita A, Coddeville B, Maes E, Sato C, Harduin-Lepers A, Kubokawa K, Kitajima K (2012) Sialome analysis of the cephalochordate Branchiostoma

belcheri, a key organism for vertebrate evolution. Glycobiology 22:479–491. doi:10.1093/glycob/cwr155

14. Petit D, Mir A-M, Petit J-M, Thisse C, Delannoy P, Oriol R, Thisse B, Harduin-Lepers A (2010) Molecular phylogeny and functional genomics of beta-galactoside alpha2,6-sialyltransferases that explain ubiquitous expression of st6gal1 gene in amniotes. J Biol Chem 285:38399–38414. doi:10.1074/jbc.M110.163931
15. Harduin-Lepers A, Mollicone R, Delannoy P, Oriol R (2005) The animal sialyltransferases and sialyltransferase-related genes: a phylogenetic approach. Glycobiology 15:805–817. doi:10.1093/glycob/cwi063
16. Felsenstein J (2002) PHYLIP: phylogeny inference package, Ver 3.6. University of Washington, Seattle, WA
17. Guindon S, Gascuel O (2003) A simple, fast and accurate algorithm to estimate large phylogenies by maximum likelihood. Syst Biol 52:696–704. doi:10.1080/10635150390235520
18. Tamura K, Peterson D, Peterson N, Stecher G, Nei M, Kumar S (2011) MEGA5: molecular evolutionary genetics analysis using maximum likelihood, evolutionary distance and maximum parsimony methods. Mol Biol Evol 28:2731–2739. doi:10.1093/molbev/msr121
19. Hammer Ø, Harper DAT, Ryan PD (2001) PAST: paleontological statistics software package for education and data analyses. Palaeontol Electron 4:9
20. Notredame C, Higgins DG, Heringa J (2000) T-coffee: a novel method for fast and accurate multiple sequence alignment. J Mol Biol 302:205–217. doi:10.1006/jmbi.2000.4042
21. Seal RL, Gordon SM, Lush MJ, Wright MW, Bruford EA (2010) Genenames.org: the HGNC resources in 2011. Nucleic Acids Res 39:D514–D519. doi:10.1093/nar/gkq892
22. Page RDM (1996) Tree view: an application to display phylogenetic trees on personal computers. Comput Appl Biosci 12:357–358. doi:10.1093/bioinformatics/12.4.357
23. Datta AK (2009) Comparative sequence analysis in the sialyltransferase protein family: analysis of motifs. Curr Drug Targets 10:483–498
24. Datta AK, Paulson JC (1997) Sialylmotifs of sialyltransferases. Indian J Biochem Biophys 34:157–165
25. Datta AK, Chammas R, Paulson JC (2001) Conserved cysteines in the sialyltransferase sialylmotifs form an essential disulfide bond. J Biol Chem 276:15200–15207. doi:10.1074/jbc.M010542200
26. Datta AK, Paulson JC (1995) The sialyltransferase « sialylmotif » participates in binding the donor substrate CMP-NeuAc. J Biol Chem 270:1497–1500
27. Datta AK, Sinha A, Paulson JC (1998) Mutation of the sialyltransferase S-sialylmotif alters the kinetics of the donor and acceptor substrates. J Biol Chem 273:9608–9614. doi:10.1074/jbc.273.16.9608
28. Geremia RA, Harduin-Lepers A, Delannoy P (1997) Identification of two novel conserved amino acid residues in eukaryotic sialyltransferases: implications for their mechanism of action. Glycobiology 7:v–vii
29. Jeanneau C, Chazalet V, Augé C, Soumpasis DM, Harduin-Lepers A, Delannoy P, Imberty A, Breton C (2004) Structure-function analysis of the human sialyltransferase ST3Gal I: role of n-glycosylation and a novel conserved sialylmotif. J Biol Chem 279:13461–13468. doi:10.1074/jbc.M311764200
30. Tsuji S, Datta AK, Paulson JC (1996) Systematic nomenclature for sialyltransferases. Glycobiology 6:v–vii
31. Patel RY, Balaji PV (2006) Identification of linkage-specific sequence motifs in sialyltransferases. Glycobiology 16:108–116. doi:10.1093/glycob/cwj046
32. Cantarel BL, Coutinho PM, Rancurel C, Bernard T, Lombard V, Henrissat B (2009) The Carbohydrate-Active EnZymes database (CAZy): an expert resource for glycogenomics. Nucleic Acids Res 37:D233–D238. doi:10.1093/nar/gkn663
33. Kent WJ (2002) BLAT—The BLAST-like alignment tool. Genome Res 12:656–664. doi:10.1101/gr.229202
34. Hubbard TJP, Aken BL, Beal K, Ballester B, Caccamo M, Chen Y, Clarke L, Coates G, Cunningham F, Cutts T, Down T, Dyer SC, Fitzgerald S, Fernandez-Banet J, Graf S, Haider S, Hammond M, Herrero J, Holland R, Howe K, Howe K, Johnson N, Kahari A, Keefe D, Kokocinski F, Kulesha E, Lawson D, Longden I, Melsopp C, Megy K, Meidl P, Ouverdin B, Parker A, Prlic A, Rice S, Rios D, Schuster M, Sealy I, Severin J, Slater G, Smedley D, Spudich G, Trevanion S, Vilella A, Vogel J, White S, Wood M, Cox T, Curwen V, Durbin R, Fernandez-Suarez XM, Flicek P, Kasprzyk A, Proctor G, Searle S, Smith J, Ureta-Vidal A, Birney E (2007) Ensembl 2007. Nucleic Acids Res 35:D610–D617. doi:10.1093/nar/gkl996
35. Huang X, Madan A (1999) CAP3: A DNA sequence assembly program. Genome Res 9:868–877. doi:10.1101/gr.9.9.868
36. Burge CB, Karlin S (1998) Finding the genes in genomic DNA. Curr Opin Struct Biol 8:346–354. doi:10.1016/S0959-440X(98)80069-9
37. Breathnach R, Chambon P (1981) Organization and expression of eucaryotic split genes coding for proteins. Annu Rev Biochem 50:349–383. doi:10.1146/annurev.bi.50.070181.002025

38. Thompson JD, Higgins DG, Gibson TJ (1994) CLUSTAL W: improving the sensitivity of progressive multiple sequence alignment through sequence weighting, position-specific gap penalties and weight matrix choice. Nucleic Acids Res 22:4673–4680. doi:10.1093/nar/22.22.4673
39. Higgins DG, Thompson JD, Gibson TJ (1996) Using CLUSTAL for multiple sequence alignments. Methods Enzymol 266:383–402
40. Morgenstern B (2004) DIALIGN: multiple DNA and protein sequence alignment at BiBiServ. Nucleic Acids Res 32:W33–W36. doi:10.1093/nar/gkh373
41. Harduin-Lepers A, Vallejo-Ruiz V, Krzewinski-Recchi M-A, Samyn-Petit B, Julien S, Delannoy P (2001) The human sialyltransferase family. Biochimie 83:727–737. doi:10.1016/S0300-9084(01)01301-3
42. Gasteiger E, Gattiker A, Hoogland C, Ivanyi I, Appel RD, Bairoch A (2003) ExPASy: the proteomics server for in-depth protein knowledge and analysis. Nucleic Acids Res 31:3784–3788. doi:10.1093/nar/gkg563
43. Finn RD, Mistry J, Tate J, Coggill P, Heger A, Pollington JE, Gavin OL, Gunasekaran P, Ceric G, Forslund K, Holm L, Sonnhammer ELL, Eddy SR, Bateman A (2010) The Pfam protein families database. Nucleic Acids Res 38:D211–D222. doi:10.1093/nar/gkp985
44. Murphy LR, Wallqvist A, Levy RM (2000) Simplified amino acid alphabets for protein fold recognition and implications for folding. Protein Eng 13:149–152. doi:10.1093/protein/13.3.149
45. Mollicone R, Moore SEH, Bovin N, Garcia-Rosasco M, Candelier J-J, Martinez-Duncker I, Oriol R (2009) Activity, splice variants, conserved peptide motifs and phylogeny of two new alpha1,3-fucosyltransferase families (FUT10 and FUT11). J Biol Chem 284:4723–4738. doi:10.1074/jbc.M809312200
46. Castresana J (2000) Selection of conserved blocks from multiple alignments for their use in phylogenetic analysis. Mol Biol Evol 17: 540–552
47. Talavera G, Castresana J (2007) Improvement of phylogenies after removing divergent and ambiguously aligned blocks from protein sequence alignments. Syst Biol 56:564–577. doi:10.1080/10635150701472164
48. Müller KF (2005) The efficiency of different search strategies in estimating parsimony jackknife, bootstrap and Bremer support. BMC Evol Biol 5:58. doi:10.1186/1471-2148-5-58
49. Battistuzzi FU, Filipski A, Hedges SB, Kumar S (2010) Performance of relaxed-clock methods in estimating evolutionary divergence times and their credibility intervals. Mol Biol Evol 27:1289–1300. doi:10.1093/molbev/msq014
50. Blair JE, Hedges SB (2005) Molecular phylogeny and divergence times of deuterostome animals. Mol Biol Evol 22:2275–2284. doi:10.1093/molbev/msi225
51. Kumar S, Hedges SB (2011) TimeTree2: species divergence times on the iPhone. Bioinformatics 27:2023–2024. doi:10.1093/bioinformatics/btr315
52. McLysaght A, Hokamp K, Wolfe KH (2002) Extensive genomic duplication during early chordate evolution. Nat Genet 31:200–204. doi:10.1038/ng884
53. Catchen JM, Conery JS, Postlethwait JH (2009) Automated identification of conserved synteny after whole-genome duplication. Genome Res 19:1497–1505. doi:10.1101/gr.090480.108
54. Muffato M, Louis A, Poisnel C-E, Crollius HR (2010) Genomicus: a database and a browser to study gene synteny in modern and ancestral genomes. Bioinformatics 26:1119–1121. doi:10.1093/bioinformatics/btq079
55. Martin R, Gallet P-F, Rocha D, Petit D (2009) Polymorphism of the prion protein in mammals: a phylogenetic approach. Recent Pat DNA Gene Seq 3:63–71
56. Petit D, Maftah A, Julien R, Petit J-M (2006) En bloc duplications, mutation rates and densities of amino acid changes clarify the evolution of vertebrate alpha-1,3/4-fucosyltransferases. J Mol Evol 63:353–364. doi:10.1007/s00239-005-0189-x
57. Yang Z (2007) PAML 4: phylogenetic analysis by maximum likelihood. Mol Biol Evol 24:1586–1591. doi:10.1093/molbev/msm088
58. Bielawski JP, Yang Z (2003) Maximum likelihood methods for detecting adaptive evolution after gene duplication. J Struct Funct Genomics 3:201–212. doi:10.1023/A:1022642807731
59. Yang Z, Nielsen R (1998) Synonymous and nonsynonymous rate variation in nuclear genes of mammals. J Mol Evol 46:409–418. doi:10.1007/PL00006320
60. Ermonval M, Petit D, Le Duc A, Kellermann O, Gallet PF (2009) Glycosylation-related genes are variably expressed depending on the differentiation state of a bioaminergic neuronal cell line: implication for the cellular prion protein. Glycoconj J 26:477–493. doi:10.1007/s10719-008-9198-5
61. Kulikova T, Aldebert P, Althorpe N, Baker W, Bates K, Browne P, van den Broek A, Cochrane G, Duggan K, Eberhardt R, Faruque N, Garcia-Pastor M, Harte N, Kanz C, Leinonen R, Lin Q, Lombard V, Lopez R, Mancuso R, McHale M,

Nardone F, Silventoinen V, Stoehr P, Stoesser G, Tuli MA, Tzouvara K, Vaughan R, Wu D, Zhu W, Apweiler R (2004) The EMBL nucleotide sequence database. Nucleic Acids Res 32:D27–D30. doi:10.1093/nar/gkh120
62. Benson DA, Karsch-Mizrachi I, Lipman DJ, Ostell J, Wheeler DL (2003) GenBank. Nucleic Acids Res 31:23–27. doi:10.1093/nar/gkg057
63. Tateno Y, Imanishi T, Miyazaki S, Fukami-Kobayashi K, Saitou N, Sugawara H, Gojobori T (2002) DNA data bank of Japan (DDBJ) for genome scale research in life science. Nucleic Acids Res 30:27–30. doi:10.1093/nar/30.1.27
64. Consortium, T. U (2007) The universal protein resource (UniProt). Nucleic Acids Res 36:D190–D195. doi:10.1093/nar/gkm895
65. Quackenbush J, Cho J, Lee D, Liang F, Holt I, Karamycheva S, Parvizi B, Pertea G, Sultana R, White J (2001) The TIGR gene indices: analysis of gene transcript sequences in highly sampled eukaryotic species. Nucleic Acids Res 29:159–164. doi:10.1093/nar/29.1.159
66. Altschul SF, Madden TL, Schäffer AA, Zhang J, Zhang Z, Miller W, Lipman DJ (1997) Gapped BLAST and PSI-BLAST: a new generation of protein database search programs. Nucleic Acids Res 25:3389–3402. doi:10.1093/nar/25.17.3389
67. Gouy M, Guindon S, Gascuel O (2010) SeaView version 4: a multiplatform graphical user interface for sequence alignment and phylogenetic tree building. Mol Biol Evol 27:221–224. doi:10.1093/molbev/msp259
68. Edgar RC (2004) MUSCLE: multiple sequence alignment with high accuracy and high throughput. Nucleic Acids Res 32:1792–1797. doi:10.1093/nar/gkh340
69. Apweiler R (2004) UniProt: the universal protein knowledgebase. Nucleic Acids Res 32:115D–119D. doi:10.1093/nar/gkh131
70. Grigoriev IV, Nordberg H, Shabalov I, Aerts A, Cantor M, Goodstein D, Kuo A, Minovitsky S, Nikitin R, Ohm RA, Otillar R, Poliakov A, Ratnere I, Riley R, Smirnova T, Rokhsar D, Dubchak I (2011) The genome portal of the Department of Energy Joint Genome Institute. Nucleic Acids Res 1–7. doi:10.1093/nar/gkr947
71. Kanehisa M, Goto S, Kawashima S, Nakaya A (2002) The KEGG databases at GenomeNet. Nucleic Acids Res 30:42–46. doi:10.1093/nar/30.1.42
72. Kanehisa M, Goto S (2000) KEGG: Kyoto encyclopedia of genes and genomes. Nucleic Acids Res 28:27–30. doi:10.1093/nar/28.1.27
73. Raman R, Venkataraman M, Ramakrishnan S, Lang W, Raguram S, Sasisekharan R (2006) Advancing glycomics: implementation strategies at the consortium for functional glycomics. Glycobiology 16:82R–90R. doi:10.1093/glycob/cwj080
74. Sprague J (2006) The Zebrafish information network: the Zebrafish model organism database. Nucleic Acids Res 34:D581–D585. doi:10.1093/nar/gkj086
75. Scheer M, Grote A, Chang A, Schomburg I, Munaretto C, Rother M, Söhngen C, Stelzer M, Thiele J, Schomburg D (2011) BRENDA, the enzyme information system in 2011. Nucleic Acids Res 39:D670–D676. doi:10.1093/nar/gkq1089
76. Kikuno R (2004) HUGE: a database for human KIAA proteins, a 2004 update integrating HUGEppi and ROUGE. Nucleic Acids Res 32:502D–504D. doi:10.1093/nar/gkh035
77. Mcwilliam H, Valentin F, Goujon M, Li W, Narayanasamy M, Martin J, Miyar T, Lopez R (2009) Web services at the European Bioinformatics Institute-2009. Nucleic Acids Res 37:W6–W10. doi:10.1093/nar/gkp302
78. Reese MG, Eeckman FH, Kulp D, Haussler D (1997) Improved splice site detection in genie. J Comput Biol 4:311–323
79. Thompson JD, Gibson TJ, Plewniak F, Jeanmougin F, Higgins DG (1997) The CLUSTAL_X Windows interface: flexible strategies for multiple sequence alignment aided by quality analysis tools. Nucleic Acids Res 25:4876–4882. doi:10.1093/nar/25.24.4876
80. Crooks GE, Hon G, Chandonia J-M, Brenner SE (2004) WebLogo: a sequence logo generator. Genome Res 14:1188–1190. doi:10.1101/gr.849004
81. Otsuka J, Sugaya N (2003) Advanced formulation of base pair changes in the stem regions of ribosomal RNAs; its application to mitochondrial rRNAs for resolving the phylogeny of animals. J Theor Biol 222:447–460. doi:10.1016/S0022-5193(03)00057-2

Chapter 8

Fluorescent Lectin Staining of *Drosophila* Embryos and Tissues to Detect the Spatial Distribution of Glycans During Development

E Tian, Liping Zhang, and Kelly G. Ten Hagen

Abstract

Glycans are the result of the coordinated activities of glycosyltransferases responsible for specific sugar additions. Glycans present on proteins can influence protein stability, transport, function, and recognition, and thus can have profound effects on cell–cell interactions, adhesion, and signaling events occurring during eukaryotic development. Lectin staining provides a useful tool to detect the spatial distribution of specific glycans in developing tissues in situ. Here we describe a method to detect diverse glycans present in developing *Drosophila* tissues and organs using fluorescently labeled lectins.

Key words Glycosylation, Glycans, Lectins, Confocal microscopy, *Drosophila*, Development

1 Introduction

Glycosyltransferases are responsible for the formation of specific glycan structures that can have diverse biological effects. Lectins are sugar-binding proteins that have been used to detect a variety of glycan structures present during eukaryotic development [1, 2]. In an effort to determine the spatial distribution of glycans during *Drosophila* development, we have stained *Drosophila* embryos and tissues with fluorescently labeled lectins and performed confocal microscopy [3]. This method is compatible with whole-mount staining, allowing one to image glycan expression in a 3D embryo or tissue at specific developmental stages.

2 Materials

Prepare all solutions using distilled water. Strictly follow all storage and waste disposal procedures.

Inka Brockhausen (ed.), *Glycosyltransferases: Methods and Protocols*, Methods in Molecular Biology, vol. 1022, DOI 10.1007/978-1-62703-465-4_8, © Springer Science+Business Media New York 2013

1. Bovine serum albumin (BSA): Albumin bovine fraction V (MP Biomedicals LLC., Solon, OH, USA). Store at 4 °C.
2. Dechorionation solution: Germicidal bleach, active ingredient is 6.15 % sodium hypochlorite (Clorox® Professional Products Company, Oakland, CA, USA). Store at room temperature.
3. Heptane (Sigma-Aldrich, St. Louis, MO, USA). Store at room temperature in flame resistant cabinet.
4. Glycerol (Life Technologies, Carlsbad, CA, USA). Make 70 % glycerol solution with 1× PBS. Store at room temperature.
5. Methanol: Methyl alcohol, anhydrous (Mallinckrodt Baker, Inc., Phillipsburg, NJ, USA.). Store at room temperature in flame resistant cabinet.
6. Paraformaldehyde: 10 % stock solution, EM grade, methanol and RNase free (Electron Microscopy Science, Hatfield, PA, USA). Store at room temperature.
7. Phosphate Buffered Saline (PBS), 10× (Quality Biological, Inc., Gaithersburg, MD, USA). Dilute to 1× with distilled water. Store at room temperature.
8. Triton®X-100 (Sigma-Aldrich). Store at room temperature.
9. Embryo Fixation Stock Buffer: 800 mM KCl, 200 mM NaCl, 150 mM PIPES (1,4-Piperazinediethanesulfonic acid, Sigma-Aldrich), and 20 mM EGTA (ethylene glycol-bis (2-aminoethylether)-*N*,*N*,*N*′,*N*′-tetraacetic acid, Sigma-Aldrich) in 400 mL. Adjust pH to 7.4 with 5 N NaOH (*see* **Note 1**) and bring to final volume of 500 mL. Autoclave to sterilize. Store at 4 °C.
10. Embryo Fixation Working Solution: Combine 2 parts distilled water, 2 parts 10 % paraformaldehyde and 1 part Embryo Fixation Stock Buffer. Make fresh before each use.
11. Larval Fixation Solution: 4 % paraformaldehyde in 1× PBS. Make fresh before each use.
12. Lectins: Fluorescent lectins purchased from EY Laboratories (San Mateo, CA, USA) included tetramethylrhodamine isothiocyanate (TRITC)-conjugated *Canavalia ensiformis* (Con A); TRITC-conjugated *Dolichos biflorus* (DBA); fluorescein isothiocyanate (FITC)-conjugated *Artocarpus integrifolia* (Jacalin); FITC-conjugated *Vicia villosa* (VVA); and FITC-conjugated *Triticum vulgare* (WGA). Lectins purchased from Molecular Probes (Eugene, OR, USA) included Alexa Fluor 488-conjugated *Arachis hypogaea* (PNA) and Alexa Fluor 568-conjugated *Glycine max* (SBA). FITC-conjugated *Helix pomatia* (HPA) (Sigma-Aldrich). Lectins were reconstituted at 1 mg/mL in distilled water and stored as 20 μL aliquots at −20 °C (*see* **Note 2**). All fluorescent lectins should be protected from light.

13. Sugars (Sigma-Aldrich) for inhibition of lectin binding: *N*-Acetyl-D-galactosamine (GalNAc), for inhibition of HPA, DBA, SBA and VVA (store at 4 °C); *N*-Acetyl-D-glucosamine (GlcNAc) for inhibition of WGA (store at −20 °C); D-(+)-Galactose (Gal) for inhibition of Jacalin and PNA (store at room temperature); D-(+)-Mannose (Man) for inhibition of Con A (store at room temperature). Make 1 M stock solution of each sugar in 1× PBS. Store stock solutions at −20 °C.
14. Hoechst 33342 nuclear counterstain, 10 mg/mL stock solution (Molecular Probes, Eugene, OR, USA). Store at 4 °C.
15. Mounting Solution: 2 % 1,4-Diazabicyclo[2.2.2]octane (Dabco, Sigma-Aldrich) in 70 % glycerol. Store at −20 °C.
16. Glass slides: 75×25 mm Superfrost®/Plus Microslides (Daigger, Vernon Hills, IL, USA).
17. Cover slips: 24×50mm microscope cover glass, No. 1 (Thermo Scientific, Portsmouth, NH, USA).
18. Cell strainer, 70 μm nylon mesh (BD Biosciences, San Jose, CA, USA).
19. Red sable paintbrush, size 2 (Thomas Scientific, Swedesboro, NJ, USA).
20. 7 mL scintillation vials with unattached cap (Daigger).
21. GyroMini™ Nutating Mixer (Labnet International, Woodbridge, NJ, USA).
22. Pyrex® Spot Test Plates, 9 well, 85×100 mm (Thomas Scientific).
23. Dissecting Dish with Plastic Dish (Electron Microscopy Science).

3 Methods

3.1 *Drosophila* Embryo Collection and Fixation

Carry out all procedures at room temperature unless specified otherwise.

1. Collect embryos at desired developmental stage from juice agar plates (*see* ref. [4] for details of Drosophila embryo collection). Transfer embryos from plate into a cell strainer using a paintbrush.
2. Wash embryos with distilled water thoroughly to remove all debris from agar plates. Immerse cell strainer containing embryos in 100 % Clorox bleach for 1–3 min to dechorionate. Dechorionation is complete when embryos become shiny and dorsal appendages are detached (*see* ref. [5] for details on dechorionation).
3. Rinse embryos thoroughly with distilled water. Transfer embryos to a 7 mL scintillation vial using a glass Pasteur

pipette. Rinse embryos with water again. Then remove as much water as possible (*see* **Note 3**).

4. Add 2.5 mL freshly made Embryo Fixation Working Solution to the vial. Then add 2.5 mL heptane. Vortex vial containing embryos at medium speed for 20 min. Embryos will then be present at the interface of the two solutions.
5. Remove all of the lower aqueous phase from vial with a glass Pasteur pipette.
6. Add 4 mL methanol and shake back and forth by hand for 30 s to 2 min to break the vitelline membrane. Embryos will then sink to the bottom of the vial. Remove heptane and most of the methanol with a glass Pasteur pipette, being careful not to aspirate embryos.
7. Rinse embryos with 100 % methanol. Transfer embryos to a 1.5 mL Eppendorf tube and rinse with methanol again. Embryos can be stored in methanol at −20 °C for several months.

3.2 *Drosophila* Larval Tissue Collection and Fixation

Carry out all steps at room temperature unless specified otherwise.

1. Collect desired larval tissues from staged larvae.
2. Immediately put the dissected larval tissues in Larval Fixation Solution on ice. Use 1 mL of Larval Fixation Solution for tissues from 4 to 5 larvae (in a 1.5 mL Eppendorf tube). Put the tube on a nutating mixer and rotate 20–30 min.
3. If larval tissues are not used immediately for lectin staining, they should be washed in methanol five times (2–5 min each) and then washed in ethanol five times (2–5 min each). Imaginal disks can then be stored in ethanol at −20 °C for 1–2 months.

3.3 Whole-Mount Lectin Staining

Carry out all steps at room temperature and on a nutating mixer.

1. Remove storage solution from fixed embryos and/or larval tissues.
2. Wash embryos and/or larval tissues with 1× PBS containing 0.1 % Triton®X-100 for 3–5 min. Remove PBS solution and repeat washing step two additional times.
3. Remove final PBS wash solution and add 0.1 % BSA in PBS containing 0.1 % Triton®X-100 for 1 h to block.
4. While blocking, dilute lectins to be used and prepare sugar inhibitor control by pre-incubating lectins with appropriate inhibitory sugars for 30 min to 1 h. Briefly, dilute lectins to 10 μg/mL in a solution of 0.1 % BSA in PBS containing 0.1 % Triton®X-100. Prepare sugar inhibitor control by incubating an aliquot of diluted lectin with 0.1–0.2 M of the appropriate inhibitory sugar. The exact concentration of sugar needed to specifically block the binding of each lectin should be determined empirically.

5. Remove the blocking solution and add 200 μL to 1 mL of diluted lectin (depending on sample size) or 200 μL to 1 mL of diluted lectin pre-incubated with sugar to samples. Incubate in the dark for 2–3 h.
6. Remove lectin solution and wash samples with 1 mL PBS containing 0.1 % Triton X-100 for 20–30 min in the dark. Repeat washing step two additional times.
7. Add 1:20,000 dilution of 10 mg/mL Hoechst 33342 nuclear counterstain to Mounting Solution.
8. Remove washing buffer and equilibrate the samples with 500 μL Mounting Solution containing nuclear counterstain overnight at 4 °C in the dark. Mount samples on glass slide with coverslip the following day (*see* **Note 4**). Use approximately 200 μL Mounting Solution to mount samples on slide. Place coverslip on the sample by holding one edge of the cover slip with forceps and letting the other edge touch the mounting media on the surface of the slide. Gently lower the cover slip to avoid trapping air bubbles beneath it. Before sealing the slide, remove any excess mounting medium at the edges of the cover slip using the edge of a paper towel. Glycerol based mounting solutions will not polymerize, so the edges of slides should be sealed with nail polish. To preserve the fluorescent signal, mounted slides should be stored at 4 °C or –20 °C and protected from light.
9. Image samples using confocal microscopy according to standard procedures (*see* **Note 5**).

4 Notes

1. PIPES does not dissolve completely until the pH approaches 7.4.
2. To avoid freeze and thaw cycles, 50 % glycerol can be added to lectin stock solutions and stored at –20 °C for up to 1 year.
3. The dechorionated embryos can stick to the inside of transfer pipettes. Caution should be taken to avoid aspiration of embryos far up into the pipette. Pipettes can be repeatedly rinsed with water to dislodge trapped embryos.
4. Embryos can be directly transferred in Mounting Solution to a glass slide for mounting. Larval tissues of interest can be dissected away from other tissues by carefully transferring the larval samples (using a plastic 1 mL pipette tip with the end cut off to enable aspiration of the tissues of interest) in a small amount of Mounting Solution to a Pyrex® Spot Test Plate or a Dissecting Dish. After dissection, tissues can then be transferred with dissection forceps to a glass slide and additional Mounting Solution can be added (if needed) prior to mounting.

5. Zeiss LSM 510 confocal laser scanning microscope was used for sample analysis. Stacked confocal images were captured. Images were processed by either Zeiss LSM Image Browser or NIH ImageJ software and assembled in Photoshop (examples are shown in Fig. 1).

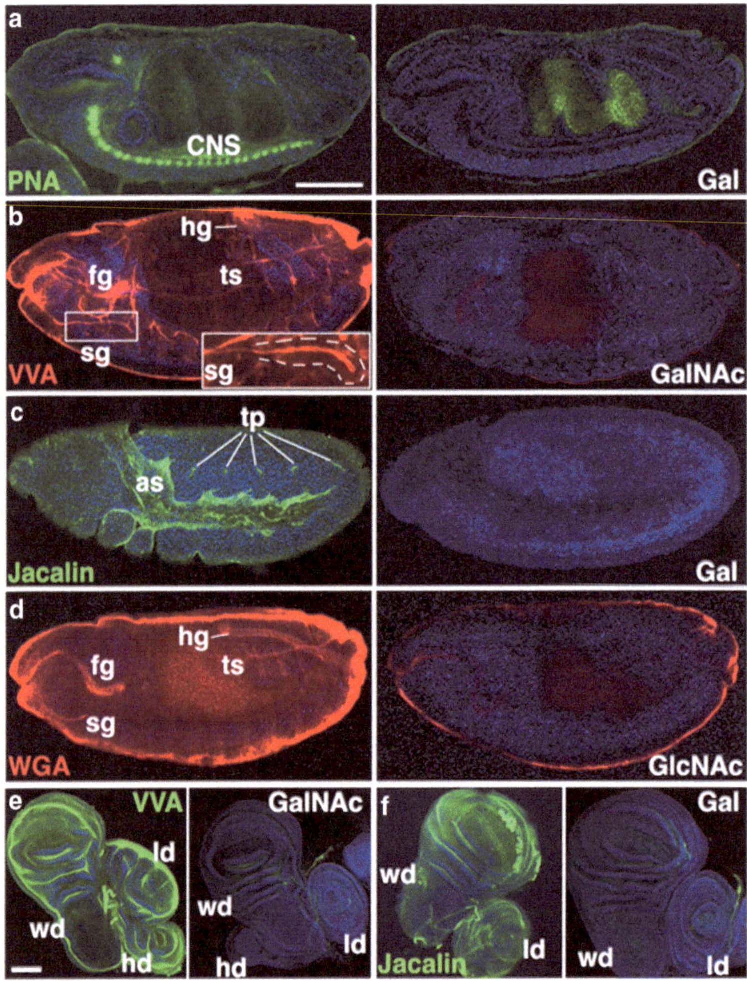

Fig. 1 Glycan expression in *Drosophila* embryos and larval tissues as detected by fluorescently labeled lectins. Fluorescent confocal stacked images are shown. Controls incubated with the competing sugar (Gal, GalNAc, or GlcNAc) to demonstrate staining specificity are shown to the *right* of each panel. (**a**) PNA detects mucin-type Core 1 O-linked glycans (Galβ1-3GalNAcα-serine/threonine) in the central nervous system (CNS) of stage 17 embryos. (**b**) VVA detects mucin-type O-linked glycans (GalNAcα-serine/threonine) in the foregut (fg), salivary glands (sg), tracheal system (ts), and hindgut (hg) in stage 15 embryos. *Inset* shows a magnified view of the salivary gland (outlined with *white dashed line*) with abundant O-linked glycans present along the apical and luminal surfaces. (**c**) Jacalin detects mucin-type Core 1 O-linked glycans in the amnioserosa (as) and tracheal placodes (tp) in stage 10 embryos. (**d**) WGA detects *N*-acetylglucosamine (GlcNAc)-containing glycans in the foregut, salivary glands, tracheal system, and hindgut in stage 15 embryos. (**e**) VVA detects mucin-type O-linked glycans (GalNAcα-serine/threonine) in the larval wing (wd), leg (ld), and haltere (hd) imaginal disks. (**f**) Jacalin stains larval wing and leg imaginal disks. Nuclear counterstaining is shown in *blue*. Scale bars = 100 μm

Acknowledgment

This work was supported by the National Institutes of Health Intramural Research Program of the NIDCR.

References

1. Fredieu JR, Mahowald AP (1994) Glycoconjugate expression during *Drosophila* embryogenesis. Acta Anat 149:89–99
2. D'Amico P, Jacobs JR (1995) Lectin histochemistry of the *Drosophila* embryo. Tissue Cell 27:23–30
3. Tian E, Ten Hagen KG (2007) O-linked glycan expression during *Drosophila* development. Glycobiology 8:820–827
4. Rothwell WF, Sullivan W (2007) *Drosophila* embryo collection. Cold Spring Harbor Protocols. Cold Spring Harbor Laboratory Press, Cold Spring Harbor, NY. doi:10.1101/pdb.prot4825
5. Rothwell WF, Sullivan W (2007) Drosophila embryo dechorionation. Cold Spring Harbor Protocol, Cold Spring Harbor Laboratory Press. Cold Spring Harbor, NY. doi:10.1101/pdb.prot4826

Chapter 9

Photoaffinity Labeling of Protein *O*-Mannosyltransferases of the PMT1/PMT2 Subfamily

Martin Loibl and Sabine Strahl

Abstract

Protein *O*-mannosylation is initiated at the endoplasmic reticulum (ER) by dolichyl phosphate-mannose: protein *O*-mannosyltransferases (PMTs). PMTs are members of the glycosyltransferase (GT) C superfamily. They are large polytopic integral membrane proteins located in the ER membrane. PMTs utilize dolichyl phosphate-activated mannose as sugar donor. Glycosyltransfer of mannose to serine and threonine residues of nascent polypeptides leads to an inversion of the stereochemistry of the glycosidic bond. Here, we describe photoaffinity labeling of yeast Pmt1p using a photo-reactive probe that is based on the artificial mannosyl acceptor peptide YATAV. Due to the high homology of PMTs, this method can also be applied to study PMT1 and PMT2 subfamily members from fungi other than baker's yeast.

Key words Mannosyltransferase, PMT1, PMT2, Protein *O*-mannosylation, Photoaffinity labeling, Photolysis, Cross-linking, Yeast

1 Introduction

In eukaryotes, a family of Dol-P-β-Man:protein α-mannosyltransferases (PMTs; EC 2.4.1.109) is initiating the essential protein *O*-mannosylation of secretory and membrane proteins in the endoplasmic reticulum (ER) [1, 2]. PMT family members are conserved throughout the fungal and animal kingdoms. They are further subdivided into the PMT1, PMT2, and PMT4 subfamilies, which include transferases closely related to *Saccharomyces cerevisiae* Pmt1p, Pmt2p, and Pmt4p, respectively [3]. In fungi, at least one member of each subfamily is present, whereas in animals only PMT2 and PMT4 subfamily members are conserved [4]. To date, PMTs have been best characterized in baker's yeast where Pmt1p, Pmt2p, and Pmt4p account for the major transferase activities although at least six PMTs (Pmt1p–Pmt6p) are present [2, 5]. Pmt1p and Pmt2p form heterodimeric protein complexes, while Pmt4p acts as homodimer [6, 7]. In addition, distinct

Inka Brockhausen (ed.), *Glycosyltransferases: Methods and Protocols*, Methods in Molecular Biology, vol. 1022, DOI 10.1007/978-1-62703-465-4_9, © Springer Science+Business Media New York 2013

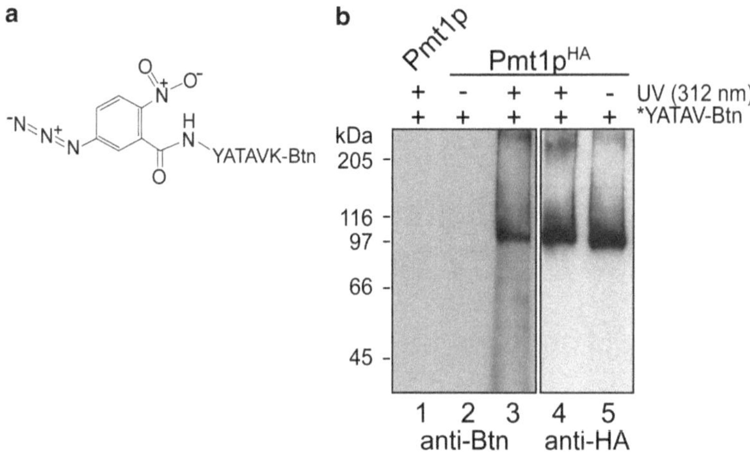

Fig. 1 Cross-linking of Pmt1p^{HA} with the photoreactive mannosyl acceptor peptide. *Left panel*: Structure of the photoreactive probe *YATAVK-Btn. *Right panel*: *YATAVK-Btn was cross-linked to Pmt1p^{HA} isolated from the yeast strain *pmt1*Δ/pSB56 (Pmt1p^{HA}). Western blots were sequentially probed with anti-biotin and anti-HA antibodies. (This research was originally published in the Journal of Biological Chemistry [12] © the American Society for Biochemistry and Molecular Biology)

mannosyl-acceptor proteins have been identified for Pmt1p/Pmt2p and Pmt4p complexes [8, 9].

PMTs are members of the glycosyltransferase (GT) C superfamily which comprises large polytopic integral membrane proteins located in the ER or the plasma membrane [10]. The majority of these enzymes utilize lipid phosphate-activated sugar donors, and glycosyltransfer leads to an inversion of the stereochemistry of the glycosidic bond. To date, very little information on three-dimensional structures is available, thus impeding access to the molecular mechanism of GT-C transferases, including PMTs.

Our previous studies showed that *S. cerevisiae* Pmt1p is an integral ER membrane protein with seven transmembrane domains (TMDs) [11]. Hydropathy profiles of PMT proteins are highly conserved suggesting that the 7-TMD topological model is applicable to all family members. N- and C-terminus of yeast Pmt1p are situated in the cytosol and the ER lumen, respectively. Two prominent hydrophilic loops are located between TMD 1 and TMD 2 (loop1), and TMD 5 and TMD 6 (loop5). They are facing the ER lumen and are crucial for transferase activity in vivo [5]. We recently established a photoaffinity labeling approach (Fig. 1) that identified the Pmt1p-loop1 region as part of the mannosyl acceptor binding and/or catalytic site (Fig. 2) [12]. In short, we developed a peptide-based photoaffinity probe that is derived from the biotinylated peptide NH_2-YATAVK-(Biotin)-COOH (YATAVK-Btn), an artificial in vitro mannosyl acceptor substrate of Pmt1p/Pmt2p complexes. After conjugation with a heterobifunctional NHS-ester and photo-activatable cross-linker, the photo-reactive peptide probe

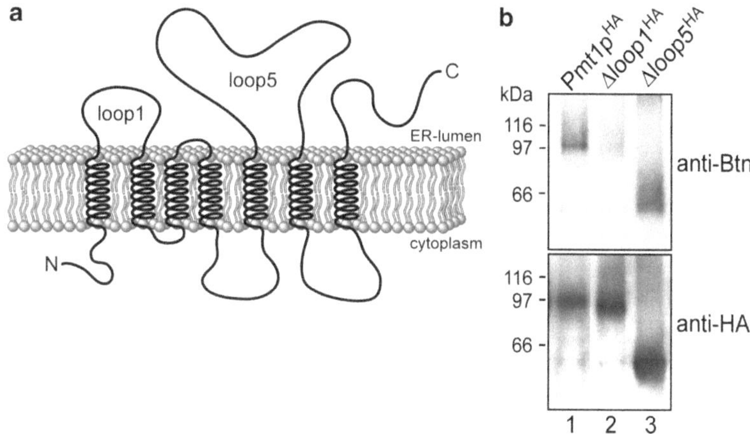

Fig. 2 Pmt1p-loop1 is crucial for cross-linking of the photoreactive peptide. *Left panel*: Schematic representation of Pmt1p membrane topology. *Right panel*: Pmt1p^{HA} mutants where the major ER-oriented loops (loop1 and loop5) have been deleted were isolated from strain *pmt1Δ* transformed with pSB56 (*lane 1*; Pmt1p^{HA}), pSB101 (*lane 2*; Δloop1^{HA}), or pVG13 (*lane 3*; Δloop5^{HA}). After photoaffinity labeling samples were analyzed by SDS-PAGE and Western blotting. Cross-linking of mutant Δloop1^{HA} to *YATAVK-Btn was lost. In contrast, deletion of loop5 did not affect binding of the photoreactive peptide. (This research was originally published in the Journal of Biological Chemistry [12] © the American Society for Biochemistry and Molecular Biology)

*YATAVK-Btn (Fig. 1) is incubated with purified Pmt1p/Pmt2p complexes. Upon photoactivation, samples are resolved on SDS-polyacrylamide (PA) gels and analyzed by Western blot using antibodies that specifically detect cross-linked products. Due to the high homology of PMTs, this method can also be applied to study PMT1 and PMT2 subfamily members from fungi other than baker's yeast.

2 Materials

Prepare all solutions with bidistilled water. Unless otherwise stated, standard laboratory equipment is used; analytical grade reagents were purchased from AppliChem GmbH (Darmstadt, Germany) and Sigma-Aldrich (Steinheim, Germany); and buffers and reagents are stored at 4 °C. Accurately regard safety precautions and follow waste disposal regulations.

2.1 Isolation of Yeast Membranes, Solubilization and Immunoprecipitation

1. Membrane buffer: 50 mM Tris-HCl (pH 7.5), 0.3 mM $MgCl_2$.
2. Protease inhibitors stock solutions: store at −20 °C.
3. 100 mM phenylmethylsulfonyl fluoride (PMSF): dissolve 174 mg in 10 mL of 100 % isopropyl alcohol.

4. 100 mM benzamidine hydrochloride hydrate: dissolve 157 mg in 10 mL of H_2O.
5. 25 mM N_α-Tosyl-L-lysine chloromethyl ketone hydrochloride (TLCK): dissolve 92 mg in 10 mL of 50 mM NaOAc (pH 5.0).
6. 15 mM *N-p*-Tosyl-L-phenylalanine chloromethyl ketone (TPCK): dissolve 53 mg in 10 mL of 100 % ethanol.
7. 33 mM antipain: dissolve 20 mg in 1 mL of H_2O.
8. 2 mM leupeptin: dissolve 5 mg in 5 mL of H_2O.
9. 1.5 mM pepstatin A: dissolve 5 mg in 5 mL of 100 % ethanol.
10. Glass beads (0.25–0.5 mm in diameter; Carl Roth, Karlsruhe, Germany).
11. Triton buffer: 50 mM Tris-HCl (pH 7.5), 150 mM NaCl, 0.3 mM $MgCl_2$, 10 % (v/v) glycerol, 0.5 % (v/v) Triton X-100.
12. Anti-HA affinity matrix from rat IgG1 (Clone 3F10; Roche, Mannheim, Germany).
13. Tris-buffered saline (TBS; 1×): 50 mM Tris-HCl (pH 7.4), 150 mM NaCl.
14. Translucent 0.2 mL thin-wall PCR microtubes (Biozym, Oldendorf, Germany).

2.2 Peptides and Cross-linking Reagents

1. Opaque 0.5 mL microtubes (neoLab, Heidelberg, Germany).
2. 50 mM NH_2-YATAVK(Btn)-COOH (YATAVK-Btn, purity >80 %; Thermo Fisher Scientific, Bonn, Germany).
3. 100 mM *N*-5-azido-2-nitrobenzoyloxysuccinimide (ANB-NOS; Thermo Fisher Scientific).
4. 0.66 M Tris-HCl (pH 7.5), 50 mM $MgCl_2$.
5. 1 % (v/v) Triton X-100.
6. 3 mg/mL (w/v) influenza hemagglutinin (HA) peptide (Roche). Dissolve 3 mg of HA-peptide in 1 mL of TBS. Store at -20 °C.
7. UV hand lamp 6 KLU (6 W; 220 V; intensity 14 μW/cm^2) (neoLab).

2.3 SDS-PAGE and Immunoblotting Components

1. 8 % SDS-PA gel electrophoresis according to [13].
2. SDS-sample buffer (5×): 312.5 mM Tris-HCl (pH 6.8), 50 % (v/v) glycerol, 25 % (v/v) β-mercaptoethanol, 10 % (w/v) SDS, 0.1 % (w/v) bromophenol blue.
3. Nitrocellulose (NC) membranes (GE Healthcare, Munich, Germany).
4. TBS/T: TBS (1×) containing 0.1 % (v/v) Tween20.
5. Blocking solution: TBS/T containing 1 % (w/v) bovine serum albumin (BSA).

6. Peroxidase-coupled goat anti-biotin antibody (Sigma-Aldrich).
7. Mouse monoclonal anti-HA antibody (16B12; Covance, Munich, Germany).
8. Peroxidase-coupled rabbit anti-mouse-IgG antibody (Sigma-Aldrich).
9. SuperSignal West Pico Chemiluminescent Substrate (Thermo Fisher Scientific).

3 Methods

3.1 Solubilization of Pmt1p^{HA} from Crude Yeast Membranes and Add-on Immunoprecipitation

Perform all procedures at 4 °C unless otherwise specified. Quantities indicated apply to one subsequent cross-linking reaction.

1. Grow yeast cells overnight to $OD_{600}=1$ (*see* **Note 1**). Harvest cells from 25 mL of culture, and wash cells once with 20 mL of membrane buffer. Suspend cell pellet in 100 μL of membrane buffer and transfer the suspension to a 1.5 mL microtube.
2. Add protease inhibitors (*see* **Notes 2** and **3**) and 100 μL of glass beads. Apply four cycles of 1-min vortexing and 1-min cooling on ice to lyse cells (*see* **Note 4**). Spin briefly, puncture top and bottom of tube with red hot 25G1 needle, and collect lysate into fresh 1.5 mL microtube by centrifugation at $250\times g_{max}$ for 30 s (*see* **Note 5**).
3. Centrifuge cell lysate at $1{,}000\times g_{max}$ for 5 min to remove cell debris. Transfer supernatant into a fresh 1.5 mL microtube and collect crude membranes by centrifugation at $48{,}000\times g_{max}$ for 30 min. Discard supernatant and suspend the membrane pellet in 500 μL of Triton buffer (*see* **Notes 6–8**).
4. Add protease inhibitors (*see* **Notes 2** and **3**), and vigorously vortex membrane suspension for 15 min to solubilize Pmt1p^{HA} (*see* **Note 9**).
5. While membranes are solubilizing, transfer 20 μL slurry of anti-HA affinity matrix into a 1.5 mL microtube (*see* **Note 10**). Add 1 mL of Triton buffer and mix by inverting the tube. Spin down affinity matrix at $250\times g_{max}$ for 2 min, and discard supernatant. Keep equilibrated affinity matrix on ice until use.
6. Following solubilization, centrifuge solubilized membrane suspension at $48{,}000\times g_{max}$ for 30 min.
7. Add approximately 300 μL of the supernatant (= Triton extract) to the equilibrated anti-HA affinity matrix (*see* **Notes 8** and **11**), seal tubes with parafilm, and incubate on an end-over-end shaker for at least 1.5 h.

8. In the meantime, prepare photoreactive peptide probe as outlined below (*see* Subheading 3.2).
9. Spin down affinity matrix at $250 \times g_{max}$ for 2 min, and remove supernatant (*see* **Note 12**).
10. Wash the pellet (= immunoprecipitate; IP) four times with 1 mL of Triton buffer, and one time with 1 mL of TBS. Mix cautiously by inverting the tube.
11. Resuspend IP in 200 μL of TBS and transfer IP quantitatively to translucent 0.2 mL thin-wall PCR microtube (*see* **Note 10** and **13**).
12. Spin down IP at $250 \times g_{max}$ for 2 min; remove supernatant; and store IP on ice (*see* **Note 14**).

3.2 Preparation of Photoreactive Peptide and Cross-linking Premix

Use opaque 0.5 mL microtubes and perform all procedures at 25 °C unless otherwise specified. Mix reagents carefully by pipetting up and down several times. Quantities indicated apply to one cross-linking reaction. We usually prepare photo-reactive probes (*see* **Note 15**) and premixes for $n+1$ reactions.

1. Pipette 1 μL of 50 mM biotinylated peptide YATAVK-Btn (*see* **Note 16**) on the bottom of opaque 0.5 mL microtube.
2. Add 1 μL of 100 mM ANB-NOS (*see* **Notes 17** and **18**), mix carefully, close tube and incubate for at least 1 h. Then, instantly use photoreactive peptide probe for photo-cross-linking (*see* **Note 19**).
3. In the meantime, pour 8 % SDS-polyacrylamide gel.
4. Prepare cross-linking premix: Mix cautiously: 16 μL of bidist. water; 5 μL of 0.66 M Tris-HCl (pH 7.5), 50 mM $MgCl_2$; 4 μL of 1 % (v/v) Triton X-100; 3 μL of 3 mg/mL HA-peptide (*see* **Notes 19–22**).
5. Assemble photolysis setup (Fig. 3).

3.3 Photo-cross-linking Reaction

Perform all procedures at 25 °C unless otherwise specified. Mix reagents cautiously by pipetting up and down several times (*see* **Note 22**).

1. Use a Hamilton syringe to fully remove TBS from the IP.
2. Suspend IP in 28 μL of cross-linking premix (*see* **Note 22**), and equilibrate for 5 min at 25 °C.
3. Add 2 μL of photoreactive peptide and mix (*see* **Notes 22** and **23**).
4. Do not close lid and place tubes in a microtube rack.

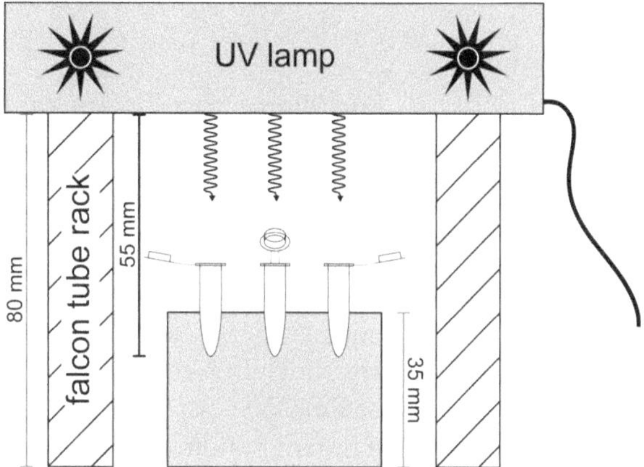

Fig. 3 Schematic representation of the experimental setup for photolysis reactions. Place UV lamp above the reaction vessel to avoid filtering of the UV light. We position the UV lamp 55 mm from the bottom of the reaction tube using Falcon tube racks

5. Immediately expose to UV light for 10 min (*see* **Note 24**). Distance between UV source and bottom of the microtubes should be approximately 55 mm. The setup we are using is outlined in Fig. 3.
6. Stop reaction by adding 5 μL of 5-times SDS-sample buffer and incubate at room temperature (RT) for 10 min (*see* **Note 25**). Briefly spin tubes at $250 \times g_{max}$, and immediately process samples further.

3.4 Analysis of Cross-linking Products

Perform all procedures at 25 °C unless otherwise specified. Incubate Nitrocellulose (NC) membranes with moderate shaking.

1. Use a Hamilton syringe to load samples on 8 % SDS-PA gel.
2. Run SDS-PAGE, and subsequently transfer resolved proteins onto NC membrane (*see* **Note 26**).
3. Incubate NC membrane in blocking solution for 1 h.
4. Replace blocking solution, add peroxidase-coupled goat anti-biotin antibody (1:5,000 dilution in TBS/T containing 0.5 % (w/v) BSA) onto the membrane, and incubate overnight at 4 °C (*see* **Note 27**).
5. Wash 6-times with 20 mL TBS/T, 30 min each time (*see* **Note 28**).
6. Visualize protein-antibody complexes by enhanced chemiluminescence using the SuperSignal West Pico Chemiluminescent Substrate according to manufacturer's instructions (*see* **Note 29**).

4 Notes

1. For photo-cross-linking experiments, we use tagged versions of Pmts. In baker's yeast, C-terminal fusions with HA-, FLAG- or GFP-epitopes do not affect PMTs mannosyltransferase activity ([7]; M. L. and S. S., unpublished data). For immunoprecipitation, we favor three to six copies of the HA epitope. Genomic integrations of $Pmt1p^{HA}$ and $Pmt2p^{HA}$ are available in our laboratory (M. L. and S. S., unpublished data). Alternatively *pmt*Δ deletion strains, expressing episomal versions of $Pmt1p^{HA}$ or $Pmt2p^{HA}$ can be used [7, 12]. Yeast strains are grown under standard conditions.
2. We learned that it is best to add protease inhibitors fresh each time. Thaw protease inhibitor stock solutions on ice. PMSF precipitates in the cold, thus warm solution prior to use. Upon thawing, mix stocks vigorously.
3. After testing a variety of different protease inhibitor combinations, we give preference to the following combination:

   PMSF, Benzamidine, TLCK and TPCK: add one-hundredth of sample volume of each stock. Antipain, leupeptin and pepstatin: add one-thousandth of sample volume of each stock. Final concentrations are 1 mM PMSF; 1 mM benzamidine; 0.25 mM TLCK; 0.15 mM TPCK; 33 μM antipain; 2 μM leupeptin; 1.5 μM pepstatin A.
4. Monitor cell lysis under the microscope. Usually the procedure results in lysis of ~80 % of the cells. If necessary, add more vortexing/cooling cycles. Alternatively, you can break cells using a homogenizer, e.g., RiboLyser (Thermo Hybaid, Ulm, Germany).
5. Use a Bunsen burner to heat a 25G1 needle. When red-hot, puncture lid of the tube at least two times to depressurize. Work speedily, since needle cools down quickly. Do not apply pressure. If necessary, heat needle again. Then, puncture bottom of the tube. Stick tube into a fresh 1.5 mL microtube (lid open). This assembly fits into most microfuge types. Open lid should face the center of the microfuge. After centrifugation, glass beads should be almost dry.
6. You may want to use a microtube pestle (e.g., ReadyPrep Mini Grinder; Bio-Rad, Munich, Germany) to suspend the sticky membrane pellet.
7. Protein concentration of suspended membranes is around 3 mg/mL.
8. To determine protein concentrations, we use the *DC* Protein Assay (Bio-Rad) that can tolerate up to 1 % (v/v) Triton X-100.

9. We use Vortex-Genie® 2 mixer, equipped with a microtube holder. But, any other vortex will do the job.
10. For pipetting the affinity matrix, we use clipped tips (~1 mm off the top).
11. Protein concentrations of Triton extracts range between 2 and 2.5 mg/mL. Process further the amount of Triton extract containing 700 μg of total protein.
12. We recommend keeping the supernatant at -20 °C in order to control efficiency of IP.
13. If you want to process more than five reactions at a time, we recommend you to use 96 well PCR plates (Bio-Rad).
14. IP of Pmt1p^{HA} under the outlined conditions results in precipitation of Pmt1p^{HA}/Pmt2p complexes that are enzymatically active [7]. You can monitor enzymatic activity using an in vitro assay as described [12, 14].
15. The photoreactive probe we developed is based on the peptide YATAVK-Btn that can be conjugated via its primary amino group with heterobifunctional cross-linkers that contain an amine-reactive and a photoactivatable group. YATAVK-Btn serves as in vitro mannosyl acceptor substrate of Pmt1p/Pmt2p and is *O*-mannosylated to a similar extent as the known acceptor peptide AcNH-YATAV-CONH$_2$ [12, 15]. If you want to develop other peptide-based photoaffinity probes, make sure that they serve as in vitro mannosyl acceptor substrate.
16. Dissolve 2 mg of YATAVK-Btn in 45 μL of the dry water-miscible organic solvent DMSO at room-temperature (RT). Store at -20 °C. Immediately before use, bring to RT and mix vigorously.
17. The NHS-ester and photo-activatable cross-linker ANB-NOS has a spacer arm length of 7.7 Å. To establish the photoprobe, we tested various heterobifunctional cross-linkers with spacer arm length varying up to 19.9 Å.
18. ANB-NOS is moisture-sensitive. Prepare immediately before use; do not use stock solutions. Equilibrate vial at RT before opening. Dissolve 1 mg of ANB-NOS in 33 μL of DMSO at RT. Discard unused material.
19. To quench non-reacted cross-linkers we perform cross-linking reactions in the presence of at least 36-fold molar excess of amine-containing buffer; final concentration: 110 mM Tris-HCl (pH 7.5). Alternatively, non-reacted cross-linkers can be removed by gel filtration or dialysis. We obtained the same results using these alternative methods. Thus for ease of handling, we prefer the quenching approach.
20. Immediately before use, thaw HA-peptide at RT and mix vigorously. HA-peptide (final concentration: 0.3 mg/mL) is

added to release Pmt1p^{HA} form the affinity matrix. It is not necessary to remove agarose beads from the reaction mixture.

21. Control peptides such as AcNH-YATAV-COOH (Eurogentec, Cologne, Germany), may be added as required (final concentrations between 1.7 and 7.5 mM). Dissolve 2 mg of peptide in 71 μL of DMSO. Store at -20 °C.
22. Avoid formation of suds caused by Triton X-100.
23. Upon addition of the photoprobe, the cross-linking mix turns cloudy. During photolysis, the reaction mixture becomes clear and takes on a yellowish color.
24. For photolysis we are using a low wattage handheld UV lamp (6 W; wavelength 366 nm). Other UV lamps such as Stratalinker 2400 or mercury vapor lamps may be used. Further detailed information concerning UV lamps is provided on ANB-NOS product information sheet. Especially helpful are the technical tips provided on the Thermo Scientific web page: Pierce Protein research product; Tech Tip #11: Light sources and conditions for photoactivation of aryl azide cross-linking reagents.
25. Pmt proteins tend to aggregate when exposed to temperatures above 60 °C.
26. For SDS-PAGE and Western blotting we follow standard protocols.
27. Overnight incubation highly enhances sensitivity of the detection of cross-linking products.
28. We found that we have to wash blots excessively to reduce unspecific background staining.
29. To monitor Pmt1p^{HA}, you can strip blots and subsequently decorated with mouse anti-HA antibody (Covance; 1:10,000 dilution in TBS/T containing 0.5 % (w/v) BSA). Incubate for 1 h at RT. Detection is accomplished with peroxidase-conjugated rabbit anti-mouse antibody (Sigma-Aldrich; 1:5,000 dilution in TBS/T containing 0.5 % (w/v) BSA).

Acknowledgments

We thank V. Hofmann for excellent technical advice. We thank A. Schott and M. Buettner for critical reading of the manuscript. This work was partially supported by the Deutsche Forschungs Gemeinschaft (SFB638). S. Strahl is a member of CellNetworks-Cluster of Excellence (EXC81).

References

1. Strahl-Bolsinger S, Immervoll T, Deutzmann R, Tanner W (1993) PMT1, the gene for a key enzyme of protein O-glycosylation in Saccharomyces cerevisiae. Proc Natl Acad Sci USA 90:8164–8168
2. Gentzsch M, Tanner W (1996) The PMT gene family: protein O-glycosylation in Saccharomyces cerevisiae is vital. EMBO J 15:5752–5759
3. Willer T, Valero MC, Tanner W, Cruces J, Strahl S (2003) O-mannosyl glycans: from yeast to novel associations with human disease. Curr Opin Struct Biol 13:621–630
4. Lommel M, Strahl S (2009) Protein O-mannosylation: conserved from bacteria to humans. Glycobiology 19:816–828
5. Girrbach V, Zeller T, Priesmeier M, Strahl-Bolsinger S (2000) Structure-function analysis of the dolichyl phosphate-mannose: protein O-mannosyltransferase ScPmt1p. J Biol Chem 275:19288–19296
6. Gentzsch M, Immervoll T, Tanner W (1995) Protein O-glycosylation in Saccharomyces cerevisiae: the protein O-mannosyltransferases Pmt1p and Pmt2p function as heterodimer. FEBS Lett 377:128–130
7. Girrbach V, Strahl S (2003) Members of the evolutionarily conserved PMT family of protein O-mannosyltransferases form distinct protein complexes among themselves. J Biol Chem 278:12554–12562
8. Gentzsch M, Tanner W (1997) Protein-O-glycosylation in yeast: protein-specific mannosyltransferases. Glycobiology 7:481–486
9. Hutzler J, Schmid M, Bernard T, Henrissat B, Strahl S (2007) Membrane association is a determinant for substrate recognition by PMT4 protein O-mannosyltransferases. Proc Natl Acad Sci USA 104:7827–7832
10. Liu J, Mushegian A (2003) Three monophyletic superfamilies account for the majority of the known glycosyltransferases. Protein Sci 12:1418–1431
11. Strahl-Bolsinger S, Scheinost A (1999) Transmembrane topology of pmt1p, a member of an evolutionarily conserved family of protein O-mannosyltransferases. J Biol Chem 274:9068–9075
12. Lommel M, Schott A, Jank T, Hofmann V, Strahl S (2011) A conserved acidic motif is crucial for enzymatic activity of protein O-mannosyltransferases. J Biol Chem 286: 39768–39775
13. Laemmli UK (1970) Cleavage of structural proteins during the assembly of the head of bacteriophage T4. Nature 227:680–685
14. Strahl-Bolsinger S, Tanner W (1991) Protein O-glycosylation in Saccharomyces cerevisiae. Purification and characterization of the dolichyl-phosphate-D-mannose-protein O-D-mannosyltransferase. Eur J Biochem 196: 185–190
15. Weston A, Nassau PM, Henly C, Marriott MS (1993) Protein O-mannosylation in Candida albicans. Determination of the amino acid sequences of peptide acceptors for protein O-mannosyl-transferase. Eur J Biochem 215: 845–849

Chapter 10

Enzymatic Analysis of the Protein *O*-Glycosyltransferase, Rumi, Acting Toward Epidermal Growth Factor-Like (EGF) Repeats

Hideyuki Takeuchi and Robert S. Haltiwanger

Abstract

Epidermal growth factor-like (EGF) repeats are found in numerous extracellular or transmembrane proteins including Notch. EGF repeats containing the appropriate consensus sequences can be modified with two unusual types of glycans: *O*-fucosylation and *O*-glucosylation. We have identified the glycosyltransferases that catalyze the addition of the first sugar to these consensus sites: protein *O*-fucosyltransferase 1 (Pofut1) and protein *O*-glucosyltransferase (Rumi/Poglut1). Recently, we have demonstrated that Rumi/Poglut1 shows protein *O*-xylosyltransferase activity as well. Here, we describe how we characterize the enzymatic activity of these enzymes, including preparation of the acceptor substrates, using bacterially expressed EGF repeats.

Key words Epidermal growth factor-like (EGF) repeats, Protein *O*-glycosyltransferase, *O*-Glucose, *O*-Xylose, Protein folding

1 Introduction

1.1 O-Linked Glycosylation of EGF Repeats

EGF repeats are small protein motifs defined by the presence of six-conserved cysteine residues spaced appropriately to allow formation of three specific disulfide bonds (Cys1-Cys3, Cys2-Cys4, Cys5-Cys6) [1]. Although there are potentially 76 different disulfide isomers for a single EGF repeat with 0-3 disulfide bonds, the correct disulfide-bonding pattern results in a distinct three-dimensional structure. During protein biosynthesis, formation of proper disulfide bonds occurs in the endoplasmic reticulum. Once EGF repeats are folded properly, they can be modified with two unusual types of glycans: *O*-fucosylation and *O*-glucosylation, which are catalyzed by specific glycosyltransferases. Protein *O*-fucosyltransferase 1 (Pofut1) transfers *O*-linked fucose from GDP-fucose [2] and protein *O*-glucosyltransferase (Rumi/Poglut1) transfers *O*-linked glucose from UDP-glucose [3]. Interestingly, both of these

Inka Brockhausen (ed.), *Glycosyltransferases: Methods and Protocols*, Methods in Molecular Biology, vol. 1022, DOI 10.1007/978-1-62703-465-4_10, © Springer Science+Business Media New York 2013

enzymes only modify properly folded EGF repeats containing the appropriate consensus sequence [4, 5]. The consensus sequences for addition of *O*-glucose (C^1-x-*S*-x-P/A-C^2) or *O*-fucose (C^2-x-x-x-x-*S*/*T*-C^3) are found in numerous cell surface and secreted proteins [6, 7], but are evolutionarily well conserved in the extracellular domain of the Notch family of receptors [8]. Indeed, many of the EGF repeats in the Notch extracellular domain are modified with these glycans [9–12]. Genetic and biochemical studies have shown that these glycans are essential for Notch function [13]. Furthermore, the early embryonic lethal phenotypes caused by elimination of *Rumi* in mice have suggested that other target(s) of Rumi, in addition to Notch, are important for embryonic development [14].

1.2 Rumi Functions as Both a Protein O-Glucosyltransferase and a Protein O-Xylosyltransferase

We originally showed that purified Rumi catalyzes the transfer of glucose from UDP-glucose to bacterially expressed (i.e., non-glycosylated) EGF repeats containing an *O*-glucose consensus sequence [3]. Activity was dependent on the concentrations of Rumi protein, donor substrate, UDP-glucose, and acceptor substrates. Product analyses showed that Rumi added a single *O*-linked glucose to the EGF repeats. Much to our surprise, we have recently discovered a very unique feature of Rumi: it is a dual specificity glycosyltransferase that can utilize either UDP-xylose or UDP-glucose as donor substrate [15]. Interestingly, the amino acid sequence surrounding the modification site influences which donor substrate (UDP-glucose or UDP-xylose) is utilized. Furthermore, we identified *O*-xylose glycans on mouse Notch2 expressed in mammalian cells, demonstrating that *O*-xylose is transferred to EGF repeats under physiological conditions [15]. These results suggest that Rumi also functions as a protein *O*-xylosyltransferase and that Notch proteins can be modified with *O*-xylose. Here we describe in detail the methods for preparing EGF repeats in bacteria for use as acceptor substrates in these assays, as well as how to characterize the products of the glycosyltransferase reactions.

2 Materials

2.1 Preparation of EGF Repeats as Acceptor Substrate

1. pET-20b expression vector encoding desired EGF repeats (the example shown here is EGF1 from human factor VII, amino acids 45–87) with C-terminal His_6-tag [2].
2. BL21 (DE3) *E. coli* cells (Invitrogen, Life Technologies, Carlsbad, CA, USA).
3. Isopropyl-β-D-thiogalactopyranoside (IPTG, Roche, Indianapolis, IN, USA).
4. 100 mg/mL ampicillin (Roche) in water.
5. 200 mM phenylmethanesulfonyl fluoride (PMSF, Sigma-Aldrich, St. Louis, MO, USA) in ethanol.

6. Ni-NTA agarose beads (Qiagen, Valencia, CA, USA).
7. Imidazole (Sigma-Aldrich, St. Louis, MO, USA).
8. Reverse phase HPLC system (1200 Series, Agilent Technologies, Santa Clara, CA, USA): Manual Injector with 2-ml Sample Loop (#G1328B), Vacuum Degasser (#G1322A), Quaternary Pump (#G1311A), Multiple Wavelength Detector (#G1365D), Agilent ChemStation for instrument control, data acquisition, and data evaluation.
9. PROTEIN&PEPTIDE C18 column (10×250 mm, VYDAC, Hesperia, CA, USA).
10. Water (HPLC grade, Pharmco-Aaper, Brookfield, CT, USA).
11. Acetonitrile (HPLC grade, Thermo Fisher Scientific, Waltham, MA, USA).
12. Trifluoroacetic acid (TFA, Thermo Fisher Scientific, Waltham, MA, USA) (*see* **Note 1**).
13. Dithiothreitol (DTT, Sigma-Aldrich).
14. Glutathione reduced form (GSH, Sigma-Aldrich).
15. Glutathione oxidized form (GSSG, Sigma-Aldrich).
16. Sephadex G-25 (GE Healthcare, Piscataway, NJ, USA).
17. Blue dextran (Sigma-Aldrich).
18. Cobalt chloride (Sigma-Aldrich).

2.2 Enzyme Assay for Protein O-Glycosyltransferase Activity of Rumi

1. Recombinant Rumi protein [15].
2. 10× Reaction buffer (500 mM HEPES pH 6.8, 100 mM $MnCl_2$) (*see* **Note 2**).
3. Purified recombinant EGF repeat expressed in *E. coli*.
4. UDP-[6-^{3}H]glucose (60 Ci/mmol, American Radiolabeled Chemicals, St. Louis, MO, USA).
5. UDP-glucose (Sigma-Aldrich).
6. UDP-[^{14}C(U)]xylose (200~250 mCi/mmol, American Radiolabeled Chemicals).
7. UDP-xylose (Complex Carbohydrate Research Center at the University of Georgia).
8. 10 % Nonidet P-40 (Fluka, St. Louis, MO, USA).
9. 100 mM EDTA (Sigma-Aldrich) pH 8.0.
10. SampliQ C18 cartridge (100 mg, Agilent Technologies).
11. Vacuum manifold (VISIPREP, SUPELCO, Sigma-Aldrich) (*see* **Note 3**).
12. Scintiverse (Thermo Fisher Scientific).
13. Beckman LS 6500 Scintillation Counter (Beckman Coulter, Brea, CA, USA).

2.3 Product Analysis of the Enzyme Assay

1. Agilent nanoflow LC system with HPLC-CHIP interface coupled to Agilent model 6340 3D-Ion Trap mass spectrometer (Agilent Technologies).
2. Acetonitrile (Mass spectrometry grade, Fisher).
3. Formic acid (Sigma-Aldrich).
4. Spin-X centrifuge tube filters (0.22 μm, Coaster, Waltham, MA, USA).

3 Methods

3.1 Preparation of EGF Repeats as Acceptor Substrate

To produce EGF repeats, *E. coli* are transformed with a pET vector encoding the EGF repeat of choice. After protein expression of EGF repeats in *E. coli* is induced by IPTG, the soluble fraction of *E. coli* contains a mixture of properly folded and mis-folded EGF repeats, as some EGF repeats fold better in *E. coli* than others. Folding isomers of EGF repeats can be separated by reverse phase HPLC. We take advantage of the fact that properly folded, but not mis-folded, EGF repeats can be modified with *O*-glucose or *O*-xylose by Rumi, to determine which peak contains the properly folded EGF repeats. We also describe the method to denature and refold EGF repeats, which is required for EGF repeats that do not fold well on their own. A typical yield of a properly folded, single EGF repeat is approximately 1 mg from 1 L *E. coli* culture with this protocol. The scale of the experiment should be determined depending on the required amount of EGF repeats. Here we describe preparation of EGF repeats from 2 L culture of *E. coli* as an example.

1. BL21 (DE3) *E. coli* transformed with pET vector encoding desired EGF repeats is cultured in approximately 60 mL of LB medium containing 100 μg/mL of ampicillin at 37 °C overnight.
2. The 60-mL culture from above is added to 2 L of LB medium containing 100 μg/mL of ampicillin and culture at 37 °C. Once OD (600 nm) reaches around 0.5 ~ 0.6, IPTG (final concentration 0.4 mM) is added to the culture. Culture at 20 °C overnight (*see* **Note 4**).
3. The culture medium is collected and centrifuged at ~1,500 × *g* at 4 °C for 20 min. Discard supernatant.
4. The cell pellet is resuspended in 40 mL of 50 mM Tris–HCl pH 8.0 containing 1 mM PMSF and sonicated using a probe sonicator ten times for 10 s on ice, and centrifuged at ~15,000 × *g* at 4 °C for 1 h.
5. The soluble fraction is filtered and applied to 0.4 mL of Ni-NTA affinity chromatography, which has been pre-equilibrated with

TBS containing 0.5 M NaCl and 10 mM imidazole, by gravity flow. The column is washed by 4 mL of TBS containing 0.5 M NaCl and 10 mM imidazole (*see* **Note 5**). The EGF repeat is eluted by TBS containing 250 mM imidazole (*see* **Note 6**).

6. Further purification of EGF repeat is performed by reverse phase HPLC. The column is eluted with a linear gradient from 10 to 90 % solvent B (0.1 % TFA in 80 % acetonitrile in water) in solvent A (0.1 % TFA in water) at a flow rate of 2 mL/min for 60 min. Eluates are monitored for absorbance at 214 nm. Individual peaks are collected, dried, and tested for ability to serve as a substrate for Rumi (*see* Subheading 3.2).
7. In case EGF repeats do not fold properly, we perform denaturing and refolding using the Ni-NTA fractions containing mis-folded EGF repeats and purify refolded EGF repeats by reverse phase HPLC as previously described with slight modification [1]. The proteins are denatured in 200 μL of 100 mM Tris–HCl pH 8.3, 10 mM EDTA, 30 mM DTT, 5 M guanidine/HCl at room temperature for 1.5 h (*see* **Note 7**).
8. The sample (200 μL) is desalted using an 8-mL Sephadex G-25 gel filtration column with gravity flow using 100 mM Tris–HCl pH 8.3 as running buffer at room temperature (*see* **Note 8**).
9. Both reduced and oxidized forms of glutathione are added to the eluate (2 mL) and the sample is incubated at room temperature for appropriate time periods (*see* **Note 9**). TFA (final concentration 2 %) is added to the sample to stop folding reaction.
10. Refolded EGF repeat is purified by reverse phase HPLC (see above for the method). Actual profiles are shown in Fig. 1.

3.2 Enzyme Assay for Protein O-Glycosyltransferase Activity of Rumi

We use both radioactive and nonradioactive assays to monitor the activity of Rumi. The radioactive assay (using UDP-[6-^{3}H]glucose or UDP-[^{14}C(U)]xylose) has high sensitivity, shorter assay time, and high-throughput, so it is appropriate for kinetic analysis [15]. The nonradioactive assay utilizes a reversed phase HPLC-based approach relying on the fact that addition of glucose or xylose to the EGF repeat causes a change in retention time [15]. The nonradioactive assay is less sensitive but allows us to monitor glycosylation status of preparative amounts of material.

3.2.1 Methods for Radioactive Assay

1. Protein *O*-glucosyltransferase assay is performed in 1× Reaction buffer containing 0.5 % Nonidet P-40, 0.1 μCi UDP-[6-^{3}H] glucose (60 Ci/mmol), 10 μM nonradioactive UDP-glucose, 1 ~ 10 μM recombinant human factor VII EGF, and 10 ng recombinant Rumi protein in a volume of 10 μL at 37 °C for 20 min. For protein *O*-xylosyltransferase assay, 10 μM of UDP-[^{14}C(U)]xylose is used (*see* **Note 10**).

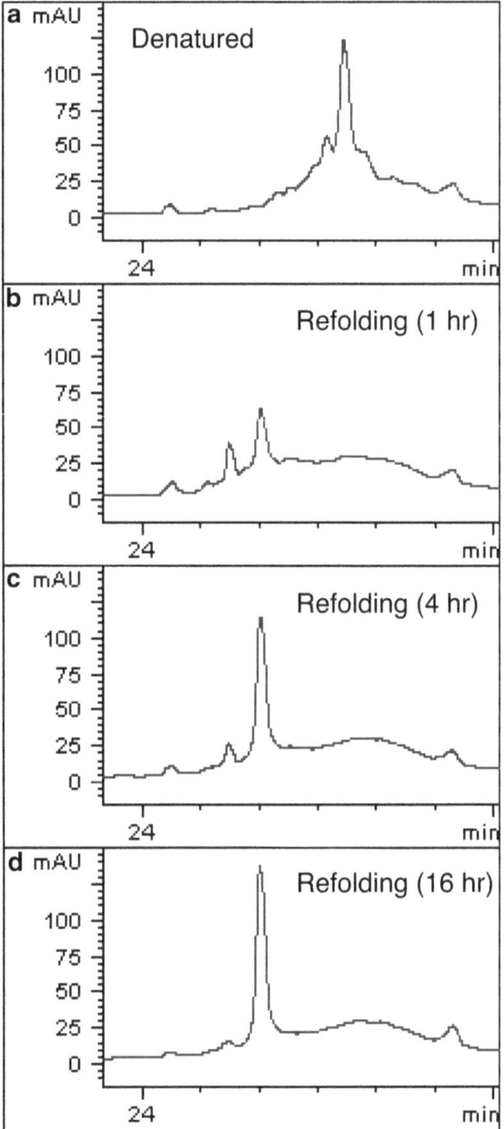

Fig. 1 Folding isomers of EGF repeats can be separated by reverse phase HPLC. Elution profiles of human factor VII EGF repeats denatured (**a**) and refolded in the presence of 0.25 mM GSSG and 0.5 mM of GSH at room temperature for 1, 4, and 16 h (**b**–**d**)

2. To stop the reaction, 900 μL of 100 mM EDTA pH 8.0 are added to the reaction mixture, which is then kept on ice until the next step.
3. The samples are applied to C18 cartridges in a vacuum manifold. Cartridges are activated with 2 mL of 100 % methanol and equilibrated with 2 mL of water. The C18 cartridges are washed with 5 mL of water and the bound samples are eluted with 1 mL of 80 % methanol.

4. The eluate (1 mL) is mixed with 4 mL of Scintiverse scintillation cocktail and vortexed, and the radioactivity is measured using a liquid scintillation counter.

3.2.2 Methods for Nonradioactive Assay

1. We typically prepare 100 μL of the reaction mixture containing 5 ~ 10 μM acceptor substrate (e.g., EGF repeats) and 100 ng ~ 1 μg recombinant glycosyltransferase (e.g., Rumi) in the presence of nonradioactive donor substrate such as UDP-glucose or UDP-xylose (final concentration of donor substrate is 200 μM), and incubate the reaction mixture at 37 °C overnight.
2. To stop the reaction, 900 μL of 100 mM EDTA pH 8.0 are added to the reaction mixture.
3. Right before HPLC analysis, the sample should be centrifuged, and then supernatant is injected into HPLC.
4. Monitor absorbance at 214 nm using a UV detector to trace elution of proteins and collect peaks as described in Subheading 3.1, **step 6**. Addition of *O*-glucose or *O*-xylose typically results in 2–3 min earlier retention times and *O*-glucose consistently causes a greater shift than *O*-xylose (Fig. 2a–c).
5. Evaporate the sample to dryness using a Speed-Vac centrifuge and store at −20 °C until use.

3.3 Product Analysis of the Enzyme Assay

The final structure of the sugars on the acceptor substrates must be characterized after the glycosyltransferase assays. For this purpose, we have developed a method which involves a series of chromatographic approaches accompanied with alkali-induced β-elimination [9]. The *O*-linked sugars were released by alkali-induced β-elimination and separated by a Superdex gel filtration chromatography. The fraction containing the sugar of interest was then analyzed using a high pH anion exchange column with pulsed amperometric detection.

Here we introduce a simple mass spectrometric method to confirm the molecular weight of the product purified by reverse phase HPLC. The resulting peptides are analyzed by MS/MS, and glycopeptides modified with *O*-glucose or *O*-xylose glycans are identified by neutral loss searches (Fig. 2d–f).

1. The samples are resuspended in 20 % solvent B (0.1 % formic acid in 95 % acetonitrile in water) in solvent A (0.1 % formic acid in water) and sonicated in a water bath for 5 min.
2. After centrifugation, the samples are filtered using Spin-X (0.22 μm) (*see* **Note 11**) and directly infused into an Agilent 6340 ion-trap mass spectrometer with a nano-HPLC CHIP-Cube interface at a rate of 18 μL/h. The MS peaks for MS/MS are chosen manually, and the data are analyzed using Agilent ChemStation data analysis software.

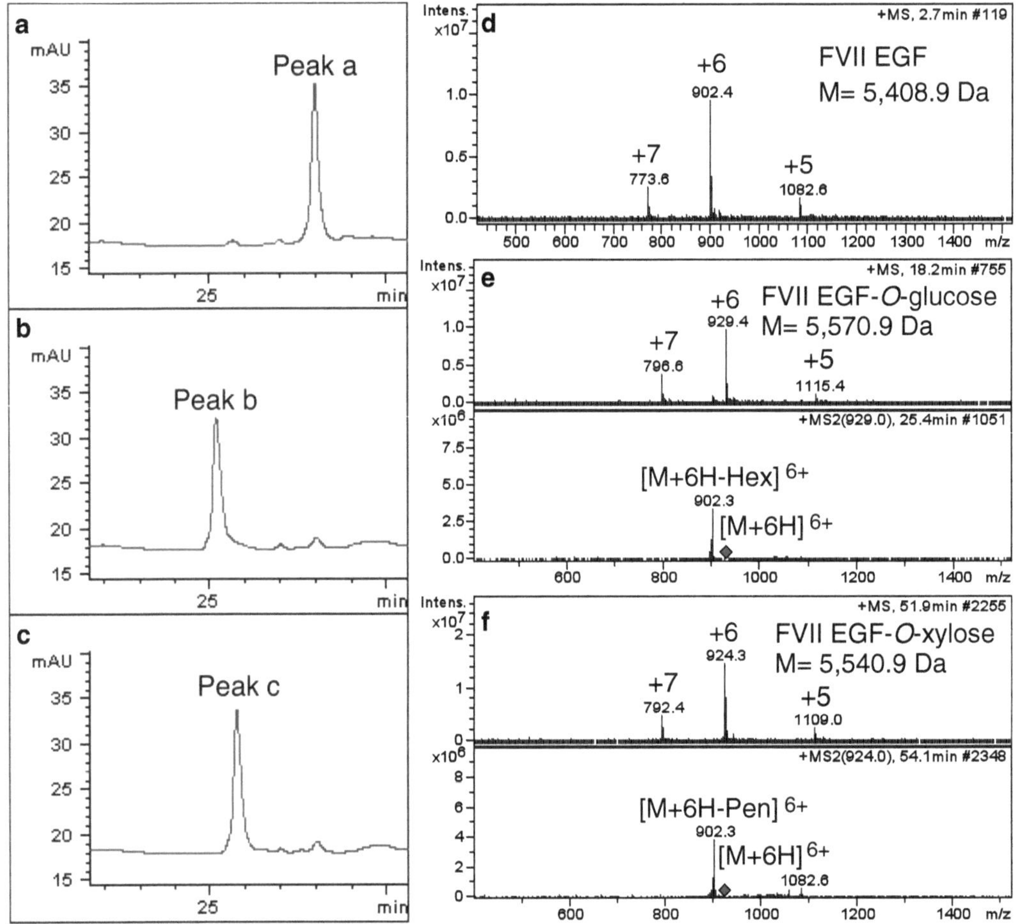

Fig. 2 Addition of glucose or xylose causes the shift of retention time of EGF repeats on reverse phase HPLC. Elution profiles of the reaction products resulting from incubation of human factor VII EGF repeat (FVII EGF) (5 μM) with Rumi in the absence of donor (**a**), or in the presence of UDP-glucose (**b**) or UDP-xylose (**c**) by reverse phase HPLC. Subsequent electrospray ionization-MS/MS analysis on peak "a" (**d**), peak "b" (**e**), and peak "c" (**f**) is shown. The theoretical average masses of FVII EGF, FVII EGF-*O*-glucose, and FVII EGF-*O*-xylose are shown. EGF repeats with different charge states were detected as indicated. MS/MS analysis indicates neutral loss of a hexose (**e**, *lower*) or a pentose (**f**, *lower*). Reproduced from [15]

4 Notes

1. We purchase 1 g of TFA in each glass vial. TFA is used as an ion pairing agent in HPLC. Since TFA is a strong acid, handling with proper caution is required.
2. To avoid precipitation of $MnCl_2$, we make 1 M HEPES pH 6.8 solution and 200 mM $MnCl_2$, and combine them at 1:1 ratio.
3. This is very useful for high throughput analysis. There are two models (12-Port or 24-Port) commercially available.

4. The concentration of IPTG should be optimized depending on the protein being expressed, although 0.4 mM has been successful for a variety of single EGF repeats.
5. We add 0.5 M NaCl and 10 mM imidazole to the sample to be applied in order to avoid nonspecific binding. The condition for washing the column should be optimized for each EGF repeat. Instructions for use of Ni-NTA agarose are available on the Qiagen website.
6. Elution of proteins should be confirmed by Western blot analysis using a specific antibody (e.g., anti-His tag antibody). The majority of the protein elutes from 0.2 to 0.8 mL.
7. The condition for denaturing proteins should be optimized. As for protein concentration, less than 1 mg/mL appears to be best.
8. We determine the volume of the fraction that contains EGF repeats but not salts by applying blue dextran and cobalt chloride to the column.
9. Conditions such as the concentrations of both forms of glutathione and incubation time should be optimized.
10. Typical conditions of the enzyme reaction are shown.
11. All samples should be filtered right before infusion. We use Spin-X (0.22 μM) for this purpose.

Acknowledgments

We would like to thank Haltiwanger lab members for helpful comments. The primary work introduced here was supported by NIH grant GM061126 (to RSH) and the research grant from Mizutani Foundation for Glycoscience (to HT).

References

1. Chang JY, Li L, Lai PH (2001) A major kinetic trap for the oxidative folding of human epidermal growth factor. J Biol Chem 276: 4845–4852
2. Wang Y, Lee GF, Kelley RF, Spellman MW (1996) Identification of a GDP-L-fucose: polypeptide fucosyltransferase and enzymatic addition of O-linked fucose to EGF domains. Glycobiology 6:837–842
3. Acar M, Jafar-Nejad H, Takeuchi H, Rajan A, Ibrani D, Rana NA, Pan H, Haltiwanger RS, Bellen HJ (2008) Rumi is a CAP10 domain glycosyltransferase that modifies Notch and is required for Notch signaling. Cell 132:247–258
4. Wang Y, Spellman MW (1998) Purification and characterization of a GDP-fucose: polypeptide fucosyltransferase from chinese hamster ovary cells. J Biol Chem 273:8112–8118
5. Shao L, Luo Y, Moloney DJ, Haltiwanger R (2002) O-Glycosylation of EGF repeats: identification and initial characterization of a UDP-glucose: protein O-glucosyltransferase. Glycobiology 12:763–770
6. Rana NA, Haltiwanger RS (2011) Fringe benefits: functional and structural impacts of O-glycosylation on the extracellular domain of Notch receptors. Curr Opin Struct Biol 21: 583–589
7. Rampal R, Luther KB, Haltiwanger RS (2007) Notch signaling in normal and disease states: possible therapies related to glycosylation. Curr Mol Med 7:427–445

8. Haines N, Irvine KD (2003) Glycosylation regulates notch signaling. Nat Rev Mol Cell Biol 4:786–797
9. Moloney DJ, Shair LH, Lu FM, Xia J, Locket R, Matta KL, Haltiwanger RS (2000) Mammalian Notch1 is modified with two unusual forms of O-linked glycosylation found on epidermal growth factor-like modules. J Biol Chem 275:9604–9611
10. Shao L, Moloney DJ, Haltiwanger RS (2003) Fringe modifies O-fucose on mouse Notch1 at epidermal growth factor-like repeats within the ligand-binding site and the Abruptex region. J Biol Chem 278:7775–7782
11. Whitworth GE, Zandberg WF, Clark T, Vocadlo DJ (2010) Mammalian Notch is modified by D-Xyl-alpha1-3-D-Xyl-alpha1-3-D-Glc-beta1-O-Ser: implementation of a method to study O-glucosylation. Glycobiology 20:287–299
12. Rana NA, Nita-Lazar NA, Takeuchi H, Kakuda S, Luther KB, Haltiwanger RS (2011) O-glucose trisaccharide is present at high but variable stoichiometry at multiple sites on mouse notch1. J Biol Chem 286: 31623–31637
13. Takeuchi H, Haltiwanger RS (2010) Role of glycosylation of Notch in development. Semin Cell Dev Biol 21:638–645
14. Fernandez-Valdivia RC, Takeuchi H, Samarghandi A, Lopez M, Leonardi J, Haltiwanger RS, Jafar-Nejad H (2011) Regulation of mammalian Notch signaling and embryonic development by the protein O-glucosyltransferase Rumi. Development 138:1925–1934
15. Takeuchi H, Fernandez-Valdivia RC, Caswell DS, Nita-Lazar A, Rana NA, Garner TP, Weldeghiorghis TK, Macnaughtan MA, Jafar-Nejad H, Haltiwanger RS (2011) Rumi functions as both a protein O-glucosyltransferase and a protein O-xylosyltransferase. Proc Natl Acad Sci USA 108:16600–16605

Chapter 11

Enzymatic Characterization of Recombinant Enzymes of *O*-GlcNAc Cycling

Eun Ju Kim and John A. Hanover

Abstract

The dynamic addition of *O*-GlcNAc to target proteins is now recognized as a major signaling paradigm impacting phosphorylation, protein turnover, gene expression, and other posttranslational modifications influencing epigenetics. Here we describe the production of and methods for assay of the recombinant enzymes of *O*-GlcNAc cycling: *O*-linked GlcNAc Transferase (OGT) and *O*-GlcNAcase (OGA).

Key words Recombinant, Fluorogenic substrates, Immunoblots, Bioorthogonal chemistry, *O*-GlcNAc

1 Introduction

The *O*-GlcNAc modification is an abundant and highly dynamic nucleocytoplasmic posttranslational modification of protein Ser and Thr residues [1, 2]. The enzymes of *O*-GlcNAc cycling play critical roles in development, signaling, gene expression, and are emerging as important players in epigenetic regulation [1, 3, 4]. The ability to produce recombinant forms of *O*-GlcNAc transferase (OGT) and *O*-GlcNAcase (OGA) has facilitated both highly informative structural studies and led to inhibitor and small molecule inhibitors of these key enzymes.

Human *O*-GlcNAc transferase was first produced in recombinant form to demonstrate the identity of this transferase as the enzyme catalyzing *O*-GlcNAc transfer [5]. Subsequent work led to identification of the 3-12 tetratricopeptide repeats (TPR) as important determinants of target specificity [6, 7]. The C-terminal domain contains a glycosyltransferase domain belonging to the GT41 family in the CAZY database and uses UDP-GlcNAc as a glycosyl donor [8]. Recombinant expression allowed us to determine the structure of the human OGT TPR domain [9]. Subsequent recombinant production of a 4 TPR version of OGT allowed crystallization of

Inka Brockhausen (ed.), *Glycosyltransferases: Methods and Protocols*, Methods in Molecular Biology, vol. 1022, DOI 10.1007/978-1-62703-465-4_11, © Springer Science+Business Media New York 2013

the human OGT catalytic domain [10]. These studies have revealed that OGT exhibits enzymatic features consistent with an ordered bi–bi reaction mechanism involving initial tight binding of sugar nucleotide followed by target peptide binding. The reaction mechanism also involves a proposed conformational entrapment by the TPR domain of the peptide linked to catalysis [10]. The strategy for the detection of OGT activity described here takes advantage of several detection strategies including radiochemical and bioorthogonal chemical approaches.

The human *O*-GlcNAcase was originally identified as hexosaminidase C and is a CAZY GH84 family member with a TIM barrel structure similar to the CAZY GH20 members of hexosaminidase A and B. When expressed in *E. coli*, the human *O*-GlcNAcase shows little activity against either GalNAc or capping GlcNAc residues and exhibits a pH optimum near pH 7. It exhibits rather high specificity for *O*-GlcNAc residues and has been shown to accommodate extension of the *N*-acetyl to longer acyl groups including *N*-pentanoyl. Such substrate flexibility is not exhibited by the CAZY 20 family members of hexosaminidases A and B. There is some evidence that the protein sequence surrounding the *O*-GlcNAc modification is also an important determinant of *O*-GlcNAcase specificity [11]. The methods we describe here take advantage of a highly sensitive fluorogenic substrate [12] which can be tailored to measuring enzyme activity in rather crude extracts.

2 Materials

Prepare all solutions using ultrapure water and analytical grade reagents unless indicated otherwise. Structures of some compounds are presented in Fig. 1.

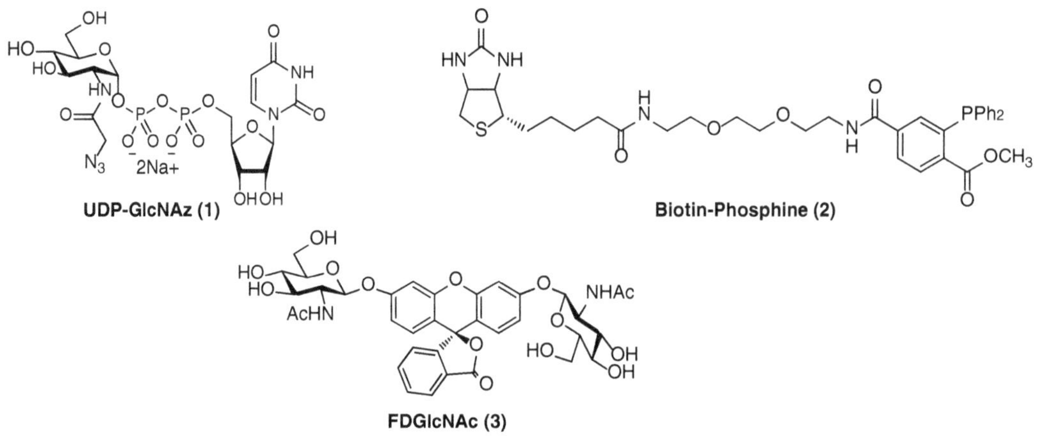

UDP-GlcNAz (1)

Biotin-Phosphine (2)

FDGlcNAc (3)

Fig. 1 Chemical structures of UDP-GlcNAz, FDGlcNAc, and Biotin-Phosphine

2.1 Components of OGT Expression in E. coli

1. Cloned plasmids of human ncOGT and mOGT inserting the coding regions in pET43.1 Ek/LIC expression vector (Novagen, San Diego, CA, USA) (*see* **Note 1**).
2. BL21(DE3) chemically competent *E. coli* (Invitrogen, Carlsbad, CA, USA).
3. Luria–Bertani (LB) broth.
4. Ampicillin solution (100 mg/mL): Weigh 0.5 g Ampicillin sodium salt (Sigma-Aldrich, St. Louis, MO, USA) and prepared in 5 mL volume by adding water to a total volume of 5 mL. Mix and filter-sterilize with a 0.22 μm filter (EMD Millipore, Billerica, MA, USA). Aliquot in 1 mL volume and store at −20 °C.
5. LB Agar plates with 50 μg/mL Ampicillin.
6. Incubator for a microbiological culture.
7. OGT lysis buffer (*see* **Note 2**): 20 mM Tris–HCl, pH 7.5, 2 mM EDTA, 1 mg/mL of lysozyme (Sigma), complete mini EDTA-free protease inhibitor cocktail (Roche Applied Science, Indianapolis, IN, USA). Prepare OGT lysis buffer right before use and store at 4 °C.
8. Lysozyme solution (100 mg/mL): Weigh 0.5 g lysozyme (Sigma) and prepare 5 mL solution by adding water to a total volume of 5 mL. Mix and filter-sterilize with a sterilized 0.22 μm filter (EMD Millipore). Aliquot in 1 mL volume and store at −20 °C.
9. Triton X-100™ (Sigma) 10 % solution: Add 1 mL of Triton X-100 to 9 mL of distilled water.
10. A centrifuge with temperature control: Sorvall Stratos Centrifuge (Thermo Scientific, Rockford, IL, USA).
11. A microcentrifuge with temperature control: Eppendorf® Refrigerated Microcentrifuge (Eppendorf, Hauppauge, NY, USA).

2.2 Components of OGT Purification

1. S-Protein Agarose (Novagen, San Diego, CA, USA).
2. Dithiothreitol (DTT) solution (1 M): Dissolve 0.787 g DTT (Sigma) in distilled water to give a total volume of 5 mL. Mix and filter-sterilize with a 0.22 μm filter device (EMD Millipore). Aliquot in 1 mL volume and store at −30 °C.
3. OGT Assay buffer (*see* **Note 3**): 50 mM Tris–HCl, pH 7.5, 1 mM DTT, 12.5 mM $MgCl_2$.
4. An orbital shaker.
5. A microcentrifuge with temperature control: Eppendorf® Refrigerated Microcentrifuge.

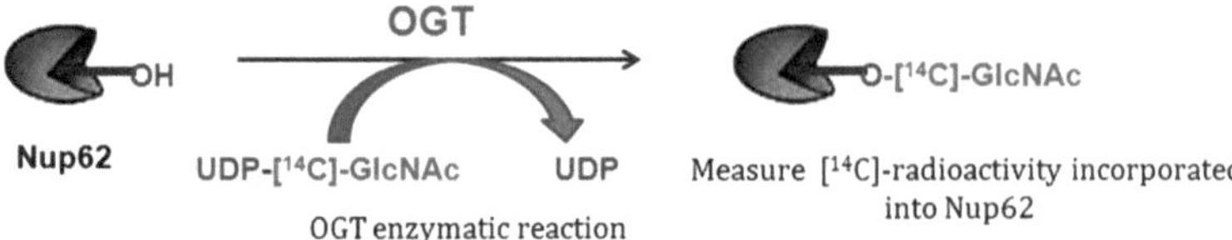

Fig. 2 Radiometric OGT activity assay. This method uses a radiolabeled sugar donor substrate and the activity is measured by quantitating radiolabeled GlcNAc incorporation into a protein such as Nup62

2.3 Components of OGT Enzymatic Activity Assay

2.3.1 Components of Radiometric OGT Activity Assay (Fig. 2)

1. Uridine diphosphate N-acetyl [1-^{14}C]D-glucosamine (UDP-[^{14}C]-GlcNAc): UDP-[^{14}C]-GlcNAc, 0.1 mCi/mL, 40–60 mCi/mmol (American Radiolabeled Chemicals, St. Louis, MO, USA).
2. Recombinant, purified Nup62 (Bioclone, Inc., San Diego, CA, USA): Concentration of Nup62 is 1 μg/μL in 20 mM sodium phosphate, pH 7.5, 0.5 M NaCl, 1 M imidazole. Recombinant and purified Nup62 can also be prepared as described previously [13].
3. Incubator for microbiological culture.
4. SDS-PAGE sample buffer: 4× NuPAGE® LDS sample buffer (Invitrogen).
5. SDS-PAGE gels: 10 % or 4–12 % NuPAGE® Bis-Tris gels (Invitrogen).
6. SDS running buffer: 20× NuPage® MOPS SDS Running Buffer (Invitrogen). Dilute it to 1× SDS Running Buffer with distilled water for the gel electrophoresis.
7. A low speed orbital shaker.
8. Simply Blue Safestain (Invitrogen).
9. En3Hance (Perkin Elmer, Wellesley, MA, USA).
10. PEG solution: 10 % PEG solution (MW 8000).
11. OGT assay buffer: 50 mM Tris–HCl, pH 7.5, 1 mM DTT, 12.5 mM $MgCl_2$.
12. A Gel Dryer equipped with a vacuum pump (Bio-Rad, Hercules, CA, USA).
13. A phosphor screen: A BAS-IIIs imaging plate (Fuji Film Co., Tokyo, Japan).
14. A phosphorimager: BAS-1500 phosphor imager (Fuji Film Co.).
15. Albumin (bovine serum) [methyl-^{14}C] methylated: [methyl-^{14}C]-BSA, 0.01 mCi/mL, 3–30 μCi/mg (American Radiolabeled Chemicals).

2.3.2 Components of Chemoenzymatic OGT Activity Assay (Fig. 3)

1. Uridine 5′-diphospho-2-azidoacetamido-2-deoxy-α-D-glucopyranose disodium salt (UDP-GlcNAz **1**, 2 mM in distilled water): UDP-GlcNAz is synthesized as described previously [14]. In order to prepare 2 mM UDP-GlcNAz, first, prepare

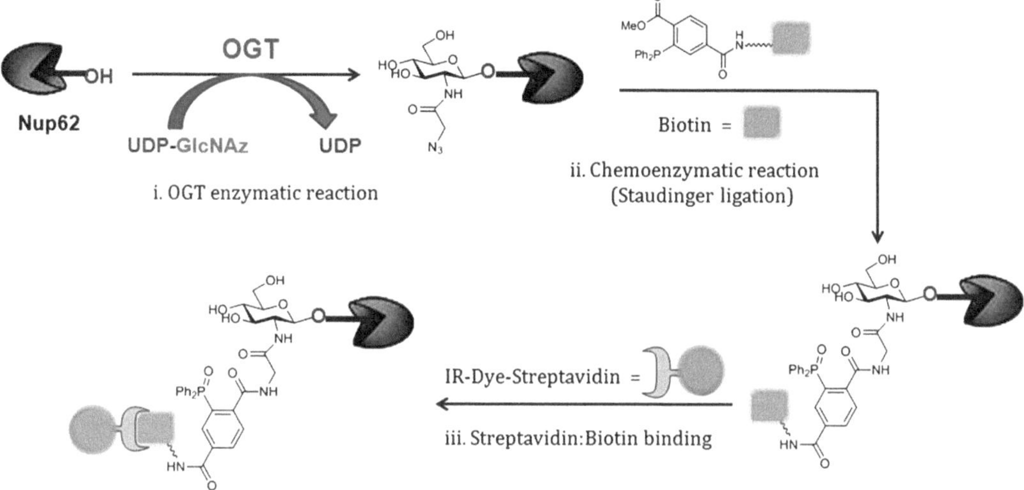

Fig. 3 Chemoenzymatic OGT activity assay. This method involves azide labeling using UDP-GlcNAz, followed by chemoenzymatic reaction either Staudinger ligation with a phosphine reagent as shown in this section, or Click reaction with a terminal alkyne reagent and Cu^{+1} ion (or strained cycloalkyne without Copper ion catalyst; not provided here)

10 mM of UDP-GlcNAz by dissolving 6.92 mg in distilled water to give a final volume of 1 mL. Transfer 100 μL of 10 mM UDP-GlcNAz solution into a new vial and add 400 μL of distilled water to make a final concentration of 2 mM solution. Store both 10 and 2 mM solutions of UDP-GlcNAz at −30 °C.

2. Recombinant, purified Nup62 (Bioclone, Inc.): Concentration of Nup62 is 1 μg/μL in 20 mM sodium phosphate, pH 7.5, 0.5 M NaCl, 1 M imidazole. Recombinant and purified Nup62 can also be prepared as described previously [13].
3. Purified OGT bound to S-protein Agarose.
4. OGT assay buffer: 50 mM Tris–HCl, pH 7.5, 1 mM DTT, 12.5 mM $MgCl_2$.
5. Incubator for microbiological culture.
6. A microcentrifuge with temperature control: Eppendorf® Refrigerated Microcentrifuge.
7. Microcon YM-30 filter device (EMD Millipore).
8. Dimethyl sulfoxide (DMSO) (Sigma).
9. A Biotin-Phosphine reagent **2**, a Staudinger reaction reagent (10 mM in DMSO): Biotin-phosphine reagent is prepared as described previously [15]. Phosphine-PEG_3-Biotin (Thermo Fisher Scientific) is also commercially available. To prepare 10 mM Biotin-Phosphine (**2**), dissolve 7.21 mg of **2** in DMSO to give a final volume of 1 mL.

10. SDS-PAGE sample buffer: 4× NuPAGE® LDS sample buffer (Invitrogen).
11. Precast SDS-PAGE gels: 10 % or 4–12 % NuPAGE® Bis–Tris gels (Invitrogen).
12. SDS running buffer: 20× NuPage® MOPS SDS Running Buffer (Invitrogen). Dilute it to 1× SDS Running Buffer with distilled water for the gel electrophoresis.
13. A low speed orbital shaker.
14. Nitrocellulose membrane (Invitrogen).
15. IR-Dye conjugated streptavidin: IRDye800CW-Streptavidin (Li-COR Biosciences, Lincoln, NE, USA).
16. An Odyssey® Infrared Imaging System (Li-COR Biosciences).

2.4 Components of OGA Expression in E. coli

1. Cloned plasmids of human OGA were used to generate inserts spanning the ORF of *O*-GlcNAcase which was ligated into pBADHisA expression vector (Novagen) (*see* **Note 4**).
2. BL21(DE3) chemically competent *E. coli* (Invitrogen).
3. Luria–Bertani (LB) broth.
4. Ampicillin solution (100 mg/mL), prepared as described above.
5. LB Agar plate with 50 μg/mL Ampicillin.
6. L-Arabinose solution (10 %): Weigh 0.5 g L-Arabinose (Sigma) and prepare 4.5 mL solution by adding water in a small falcon tube. Mix and filter-sterilize with a 0.22 μm filter (EMD Millipore). Aliquot in 1 mL volume and store at −20 °C.
7. Lysozyme solution (100 mg/mL): Weigh 0.5 g lysozyme (Sigma) and prepare 5 mL solution by adding water to a total volume of 5 mL. Mix and filter-sterilize with a sterilized 0.22 μm filter (EMD Millipore). Aliquot in 1 mL volume and store at −20 °C.
8. OGA lysis buffer (*see* **Note 5**): 20 mM Tris–HCl, pH 7.5, 0.1 mg/mL of lysozyme, and Complete mini EDTA-free protease inhibitor cocktail (Roche Applied Science). Place on ice until it is used.
9. Triton X-100™ (Sigma) 10 % solution.
10. A centrifuge: Sorvall_RC 5C Plus centrifuge (DuPont, Delaware City, DE, USA)
11. A sonicator: Misonix Sonicator® S-4000 Ultrasonic Processor (Cole-Parmer, Vernon Hills, IL, USA).

2.5 Components of OGA Purification

1. HisTrap HP column (GE Healthcare Biosciences, Pittsburgh, PA, USA): Column is pre-charged with Ni^{2+}.
2. 8× Phosphate buffer stock solution, pH 7.4 (GE Healthcare Biosciences).

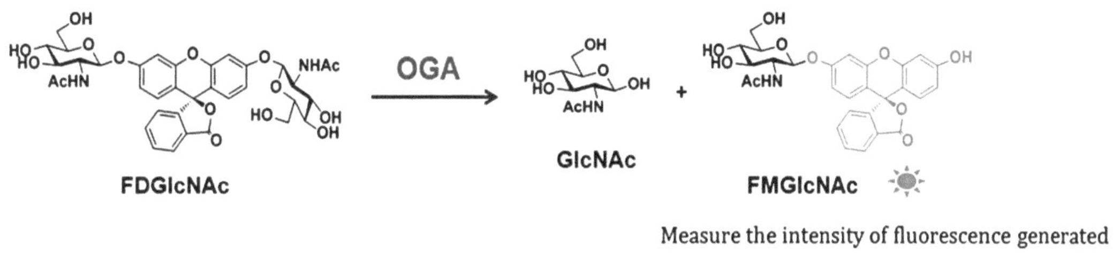

Fig. 4 Fluorogenic OGA activity assay. This method uses nonfluorescent FDGlcNAc as a OGA substrate. Upon enzymatic cleavage of FDGlcNAc by OGA, fluorescent FMGlcNAc is generated

3. 2 M imidazole, pH 7.4 (GE Healthcare Biosciences).
4. A 5-mL syringe.
5. A 0.45 μm filter (EMD Millipore).
6. A UV–vis Spectrophotometer: NanoDrop 2000 (Thermo Scientific).
7. A fluorescence microplate reader: Victor 2 Microplate Reader (Perkin-Elmer Life Sciences, Waltham, MA, USA).

2.6 Components of OGA Enzymatic Assay

2.6.1 Components of Standard OGA Activity Assay (Fig. 4)

1. A fluorogenic substrate (FDGlcNAc **3**): FDGlcNAc is synthesized as described previously [12]. In order to prepare 1 mM FDGlcNAc, first, prepare 10 mM of FDGlcNAc by dissolving 7.38 mg in distilled water to give a final volume of 1 mL. Then transfer 50 μL from 10 mM FDGlcNAc solution into a new vial and add 450 μL of distilled water to make a final concentration of 1 mM solution. Store both 10 and 1 mM solutions of FDGlcNAc at −30 °C.
2. OGA assay buffer: 0.5 M citrate–phosphate buffer, pH 6.5 (*see* **Note 6**).
3. *N*-Acetyl-D-galactosamine (GalNAc) solution (0.1 M): Dissolve 22.6 mg of GalNAc (Sigma) in distilled water to a final volume of 1 mL. Store at −30 °C.
4. Na_2CO_3 solution (0.5 M): Dissolve 5.30 g of sodium carbonate (Sigma) in distilled water to a final volume of 100 mL. Filter-sterilize with a sterilized 0.22 μm filter (EMD Millipore). Store at room temperature.
5. A fluorescence microplate reader: Victor 2 Microplate Reader (Perkin-Elmer Life Sciences).

2.6.2 Components of OGA Activity Assay in a 96-Well Plate Format

1. A fluorogenic substrate (FDGlcNAc **3**): FDGlcNAc is synthesized as described above in Subheading 2.6.1.
2. OGA assay buffer, 0.1 M GalNAc and 0.5 M Na_2CO_3 solutions: as described in Subheading 2.6.1.
3. A 96-well plate (BD Biosciences, Bedford, MA, USA).
4. A fluorescence microplate reader: Victor 2 Microplate Reader (Perkin-Elmer Life Sciences).

3 Methods

3.1 OGT Expression in E. coli

1. pET43.1 Ek/LIC/ncOGT or mOGT is transformed into competent BL21(DE3) cells according to the manufacturer's instructions. Grow overnight on a LB Agar plate with 50 μg/mL Ampicillin (LB Agar/Amp plate) at 37 °C in the incubator.
2. Inoculate a single colony on the LB Agar/Amp plate in a 20 mL culture of LB broth supplemented with Ampicillin (50 μg/mL).
3. Grow overnight at room temperature (*see* **Note 7**).
4. Centrifuge the cells at 2,500 × *g* at 4 °C for 10 min.
5. Remove the supernatant and resuspend the pellets in 0.99 mL of OGT lysis buffer containing 20 mM Tris–HCl, pH 7.5, 2 mM EDTA, 1 mg/mL of lysozyme, and complete mini EDTA-free protease inhibitor cocktail.
6. Incubate at room temperature for 5 min to perform the lysozyme digestion.
7. Add 10 μL of 10 % Triton X-100 to give 0.1 % final concentration of Triton X-100 and vortex.
8. Sonicate the lysate for 15 s on ice with 30 s pause in between each until DNA is completely sheared (*see* **Note 8**).
9. Centrifuge lysate at 14,000 × *g* for 10 min and aliquot supernatant (~200 μL) in a clean tube and store at −80 °C.
10. Determine the OGT expression level in the supernatant (*see* **Note 9**).

3.2 Purification of Expressed OGT in Lysate

Perform the OGT purification procedure in the cold room (~4 °C).

1. Gently suspend S-protein Agarose by inversion and transfer 240 μL of the slurry (equivalent to 120 μL settled resin) to a clean tube (*see* **Note 10**).
2. Add 600 μL of OGT assay buffer to the resin and gently mix by inversion to wash the resin. Centrifuge at 500 × *g* for 5 min and carefully discard the supernatant.
3. Repeat **step 2**.
4. Add 120 μL of *E. coli* lysate containing OGT to the pre-washed S-protein Agarose.
5. Mix thoroughly and incubate at 4 °C on an orbital shaker for 1–2 h.
6. Centrifuge the entire volume at 500 × *g* at 4 °C for 5 min and carefully decant supernatant.
7. Resuspend the S-protein Agarose, which now contains bound S-tag fusion OGT enzyme, in 600 μL OGT assay buffer. Mix thoroughly by inverting the tube 5–7 times.

8. Repeat **steps 6** and 7 twice more to wash away unbound proteins.
9. Centrifuge at 500 × *g* at 4 °C for 5 min and carefully remove the final supernatant.
10. Resuspend OGT bound to S-protein Agarose in 240 μL of OGT assay buffer and use 10–15 μL for each reaction.

3.3 OGT Enzymatic Activity Assay

3.3.1 Radiometric OGT Activity Assay

Keep UDP-[^{14}C]-GlcNAc, *E. coli* lysate containing active OGT enzyme, and Nup62 on ice at all times during the experiment. Experiment has been performed at room temperature unless indicated otherwise.

1. Calculate the volume of OGT assay buffer to make a total reaction volume be 40 μL: If there are UDP-[^{14}C]-GlcNAc 0.4 μL (20 μM), *E. coli* lysate containing active OGT enzyme 8 μL, and Nup62 1 μL (1 μg), then amount of OGT assay buffer needed is 30.6 μL.
2. Place appropriate amount of OGT assay buffer calculated from the **step 1** in a clean tube and add 10–20 μM of UDP-[^{14}C]-GlcNAc and 1 μg of recombinant, purified Nup62.
3. Add 8 μL of the lysate containing OGT enzyme into reaction mixture and vortex briefly.
4. Incubate reactions at 37 °C and rotate at 220 rpm for 1–2 h.
5. Stop the enzymatic reaction by adding 20 μL of 4× NuPAGE® LDS sample buffer and boil samples for 3 min.
6. Load samples onto a SDS-PAGE gel (10 % or 4–12 % NuPAGE® Bis-Tris gel) and run with 1× NuPage® MOPS SDS Running Buffer for 50 min at 200 V.
7. After finishing gel electrophoresis, rinse the gel with distilled water by transferring the gel into a tray containing distilled water and incubating on a shaker for 5 min.
8. Carefully pour off distilled water from the tray. Repeat the **step 7** twice more.
9. Carefully pour off distilled water and pour stain solution, e.g., Simply Blue Safestain solution to cover the gel in the tray, and incubate the gel in the stain solution on a shaker for 1 h.
10. Carefully pour off the stain solution and replace with distilled water to de-stain the gel and incubate the gel in distilled water to de-stain on a shaker for 1 h.
11. Pour off distilled water and soak the gel in En3Hance in a fume hood and incubate on a shaker for 1 h.
12. Carefully pour off the En3Hance solution and rinse the gel with a 10 % PEG (MW8000) solution for 30 min.
13. After rinsing, dry the gel using a gel dryer (*see* **Note 11**).

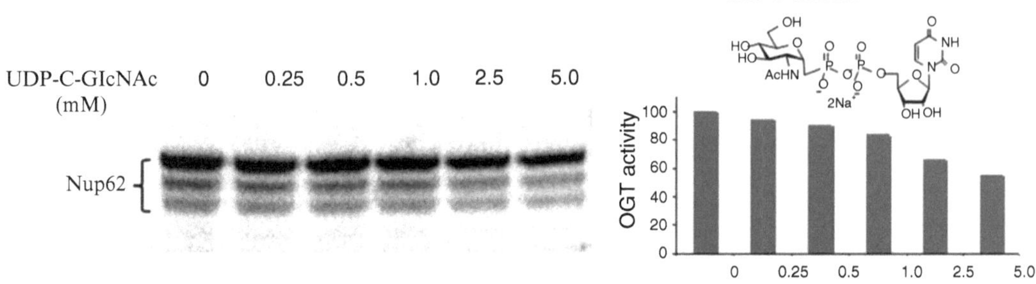

Fig. 5 Radiometric ncOGT activity assay in the absence and presence of UDP-C-GlcNAc. Inhibition of ncOGT by UDP-C-GlcNAc can be evaluated using Nup62 and UDP-[^{14}C]-GlcNAc substrate [16]

14. Expose the gel to a phosphor screen. Scan and analyze the imaging plate on the Fujifilm BAS-1500 phosphorimager (Fig. 5, *see* **Note 12**). Densitometry of the phosphoimage data is performed with Image Gauge 3.0 software (*see* **Note 13**).

3.3.2 Chemoenzymatic OGT Activity Assay

Keep UDP-GlcNAz, purified OGT bound to S-protein Agarose, and Nup62 on the ice all the time during the experiment. Experiment has been performed at room temperature unless indicated otherwise.

1. Calculate the volume of OGT assay buffer to make a total reaction volume of 40 μL: If UDP-GlcNAz solution is required, add 1 μL (50 μM), Purified OGT bound to S-protein Agarose 15 μL, and Nup62 2 μL (2 μg), then OGT assay buffer needed is 22 μL.
2. Place appropriate amount of OGT assay buffer calculated from the **step 1** in a clean tube and add 10–50 μM of UDP-GlcNAz and 2 μg of recombinant, purified Nup62.
3. Add 10–15 μL of purified OGT bound to S-protein Agarose into reaction mixture and vortex briefly.
4. Incubate reactions at 37 °C with frequent mixing for 1–2 h.
5. Centrifuge at 500×*g* for 5 min and carefully transfer the supernatant to the new clean tube.
6. Wash the resin with 50 μL of OGT assay buffer. Centrifuge at 500×*g* for 5 min and combine the wash solution with the supernatant in the tube obtained from **step 5**.
7. Remove excess, unreacted UDP-GlcNAz and buffer-exchange into 1× phosphate buffer using a Microcon YM-30 from the combined solution (*see* **Note 14**).
8. To the buffer-exchanged filtrate (~40 μL) containing Nup62, add 1 μL of Biotin-Phosphine reagent (10 mM stock solution) and vortex.
9. Perform the Staudinger ligation by incubating reaction mixture at 37 °C at 200 rpm for 2 h (*see* **Note 15**).

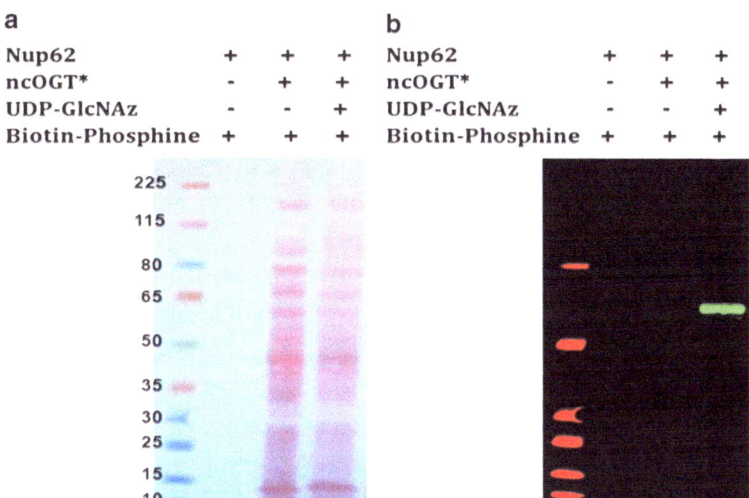

Fig. 6 Ponce S. Stained (**a**) and Western blot (**b**) of Nup62. ncOGT utilizes UDP-GlcNAz to label Nup62 giving a distinctive signal after chemoselective reaction with Biotin-Phosphine. Nitrocellulose membrane was probed with IRDye800CW-Streptavidin. *Asterisk* (*) represents ncOGT in a crude extract of *E. coli*

10. Add 20 μL of 4× NuPAGE® LDS sample buffer and boil samples for 3 min.
11. Load samples onto a SDS-PAGE gel (10 % or 4–12 % NuPAGE® Bis-Tris gel) and run with 1× NuPage® MOPS SDS Running Buffer for 50 min at 200 V.
12. Electrophoretically transfer the protein from the gel onto the nitrocellulose membrane and probe the membrane with IR-Dye conjugated Streptavidin (Fig. 6).

3.4 OGA Expression in E. coli

Carry out all procedures at room temperature unless otherwise specified.

1. pBADHisA/MEGE5 plasmid is transformed into competent BL21(DE3) cells according to the manufacturer's instruction. Grow overnight on a LB Agar plate with 50 μg/mL Ampicillin (LB Agar/Amp plate) at 37 °C in the incubator.
2. Pick a single colony on the LB Agar/Amp plate in 5 mL LB medium containing Ampicillin (50 μg/mL).
3. Grow overnight at 37 °C at 200 rpm.
4. Inoculate 1 mL of overnight culture into 99 mL of fresh LB medium supplemented with Ampicillin (100 μg/mL) and cultivate at 37° at 200 rpm until they reach mid-log phase (OD600 ~ 0.5; 2.5–3.5 h).
5. Induce the culture by adding 200 μL of 10 % arabinose solution to a final concentration of 0.02 % and culture for an additional 3–4 h at 30 °C at 200 rpm.

6. Harvest cells by centrifugation at 3,000 × *g* for 10 min. Discard the supernatant. Store the cell pellet at −80 °C until the cells are lysed.
7. Freeze-thaw the cell pellet prepared in the previous step and suspend in 1.8 mL OGA lysis buffer (add 0.9 mL OGA breaking buffer for 50 mL culture volume) containing 20 mM Tris–HCl, pH 7.5, 100 μg/mL of lysozyme, and complete mini EDTA-free protease inhibitor cocktail.
8. Add 18 μL of 10 % Triton X-100™ solution and vortex.
9. Incubate at room temperature for 15 min and sonicate lysate for 15 s on ice with 30 s pause in between each until DNA is completely sheared.
10. Centrifuge lysate at 14,000 × *g* for 10 min and aliquot supernatant (~200 μL) in a clean tube and store at −80 °C.
11. Determine the OGA expression level of the supernatant (*see* **Note 16**).

3.5 Purification of His-Tag Fused OGA

To avoid clogging of the column it is recommended to filter lysate containing recombinant OGA through a 0.45 μm filter. Carry out the purification procedure in the cold room (4 °C). For 2 mL volume of lysate, 1-mL size of HisTrap HP column can be used. Use a buffer containing 1× Phosphate buffer (20 mM sodium phosphate, 500 mM NaCl, pH 7.4) and 40 mM imidazole as binding and wash buffer and a flow rate of 1 mL/min.

1. Prepare the column for His-tagged OGA's purification (*see* **Note 17**) at room temperature. But perform the purification in a cold room.
2. Thaw the lysate (2 mL) and add 40 μL of 2 M imidazole solution to a final concentration of 40 mM (*see* **Note 18**).
3. Apply the lysate containing 40 mM imidazole to the column using a syringe with a flow rate of 1 mL/min.
4. Collect the flowthrough fraction in a falcon tube.
5. Wash with 20 mL binding buffer.
6. Start elution with 5 mL of 1× phosphate buffer containing 100 mM imidazole and collect the eluate in five 1-mL fractions.
7. Next add 5 mL of 1× phosphate buffer containing 300 mM imidazole and collect the eluate in five 1-mL fractions (*see* **Note 19**).
8. Finally add 3 mL of 1× phosphate buffer containing 500 mM imidazole and collect the eluate in three 1-mL fractions (*see* **Note 20**).
9. Check the different fractions for protein by measuring the absorbance of eluate at 280 nm (A_{280nm}) for protein assays

(*see* **Note 21**), and by performing an OGA activity assay using a protocol described below in this chapter.

10. Pool the fractions that exhibit OGA activity. Perform buffer exchange into a buffer containing 20 mM Tris–HCl, pH 7.5 using a Microcon YM-100.
11. Aliquot purified OGA and store at −80 °C.
12. Concentration of purified OGA can be determined using a BCA assay and the purity of OGA can be determined by Western blot analysis using anti-His-tag antibody.

3.6 OGA Activity Assay

3.6.1 Standard OGA Activity Assay

1. First, calculate distilled water required to make the total volume of assay be 100 μL. (If there are enzyme volume 2 μL, FDGlcNAc 4 μL, OGA assay buffer 20 μL, GalNAc solution 10 μL, then volume of distilled water needed is 72 μL).
2. Place distilled water that is calculated from **step 1** in a clean tube.
3. Add 20 μL of OGA assay buffer.
4. Add 10 μL of 0.1 M GalNAc solution and briefly vortex.
5. Add 10 μL of 1 mM FDGlcNAc solution and briefly vortex.
6. Carefully, add 2 μL of OGA overexpressed lysate and briefly vortex. Upon addition of the lysate, OGA enzymatic reaction starts.
7. Incubate assay solution at 37 °C for 20 min.
8. Terminate the enzymatic reaction by adding 900 μL of 0.5 M Na_2CO_3 solution and vortex.
9. Measure the fluorescence generated from the enzymatic reaction on a fluorescence spectrofluorometer or on a fluorescence microplate reader with the excitation wavelength of 485 nm and the emission wavelength of 535 nm. For fluorescence measurement on a fluorescence microplate reader, transfer 200 μL of OGA assay reaction solution to a 96-well plate and read the fluorescence at the excitation wavelength of 485 nm and the emission wavelength of 535 nm.

3.6.2 OGA Activity Assay in a 96-Well Plate Format

1. First, calculate distilled water required to make the total volume of assay be 50 μL (If there is enzyme volume 1 μL, FDGlcNAc 2 μL, OGA assay buffer 10 μL, GalNAc solution 5 μL, then volume of distilled water needed is 32 μL).
2. Place distilled water that is calculated from **step 1** in a clean tube.
3. Add 10 μL of OGA assay buffer.
4. Add 5 μL of 0.1 M GalNAc solution and briefly vortex.
5. Add 2 μL of 1 mM FDGlcNAc solution and briefly vortex.

6. Carefully, add 1 μL of OGA overexpressed lysate and briefly vortex. Upon addition of the lysate, OGA enzymatic reaction starts.
7. Incubate assay solution at 37 °C for 20 min.
8. Terminate the enzymatic reaction by adding 200 μL of 0.75 M Na_2CO_3 solution.
9. Read the fluorescence on a fluorescence microplate reader at the excitation wavelength of 485 nm and the emission wavelength of 535 nm.

4 Notes

1. Recombinant human ncOGT and mOGT have both His-Tag and S-Tag.
2. To prepare a10 mL volume of OGT lysis buffer, place 200 μL of 1 M Tris–HCl, pH 7.5, 40 μL of 0.5 M EDTA, 100 μL of 100 mg/mL lysozyme stock solution, and a complete mini EDTA-free protease inhibitor cocktail tablet and add water to give a final volume of 10 mL. Mix thoroughly and place the lysis buffer on ice while cells are lysed.
3. To prepare a 10 mL volume of OGT assay buffer, place 500 μL of 1 M Tris–HCl, pH 7.5, 125 μL of 1 M $MgCl_2$, and 10 μL of 1 M DTT and add water to give a final volume of 10 mL. Mix thoroughly and store at 4 °C.
4. Recombinant human OGA has a His-tag.
5. To prepare a 10 mL volume of OGA lysis buffer, place 200 μL of 1 M Tris–HCl, pH 7.5, 10 μL of 100 mg/mL lysozyme stock solution, and a complete mini EDTA-free protease inhibitor cocktail tablet and add water to give a final 10 mL volume. Mix thoroughly and place the OGA breaking buffer on ice while cells are lysed.
6. To make 0.5 M citrate–phosphate buffer, first dissolve 15.0 g of anhydrous sodium dihydrogen phosphate (NaH_2PO_4) in 200 mL of distilled water. Prepare 0.5 M citric acid solution by dissolving 7 mL (10.51 g) of citric acid monohydrate in distilled water to a final volume of 100 mL. Adjust sodium phosphate buffer's pH to be 6.5 by adding 0.5 M citric acid solution. When pH of phosphate buffer gets 6.5, add distilled water to give a final volume of 250 mL and filter-sterilize with a 0.22 μm filter.
7. Temperature control is critical to obtain active OGT enzyme. No not allow the temperature of cell culturing over 30 °C or an OGT enzymatic activity will be dramatically decreased.
8. Perform sonication of the lysate on ice all the time because heat generated during sonication can cause the loss of an OGT activity.

9. Total protein level in the supernatant can be determined using the Pierce® bicinchoninic acid (BCA) protein assay protocol as described by the manufacturer (Thermo Fisher Scientific). OGT expression can be determined by detecting His-tag fusion protein or S-tag fusion protein by immunoblotting with anti-His-tag antibody (Abcam, Cambridge, MA, USA) or anti-S-tag antibody (Abcam).
10. The resin is most conveniently transferred with a 1 mL wide-mouth pipet tip.
11. For gel drying experiment, empty gel dryer cold trap flask before turning on the dryer to prevent solvent accumulation and potential vacuum cutoff. Prechill gel dryer cold trap. Place gel onto a Whatman 3 MM filter paper and cover with plastic wrap. Put gel sandwich (paper, gel, and plastic wrap) on dryer. Cover with flexible plastic membrane and turn on vacuum. Then turn on heat for 2 h at 75 °C. When the gel is completely dried, turn off the vacuum and trap chiller and remove gel sandwich. Remove plastic wrap.
12. For phosphorimaging of dried gel, first, blank phosphor screen (BAS III) on light source for about 20 min. Put the gel in the phorphor cassette (BAS cassette 2040, Fuji Film Co.) and carefully place the erased screen phosphor side down on the gel. Close cassette and leave to expose at room temperature for 16–24 h. When ready to read, take the screen to the phosphorimager.
13. Stoichiometry of *O*-GlcNAc modified Nup62 can be determined using phosphorimager quantitation against [methyl-^{14}C]-BSA internal control.
14. Insert Microcon sample reservoir into vial. Pipette the combined solution into sample reservoir without touching the filter membrane with the pipette tip. Seal with attached cap and centrifuge at 14,000 × *g* at 4 °C. Place 100 μL of 1× PBS and centrifuge at 14,000 × *g* at 4 °C. Add 100 μL of 1× PBS and centrifuge at 14,000 × *g* at 4 °C one more time. Separate vial from sample reservoir and place sample reservoir upside down in a new vial, then spin 3 min at 1,000 × *g* at 4 °C to transfer concentrate to vial. Adjust the concentrate volume to be about 40 μL.
15. Azide can be selectively reacted with phosphine reagent by Staudinger ligation or with terminal alkyne and copper catalyst by Click chemistry. In this protocol, we demonstrated the Staudinger ligation method.
16. Total protein level in the supernatant can be determined using the Pierce® bicinchoninic acid (BCA) protein assay protocol as described by the manufacturer. OGA expression can be determined by detecting His-tag fusion protein by immunoblotting with anti-His-tag antibody (Abcam).

17. Preparation of HisTrap HP column for His-tagged protein's purification can be performed at room temperature. Remove the snap-off end of the column and wash the column with 5 mL distilled water at a rate of 1 mL/min. Equilibrate the column with 10 mL binding buffer (20 mM sodium phosphate, 500 mM NaCl, pH 7.4, and 40 mM imidazole) using the syringe.
18. To minimize nonspecific binding of host cell proteins, lysate to be purified is adjusted to have the same concentration of imidazole as the binding and wash buffer (40 mM) before it is applied to the column.
19. The purified OGA is most likely found in the fractions eluted with a 1× Phosphate buffer containing 300 mM imidazole.
20. After the OGA protein has been eluted, the column can be re-equilibrated with 10 mL of binding buffer. The column is ready for a new purification.
21. For A_{280} measurement, use the elution buffer as a blank.

Acknowledgments

This work was supported by NIDDK intramural funds (NIH) and the National Research Foundation of Korea (2011-0027257).

References

1. Hanover JA, Krause MW, Love DC (2010) The hexosamine signaling pathway: O-GlcNAc cycling in feast or famine. Biochim Biophys Acta 1800:80–95
2. Butkinaree C, Park K, Hart GW (2010) O-linked beta-N-acetylglucosamine (O-GlcNAc): extensive crosstalk with phosphorylation to regulate signaling and transcription in response to nutrients and stress. Biochim Biophys Acta 1800:96–106
3. Hart GW, Slawson C, Ramirez-Correa G, Lagerlof O (2011) Cross talk between O-GlcNAcylation and phosphorylation: roles in signaling, transcription, and chronic disease. Annu Rev Biochem 80:825–858
4. Love DC, Krause MW, Hanover JA (2010) O-GlcNAc cycling: emerging roles in development and epigenetics. Semin Cell Dev Biol 21:646–654
5. Lubas WA, Frank DW, Krause M, Hanover JA (1997) O-Linked GlcNAc transferase is a conserved nucleocytoplasmic protein containing tetratricopeptide repeats. J Biol Chem 272:9316–9324
6. Kreppel LK, Hart GW (1999) Regulation of a cytosolic and nuclear *O*-GlcNAc transferase. Role of the tetratricopeptide repeats. J Biol Chem 274:32015–32022
7. Lubas WA, Hanover JA (2000) Functional expression of O-linked GlcNAc transferase. Domain structure and substrate specificity. J Biol Chem 275:10983–10988
8. Coutinho P, Deleury E, Davies GJ, Henrissat B (2003) An evolving hierarchical family classification for glycosyltransferases. J Mol Biol 328:307–317
9. Jinek M, Rehwinkel J, Lazarus BD, Izaurralde E, Hanover JA, Conti E (2004) The superhelical TPR-repeat domain of *O*-linked GlcNAc transferase exhibits structural similarities to importin alpha. Nat Struct Mol Biol 11:1001–1007
10. Lazarus MB, Nam Y, Jiang J, Sliz P, Walker S (2011) Structure of human O-GlcNAc transferase and its complex with a peptide substrate. Nature 469:564–567
11. Schimpl M, Borodkin VS, Gray LJ, van Aalten DM (2012) Chem Biol 19:173–178
12. Kim EJ, Kang DO, Love DC, Hanover JA (2006) Enzymatic characterization of *O*-GlcNAcase isoforms using a fluorogenic GlcNAc substrate. Carbohydr Res 341:971–982

13. Lubas WA, Smith M, Starr CM, Hanover JA (1995) Analysis of nuclear pore protein p62 glycosylation. Biochemistry 34: 1686–1694
14. Vocadlo DJ, Hang HC, Kim EJ, Hanover JA, Bertozzi CR (2003) A chemical approach for identifying *O*-GlcNAc modified proteins in cells. Proc Natl Acad Sci USA 100: 9116–9121
15. Saxon E, Bertozzi C (2000) Cell surface engineering by a modified Staudinger reaction. Science 287:2007–2010
16. Hajduch J, Nam G, Kim EJ, Fröhlich R, Hanover JA, Kirk KL (2008) A convenient synthesis of the C-1-phosphonate analogue of UDP-GlcNAc and its evalution as an inhibitor of O-linked GlcNAc transferase (OGT). Carbohydr Res 343:189–195

Chapter 12

Antibodies and Activity Measurements for the Detection of *O*-GlcNAc Transferase and Assay of its Substrate, UDP-GlcNAc

Tony Lefebvre, Ludivine Drougat, Stephanie Olivier-Van Stichelen, Jean-Claude Michalski, and Anne-Sophie Vercoutter-Edouart

Abstract

Since the discovery of *O*-GlcNAc modification (*O*-GlcNAcylation) 20 years ago, much attention has been given to OGT (*O*-GlcNAc transferase), the unique enzyme responsible for the nuclear and cytosolic *O*-GlcNAcylation processes. This review focuses on protocols that are routinely used to analyze OGT expression and activity. First are detailed techniques using rabbit polyclonal anti-OGT antibodies, namely, Western blot, (co-)immunoprecipitation, and immunofluorescence. We also describe the measurement of OGT activity by using synthetic peptides as acceptors and radiolabeled UDP-GlcNAc. Finally, a sensitive HPAEC-based technique to measure the cellular content of UDP-GlcNAc, the donor substrate of OGT, is described in detail.

Key words *O*-GlcNAc transferase, Polyclonal anti-OGT antibodies, OGT activity assay, UDP-GlcNAc content assay

1 Introduction

OGT is a nucleocytoplasmic glycosyltransferase (uridine diphospho-N-acetylglucosamine:polypeptide β-N-acetylglucosaminyltransferase or *O*-GlcNAc transferase; EC. 2.4.1.255) assigned to the GT41 family in the CAZY (Carbohydrate-Active enZYme) database [1]. Three isoforms of OGT have been described: the 110 and 78 kDa forms of OGT are localized in the nuclear and cytoplasmic compartments whereas the 103 kDa form is localized in the mitochondria [2, 3]. Using UDP-GlcNAc as the donor substrate, this enzyme modifies thousands of proteins by adding a unique *N*-acetylglucosamine residue onto acceptor substrates mainly confined within cytosol and nucleus. The detection of OGT is quite easy to do since reliable antibodies have been developed, first by

Inka Brockhausen (ed.), *Glycosyltransferases: Methods and Protocols*, Methods in Molecular Biology, vol. 1022, DOI 10.1007/978-1-62703-465-4_12, © Springer Science+Business Media New York 2013

G. W. Hart's group which has discovered the *O*-GlcNAc modification [4], characterized and cloned OGT [5, 6], and by companies such as Sigma, which commercialize a panel of anti-OGT antibodies. The advisable utilization of these polyclonal antibodies allows OGT detection by Western blot (WB)—in different cell types, tissue, or organisms [7, 8]—and by immunofluorescence (IF) [9]. OGT antibodies can also be used in immunoprecipitation (IP), co-IP experiments to identify protein partners and in microinjection experiments to neutralize the enzyme [7].

Within the cell, OGT is not the only *N*-acetylglucosaminyltransferase using the nucleotide sugar UDP-GlcNAc; as an example, the UDP-*N*-acetyl-D-glucosamine:*N*-acetyl-D-glucosaminyldiphosphodolichol *N*-acetyl-D-glucosaminyltrans-ferase, EC. 2.4.1.141 is anchored within the endoplasmic *reticulum* membrane, participates in the biosynthesis of *N*-glycans and consequently competes with OGT for the use of UDP-GlcNAc since its catalytic domain is localized in the cytosol. Measurements of the enzymatic activity of OGT require specific substrates such as Nup62/p62 or casein kinase II (CKII) for macromolecular substrates, or c-Myc, Nup62/p62 or CKII-derived synthetic peptides [10]. The use of full-length proteins as substrate can be followed by the separation of [^{3}H]-GlcNAc-labeled proteins by SDS-PAGE and autoradiography of the dried gel. We show in this chapter that incorporation of radioactivity into peptide substrates is quantified by scintillation counting after fractionation by ion-exchange, gel-filtration, or HPLC.

The intracellular concentration of UDP-GlcNAc is closely dependent upon nutrient availability. Consequently OGT activity tightly correlates with the cellular nutritional status. Therefore, knowing the UDP-GlcNAc level can help in the understanding of *O*-GlcNAcylation processes. We have developed a two-step procedure, based on cation- and anion-exchange chromatography to measure cellular and tissue-derived UDP-GlcNAc contents [11, 12]. This highly sensitive method is presented in the last part of this chapter.

2 Materials

All buffers should be prepared using ultrapure water (18 MΩ water) except for electrophoresis (running) and electroblot buffers.

2.1 SDS-PAGE, Electro-transfer and Western Blot

For electrophoresis and Western blotting, refer to suppliers' recommendations.

2.2 Immunoblot

1. Rabbit polyclonal anti-OGT antibodies, DM17, SQ17 and TI14 were developed using synthetic peptides corresponding respectively to residues 740–756, 833–849 and 1024–1037 of human OGT (*see* **Note 1**).

2. HRP (Horseradish peroxidase)-linked whole Ab (from sheep) and ECL (Enhanced chemiluminescence) Plus (GE Healthcare, Velizy, France).
3. Cell lysis buffer, see IP lysis buffer (*see* Subheading 2.3).
4. TBS-Tween (TBS-T): 150 mM Tris–HCl, pH 8.0, 140 mM NaCl, 0.05 % (v/v) Tween-20 (Sigma, Saint Quentin Fallavier, France).
5. Blocking solution: 5 % (w/v) nonfat milk in TBS-T.
6. Hyperfilm (GE Healthcare).
7. Developer and fixer (Sigma).
8. CCD camera (ChemiGenius^2 bio imaging system, Syngene, Ozyme, Montigny le Bretonneux, France) (*see* **Note 2**).

2.3 Immunoprecipitation (IP) and Co-immunoprecipitation (co-IP) (See Note 3)

1. Protein A Sepharose™ 4 Fast Flow (GE Healthcare).
2. IP lysis buffer: 10 mM Tris–HCl, pH 7.4, 150 mM NaCl, 1 % (v/v) Triton X-100, 0.5 % (w/v) sodium deoxycholate, 0.1 % (w/v) sodium dodecyl sulfate (SDS), Inhibitor cocktail tablets (Roche Diagnostics, Meylan, France).
3. TNE: 10 mM Tris–HCl, pH 7.4, 150 mM NaCl, 1 mM EDTA.
4. IP lysis buffer/NaCl: 0.5 M NaCl in IP lysis buffer.
5. IP lysis buffer/TNE: 50 % IP lysis buffer/50 % TNE (v/v).
6. Co-IP lysis buffer: 20 mM Tris–HCl, pH 7.5, 150 mM NaCl, 1 % (v/v) NP-40, 0.25 % (w/v) sodium deoxycholate, 0.1 % (w/v) sodium dodecyl sulfate (Sigma).
7. Laemmli buffer: 50 mM Tris–HCl, pH 6.5, 2.5 % (w/v) SDS, 5 % (v/v) 2-mercaptoethanol, 10 % (v/v) glycerol, 0.05 % (m/v) bromophenol blue (Sigma).

2.3.1 Silver-Staining of Proteins

1. Fixing Solution: 30 % (v/v) Ethanol, 5 % (v/v) acetic acid in H_2O.
2. Sensitizer Solution: 0.02 % (w/v) Sodium Thiosulphate (S6672, Sigma). Prepare 10 % (w/v) Sodium Thiosulphate and dilute in H_2O to obtain the final solution.
3. Silver Solution: 0.1 % (w/v) $AgNO_3$ (Ultrapure, Sigma) and 0.028 % (v/v) of 37 % formaldehyde solution (Sigma).
4. Developer Solution: 2.4 % (w/v) anhydrous Sodium Carbonate, 0.028 % (v/v) of 37 % formaldehyde solution, 0.0125 % (v/v) of 10 % (w/v) = Sodium Thiosulphate.
5. Stop solution: 4 % (w/v) Tris-base (Sigma), 2 % (v/v) acetic acid.
6. Destaining solution: 1.6 % (w/v) Sodium Thiosulphate and 1 % (w/v) potassium ferricyanide (Sigma). It has to be prepared just before use.

2.3.2 In-Gel Trypsin Digestion

1. Acetonitrile (ACN) (HPLC-grade).
2. Ammonium bicarbonate Solution: 50 mM NH_4HCO_3 (Sigma).
3. 25 mM NH_4HCO_3.
4. 50 % (v/v) ACN in 50 mM NH_4HCO_3.
5. Reduction Solution: 20 mM DTT in 50 mM NH_4HCO_3.
6. Alkylation Solution: 100 mM iodoacetamide in 50 mM NH_4HCO_3.
7. Trypsin solution: 10 ng/μL trypsin in 25 mM NH_4HCO_3 (20 μg trypsin/2 mL 25 mM NH_4HCO_3) (Trypsin, mass spectrometry sequence grade, G-BioSciences, Agro-Bio, La Ferté Saint Aubin, France).
8. Extraction solution: 45 % (v/v) ACN, 10 % (v/v) formic acid.

2.4 Indirect Immunofluorescence

1. PBS (Phosphate-buffered saline): 0.02 M Phosphate, pH 7.5, 150 mM NaCl.
2. Fixation solution: 3 % (w/v) PFA (paraformaldehyde) in PBS.
3. Neutralization solution: 50 mM NH_4Cl (ammonium chloride) in PBS.
4. Permeabilization solution: 0.1 % (v/v) Triton X-100 in PBS.
5. IF blocking solution: 10 % (v/v) goat serum in PBS (Lonza, Verviers, Belgium).
6. Texas Red-, FITC- or Alexa-conjugated secondary antibodies (Molecular Probes, Life Technologies, Invitrogen, Saint Aubin, France).
7. DAPI solution: Prepare a 10 mg/L stock solution of DAPI (4,6-diamino-2-phenylindole) in ultrapure water and store at 4 °C in the dark. Before use, dilute the stock solution to reach a final concentration of 50 ng/mL (1/200) in 10 mM Tris–HCl, pH 7.4, 100 mM NaCl, 10 mM EDTA, 10 mM mercaptoethylamine.
8. Mounting solution. Mowiol solution (Calbiochem, Merck chemicals, Nottingham, UK).

2.5 OGT Activity Measurement

1. Cell lysis buffer: 50 mM Tris–HCl pH 7.8, 150 mM NaCl, 1 % (v/v) NP-40, EDTA-free inhibitor cocktail (Roche Diagnostics).
2. OGT activity buffer [2×]: 100 mM Sodium Cacodylate, pH 6.5, 10 mM $MnCl_2$, 5 mM 5′-AMP.
3. Peptide-substrate YSDSPSTST (100 nmol/μL).
4. UDP-[^{3}H]-GlcNAc (50 μCi/500 μL).
5. Stop Solution: 100 mM formic acid.
6. Methanol (HPLC grade).

7. C18 phase Equilibration buffer: 50 mM formic acid.
8. C18 phase wash buffer: 1 M NaCl in 50 mM formic acid.
9. Elution buffer: 50 % ACN.
10. Sep-Pak C18 cartridges (200 mg Sorbent per Cartridge, 37–55 μm Particle Size, Waters, Milford, MA, USA).

2.6 UDP-GlcNAc Assay

This procedure is based on HPAEC (High pH Anion Exchange Chromatography).

1. Hypotonic lysis buffer: 10 mM Tris–HCl, pH 7.2, 10 mM NaCl, 15 mM 2-mercaptoethanol, 1 mM $MgCl_2$, protease inhibitors.
2. 1 M HCl.
3. Dowex 50WX2-400 in its H^+ form.
4. ProPAC-PA1 column (4×250 mm) (Dionex, Jouy en Josas, France).
5. Dionex HPLC system (Dionex).
6. UV spectrophotometer (Shimadzu, Marne la Vallée, France).
7. Neutralization buffer: 1 M Tris–HCl, pH 8.0.
8. Solution A: 20 mM Tris–HCl, pH 9.2.
9. Solution B: 2 M NaCl.

3 Methods

3.1 Western Blotting

1. Lyse cells in IP lysis buffer on ice (alternatively, cells may be directly homogenized in Laemmli buffer).
2. Centrifuge the cell extract at 20,000 × *g* for 10 min at 4 °C.
3. Discard the membrane pellet.
4. Perform a Bradford, Lowry or BCA assay to determine protein concentration. Bovine serum albumin (BSA) is a frequently used protein standard. You can freeze samples at −20 °C (short time) or −80 °C for later use (protein concentration cannot be determined if cell lysis is performed in Laemmli buffer).
5. Add a volume of Laemmli buffer.
6. Boil the samples for 10 min.
7. Run proteins by SDS-PAGE (load the equivalent of 10–40 μg proteins per lane for a mini-gel).
8. Electrotransfer proteins onto nitrocellulose membrane.
9. Control the equal loading and transfer efficiency using Ponceau red staining (0.1 % (m/v) Ponceau S, 5 % (v/v) acetic acid).
10. Saturate the membrane and block unspecific sites with blocking buffer for 30 min at room temperature. It is not necessary to wash.

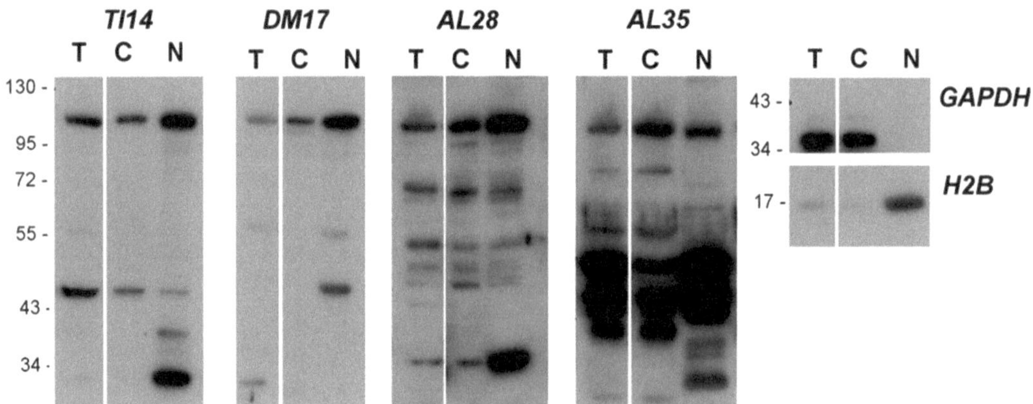

Fig. 1 Immunoblots obtained with anti-OGT. Cytosol and nuclei were prepared from MCF7 cells (not described). Cytosolic (C), nuclear (N), and total proteins (T) were run on a 10 % SDS-PAGE, electroblotted onto nitrocellulose sheet and analyzed by Western blot using anti-OGT antibodies (TI14, DM17, AL28, and AL35). GAPDH was used as a marker of cytosol and H2B (Histone H2B) as a nuclear control. Molecular mass markers are indicated at the *left* (kDa)

11. Dilute the anti-OGT into the blocking buffer at a final dilution of 1:1,000–1:2,000.
12. Incubate the solution with the membranes either 1 h at room temperature or overnight at 4 °C.
13. Discard the solution containing the primary antibody and store it at −20 °C (anti-OGT may be reused at least two or three times).
14. Wash membranes three times for 10 min in TBS-T.
15. Incubate membranes with anti-rabbit IgG HRP-labeled secondary antibody at a dilution of 1:10,000.
16. Wash membranes three times for 10 min in TBS-T.
17. Prepare ECL reagent just before use according to the manufacturer's recommendations.
18. Perform the development of the reaction using Hyperfilms, BioMax films of a CCD camera (Fig. 1).

3.2 Immunoprecipitation

All procedures should be carried out on ice (*see* **Note 4**).

3.2.1 Cell Lysis

1. Wash cells twice with ice-cold PBS (10 mL each wash for a 100 mm diameter Petri dish).
2. Place the cell culture dishes on ice and proceed to cell lysis by adding IP lysis buffer (0.5–1 mL for a 100 mm diameter Petri dish) for 10 min. Avoid exaggerated shaking.
3. Centrifuge cell lysates at 20,000 × *g* for 10 min at 4 °C.

4. Collect supernatants.
5. Perform a Bradford, Lowry or BCA assay to determine protein concentration. You can freeze samples at −20 °C (short time) or −80 °C for later use.

3.2.2 Preclearing the Lysates (See Note 5)

1. Prepare the Sepharose beads: Rinse the beads twice in PBS (to remove ethanol) and finally dilute them 1:1 (v/v) in IP lysis buffer.
2. Add 50 μL of protein A-coupled Sepharose to each sample (0.5–1 mg) for 1 h at 4 °C, under rotation.
3. Spin in micro-centrifuge at 1,500 rpm at 4 °C for 5 min.
4. Discard bead pellet and keep supernatant for immunoprecipitation.

3.2.3 OGT Immunoprecipitation

1. On ice, add 10 μL of rabbit polyclonal anti-OGT antibody to the precleared sample (final dilution of 1:100–1:500).
2. Place the tubes at 4 °C overnight under gentle agitation or rotation.
3. Mix the slurry well and add 30 μL of the beads to each sample. Incubate the lysate beads mixture at 4 °C under rotary agitation for 1 h.
4. Gently centrifuge beads for 1 min at 4 °C.
5. Carefully discard the supernatant using a vacuum water pump.
6. Wash the beads by adding successively 1 mL IP lysis buffer, IP lysis buffer/NaCl, IP lysis buffer/TNE and TNE alone and by vortexing for 1 min. Repeat **steps 4** and **5** between each wash.
7. Remove the last supernatant and add 25–50 μL of 2× Laemmli buffer. Boil for 5 min. You can then freeze the samples or run them on a SDS-PAGE gel.

3.3 Co-immunoprecipitation

All procedures should be carried out on ice (*see* **Note 4**). Wear powder-free gloves at all stages to avoid keratin contamination of the samples.

3.3.1 Cell Lysis

1. Wash cells twice with ice-cold PBS (10 mL each wash for a 100 mm diameter Petri dish).
2. Place the cell culture dishes on ice and proceed to cell lysis by adding co-IP lysis buffer (0.5–1 mL for a 100 mm diameter Petri dish) for 10 min. Avoid shaking.
3. Centrifuge cell lysates at 20,000 × g for 10 min at 4 °C.
4. Collect supernatants.
5. Perform a Bradford, Lowry, or BCA assay to determine protein concentration. Bovine serum albumin (BSA) is a frequently used protein standard. You can freeze samples at −20 °C (short time) or −80 °C for later use.

3.3.2 Preclearing the Lysates, Co-immunoprecipitation of OGT and Partners

This is the same procedure than described in the Subheading 3.2.

1. On ice, add 10 μL of rabbit polyclonal anti-OGT antibody to the precleared sample.
2. Place the tubes at 4 °C overnight under gentle agitation or rotation.
3. Mix the slurry well and add 30 μL of the beads to each sample. Incubate the lysate beads mixture at 4 °C under rotary agitation for 1 h.
4. Gently centrifuge beads for 1 min at 4 °C.
5. Carefully discard the supernatant using a vacuum water pump.
6. Wash beads four times by adding 1 mL co-IP lysis buffer and by vortex for 1 min very gently. Repeat **steps 4** and **5** between each wash.
7. Remove the last supernatant and add 25–50 μL of 2× Laemmli buffer. Boil at 95–100 °C for 5 min. You can then freeze the samples or run them on a SDS-PAGE gel.
8. Resolve proteins by SDS-PAGE (*see* **Note 6**).

3.3.3 Silver-Staining

Silver staining is used for sensitive detection of proteins separated by SDS-PAGE with detection limits from 0.5 to 5 ng. However its linearity is limited only over a short detection range. Gently agitate. Wear powder-free gloves at all stages and use clean staining trays to avoid keratin contamination of the gel. Use ONLY ultrapure water for all the solutions as well as for all the wash steps. Make Sensitizer, Stain and Developer Solutions *fresh* before each use.

1. Remove gel from glass and place in Fixing Solution for 10 min.
2. Replace Fixing solution and shake for at least 1 h.
3. Wash the gel in ultrapure water for 15 min. Repeat twice.
4. Sensitize the gel in Sensitizer Solution for only 1 min.
5. Wash the gel in ultrapure water for 1 min. Repeat once.
6. Incubate the gel for 45 min to 1 h in silver nitrate solution (*see* **Note 7**).
7. Wash the gel with ultrapure water for 30 s.
8. Discard water and add Developer solution. Shake for 30 s to remove excess of silver nitrate.
9. Change Developer solution and agitate until the staining is sufficient (5–10 min). Do not develop for more than 10 min.
10. Discard developer solution and add Stop solution. Shake for at least 10 min.
11. Wash the gel with ultrapure water.

3.3.4 In-Gel Trypsin Digestion

Wear powder-free gloves at all stages and work on a clean surface to avoid keratin contamination. Use only ultra pure water for all the solutions that have to be prepared freshly. It is advisable to prepare small volumes of solutions in a small flask rather than in plastic tubes to avoid contaminations with plastic polymers that interfere with mass spectrometry analysis. For low-level proteins (<1 pmol), reduction and alkylation is recommended to increase the sequence recovery by mass spectrometry.

1. Add 100 μL (enough to cover gel pieces) of 50 mM NH_4HCO_3 and vortex for 10 min.
2. Discard the supernatant and add 100 μL of 50 mM NH_4HCO_3/50 % ACN. Vortex for 10 min.
3. Discard and add ACN. Vortex for 10 min.
4. Discard ACN and dry the gel pieces by SpeedVac centrifugation to complete dryness (5–10 min).
5. Add 100 μL (or enough to cover) of Reduction Solution to the dried gel pieces and incubate the tubes at 56 °C for 1 h.
6. Discard supernatant and add 100 μL of Alkylation Solution. Allow the reaction to proceed in the dark for 45 min at room temperature. Vortex every 15 min.
7. Discard supernatant. Wash gel pieces with 100 μL of 50 mM NH_4HCO_3. Vortex for 10 min.
8. Discard supernatant and repeat successively with 50 mM NH_4HCO_3/50 % ACN and ACN to dehydrate gels.
9. Dry (SpeedVac) the gel pieces to complete dryness.
10. Add trypsin solution to just barely cover the gel pieces (~10–30 μL). Rehydrate the gel pieces on ice or at 4 °C for 20–30 min.
11. Remove excess of trypsin solution and add 25 mM NH_4HCO_3 as needed to cover the gel pieces.
12. Incubate at 37 °C overnight.

3.3.5 Extraction of Peptides

1. Spin briefly and Transfer the digest solution (aqueous extraction) into a clean 0.5 mL tube.
2. Add 50–70 μL (enough to cover) of extraction solution (45 % ACN/10 % formic acid) and vortex for 20 min.
3. Transfer the extracted solution into the 0.5 mL tube. Repeat once.
4. Vortex the extracted digests and dry (SpeedVac) the extracted peptides to complete dryness (~2 h).
5. Analyze with LC-MS/MS.

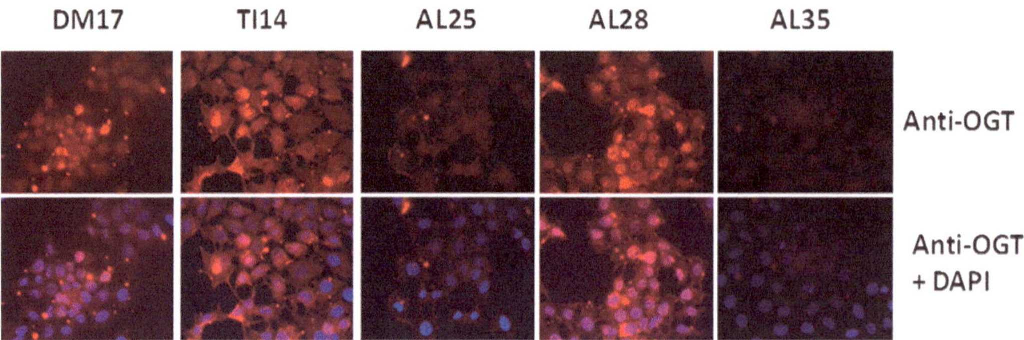

Fig. 2 Indirect IF pictures obtained with anti-OGT. MCF7 cells were prepared as described in Subheading 3.3 and analyzed with a panel of anti-OGT antibodies (TI14, DM17, AL25, AL28, and AL35). A DAPI staining was performed to specifically visualize the nuclei. Note that TI14, DM17, and AL28 are particularly efficient for staining OGT by IF

3.4 Immunofluorescence

Glass coverslips must be handled with care.

1. Cells are grown on glass coverslip in 60-mm plate dishes or on a six-well plate.
2. Gently rinse cells three times with ice-cold PBS.
3. Fix cells in fixative solution for 15 min at room temperature.
4. Wash with PBS.
5. Eliminate excess of PFA by incubating cells with neutralization solution for 10 min.
6. Wash with PBS.
7. Incubate cells with the permeabilization buffer for 5 min.
8. Block nonspecific sites with IF blocking solution for 30 min at room temperature.
9. Incubate coverslips with polyclonal anti-OGT antibody AL28 at a dilution of 1:100 in the IF blocking solution for 1 h. Alternatively, incubation can be performed at 4 °C overnight.
10. Wash three times with PBS.
11. Incubate coverslips with Texas Red-, FITC- or Alexa-conjugated secondary antibodies at a dilution of 1:100 in the IF blocking solution for 1 h at room temperature, in the dark.
12. Wash with PBS.
13. Stain nuclei with the DAPI solution for 30 min. Do not wash.
14. Perform mounting of coverslips in Mowiol solution (5 μL) onto the microscope slides (Fig. 2).

3.5 OGT Activity Assay

3.5.1 Preparation of Lysate from Cell Culture

1. Wash the cells with ice-cold PBS. Repeat once.
2. Place the cell culture dish in ice and add ice-cold lysis buffer. Allow the lysis to proceed for 15 min.
3. Scrape cells and transfer the lysate into a precooled microcentrifuge tube.
4. Spin at 20,000 × *g* for 20 min in a 4 °C precooled microcentrifuge.
5. Transfer the supernatant to a fresh tube kept on ice and discard the pellet.
6. Perform a Bradford, Lowry, or BCA assay to determine protein concentration of samples.

3.5.2 Labeling of OGT Peptide Substrate

1. Transfer 150 μL of protein extract (250–500 μg) into a precooled microcentrifuge tube in ice.
2. Add 160 μL of OGT activity buffer to 150 μL of protein extract (200–500 μg). Check the pH of the mix and adjust to pH 6.5 if necessary.
3. Add 1–10 μg of peptide substrate and 0.5 μCi UDP-[^{3}H]-GlcNAc. As a negative control of the assay, perform the reaction without any peptide substrate.
4. Allow the reaction to proceed for 1 h at room temperature.
5. Stop the reaction by adding 288 μL of 100 mM formic acid.
6. Spin at 10,000 × *g* for 5 min to clarify the mixture.
7. Transfer the supernatant into a clean microcentrifuge tube.

3.5.3 Removing Excess of UDP-[^{3}H]-GlcNAc and Desalting Labeled Peptide Substrate

1. Prepare the C18 phase by loading 10 volumes of methanol onto the top of the column. Methanol can be pushed through the C18 material by using a plastic syringe (e.g., 10 mL plastic disposable syringe).
2. Equilibrate the C18 phase by loading 10 volumes of 50 mM formic acid.
3. Load the sample onto the C18 phase. Slowly push the sample over the C18 phase.
4. Collect the flow-through fraction in a clean Eppendorf tube. Capture efficiency can be improved by passing the flowthrough over the C18 phase again.
5. Desalt the labeled peptide by loading successively 5 volumes of wash buffer, 5 volumes of equilibration buffer, and 5 volumes of H_2O onto the C18 tip. Buffers can be pushed through the C18 material by using a plastic syringe.
6. Elute the desalted peptide by loading 5 volumes of elution buffer onto the C18 column and slowly passing it over the C18 phase. Collect eluate in clean microcentrifuge tubes.

7. Dry the sample in a SpeedVac centrifuge. Alternatively, eluates can be frozen in a 15-mL conical tube and lyophilized.
8. Resuspend in H_2O and combine eluates in a microcentrifuge tube (200 μL).
9. Measure the incorporation of radioactive GlcNAc into the peptide substrate by scintillation counting.

3.6 UDP-GlcNAc Assay

1. Wash cells in ice-cold PBS.
2. Lyse cells in 1 mL of hypotonic buffer.
3. Add 50 μL of 1 M HCl to acidify the protein lysate. Perform a protein assay using the Bradford procedure (not BCA due to oxydoreduction interference with 2-mercaptoethanol).
4. Prepare a column using a Pasteur pipette with 1.5 mL of Dowex 50WX2-400 (H^+-activated).
5. Wash with 5 mL of ultrapure water.
6. Load the cell lysis onto the column.
7. Keep the unbound fraction into a 15 mL Falcon tube (polypropylene) in which 0.5 mL of neutralization buffer was previously added. Perform the collect on ice.
8. Wash the column with ultrapure water until the final volume reaches 10 mL in the Falcon tube.
9. Inject 250–1,000 μL of the unbound fraction using the ProPAC-PA1 column.
10. Use the following elution program: solution A for 1 min; elution gradient for 29 min with 85 % solution A and 15 % solution B; plateau of 5 min in these conditions; an elution gradient of 10 min until 100 % solution B is reached; maintain a plateau at 100 % solution B for 5 min.
11. Re-equilibrate the column in 100 % solution A.
12. Use a flow rate of 1 mL/min.
13. Detection is performed at 256 nm.

4 Notes

1. Alternatively rabbit polyclonal AL25, AL28, and AL35 antibodies produced in Dr. Gerald Hart's laboratory work also very well [6].
2. Films are usually more sensitive than CCD cameras. However regarding grey levels, the dynamic range is near 3.5–4.5 logs for CCD cameras against 2 logs for films.
3. Add protease inhibitors to all buffers (Inhibitor cocktail tablets, Roche Diagnostics, or Sigma).

4. Controls for the immunoprecipitation specificities of the polyclonal antibodies were performed with normal rabbit IgG (Santa Cruz Biotechnology, Heidelberg, Germany).
5. Preclearing the lysate is recommended to reduce non specific binding of proteins to agarose or Sepharose beads. It is advisable to use pipette tips with the end cut off to prevent damage to the beads.
6. A gradient gel will provide a greater separation of proteins. Usually 7.5–15 % SDS-PAGE allows separation of proteins ranging from 250 to 10 kDa.
7. Staining is enhanced with cold $AgNO_3$.

References

1. Cantarel BL, Coutinho PM, Rancurel C, Bernard T, Lombard V, Henrissat B (2009) The Carbohydrate-Active EnZymes database (CAZy): an expert resource for Glycogenomics. Nucleic Acids Res 37:D233–D238
2. Hanover JA, Yu S, Lubas WB, Shin SH, Ragano-Caracciola M, Kochran J, Love DC (2003) Mitochondrial and nucleocytoplasmic isoforms of O-linked GlcNAc transferase encoded by a single mammalian gene. Arch Biochem Biophys 409:287–297
3. Hu Y, Suarez J, Fricovsky E, Wang H, Scott BT, Trauger SA, Han W, Hu Y, Oyeleye MO, Dillmann WH (2009) Increased enzymatic O-GlcNAcylation of mitochondrial proteins impairs mitochondrial function in cardiac myocytes exposed to high glucose. J Biol Chem 284:547–555
4. Torres CR, Hart GW (1984) Topography and polypeptide distribution of terminal N-acetylglucosamine residues on the surfaces of intact lymphocytes. Evidence for O-linked GlcNAc. J Biol Chem 259:3308–3317
5. Haltiwanger RS, Blomberg MA, Hart GW (1992) Glycosylation of nuclear and cytoplasmic proteins. Purification and characterization of a uridine diphospho-N-acetylglucosamine: polypeptide beta-N-acetylglucosaminyltransferase. J Biol Chem 267:9005–9013
6. Kreppel LK, Blomberg MA, Hart GW (1997) Dynamic glycosylation of nuclear and cytosolic proteins. Cloning and characterization of a unique O-GlcNAc transferase with multiple tetratricopeptide repeats. J Biol Chem 272: 9308–9315
7. Dehennaut V, Hanoulle X, Bodart JF, Vilain JP, Michalski JC, Landrieu I, Lippens G, Lefebvre T (2008) Microinjection of recombinant O-GlcNAc transferase potentiates Xenopus oocytes M-phase entry. Biochem Biophys Res Commun 369:539–546
8. Perez-Cervera Y, Harichaux G, Schmidt J, Debierre-Grockiego F, Dehennaut V, Bieker U, Meurice E, Lefebvre T, Schwarz RT (2011) Direct evidence of O-GlcNAcylation in the apicomplexan Toxoplasma gondii: a biochemical and bioinformatic study. Amino Acids 40:847–856
9. Dehennaut V, Slomianny MC, Page A, Vercoutter-Edouart AS, Jessus C, Michalski JC, Vilain JP, Bodart JF, Lefebvre T (2008) Identification of structural and functional O-linked N-acetylglucosamine-bearing proteins in Xenopus laevis oocyte. Mol Cell Proteomics 7:2229–2245
10. Lubas WA, Hanover JA (2000) Functional expression of O-linked GlcNAc transferase. Domain structure and substrate specificity. Domain structure and substrate specificity. J Biol Chem 275:10983–10988
11. Guinez C, Mir AM, Leroy Y, Cacan R, Michalski JC, Lefebvre T (2007) Hsp70-GlcNAc-binding activity is released by stress, proteasome inhibition, and protein misfolding. Biochem Biophys Res Commun 361: 414–420
12. Dehennaut V, Lefebvre T, Leroy Y, Vilain JP, Michalski JC, Bodart JF (2009) Survey of O-GlcNAc level variations in Xenopus laevis from oogenesis to early development. Glycoconj J 26:301–311

Chapter 13

In Vitro Glycosylation Assay for Bacterial Oligosaccharyltransferases

Matias A. Musumeci, Maria V. Ielmini, and Mario F. Feldman

Abstract

Oligosaccharyltransferases (OTases) constitute a family of glycosyltransferases that catalyze the transfer of an oligosaccharide from a lipid donor to an acceptor molecule, commonly a protein. These enzymes can transfer a variety of glycan structures, including polysaccharides, to different protein acceptors. Therefore, this property endows the OTases with great biotechnological potential as these enzymes could be applied to produce several glycoconjugates relevant to the pharmaceutical industry. Furthermore, bacterial OTases are thought to be involved in pathogenesis mechanisms. Here we describe how to purify a representative OTase and its protein acceptor and glycan donor to perform in vitro glycosylation studies.

Key words *O*-oligosaccharyltransferase, *O*-linked glycosylation, Glycoconjugates, In vitro glycosylation, PglL from *Neisseria meningitidis*

1 Introduction

Glycosylation is the reaction in which a carbohydrate or glycan is attached to functional groups of another molecule [1]. Although different mechanisms for protein glycosylation have been described, here we focus on the "*en bloc* glycosylation" process, in which the sugars are preassembled onto a lipid carrier such as dolichol and undecaprenol pyrophosphate before being linked to the protein that acts as glycan acceptor. Such protein glycosylation processes occur in the lumen of the endoplasmic reticulum within eukaryotes or in the periplasm of bacteria like *Campylobacter* and *Neisseria* [2–4]. The *en bloc* glycosylation reactions are catalyzed by enzymes named oligosaccharyltransferases (OTases) [4]. In *N*-linked glycosylation the glycans are linked to amine groups present in asparagines [5], whereas the *O*-linked glycosylation involves the attachment of glycans to hydroxyl groups of serine or threonine residues [6, 7]. By employing OTases it is possible to produce a wide repertoire of glycoconjugates, which could be potentially

Inka Brockhausen (ed.), *Glycosyltransferases: Methods and Protocols*, Methods in Molecular Biology, vol. 1022, DOI 10.1007/978-1-62703-465-4_13, © Springer Science+Business Media New York 2013

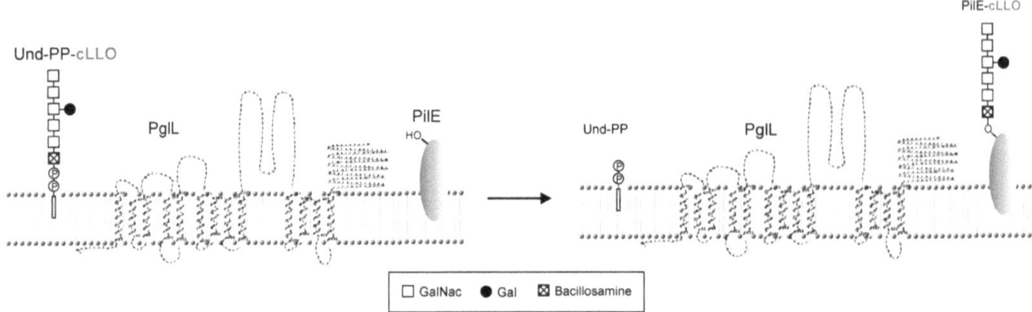

Fig. 1 The reaction catalyzed by the *O*-Oligosaccharyltransferase PglL which is the subject of this chapter. The enzyme was modeled by using TMHMM server. Und-PP-cLLO: Heptasaccharide from *Campylobacter jejuni* linked to undecaprenyl pyrophosphate

applied as vaccines and in diagnostics [8–10]. Understanding the mechanism of these reactions could have an important impact on biotechnology.

In this chapter we describe how to perform an in vitro *O*-linked glycosylation assay by studying the reaction catalyzed by PglL, the *O*-OTase from *Neisseria meningitidis* responsible for the glycosylation of several proteins in the bacterium (Fig. 1) [6, 11]. As an acceptor we will employ the first protein known to be recognized by this enzyme, the type IV pilin, encoded by the *pilE* gene. Type IV pilins form a pilus structure at the bacterial surface and are important for host cell adhesion and virulence [12]. When the protein is expressed in *E. coli* in the absence of the machinery that assembles the pilus, the pilins insert spontaneously into the inner membrane on the basis of hydrophobic interactions [13]. In the in vitro assay we use the purified enzyme (PglL), the protein acceptor pilin (PilE), and the glycan donor is the heptasaccharide employed in *Campylobacter jejuni* *N*-linked glycosylation, which is obtained as lipid-linked oligosaccharide (cLLO). However, the assay can be easily adapted for other *O*- and *N*-OTases, diverse lipid-linked glycan donors, and different protein acceptors. Moreover, by using a similar protocol, it is possible to study other related enzymes. For example, we have performed similar in vitro assays for studying the WaaL ligase, which constitutes a special class of OTase, evolutionarily related to PglL, that transfers a variety of glycans from undecaprenol-pyrophosphate to a lipid A core acceptor during lipopolysaccharide (LPS) biosynthesis [14].

Here, we describe the protocols for the purification of both, OTase and pilin, as well as for the isolation of the glycan donor, the conditions for the in vitro glycosylation reaction and methods to analyze the generation of the glycosylated protein. The detection of glycosylated product is based on changes in the properties of the glycosylated protein acceptor. In most cases, the glycosylated

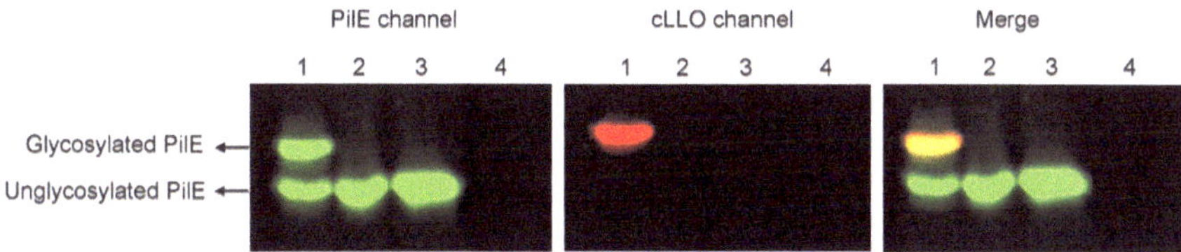

Fig. 2 Result of in vitro glycosylation assay. Shift in the electrophoretic mobility of glycosylated and non-glycosylated forms of protein acceptor. *Lane* 1 shows PilE monoglycosylated (*yellow band*). The *red channel* corresponds to glycan signal, *green* to the unglycosylated protein acceptor (PilE), and the *yellow band* to the merge of both signals. *Lanes* 2, 3, and 4 correspond to controls in the absence of enzyme (PglL), glycan donor (cLLO), and glycan acceptor (PilE) respectively

protein displays a lower electrophoretic mobility with respect to non-glycosylated protein. Therefore, both glycosylated and unglycosylated acceptor forms can be separated on SDS-PAGE and analyzed by Western blot. For the detection of the proteins in glycosylation assays we use an infrared fluorescent imaging system. This system allows the simultaneous detection of protein and glycan in separate fluorescent channels, red and green, for which anti-glycan and anti-protein antibodies are required. The co-localization of both signals is visualized in yellow (Fig. 2). The formation of the glycosylated product usually requires further confirmation using mass spectrometry methods. This kind of analysis can be found in several reports [6, 7, 15, 16].

2 Materials

2.1 Materials Required for Purification of PglL and PilE

1. Luria–Bertani (LB) medium: 1.0 % Tryptone, 0.5 % Yeast Extract, 0.5 % Sodium Chloride (NaCl), pH 7.0.
2. Antibiotics stock: 100 mg/mL ampicillin (Sigma-Aldrich, St. Louis. MO, USA) and 100 mg/mL trimethoprim (Sigma-Aldrich).
3. *E. coli* Strain CLM24 [9] (*see* **Note 1**).
4. *E. coli* CLM24 cells containing the plasmid pAMF10 [7] (*see* **Note 2**).
5. *E. coli* CLM24 cells containing the plasmid pAMF16 [7] (*see* **Note 2**).
6. Inductors stock: 1 M Isopropyl β-D-1-thiogalactopyranoside (IPTG) (Calbiochem, La Jolla, CA, USA) and 20 % (w/v) L-(+)-arabinose (MP Biomedicals, Solon, OH, USA).
7. Buffer A: 50 mM Tris–HCl (pH 8.0), 0.25 M NaCl, 5 % (v/v) glycerol.

8. EDTA-free protease inhibitor mixture (Roche, Mannheim, Germany).
9. DNAseI (Sigma-Aldrich) 100 mg/L.
10. Buffer B: 50 mM Tris–HCl (pH 8.0), 0.25 M NaCl, 5 % (v/v) glycerol, 1 % (w/v) *n*-dodecyl-β-D-Maltopyranoside detergent (Affymetrix, Santa Clara, CA, USA).
11. Ni-NTA affinity resin (QIAGEN, Montreal, QC, Canada).
12. Buffer C: 50 mM Tris–HCl (pH 8.0), 0.25 M NaCl, 5 % (v/v) glycerol, 0.5 % (w/v) *n*-dodecyl-β-D-Maltopyranoside detergent, and 25 mM imidazole (Sigma-Aldrich).
13. Buffer D: 50 mM Tris–HCl (pH 8.0), 0.25 M NaCl, 5 % (v/v) glycerol, 0.5 % (w/v) *n*-dodecyl-β-D-Maltopyranoside detergent, and 50 mM imidazole.
14. Buffer E: 50 mM Tris–HCl (pH 8.0), 0.25 M NaCl, 5 % (v/v) glycerol, 0.5 % (w/v) *n*-dodecyl-β-D-Maltopyranoside detergent, and 250 mM imidazole.
15. All solutions, buffers, and equipment needed to carry out SDS-PAGE.
16. Coomassie stain solution: 0.1 % (w/v) Coomassie R250 (Santa Cruz Biotechnology, Santa Cruz, CA, USA), 10 % (v/v) acetic acid, 40 % (v/v) methanol.
17. Stock of 0.1 M β-mercaptoethanol (Sigma-Aldrich).
18. Ultracentrifuge.

2.2 Materials Required for Isolation of *Campylobacter jejuni* Lipid-Linked Heptassacharide

1. Luria-Bertani (LB) medium.
2. Antibiotics stock: 30 mg/mL chloramphenicol (Sigma-Aldrich).
3. *E. coli* CLM24 cells containing the plasmid pACYC$pglB_{mut}$ [17] (*see* **Note 3**).
4. Solution chloroform–methanol–water 10:20:3 (v/v).
5. Rotary evaporator.

2.3 Materials Required for In Vitro Glycosylation Assay

1. All solutions, buffers, and equipment needed to carry out SDS-PAGE.
2. BSA standard stock 1 mg/mL (Bio-Rad, Hercules, CA, USA).
3. Coomassie stain solution.
4. Software to quantify band intensities: Odyssey® Software (Li-cor Biosciences, Lincoln, NE, USA); Gel-Pro Analyzer (MediaCybernetics, Bethesda, MD, USA).
5. Reaction Buffer 10×: 0.5 M Tris–HCl (pH 6.5), 1 M sucrose, 100 mM $CaCl_2$, 100 mM $MgCl_2$, 10 mM $MnCl_2$.
6. Deionized water.

2.4 Materials Required for Determination of Product Formation

1. All solutions, buffers, and equipment needed to carry out Western Blot.
2. Odyssey blocking buffer (Li-cor Biosciences).
3. Clean 50 mL falcon tubes.
4. Buffer PBST 1×: 137 mM NaCl, 2.7 mM KCl, 10.0 mM Na_2HPO_4, 1.8 mM KH_2PO_4 (pH 7.4), 0.1 % (v/v) Tween 20 (Fisher Scientific, Waltham, MA, USA).
5. Antibody anti-Heptasaccharide of *C. jejuni* from rabbit.
6. Antibody anti-PilE from mouse.
7. Antibody anti-IgG of rabbit coupled to fluorescent probe. We use Goat anti-Rabbit-IgG IRDye 800 (Li-cor Biosciences).
8. Antibody anti-IgG of mouse coupled to fluorescent probe. We use Goat anti-Mouse-IgG IRDye 680 (Li-cor Biosciences).
9. Buffer PBS 1×: 137 mM NaCl, 2.7 mM KCl, 10.0 mM Na_2HPO_4, 1.8 mM KH_2PO_4 (pH 7.4).
10. System to detect fluorescent signals at 685 and 785 nm (Odyssey® Infrared Imaging System, Li-cor Biosciences).

3 Methods

3.1 Purification of PglL and PilE

These protocols were adapted from [7].

Day 1

1. Grow overnight at 37 °C, 25 mL of LB medium (containing ampicillin 100 μg/mL to PglL expression or trimethoprim 100 μg/mL to PilE expression) inoculated with a colony of CLM24 *E. coli* cells transformed with the vector expressing PglL (pAMF10) or PilE (pAMF16).

Day 2

2. Reinoculate into 1 L of fresh LB media (1/40 dilution) containing the same antibiotic concentration and grow with shaking at 37 °C until an OD between 0.4 and 0.6 at 600 nm is reached.
3. Induce with 0.1 mM IPTG (in the case of PglL) or L-(+)-arabinose 0.2 % (w/v) (in the case of PilE).
4. Grow with shaking at 30 °C. After 6 h of the initial induction add again L-(+)-arabinose 0.2 % (w/v) to the culture of PilE.
5. Incubate overnight at 30 °C with shaking.

Day 3

6. Harvest cells by centrifuging at 6,000 ×*g*, 10 min at 4 °C.
7. Wash pellet with 25 mL of Buffer A and centrifuge at 6,000 ×*g* for 10 min at 4 °C. Resuspend in same buffer (25 mL) and add

one EDTA-free protease inhibitor mixture tablet to the solution.

8. Disrupt cells by using a French press.
9. Add DNAseI 1 mg/L of culture and incubate 15 min on ice.
10. Centrifuge suspension (10,000 × *g*, 10 min) to remove large particles and cell debris. Save supernatant.
11. Isolate the membrane fraction by ultracentrifugation of supernatant for 1 h at 100,000 × *g* and 4 °C.
12. Remove supernatant and resuspend pellet in 7.5 mL of Buffer B. Add 1 EDTA-free protease inhibitor mixture tablet to the buffer and solubilize the pellet overnight at 4 °C while rolling (*see* **Note 4**).

Day 4

13. Ultracentrifuge the sample again with the same parameters described in **step 11**.
14. The supernatant contains solubilized PglL or PilE. Keep it on ice.
15. All of the following procedures must be carried out at 4 °C. Equilibrate a Ni-NTA column by washing with 5 volumes of Buffer C.
16. Load sample at flow rate of 0.5 mL/min.
17. Wash column with 5 volumes of Buffer D.
18. Elute sample from column with 7 volumes of elution Buffer E. Harvest aliquots of 500 μL and keep them on ice.
19. Determine the degree of purity by SDS-PAGE electrophoresis and subsequent Coomassie staining of all collected aliquots. A 15 % polyacrylamide gel is required to determine the purity of PilE (Fig. 3).
20. Store the aliquots containing PglL at −70 °C in a final 15 % (v/v) glycerol concentration.

The purification of PilE can be performed using the same protocol with the exception that 15 mM β-mercaptoethanol is added to all buffers and the purified PilE is stored at 4 °C.

3.2 Isolation of *Campylobacter jejuni* Lipid-Linked Heptassacharide

The protocol of *C. jejuni* lipid-linked heptassacharide isolation was adapted from Ielpi et al. [18].

Day 1

1. Inoculate 25 mL of LB medium containing 30 μg/mL of chloramphenicol with a colony of CLM24 *E. coli* cells containing the plasmid pACYC$pglB_{mut}$ (which expresses the enzymes involved in the synthesis of *Campylobacter jejuni* lipid-linked heptassacharide). Grow overnight at 37 °C.

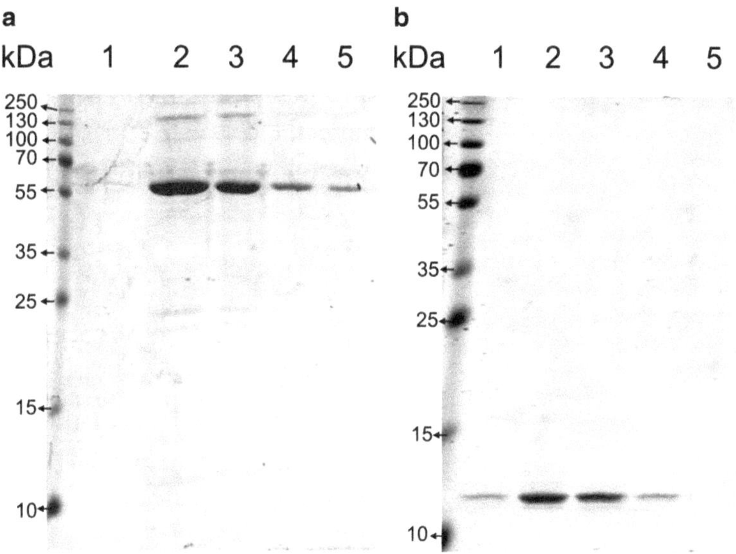

Fig. 3 Result of purification of PglL (**a**) and PilE (**b**). *Lanes* 1–5 correspond to the aliquots that were collected at the elution step during the purification. 10 μL of the different aliquots were loaded in a 15 % polyacrylamide gel

Day 2

2. Reinoculate into 1 L of LB medium (with 30 μg/mL chloramphenicol) and grow with shaking 5 h at 37 °C.
3. Harvest cells by centrifuging at 6,000 × *g*, 10 min and 4 °C.
4. Wash the pellet with 20 mL of water and centrifuge at 6,000 × *g* for 10 min at 4 °C.
5. Solubilize the cell pellet in 20 mL of chloroform–methanol–water (10:20:3 (v/v)).
6. Extract *C. jejuni* lipid-linked heptassacharide by mild tumbling at room temperature for 30 min.
7. Centrifuge at 10,000 × *g*, 10 min at 4 °C. Collect supernatant.
8. Repeat extraction 5–7.
9. Evaporate chloroform and methanol with a rotary evaporator at 40 °C under reduced pressure. The aqueous fraction contains *C. jejuni* lipid-linked heptasaccharide.
10. Centrifuge the liquid solution at maximum speed, for 10 min at 4 °C to precipitate insoluble aggregates. Store the solution of *C. jejuni* lipid-linked heptasaccharide at 4 °C.

3.3 In Vitro Glycosylation Assay

1. Quantify PglL and PilE samples by applying SDS-PAGE (*see* **Note 5**). In a 15 % polyacrylamide gel load 1, 1.5, 2, 3, and 5 μg of BSA standard stock. In the same gel load 5 or 10 μL of PglL and PilE samples.

2. After Coomassie staining measure the band intensities of the BSA standards.
3. Graph band intensity versus μg of BSA. Through interpolation determine the concentration of PglL and PilE samples.
4. To a 1.5 mL eppendorf tube add the following components (*see* **Note 6**):

   2 μg of PglL (~2 μL of the most concentrated aliquot).

   0.3 μg of PilE (~5 μL of the most concentrated aliquot).

   10 μL of extract of *C. jejuni* heptassacharide.

   5 μL reaction buffer 10×.

   Bring to 50 μl final volume with deionized water.
5. In parallel, three controls must be prepared: without glycan donor, without protein acceptor, and without enzyme (*see* **Note 7**).
6. Incubate at 30 °C for 30 min to 2 h.

The reaction conditions and purification procedures can be modified depending on the system upon study, the OTase, protein acceptor, and glycan donor.

3.4 Determination of Product Formation

1. Load 20 μL of the reaction on a 15 % polyacrylamide gel and run at 110 V, 40 mA for 3.5 h.
2. Transfer the proteins onto nitrocellulose membrane by Western blot.
3. Apply Western blot analysis:
   (a) Block the membrane with Odyssey blocking buffer for 1 h.
   (b) Insert the membrane into a 50 mL clean falcon tube. The proteins must be inward facing. Incubate via rolling for 1 h with 1 mL of Odyssey buffer, 1 mL of PBST 1× and the primary antibodies (anti-Glycan from Rabbit and anti-PilE from Mouse) while rolling.
   (c) After incubation, discard the solution and wash three times with 25 mL of PBST 1× for 5 min. Roll the tube during each wash step.
   (d) Incubate the membrane for 1 h with 1 mL of Odyssey buffer, 1 mL of PBST 1× and the secondary antibodies coupled to fluorescent probe while rolling (*see* **Note 8**).
   (e) Discard the solution and wash three times with 25 mL of PBST for 5 min. Roll the tube during each wash.
   (f) Finally wash with 25 mL of PBS 1× for 5 min (rolling).
   (g) Membrane is ready for scanning. Use a fluorescent imaging system to detect both glycan and protein components (*see* **Note 9**).

The results of in vitro glycosylation assays are shown in Fig. 2. To confirm glycosylation of the protein acceptor it is often necessary to apply mass spectrometry techniques, which can provide additional information about glycosylation sites and glycan structures.

4 Notes

1. The *E. coli* strain CLM24 has a mutation in the gene *waaL* that prevents the synthesis of the WaaL ligase resulting in the suppression of the transfer of oligosaccharide from undecaprenyl pyrophosphate to lipid A, and therefore enabling better yields of glycans linked to undecaprenyl pyrophosphate. This strain is also preferred for in vivo glycosylation assays because the cellular pool of glycans linked to undecaprenyl pyrophosphate available for the OTase is maximized [9].
2. The plasmid pAMF10 overexpresses PglL linked with ten histidines at its C-terminus [7]. This vector is inducible with IPTG and resistant to 100 μg/mL ampicillin.

   The plasmid pAMF16 overexpresses PilE linked linked with ten histidines at the C-terminus [7]. This vector is inducible with L-(+)-arabinose and resistant to 100 μg/mL trimethoprim.

   We have found that tagging membrane proteins like the OTase and pilin with a C-terminal Deca-His tag improves the purification of these proteins via Ni-NTA resin compared to the more commonly used Hexa-His tag.
3. The plasmid pACYC*pglB*$_{mut}$ carries *C. jejuni* protein glycosylation locus (pgl) containing mutations W458A and D459A in PglB and express the *C. jejuni* heptassacharide linked to undecaprenyl-pyrophosphate (resistance chloramphenicol 30 μg/mL).
4. The solubilization of PglL is one of the most critical steps in this protocol. During the solubilization stage, the transmembrane enzymes are extracted from their natural environment, the lipid membrane, to an aqueous environment by the use of detergents. Some transmembrane enzymes require the interaction with native lipids from the lipid bilayer to remain in their active conformation. Harsh solubilization and purification procedures may lead to the removal of these essential lipids, and hence inactivation of the protein. Therefore, we suggest not extending the solubilization time for more than 10 h. Addition of glycerol to the samples helps to stabilize the proteins.
5. We advise that the Bradford or other colorimetric methods not be utilized to quantify the samples because the detergent

present in the buffer interferes with the accuracy in these measurements.

6. The quantity of the components, reaction buffer composition, and reaction time can be adjusted taking into account the desired yield of enzyme to be obtained.
7. Negative controls are needed to guarantee that the OTase activity comes from the purified OTase and to rule out the presence of any potential contaminants that may yield false positive results.
8. We employed the following antibodies, from goat: anti-Rabbit-IgG IRDye 800 and anti-Mouse-IgG IRDye 680 (LI-COR Biosciences). These antibodies are labile to light exposure and need to be protected by covering the falcon tube with aluminum foil.
9. The system employed in our laboratory is the Odyssey® Infrared Imaging System (Li-cor Biosciences). This system can detect two signals simultaneously at 685 and 785 nm. Secondary antibodies linked to fluorescent probes that emit in the infrared region are available commercially.

Acknowledgments

We thank Ken McMahon and Nancy Price for proofreading the manuscript. This work was supported by funds from the Alberta Glycomics Centre to MFF. MFF is a CIHR New investigator and an AHFMR scholar.

References

1. Tennant-Eyles RJ, Davis BJ, Fairbanks AJ (2000) Peptide templated glycosylation reactions. Tetrahedron: Asymmetry 11:231–243
2. Weerapana E, Imperiali B (2006) Asparagine-linked protein glycosylation: from eukaryotic to prokaryotic systems. Glycobiology 16:91R–101R
3. Dell A, Galadari A, Sastre F, Hitchen P (2010) Similarities and differences in the glycosylation mechanisms in prokaryotes and eukaryotes. Int J Microbiol 2010:148178
4. Hug I, Feldman MF (2011) Analogies and homologies in lipopolysaccharide and glycoprotein biosynthesis in bacteria. Glycobiology 21(2):138–151
5. Kowarik M, Numao S, Feldman MF, Schulz BL, Callewaert N, Kiermaier E, Catrein I, Aebi M (2006) *N*-linked glycosylation of folded proteins by the bacterial oligosaccharyltransferase. Science 314:1148–1149
6. Vik Å, Aas FE, Anonsena JH, Bilsborough S, Schneider A, Egge-Jacobsena W, Koomey M (2009) Broad spectrum *O*-linked protein glycosylation in the human pathogen *Neisseria gonorrhoeae*. Proc Natl Acad Sci USA 106(11):4447–4452
7. Faridmoayer A, Fentabil MA, Haurat MF, Yi W, Woodward R, George R, Wang P, Feldman MF (2008) Extreme substrate promiscuity of the *Neisseria* oligosaccharyl transferase involved in protein *O*-glycosylation. J Biol Chem 283(50):34596–34604
8. Nothaft H, Szymanski CM (2010) Protein glycosylation in bacteria: sweeter than ever. Nat Rev Microbiol 8(11):765–778
9. Feldman MF, Wacker M, Hernandez M, Hitchen PG, Marolda CL, Kowarik M, Morris HR, Dell A, Valvano MA, Aebi M (2005) Engineering *N*-linked protein glycosylation with diverse *O*-antigen lipopolysaccharide

structures in *Escherichia coli*. Proc Natl Acad Sci 102(8):3016–3021
10. Iwashkiw JA, Fentabil MA, Faridmoayer A, Mills DC, Peppler M, Czibener C, Ciocchini AE, Comerci DJ, Ugalde JE, Feldman MF (2012) Exploiting the *Campylobacter jejuni* protein glycosylation system for glycoengineering vaccines and diagnostic tools directed against brucellosis. Microb Cell Fact 11:13. doi:10.1186/1475-2859-11-13
11. Power PM, Seib KL, Jennings MP (2006) Pilin glycosylation in *Neisseria meningitidis* occurs by a similar pathway to wzy-dependent *O*-antigen biosynthesis in *Escherichia coli*. Biochem Biophys Res Commun 347:904–908
12. Craig L, Li J (2008) Type IV pili: paradoxes in form and function. Curr Opin Struct Biol 18(2):267–277
13. Engelman DM, Steitz TA (1981) The spontaneous insertion of proteins into and across membranes: the helical hairpin hypothesis. Cell 23:411–422
14. Hug I, Couturier MR, Rooker MM, Taylor DE, Stein M, Feldman MF (2010) *Helicobacter pylori* lipopolysaccharide is synthesized via a novel pathway with an evolutionary connection to protein *N*-glycosylation. PLoS Pathog 6(3):e1000819
15. Glover KJ, Weerapana E, Imperiali B (2005) In vitro assembly of the undecaprenyl pyrophosphate linked heptasaccharide for prokaryotic *N*-linked glycosylation. Proc Natl Acad Sci 102(40):14255–14259
16. Aas FE, Vik Å, Vedde J, Koomey M, Jacobsen JE (2007) *Neisseria gonorrhoeae O*-linked pilin glycosylation: functional analyses define both the biosynthetic pathway and glycan structure. Mol Microbiol 65(3):607–624
17. Wacker MD, Linton PG, Hitchen M, Nita-Lazar SM, Haslam SJ, North M, Panico HR, Morris A, Dell B, Wren W, Aebi M (2002) *N*-linked glycosylation in *Campylobacter jejuni* and its functional transfer into *E. coli*. Science 298:1790–1793
18. Ielpi L, Couso RO, Dankert MA (1993) Sequential assembly and polymerization of the polyprenol-linked pentasaccharide repeating unit of the xanthan polysaccharide in *Xanthomonas campestris*. J Bacteriol 175(9): 2490–2500

Chapter 14

In Vitro UDP-Sugar:Undecaprenyl-Phosphate Sugar-1-Phosphate Transferase Assay and Product Detection by Thin Layer Chromatography

Kinnari B. Patel and Miguel A. Valvano

Abstract

In vitro assays are invaluable for the biochemical characterization of UDP-sugar:undecaprenyl-phosphate sugar-1-phosphate transferases. These assays typically involve the use of a radiolabeled substrate and subsequent extraction of the product, which resides in a lipid environment. Here, we describe the preparation of bacterial membranes containing these enzymes, a standard in vitro transferase assay with solvents containing chloroform and methanol, and two methods to measure product formation: scintillation counting and thin layer chromatography.

Key words Lipopolysaccharide, WbaP, WecA, Glycosyltransferase, Sugar-1-phosphate transferase, Membrane protein, O-antigen, Undecaprenyl-diphosphate, Lipid A-core oligosaccharide

1 Introduction

UDP-sugar:undecaprenyl-phosphate sugar-1-phosphate transferases are bacterial enzymes involved in the synthesis of glycans of medical, agricultural and industrial importance. These integral membrane proteins catalyze the transfer of sugar-1-phosphate from a uridine diphosphate (UDP) sugar to the isoprenoid undecaprenyl-phosphate (Und-P) resulting in the formation of an Und-PP-sugar product [1, 2]. Additional glycosyltransferases continue synthesis of the polymer, or a glycan subunit that is subsequently polymerized [1, 2]. Certain glycans can have advantageous properties, such as the emulsifying and thickening agent xanthan gum produced by *Xanthomonas campestris* [3], however many microbial polysaccharides are also produced by pathogenic bacteria and are important in virulence. For example, in *Salmonella enterica* the surface exposed O antigen glycan (a component of lipopolysaccharide) has been shown to protect against killing by serum complement by preventing proper insertion of the membrane attack

Inka Brockhausen (ed.), *Glycosyltransferases: Methods and Protocols*, Methods in Molecular Biology, vol. 1022, DOI 10.1007/978-1-62703-465-4_14, © Springer Science+Business Media New York 2013

complex in the outer membrane [4, 5]. Similarly, *Streptococcus pneumoniae* capsular polysaccharides have been extensively studied and have been shown to be a major factor in biofilm formation and virulence [6]. Since most UDP-sugar:undecaprenyl-phosphate sugar-1-phosphate transferases are involved in the initiation reaction, they are attractive targets for antimicrobials.

Studying these enzymes in vivo has provided information on the function of these proteins, their membrane topology, and potential functional domains. However, in vitro assays are also required to biochemically characterize these enzymes. In vitro assays have been used to study UDP-sugar:undecaprenyl-phosphate sugar-1-phosphate transferases from *Campylobacter jejuni* (PglC; UDP-bacillosamine) [7], *Geobacillus stearothermophilus* (WsaP; UDP-galactose) [8], *Escherichia coli* (*E. coli*) and *Thermatoga maritima* (WecA; UDP-*N*-Acetyl-Glucosamine) [9, 10], and *S. enterica* and *E. coli* (WbaP; UDP-galactose) [11–14]. In these experiments bacterial membranes or pure protein is incubated with radiolabeled sugar substrate, and Und-P as required. After the reaction occurs, the lipid phase containing the Und-PP-sugar product is extracted and radioactivity is measured or detected.

In our work we have found that certain extraction methods work better than others. For example, butanol was routinely used to extract the lipid fraction after reactions with *E. coli* WecA and *S. enterica* WbaP were complete [10, 12, 13]. Although we could determine the radioactivity in this extraction by scintillation counting, only product from the WecA reaction (Und-PP-GlcNAc) would migrate by thin layer chromatography (TLC). Performing the lipid extraction with solvents containing chloroform and methanol has enhanced our in vitro assays. We have modified a protocol from Schäffer et al. [14] to accommodate the use of 1.5 mL microfuge tubes and have given instructions to measure product formation by scintillation counting and TLC. We also provide here detailed instructions for preparing bacterial membranes for the in vitro assay.

2 Materials

2.1 Total Membrane Preparation Components

1. Resuspension buffer: 50 mM Tris–HCl (pH 8), 150 mM NaCl (*see* **Note 1**).
2. Protease inhibitor cocktail (Roche Diagnostics, Indianapolis, IN, USA).
3. Protein Assay Dye Reagent Concentrate (Bio-Rad, Hercules, CA, USA) or a commercial kit to measure protein concentration.

2.2 In Vitro Assay Components

1. 1 M Tris–HCl, pH 8 (*see* **Note 1**).
2. 100 mM $MgCl_2$.

3. 40 mM $MgCl_2$.
4. ^{14}C-labeled UDP-sugar: UDP-glucose (Perkin-Elmer, Waltham, MA, USA), UDP-galactose and UDP-GlcNAc (American Radiolabeled Chemicals, Inc., St Louis, MO, USA) (*see* **Note 2**).
5. β-Mercaptoethanol: 10 % solution in water (*see* **Note 3**).
6. Undecaprenyl-phosphate*: Prepared in 1 % 3-[(3-cholamidopropyl)dimethylammonio]-1-propanesulfonate (CHAPS) (Obtained from Ewa Swiezewska, Institute of Biochemistry and Biophysics of the Polish Academy of Sciences, Warsaw, Poland) (*see* **Note 4**).
7. CHAPS*: 10 % solution in water.
8. Solvent 1: Chloroform–Methanol/3:2 (*see* **Note 5**). 500 μL per sample plus additional solvent is required to pre-run the TLC plate (*see* **Note 6**).
9. Solvent 2: Chloroform–Methanol–Water/3:2:0.4 (*see* **Note 5**). 600 μL per sample is required.
10. Pure solvent upper phase (PSUP): Chloroform–Methanol–1 M $MgCl_2$–Water/18:294:293:1 (*see* **Note 5**). 400 μL per sample.

**These materials are only required when working with purified protein*

2.3 Scintillation Counting Components

1. Scintillation fluid (Ecolume, MP Biomedicals, Solon, OH, USA).
2. Scintillation Vials, Polyethylene, with Screw Cap (VWR International, Mississauga, Ontario Canada).
3. Scintillation counter (Beckman Coulter Canada, Inc., Mississauga, Ontario Canada).

2.4 Thin Layer Chromatography Components

1. Glass vials, 1–3 mL with diameter of ≤1 cm (for collecting extracted lipids).
2. Polyester-backed Flexible Silica TLC Plates, 250 μm, 20×20 cm (Whatman, GE Health Care Life Sciences, Baie d'Urfe, Quebec,Canada).
3. TLC rack (to hold plate(s)).
4. TLC glass chamber with lid (30*h*×30*w*×10*d*cm) (may use chamber with different dimensions if it accommodates TLC rack/plate(s)) (Sigma-Aldrich, St. Louis, MO, USA).
5. Petroleum jelly (e.g., Vaseline).
6. Solvent 1 (*see* Subheading 2.2).
7. Solvent 3: Chloroform–Methanol–Water/65:25:4 (*see* **Note 5**).
8. Hair dryer.

9. Phosphor screen and cassette, 20×25 cm (GE Healthcare).
10. Phosphor imaging system (e.g., Typhoon, Storm, or PhosphorImager) and ImageQuant software (GE Health Care Life Sciences).

3 Methods

3.1 Total Membrane Preparation

1. Grow bacteria cells expressing protein of interest (the size of the culture will depend on the number of samples required and the level of protein expression). Grow a starter culture overnight. For constitutively expressed proteins dilute the culture to an initial OD_{600} of 0.2 and incubate at 37 °C for 5 h. Harvest cells by centrifugation at 10,000×*g* for 10 min at 4 °C. For proteins requiring induction, dilute the culture to an initial OD_{600} of 0.2 and incubate at 37 °C for 2 h until reaching an OD_{600} of 0.6. At this point, add the inducing agent to a final concentration that is optimal for your protein of interest. Incubate cells for 5 h at 30 °C or as required.
2. Harvest cells by centrifugation at 10,000×*g* for 10 min at 4 °C. Continue with cell protocol or store pellet at −20 °C.
3. Resuspend cell pellet in resuspension buffer (*see* **Note 7**) and add protease inhibitor cocktail (Roche). From this point on samples should always be kept cool on ice.
4. Lyse bacteria using a method appropriate for membrane proteins (*see* **Note 8**). Once the cells are lysed continue immediately with the membrane preparation protocol.
5. Remove cell debris by centrifugation at 15,000×*g* for 15 min at 4 °C. Recover supernatant and centrifuge at 1,00,000×*g* for 45 min at 4 °C. Recover pellet and homogenize in resuspension buffer (*see* **Note 9**).
6. Measure protein concentration using the Bradford assay protocol (*see* **Note 10**) or with a commercial kit.
7. Continue with in vitro assay. Some proteins will tolerate storage at −80 °C and subsequent freeze-thaw, however others such as WbaP will not. In these cases, working with freshly prepared membranes is always optimal. If you would like to do the assay the next day, store your samples at 4 °C (*see* **Note 11**).

3.2 In Vitro Assay

1. This assay requires the use of ^{14}C-labeled UDP-sugars. Follow all protocols required by your institution when handling samples containing radioactivity. Use conical 1.5 mL microfuge tubes with a screw cap lid to prevent leakage of the samples during shaking, vortexing and/or centrifugation steps.
2. Set up the reaction. Always add the UDP-sugar last as the reaction will commence once this component is added. *For assays*

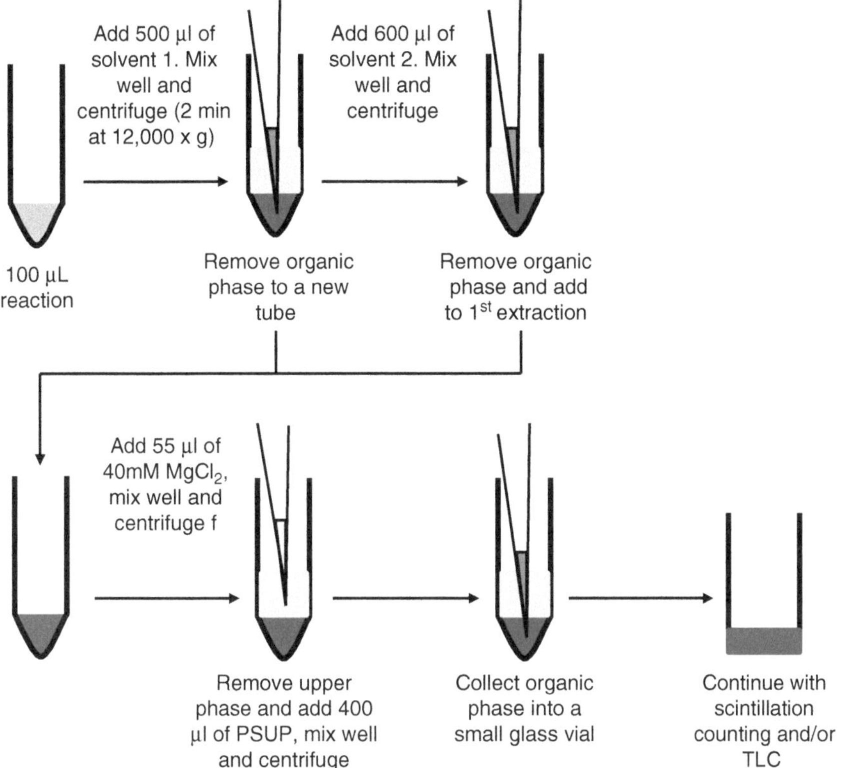

Fig. 1 Procedure for in vitro assay and lipid extraction

with total membranes set up the reaction in a total volume of 100 μL including 100 μg of total membrane protein, 5 μL 1 M Tris–HCl at pH 8.5 (final concentration 50 mM), 25 μL of 100 mM $MgCl_2$ (final 25 mM), 1 μL of 10 % β-mercaptoethanol (final 1 %), and the ^{14}C-labeled UDP-sugar to a final concentration of 0.33 μM (0.01 μCi) (or more/less depending on the experiment). The volume of ^{14}C-labeled UDP-sugar added will depend on the concentration of the stock solution (*see* **Note 2**). Cold UDP-sugar may also be added. Top up with water. *For assays with purified protein* set up the reaction in a total volume of 50 or 100 μL. For a 100 μL reaction include 200 ng of protein and, in addition to the reagents listed above, 10 μL of 10 % CHAPS (final 1 %), and 25 μL of 400 μM Und-P (final 100 μM); top up with water.

3. Vortex samples and incubate at 37 °C in a water bath. The length of incubation will depend on the nature of the experiment. For purified WbaP, the reaction is linear up to 15 min for concentrations of protein up to 4 ng/μL [11].

4. Extract the lipid phase (*see* Fig. 1). Terminate the reaction by adding 500 μL of solvent 1. Shake the tubes vigorously for 2 min (*see* **Note 12**). Centrifuge for 2 min at 14,000×*g*.

Remove tubes gently. You will see two phases. Carefully remove the bottom phase (organic phase) to a new tube by slowly pushing the pipette tip through the upper phase and through the interface. Re-extract the pellet by adding 600 μL of solvent 2 and shake and centrifuge as before. Remove the organic phase and add to the first extraction. Add 55 μL of 40 mM $MgCl_2$, shake and centrifuge. Remove the upper phase and wash by adding 400 μL of PSUP, shake and centrifuge. Collect the bottom (organic) phase in glass vials (*see* **Note 13**).

3.3 Scintillation Counting

1. Add lipid extract (or half of the extract if also doing TLC) to 5 mL of scintillation fluid. Swirl vials to mix sample thoroughly.
2. Count radioactivity (5 min per sample) using a scintillation counter.
3. Calculate the incorporation of UDP-sugar into the lipid fraction (*see* **Note 14**).

3.4 Thin Layer Chromatography

1. Dry the lipid extract under nitrogen (*see* **Note 15**). If there is no nitrogen available, air-dry the samples overnight in a fume hood.
2. The day before you wish to perform the TLC experiment pre-run the plate to remove impurities. Add solvent 1 (*see* **Note 6**) to the TLC glass chamber (*see* **Note 16**) and leave for a minimum of 3 h to saturate the tank (*see* **Note 17**). Place the plate into the TLC chamber and run until solvent reaches the top (*see* **Note 18**). Dry the plate with a hair dryer or overnight.
3. Remove solvent 1 from the chamber and replace with solvent 3 (*see* **Note 6**). Saturate the chamber for a minimum of 3 h or overnight (*see* **Note 17**).
4. Using a pencil and ruler carefully draw a line across the plate approximately 1 in. from the bottom (*see* **Note 19**). This will represent the origin. Draw equally spaced spots on the origin that will indicate where to spot your samples.
5. Resuspend the lipid extract in 10 μL of solvent 1 and spot onto the plate (*see* **Note 20**).
6. Run the plate in solvent 3 (*see* **Notes 6** and **18**). When the solvent front is approximately 2 cm from the top, remove the plate from the chamber and trace the solvent front with a pencil before it dries.
7. Dry the plate with a hair dryer or overnight.
8. Place the plate in a plastic cover (*see* **Note 21**) and into a Phosphor screen (*see* **Note 22**). Leave to expose overnight or longer.
9. Detect the sample spots with a Phosphorimager (Fig. 2) (*see* **Note 23**) and scan the entire plate.

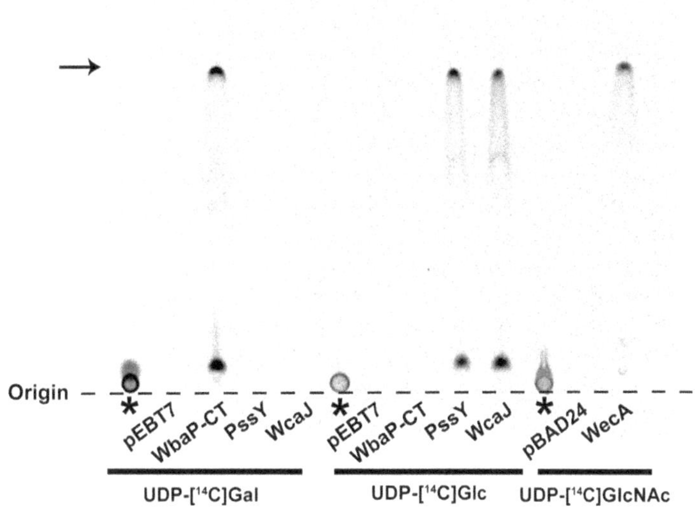

Fig. 2 Thin layer chromatography. Total membranes from *E. coli* C43(DE3) cells expressing vector control (pEB-T7), the C-terminal domain of *S. enterica* WbaP (WbaP-CT), *E. coli* WcaJ or *Caulobacter crescentus* PssY were incubated with ^{14}C-labeled UDP-galactose and UDP-glucose. Similarly total membranes from *E. coli* MV501 cells expressing a vector control (pBAD24) and *E. coli* WecA were incubated with ^{14}C-labeled UDP-GlcNAc. The lipid extractions and the radiolabeled sugars alone (0.025 μCi) were also spotted along the origin. All of the Und-PP-sugar products (*arrow*) migrated a similar distance, which has been previously observed with *E. coli* WecA and WbaP expressed in *Bacillus subtilis* [14]. Reproduced from [17] with permission from the Journal of Bacteriology

10. Calculate the migration of the spot, or the Retention factor (R_f), to aid in the identification of the product. The R_f is the distance the spot travelled from the origin, divided by the distance between the origin and the solvent front (*see* **Note 24**).

4 Notes

1. The pH of the resuspension buffer and the buffer used in the in vitro assay will depend on the isoelectric point (pI) of your protein of interest. Test different buffers and pH conditions to determine what works best. For example, a pH of 8.5 is optimal for *S. enterica* WbaP [10, 11].
2. The ^{14}C-labeled UDP-sugars come in very small volumes and are expensive; therefore care must be taken to store them properly. Also, diluting the stock will allow proper dispensing of the sugars when adding them to the reaction. For example, 10 μCi of ^{14}C-labeled UDP-galactose arrives in a total volume of 100 μL in ethanol–water (7:3). Evaporate the stock by air (or dry under nitrogen) and resuspend in 1,000 μL of water. Each μL will represent 0.01 μCi. Remember to account for the change

in molarity when performing your calculations. Always store at −20 °C in a tube with a screw cap lid. Before opening the tubes centrifuge for 5 s at max speed to prevent the sugars from sticking to the lid.

3. When dispensing β-mercaptoethanol always work under the fume hood as this reagent has an offensive odor. Store the 10 % solution at 4 °C in a bottle with a tight lid to prevent exposure to the air.
4. Undecaprenyl-phosphate is received in sealed glass vials stored in a solution of chloroform–methanol. This must be dried under nitrogen and reconstituted in a lipid environment, provided by 1 % CHAPS. Using a glass syringe remove the Und-P from the vials and decant into a new glass vial under a stream of nitrogen. After the Und-P is completely dry add 1 % CHAPS (the amount will depend on the concentration of the Und-P and the desired concentration of the stock solution). To efficiently produce micelles with Und-P the solution must be sonicated or placed in a sonication bath.
5. All solutions containing chloroform must be prepared and stored in glass containers with glass stoppers, never metal or plastic. Also, any solution made with chloroform should be prepared fresh every time. Chloroform should be measured and dispensed using a glass cylinder or glass pipettes. Plastic pipettes should never be placed into chloroform as they will breakdown and contaminate your chloroform stock. Also, a micropipette should not be used, as the chloroform vapors will damage its internal mechanism.
6. A volume of 100 mL of solvent is required for a chamber with the dimensions above. If the TLC chamber is smaller or larger, make enough solvent to fill the chamber a centimeter from the bottom.
7. The volume of resuspension buffer to use will depend on the mass of your cell pellet. Generally, for a 100 mL culture, resuspend the pellet in 5 mL. A transfer pipette is useful for this step.
8. Techniques suitable to yield high amounts of membrane proteins include liquid shear pressure, such as the French press system by Thermo Scientific (which is now discontinued), the One Shot by Constant Systems Inc., or ultrasonication. Liquid shear pressure is preferable, as ultrasonication can destroy proteins by shearing. Both methods generate heat therefore care must be taken to keep samples cool on ice.
9. Total membrane pellets will appear amber in color and may be difficult to homogenize. With small sample volumes requiring centrifugation in microfuge tube(s), use a high speed centrifuge with a microfuge tube rotor (e.g., Beckman rotor JA 18.1) and centrifuge recovered supernatant at 40,000 × *g* for 45 min at

4 °C. A polypropylene pestle may help to resuspend the pellet. A properly homogenized sample is important when measuring protein concentration and performing the in vitro assay.

10. The Bradford assay requires the use of standard samples at known concentrations and a colorimetric reagent that detects proteins. Make the protein standards (0, 2, 4, 6, 8, 10 μg/mL) using bovine serum albumin (BSA) (1 mg/mL) and Bio-Rad Protein Assay Dye Reagent Concentrate. The standards and samples should be diluted in a total volume of 1 mL, which includes 200 μL of Bradford reagent and 800 μL of the sample buffer. Mix the standards and samples well and let stand at room temperature for 5 min, but never longer than 1 h. Measure the absorbance at 595 nm making sure to always correct with the blank (0 mg/mL BSA). Produce a standard curve to determine concentrations of samples.

11. Determining the stability of a protein will require some trial and error. Freezing may cause changes in pH affecting the structure of the protein making it more susceptible to degradation [15]. Short-term storage at 4 °C with sufficient protease inhibitor is usually optimal and proteins may remain stable in these conditions for days. To enhance stability, reducing agents such as dithiothreitol (DTT) and β-mercaptoethanol may help maintain proteins in a reduced state. Metal chelators such as EDTA prevent metal-induced oxidation of –SH groups by preventing oxidation of cysteines (*see* [16] for information on storing pure proteins which can be applied to total membrane preparations).

12. If there is no vortex with the necessary attachments available to shake the samples, simply place the microfuge tubes in a flat microfuge rack. Secure them with tape or place another rack on top and manually shake them up and down.

13. To continue with the TLC protocol it is important that the diameter of the glass vials be 1 cm or less as the sample will be resuspended in a final volume of 10 μL before spotting. If only performing the scintillation counting, the extraction can be dispensed directly into the scintillation vial instead of the glass vial.

14. To calculate the incorporation of the UDP-sugar per unit of protein (mmol/mg) use the following formula:

$$\frac{\text{CPM}}{\text{Assay volume(mL)}} \times \frac{\text{Efficiency}}{(\text{CPM / DPM})} \times \frac{\text{Ci}}{2.2 \times 10^{12}\,\text{DPM}} \times \frac{\text{mmol}}{\text{Ci in assay}} \times \frac{1{,}000\ \text{mL}}{1\text{L}} \times \frac{1\ \text{mol}}{1{,}000\ \text{mmol}} \times \frac{10^{12}}{1\ \text{mol}}$$

The CPM and DPM values will be provided by the scintillation counter.

15. Prior to placing the sample(s) under the stream of nitrogen insure that the force of the nitrogen flow is not too high as this can cause the sample to dry to the side of the glass vial and will make it difficult to resuspend.
16. To insure a tight seal between the lid and the chamber apply a generous amount of petroleum jelly to the lip of the chamber and press the lid down on top. If desired, add weights to the top of the lid to maintain a seal.
17. Allowing the solvent to sit in the sealed chamber will saturate the chamber with solvent vapors, which will result in reproducible R_f values (*see* **Note 22**).
18. It is very important that the TLC plate is sitting properly in the rack and level with the bottom of the rack. Also, when placing the rack into the chamber ensure that it is level with the bottom of the chamber. If the plate enters the solvent on an angle it will affect the migration of the sample(s).
19. Do not use a lot of force when drawing the origin as it may disrupt the silica matrix and affect the migration of the spots.
20. Do not spot all 10 μL of the sample at once. Spot 2–3 μL and dry with a hair dryer before spotting the next volume on top of the dried sample. Also, remember that the solvent evaporates quickly; therefore the total amount spotted may be less than 10 μL.
21. A clear plastic sealable bag or plastic wrap may be used. The plastic should cover the whole plate to contain the radiation.
22. Place the plate in the phosphor cassette. Cover with the phosphor screen and take note of the coordinates of the plate on the cassette.
23. When placing the phosphor screen in the imaging equipment, line up the screen with the grid and scan the whole area of the screen that the plate was exposed to. The ImageQuant program will display these coordinates with the detected radioactivity. This is important as the coordinates can then be correlated to the solvent front, the origin and the detected samples. These measurements are required to calculate the Rf value (below).
24. Retardation factor, or Rf value, is the ratio of the distance the substance moved to the distance the solvent moved.

Acknowledgments

This work was supported by a grant from the Canadian Institutes of Health Research to M.A.V., who also holds a Canada Research Chair in Infectious Diseases and Microbial Pathogenesis.

References

1. Valvano MA (2003) Export of O-specific lipopolysaccharide. Front Biosci 8:452–471
2. Valvano MA, Furlong SE, Patel KB (2011) Genetics, biosynthesis and assembly of O-antigen. In: Knirel YA, Valvano MA (eds) Bacterial lipopolysaccharides. Structure, chemical synthesis, biogenesis and interaction with host cells. Springer, Vienna, Austria, pp 275–310
3. Ielpi L, Couso RO, Dankert MA (1993) Sequential assembly and polymerization of the polyprenol-linked pentasaccharide repeating unit of the xanthan polysaccharide in Xanthomonas campestris. J Bacteriol 175:2490–2500
4. Murray GL, Attridge SR, Morona R (2003) Regulation of Salmonella typhimurium lipopolysaccharide O antigen chain length is required for virulence; identification of FepE as a second Wzz. Mol Microbiol 47: 1395–1406
5. Murray GL, Attridge SR, Morona R (2005) Inducible serum resistance in Salmonella typhimurium is dependent on wzz(fepE)-regulated very long O antigen chains. Microbes Infect 7:1296–1304
6. AlonsoDeVelasco E, Verheul AF, Verhoef J, Snippe H (1995) *Streptococcus pneumoniae*: virulence factors, pathogenesis, and vaccines. Microbiol Rev 59:591–603
7. Glover KJ, Weerapana E, Chen MM, Imperiali B (2006) Direct biochemical evidence for the utilization of UDP-bacillosamine by PglC, an essential glycosyl-1-phosphate transferase in the *Campylobacter jejuni N*-linked glycosylation pathway. Biochemistry 45:5343–5350
8. Steiner K, Novotny R, Patel K, Vinogradov E, Whitfield C, Valvano MA, Messner P, Schaffer C (2007) Functional characterization of the initiation enzyme of S-layer glycoprotein glycan biosynthesis in *Geobacillus stearothermophilus* NRS 2004/3a. J Bacteriol 189:2590–2598
9. Al-Dabbagh B, Mengin-Lecreulx D, Bouhss A (2008) Purification and characterization of the bacterial UDP-GlcNAc:undecaprenyl-phosphate GlcNAc-1-phosphate transferase WecA. J Bacteriol 190:7141–7146
10. Lehrer J, Vigeant KA, Tatar LD, Valvano MA (2007) Functional characterization and membrane topology of *Escherichia coli* WecA, a sugar-phosphate transferase initiating the biosynthesis of enterobacterial common antigen and O-antigen lipopolysaccharide. J Bacteriol 189:2618–2628
11. Patel KB, Ciepichal E, Swiezewska E, Valvano MA (2012) The C-terminal domain of the Salmonella enterica WbaP (UDP-galactose:Und-P galactose-1-phosphate transferase) is sufficient for catalytic activity and specificity for undecaprenyl monophosphate. Glycobiology 22:116–122
12. Patel KB, Furlong SE, Valvano MA (2010) Functional analysis of the C-terminal domain of the WbaP protein that mediates initiation of O antigen synthesis in *Salmonella enterica*. Glycobiology 20:1389–1401
13. Saldías MS, Patel K, Marolda CL, Bittner M, Contreras I, Valvano MA (2008) Distinct functional domains of the *Salmonella enterica* WbaP transferase that is involved in the initiation reaction for synthesis of the O antigen subunit. Microbiology 154:440–453
14. Schäffer C, Wugeditsch T, Messner P, Whitfield C (2002) Functional expression of enterobacterial O-polysaccharide biosynthesis enzymes in *Bacillus subtilis*. Appl Environ Microbiol 68:4722–4730
15. Pikal-Cleland KA, Rodriguez-Hornedo N, Amidon GL, Carpenter JF (2000) Protein denaturation during freezing and thawing in phosphate buffer systems: monomeric and tetrameric beta-galactosidase. Arch Biochem Biophys 384:398–406
16. Ó'Fágáin C (2011) Storage and lyophilisation of pure proteins protein chromatography. In: Walls D, Loughran ST (eds) Methods in molecular biology, vol 681. Humana Press, New York, pp 179–202
17. Patel KB, Toh E, Fernandez XB, Hanuszkiewicz A, Hardy GG, Brun YV, Bernards MA, Valvano MA (2012) Functional characterization of UDP-glucose:undecaprenyl-phosphate glucose-1-phosphate transferases of Escherichia coli and Caulobacter crescentus. J Bacteriol 194:2646–2657

Chapter 15

In Vitro O-Antigen Ligase Assay

Xiang Ruan and Miguel A. Valvano

Abstract

WaaL is a membrane enzyme that catalyzes the glycosidic bonding of a sugar at the proximal end of the undecaprenyl-diphosphate (Und-PP)-O-antigen with a terminal sugar of the lipid A-core oligosaccharide (OS). This is a critical step in lipopolysaccharide synthesis. We describe here an assay to perform the ligation reaction in vitro utilizing native substrates.

Key words Lipopolysaccharide, WaaL, Ligase, Oligosaccharyl transferase, Glycosyltransferase, Membrane protein, O-antigen, Undecaprenyl-diphosphate, Lipid A-core oligosaccharide

1 Introduction

The lipopolysaccharide (LPS) is a major component of the outer leaflet of the outer membrane of gram-negative bacteria, and consists of lipid A, core oligosaccharide (OS), and O-specific polysaccharide (O antigen) [1]. LPS plays a major role as an elicitor of innate immune responses ranging from localized inflammation to disseminated sepsis [2]. The O antigen, which is the most surface-exposed LPS moiety, also contributes to bacterial pathogenicity by influencing macrophage recognition and resistance to the lytic action of the complement system, epithelial cell invasion, and intracellular survival [3–9]. O antigens are synthesized as lipid-linked sugar precursors; typically using a C_{55} polyisoprenyl known as undecaprenyl phosphate (Und-P), while the lipid A core is synthesized by an independent pathway [10], and eventually both components are joined at the periplasmic side of the inner membrane [11] by the WaaL (formerly RfaL) ligase [1, 12].

WaaL is an integral membrane protein that catalyzes the glycosidic bonding of a sugar at the proximal end of the undecaprenyl-diphosphate (Und-PP)-O-antigen with a terminal sugar of the lipid A-core oligosaccharide (OS). This is a specific glycosyl transfer reaction resulting in the release of the Und-PP attached to the

Inka Brockhausen (ed.), *Glycosyltransferases: Methods and Protocols*, Methods in Molecular Biology, vol. 1022, DOI 10.1007/978-1-62703-465-4_15,

O antigen polysaccharide. Mutant strains devoid of a functional *waaL* gene cannot ligate O antigen molecules to lipid A-core OS and produce LPS lacking O antigen polysaccharide, while accumulating membrane embedded Und-PP-linked O antigen molecules [11, 13]. WaaL proteins do not share any similarities with glycosyltransferases that use sugar nucleotides and the mechanism of ligation is unknown. A requirement for a specific lipid A-core OS acceptor structure has been established in several model systems [14–17], which has led to the generalized notion that a specific WaaL protein can recognize a cognate lipid A-core OS terminal structure [18, 19]. However, WaaL lacks specificity for the lipid-linked O antigen [1, 18, 20]. Non-O-antigen molecules originated from various biosynthesis pathways, such as colanic acid [21, 22] and enterobacterial common antigen (ECA) [23] can also be substrates for the ligase.

It is unclear how WaaL recognizes the lipid-linked O antigen and in particular, which part of the donor molecule participates in the enzymatic reaction. Conceivably, WaaL activity requires amino acids exposed to the periplasmic space where they could interact with the donor and acceptor molecules. A critical histidine, which is conserved in many WaaL proteins, was identified in a periplasmic loop of several WaaL proteins [14, 17, 24]. All ligases possess a relatively large periplasmic loop [17, 24–26]. A tri-dimensional structural model of this loop in the *Escherichia coli* WaaL protein was recently proposed, which may be part of a putative catalytic region [24]. More recently, Ruan et al. [27] utilized an in vitro assay to demonstrate that ligation with purified *Escherichia coli* WaaL occurs without ATP and magnesium ions, and confirmed the critical role of conserved, periplasmic arginine and histidine residues. Only one report suggested that the O antigen ligase requires ATP hydrolysis [14], but others could not confirm these results [27–29]. Current biochemical and in silico data strongly suggest that WaaL proteins use a common reaction mechanism that shares features of metal ion-independent inverting glycosyltransferases [27]. This chapter provides detailed protocols for WaaL purification and a ligase assay (*see* **Note 1**) with native substrate (Fig. 1a) that can also be adapted for use with synthetic substrates.

2 Materials

2.1 LPS Preparation

1. SCM3(pJHCV32): SCM3 is an *Escherichia coli* K-12 derived from strain SØ874 [30] [*lacZ trp* Δ(*sbcB-rfb*) *upp rel rpsL*] that contains a deleted O antigen ligase gene (Δ*waaL*) [24]. pJHCV32 is a cosmid encoding the gene cluster for the synthesis of O7 antigen [31].
2. Phosphate-buffered saline, pH 7.2: tenfold dilute the Premixed PBS Buffer (10×) (Roche Diagnostics, Indianapolis, IN, USA) with deionized water and autoclave.

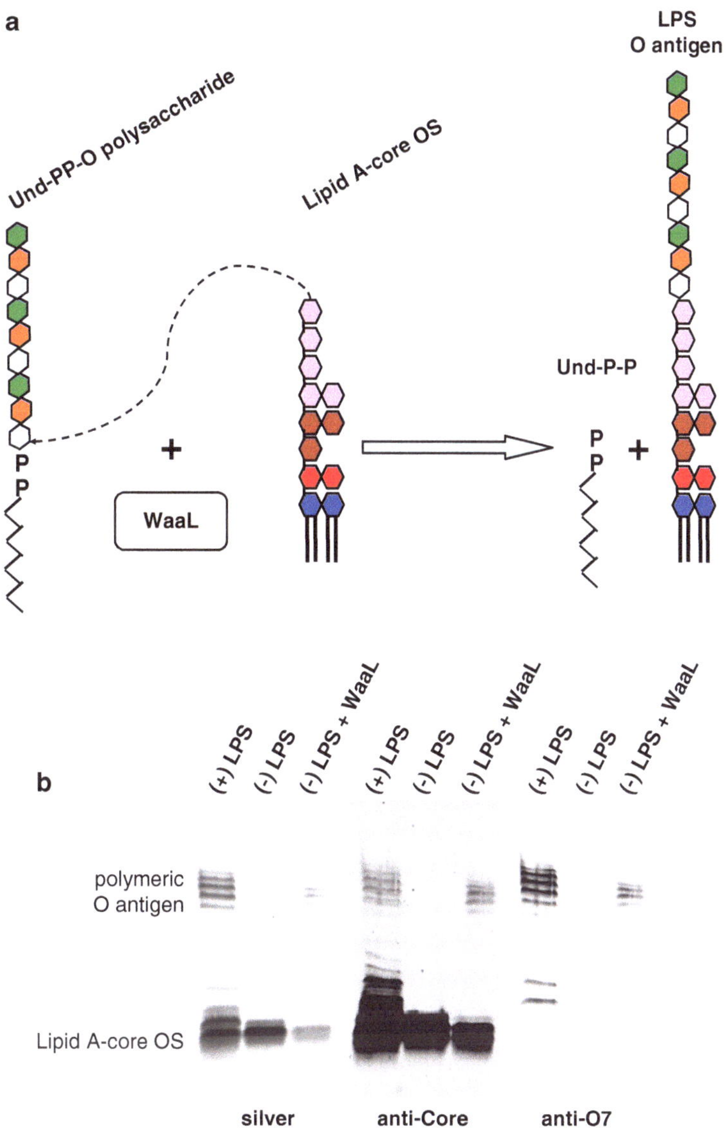

Fig. 1 (**a**) Diagram of the ligation reaction. The components are Und-PP-linked O7 antigen produced in SCM3 with pJHCV32, lipid A-core OS acceptor provided by the same strain, and purified WaaL. Purified WaaL was obtained from the strain JM105v, which has a truncated lipid A-core OS. (**b**) Analyses of in vitro ligase assays (modified from Ruan et al. [27]). (+) LPS, indicates LPS produced in vivo by the SCM3(pJHCV32/pXR1) strain expressing parental WaaL; (–) LPS, indicates LPS produced in vivo by SCM3(pJHCV32) in the absence of WaaL; (–) LPS + WaaL, indicates the product of in vitro ligation after 16 h incubation of purified WaaL and LPS prepared from SCM3(pJHCV32). The in vitro ligation was carried out with 0.67 μM of purified WaaL in 50 mM phosphate buffer, pH 6.0, at 37 °C. Samples were analyzed by SDS-PAGE followed by silver staining (silver), and immunoblot with anti-lipid A-core OS-specific mAb (anti-core) or polyclonal anti-O7-specific antibodies (anti-O7), as indicated. The regions of the gel corresponding to O antigen polymeric bands and the lipid A-core OS bands are also indicated

3. Lysis buffer: 2 % sodium dodecylsulfate (SDS), 4 % β-mercaptoethanol, 10 % glycerol, 1 M Tris–HCl, pH 6.8.
4. Proteinase K: stock solution of 20 mg/mL in deionized water (Roche Diagnostics).
5. 90 % phenol: 90 % phenol containing 0.1 % β-mercaptoethanol and 0.2 % 8-hydroxyquinoline (Fisher Scientific, Waltham, MA, USA).
6. Ethyl ether saturated with 10 mM of Tris–HCl and 1 mM of ethylenediamine tetraacetic acid (EDTA), pH 8.0.
7. Microcentrifuge (Model 5415D, Eppendorf, Hauppauge, NY, USA).

2.2 Purification of WaaL Protein

1. JM105v(pXR1): JM105v is an *E. coli* JM105 derivative with a spontaneous mutation resulting in a truncated lipid A-core OS, and pXR1 plasmid is pBAD24 derivative encoding $WaaL_{FLAG\text{-}10xHis}$ [27].
2. L-Arabinose: stock solution of 20 % (w/v) in deionized water, filtered through a 0.22-μm filter (Sigma-Aldrich, St. Louis, MO, USA).
3. 0.9 % saline: dissolve 9 g NaCl in 1 L of deionized water and autoclave.
4. 50 mM Tris–HCl, pH 7.5: 20-fold dilute 1 M Tris–HCl, pH 7.5 with deionized water.
5. 1 M Tris–HCl, pH 7.5: dissolve 60.6 g Tris–HCl in 450 mL of deionized water, drop HCl to adjust pH to 7.5, add more deionized water to a total volume of 500 mL, and autoclave.
6. 25× protease inhibitor: dissolve 1 tablet of complete, EDTA-free Protease Inhibitor Cocktail Tablets (Roche Diagnostics) in 2 mL of deionized water.
7. Equilibration buffer (300 mL): 50 mM phosphate buffer, pH 8.0, 150 mM NaCl, 10 % glycerol, 0.05 % (w/v) *n*-Dodecyl-β-D-Maltoside (DDM) (Sigma-Aldrich).
8. Ni^{2+} resin: charge the Chelating Sepharose Fast Flow (Amersham Biosciences, Pittsburgh, PA, USA) with nickel (Ni^{2+}) as described by the manufacturer's manual with minor modifications (*see* **Note 2**).
9. 1 M imidazole: dissolve 1.36 g of imidazole in 20 mL of Equilibration buffer.
10. Washing buffer (100 mL): mix 95 mL of Equilibration buffer with 5 mL of 1 M imidazole.
11. Elution buffer (40 mL): mix 30 mL of Equilibration buffer with 10 mL of 1 M imidazole.
12. 20-mL 10-KDa Vivascience filter (Fisher Scientific).

13. JA-10 rotor, 70 Ti rotor (Beckman Coulter Canada, Inc., Mississauga, Ontario Canada).
14. Rocking Platform (VWR International, Mississauga, Ontario Canada).
15. Roller (Labquake Shaker, Barnstead Thermolyne, Sigma-Aldrich).

2.3 In Vitro Ligase Assay

1. LPS substrates: LPS prepared from SCM3(pJHCV32) as described in Subheading 3.1.
2. Purified WaaL: WaaL protein purified from JM105v/(pXR1) as described in Subheading 3.2.
3. 0.2 M phosphate buffer, pH 6.0.
4. Proteinase K: stock solution of 2 mg/mL in deionized water.
5. 90 % phenol: 90 % phenol containing 0.1 % β-mercaptoethanol, 0.2 % 8-hydroxyquinoline (Fisher Scientific), and 10 % 0.1 M Tris–HCl, pH 8.0.

2.4 LPS Detection

1. Acrylamide mixture (49.5 % T, 6 % C): dissolve 23.25 g of acrylamide and 1.5 g of bis-acrylamide in deionized water to a final volume of 50 mL (Sigma-Aldrich).
2. Acrylamide mixture (49.5 % T, 3 % C): dissolve 24.0 g of acrylamide and 0.75 g of bis-acrylamide in deionized water to a final volume of 50 mL.
3. Gel buffer: 3.0 M Tris, 0.3 % SDS, pH 8.45.
4. Cathode buffer: 0.1 M Tris, 0.1 M Tricine, 0.1 % SDS, pH 8.25 (loaded in the upper chamber).
5. Anode buffer: 0.2 M Tris, pH 8.9 (loaded in the bottom chamber).
6. 3× loading dye: 0.825 M Tris–HCl, pH 6.8, 6 % (w/v) SDS, 30 % (v/v) glycerol, 15 % (v/v) β-mercaptoethanol, and 0.03 % (w/v) bromophenol blue (Sigma-Aldrich).
7. Fixing solution (200 mL): Mix 120 mL of methanol, 20 mL of acetic acid, and 60 mL of deionized water.
8. 7.5 % acetic acid (200 mL): Mix 15 mL of acetic acid with 185 mL deionized water.
9. 0.2 % periodic acid: dissolve 0.4 g of periodic acid in 200 mL of deionized water in a glass flask.
10. Staining solution: Dissolve 1 g of $AgNO_3$ in 5 mL of deionized water. Add 2.8 mL of 1 M NaOH and then 1 mL of ammonium into 120 mL of deionized water in a glass cylinder. In a fume hood, add slowly the ammonium and the silver nitrate simultaneously in the above glass cylinder, and swirl the cylinder to clear the precipitation. Keep adding until all the 5 mL of $AgNO_3$ are dissolved in the cylinder. Add deionized water to a total volume of 150 mL.

11. Developer: dissolve 10 mg of citric acid in 180 mL of deionized water. Add 60 μL of 37 % formaldehyde. Mix with 20 mL of methanol.
12. Nitrocellulose membrane (Whatman, GE Health Care Life Sciences, Baie d'Urfe, Quebec,Canada).
13. TBS (Tri-buffered saline): 50 mM Tris, 150 mM NaCl, pH 7.5.
14. 10 % Western Blocking Solution: dilute the Western Blocking Solution (Roche Diagnostics) with TBS to 10 % (v/v).
15. Antibody dilution buffer: dilute the Western Blocking Solution with TBS to 5 % (v/v).
16. Anti-O7 polyclonal rabbit antiserum: 1:1,000 (v/v) dilute the antiserum with the antibody dilution buffer (The antiserum was obtained by immunization at the University of Western Ontario following protocols approved for the ethical use of animals).
17. Anti-core LPS WN1 222-5 mAb (HyCult Biotechnology, Hycult Biotech, Uden, The Netherlands): monoclonal antibody of mouse anti-core LPS epitope. 1:5,000 (v/v) dilute the antibody with the antibody dilution buffer.
18. IRDye800CW affinity-purified anti-rabbit IgG antibodies (Rockland, Gilbertsville, PA, USA): 1:10,000 (v/v) dilute the antibodies with the antibody dilution buffer.
19. Alex Fluor® 680 anti-mouse IgG (Invitrogen, Carlsbad, CA, USA): 1:10,000 (v/v) dilute the antibodies with the antibody dilution buffer.
20. Odyssey infrared imaging system (Li-cor Biosciences, Lincoln, NE, USA).

3 Methods

3.1 Preparation of LPS Substrates for the Ligase Assay

LPS was prepared as described previously [32] with minor modifications, as follows:

1. Grow SCM3(pJHCV32) on LB agar plates containing 20 μg/mL of tetracycline at 37 °C overnight.
2. With a sterile cotton swab suspend bacteria in 2 mL phosphate-buffered saline, pH 7.2.
3. Measure the optical density of the suspension at 600-nm (OD600) and adjust turbidity to 3.0 in 1.5 mL cell suspension.
4. Centrifuge in a microcentrifuge at 16,000 × *g* for 1 min.
5. Resuspend cell pellets in 150 μL of lysis buffer and heat at 100 °C for 10 min.

6. Add 10 μL of Proteinase K stock solution. Vortex and incubate at 60 °C for 180 min (*see* **Note 3**).
7. Add 150 μL of pre-warmed (55 °C) 90 % phenol and incubate for 15 min at 70 °C with vortexing every 5 min.
8. Place on ice for 10 min.
9. Centrifuge in a microcentrifuge at 16,000 × *g* for 10 min and then transfer the upper aqueous phase to a clean tube.
10. Add 500 μL of ethyl ether saturated with Tris–EDTA to the clear phase and mix (*see* **Note 4**).
11. Centrifuge in a microcentrifuge at 16,000 × *g* for 1 min and aspirate the top ether phase (*see* **Note 5**).
12. Quantify the isolated LPS by Purpald assay for 2-keto-3-deoxy-octulosonic acid (Kdo) [32, 33].

3.2 Purification of WaaL Protein from E. coli

1. Grow JM105v(pXR1) in LB agar plates at 37 °C overnight. Inoculate one colony into 5 mL of LB containing 100 μg/mL of ampicillin (Ap) and incubate overnight with shaking at 37 °C.
2. Inoculate the 5 mL culture into 1 L of LB/Ap to reach an initial OD_{600} of 0.02, and incubate at 37 °C for 4 h with shaking until culture reaches an OD_{600} of 0.5. Add 10 mL of 20 % (w/v) L-arabinose to a final concentration of 0.2 % (w/v), and continue incubation with shaking at 37 °C for another 3 h until the culture reaching an OD_{600} of approximately 0.9.
3. Incubate culture on ice for 10 min. Use JA-10 rotor to spin down the 1 L of cells in four centrifuge tubes at 10,000 × *g* for 10 min at 4 °C.
4. Resuspend cell pellets with 30 mL of ice-cold 0.9 % saline. Spin down at 10,000 × *g* for 10 min at 4 °C (*see* **Note 6**).
5. Resuspend the cell pellets with 15 mL 50 mM Tris–HCl, pH 7.5, containing 0.6 mL 25× protease inhibitor.
6. Lyse cells in a French press at 1,500 psi for three times and spin down the lysate at 10,000 × *g* for 20 min at 4 °C.
7. Transfer the supernatant into one ultracentrifuge tube. Add more 50 mM Tris–HCl, pH 7.5 (~7 mL) to fill the tubes. Ultracentrifuge in a 70 Ti rotor at 1,00,000 × *g*, 4 °C, for 1 h.
8. Remove the supernatant. Add 1 mL 50 mM Tris–HCl, pH 7.5, to wash the membrane pellets. Repeat washing five times.
9. Resuspend the membrane pellets with 1 mL of 50 mM Tris–HCl, pH 7.5.
10. Mix all the above membrane suspension with 8 mL Equilibration buffer containing 1 % DDM.
11. Gently mix with a rocker in a cold room for 4 h.

12. Aliquot the mix in six microcentrifuge tubes. Spin at 16,000 × *g* at room temperature for 10 min (*see* **Note 7**). Combine the supernatant in one 15-mL Falcon tube. Leave on ice to wait for mixing with Ni^{2+}resin. (This is called DDM-membrane extract).
13. Mix the DDM-membrane extract with the Ni^{2+}-resin with a roller in a cold room for 2 h.
14. In a cold room, load the DDM-membrane extract/Ni^{2+}-resin mixture into a 1.5-cm diameter, 20-mL glass column. Let the resin pack by gravity to reach a bed volume of 1.5 mL. Remove the bottom cap of the column. Collect about 9 mL of flow-through. Add 100 mL of Washing buffer containing 50 mM imidazole to wash the resin. Add 40 mL of Elution buffer containing 250 mM imidazole to elute the protein.
15. Protein concentration. Add 10 mL of Equilibration buffer in a 20-mL 10-kDa Vivascience filter. Centrifuge at 3,000 × *g* for 10 min at 4 °C. Discard the filtrate (around 5 mL) and un-filtered Equilibration buffer. Add the above eluate in this pretreated Vivascience filter. Centrifuge at 3,000 × *g* for about 140 min until the concentrated volume is less than 500 μL.
16. Measure protein concentration by Bradford assay. The total amount is about 1 mg. Prepare 20-μL aliquots in microcentrifuge tubes. Store at −80 °C for usage (*see* **Note 8**).

3.3 *In Vitro* Ligase Assay

1. Mix 6 μL of the LPS prepared from SCM3(pJHCV32) which corresponds to 10 μmol of Kdo equivalents, 2 μg of purified WaaL (*see* **Note 9**), and 15 μL of 0.2 M phosphate buffer, pH 6.0 (*see* **Note 10**). Add water to reach a reaction volume of 60 μL.
2. Incubate at 37 °C for 16 h.
3. Terminate the reaction by adding 1 μL of 2 mg/mL of Proteinase K to a final concentration of 0.03 mg/mL. Vortex and incubate at 60 °C for 20 min (*see* **Note 11**).
4. Add 60 μL of pre-warmed (55 °C) 90 % phenol and incubate at 70 °C for 15 min with vortexing every 5 min.
5. Place on ice for 10 min.
6. Centrifuge in a microcentrifuge at 16,000 × *g* for 10 min. Transfer the upper aqueous phase to a clean tube (*see* **Note 12**). Typically, the collected volume is 40 μL, from which 20 μL is used for silver stain and the rest for Western blotting.

3.4 LPS Detection

The LPS products can be detected with Tricine-SDS-PAGE, silver staining and Western blotting.

3.4.1 Tricine-SDS- PAGE

1. Prepare 14 % Running gel. Mix 2.84 mL of acrylamide mixture (49.5 % T, 6 % C), 3.32 mL of Gel buffer, 1.04 mL of glycerol, and 2.78 mL of water. Add 20 μL of 10 % ammonium persulfate and 12 μL of TEMED, and cast gel in an 8 cm × 10 cm × 1 mm gel cassette. Add a layer of isopropanol. Let polymerize for 40 min.
2. Prepare the stacking gel. Mix 0.25 mL of acrylamide mixture (49.5 % T, 3 % C), 0.775 mL of Gel buffer, and 2.1 mL of water. Add 38 μL of 10 % ammonium persulfate and 10 μL of TEMED. Layer the running gel with the stacking solution. Place the comb and let polymerize for 30 min.
3. Set up the gel apparatus and add the cathode buffer to the upper chamber and the anode buffer to the lower chamber.
4. Mix 20 μL of the upper aqueous phase with 5 μL of 3× loading dye (*see* **Note 13**). Load all the 25 μL of sample mixture into each well and run at 50 V until the dye reaches the running gel (30–40 min), then switch to 130 V. Run the samples for 30 min after the dye has left the gel.

3.4.2 Silver Stain

Silver stain for LPS follows the procedure as described previously [32].

1. Soak the gel in 200 mL of fresh fixing solution overnight.
2. Discard the fixing solution and wash the gel with 7.5 % acetic acid for 30 min.
3. Discard the acetic acid, add 200 mL of 0.2 % periodic acid and rock for 30 min.
4. Wash the gel with deionized water for an hour, changing the water every 15 min.
5. Add the staining solution and rock for 15 min.
6. Pour out the staining solution and wash the gel with deionized water for 45 min, changing the water every 15 min.
7. Add the developer and rock until brown bands appear (around 5 min).
8. Stop the development with several changes of deionized water.
9. Place the gel in a plastic sleeve to scan.

3.4.3 Western Blotting

1. Transfer the SDS-PAGE gel onto nitrocellulose transfer membrane under 250 mA for 70 min.
2. Place the membrane into 10 % Western Blocking Solution and rock for 1 h at room temperature.
3. Incubate the membrane overnight at 4 °C with either of the following primary antibodies: anti-O7 polyclonal rabbit antiserum at a 1:1,000 dilution to detect O7 polysaccharide, or anti-core

LPS WN1 222-5 mAb at a 1:5,000 dilution to detect LPS core (*see* **Note 14**).

4. Wash the membrane with 20 mL of TBS. Repeat five times.
5. Incubate the membrane at room temperature for 2 h with either of the following secondary antibodies: IRDye800CW affinity-purified anti-rabbit IgG antibodies at a 1:10,000 dilution to detect O7 polysaccharide, or Alex Fluor® 680 anti-mouse IgG at a 1:10,000 dilution to detect LPS core (*see* **Note 14**).
6. Wash the membrane with 20 mL of TBS. Repeat five times.
7. Detect reacting bands by fluorescence with an Odyssey infrared imaging system.

4 Notes

1. The assay and procedures described here were initially used with the *E. coli* K-12 WaaL protein but they can also be employed for other ligases. For example, the authors have used similar protocols to purify and assay the *Pseudomonas aeruginosa* WaaL protein.
2. Take 2.5 mL of Chelating Sepharose Fast Flow in one 15-mL Falcon tube. Add 12 mL of deionized water; resuspend and spin at 1,000 × *g* for 2 min. Keep resin pellets. Repeat three times. The resulting resin pellets are in 1.5 mL. Add 1.5 mL of 0.2 M $NiSO_4$ in each tube. Mix with a roller at room temperature for 40 min. Spin at 1,000 × *g* for 2 min. Keep resin pellets. Add 12 mL of deionized water; resuspend and spin at 1,000 × *g* for 2 min. Keep resin pellet. Repeat three times. Add 10 mL of Equilibration buffer; resuspend and spin at 1,000 × *g* for 2 min. Keep resin pellet. Repeat three times. The resulting resin is called Ni^{2+}-resin.
3. The incubation step at 60 °C can last longer, even overnight.
4. This step will remove any phenol left in the aqueous phase.
5. The prepared LPS can be kept at −20 °C for as long as 1 year.
6. The cell pellets can be stored at −20 °C for 2–3 days before being lysed.
7. We find that spinning in a microcentrifuge at room temperature for 10 min gives a clear separation between the brown colored supernatant and the white colored pellet. The protein prepared by this method is functional despite of such a treatment at room temperature.
8. The WaaL proteins prepared by this method can be stored at −80 °C for several months. When we use a tube of proteins, which have been stored at −80 °C over 1 year for in vitro ligase

assay, we find that the protein is still functional, close to a freshly prepared one.

9. The final concentration of WaaL equals to 0.67 μM. The molecular weight (MW) of the WaaL construct with a C-terminal FLAG-His10× tag is 50,794.8 Da as determined by mass spectrometry [26, 34]. For simplification, we consider the MW of WaaL as 50 kDa. The concentration of WaaL with 2 μL in 60 μL of reaction is calculated to be 0.67 μM.
10. The ligase assay also works well at pH 7.0 and pH 8.0. The amounts of the product LPS at pH 6.0, pH 7.0, and pH 8.0 after 16 h of incubation at 30 or 37 °C are at the same level. However, at pH 6.0, the amount of the product LPS after 30 min at 37 °C is significantly higher than that at pH 7.0 and pH 8.0, which indicates that a pH with 6.0, similar to the periplasmic pH, favors the ligation reaction [27].
11. We use a 2 mg/mL of Proteinase K in the in vitro assay instead of 20 mg/mL in the preparation of LPS from whole cells. Adding 1 μL of 2 mg/mL of Proteinase K in the 60 μL of reaction and then incubating at 60 °C for 20 min are enough to destroy the WaaL protein, since we cannot see any protein bands in the silver staining which is also sensitive to proteins [27].
12. Typically, from 60 μL of reaction, we can collect 40 μL from the upper phase, all of which is used for further electrophoresis analysis. So, to save samples, we set a 60 μL of reaction volume.
13. Mixing 20 μL of the upper aqueous phase with 5 μL of 3× loading dye results in a sample mixed in 0.6× loading dye. The maximum loading volume in a well is 25 μL. We load all the 25 μL of sample mixture into each well. So in each well, we actually load 20 μL of the upper aqueous phase in a purpose to maximize the sample amount. Since this is a full loading, it is very crucial to run first at 50 V until the dye reach the running gel.
14. The LPS product contains the LPS core and O7 polysaccharide; therefore, both antibodies can recognize the product. To avoid the binding advantage of the polyclonal antibodies over the monoclonal antibody, we incubate the membrane with the two primary antibodies separately. First, we incubate the membrane with anti-core LPS mAb overnight at 4 °C and then after washing incubate the membrane at room temperature for 2 h with Alex Fluor® 680 anti-mouse IgG. Second, after washing we incubate the same membrane with anti-O7 polyclonal rabbit antiserum overnight at 4 °C and then after washing incubate the membrane at room temperature for 2 h with IRDye800CW anti-rabbit IgG antibodies before imaging.

Acknowledgments

This work was supported by a grant from the Canadian Institutes of Health Research to M.A.V., who also holds a Canada Research Chair in Infectious Diseases and Microbial Pathogenesis.

References

1. Raetz CRH, Whitfield C (2002) Lipopolysaccharide endotoxins. Annu Rev Biochem 71:635–700
2. Opal SM (2007) The host response to endotoxin, antilipopolysaccharide strategies, and the management of severe sepsis. Int J Med Microbiol 297:365–377
3. Duerr CU, Zenk SF, Chassin C, Pott J, Gutle D, Hensel M, Hornef MW (2009) O-antigen delays lipopolysaccharide recognition and impairs antibacterial host defense in murine intestinal epithelial cells. PLoS Pathog 5:e1000567
4. Murray GL, Attridge SR, Morona R (2003) Regulation of *Salmonella typhimurium* lipopolysaccharide O antigen chain length is required for virulence; identification of FepE as a second Wzz. Mol Microbiol 47:1395–1406
5. Murray GL, Attridge SR, Morona R (2006) Altering the length of the lipopolysaccharide O antigen has an impact on the interaction of *Salmonella enterica* serovar Typhimurium with macrophages and complement. J Bacteriol 188:2735–2739
6. Saldías MS, Ortega X, Valvano MA (2009) *Burkholderia cenocepacia* O antigen lipopolysaccharide prevents phagocytosis by macrophages and adhesion to epithelial cells. J Med Microbiol 58:1542–1548
7. Paixao TA, Roux CM, den Hartigh AB, Sankaran-Walters S, Dandekar S, Santos RL, Tsolis RM (2009) Establishment of systemic *Brucella melitensis* infection through the digestive tract requires urease, the type IV secretion system, and lipopolysaccharide O antigen. Infect Immun 77:4197–4208
8. Bengoechea JA, Najdenski H, Skurnik M (2004) Lipopolysaccharide O antigen status of *Yersinia enterocolitica* O:8 is essential for virulence and absence of O antigen affects the expression of other *Yersinia* virulence factors. Mol Microbiol 52:451–469
9. West NP, Sansonetti P, Mounier J, Exley RM, Parsot C, Guadagnini S, Prevost MC, Prochnicka-Chalufour A, Delepierre M, Tanguy M, Tang CM (2005) Optimization of virulence functions through glucosylation of *Shigella* LPS. Science 307:1313–1317
10. Valvano MA (2011) Common themes in glycoconjugate assembly using the biogenesis of O-antigen lipopolysaccharide as a model system. Biochemistry (Mosc) 76:729–735
11. McGrath BC, Osborn MJ (1991) Localization of the terminal steps of O-antigen synthesis in *Salmonella typhimurium*. J Bacteriol 173: 649–654
12. Valvano MA, Furlong SE, Patel KA (2011) Genetics, biosynthesis and assembly of O antigen. In: Knirel Y, Valvano MA (eds) Bacterial lipopolysaccharides: structure, chemical synthesis, biogenesis and interaction with host cells. Springer, Vienna, Austria, pH 275–310
13. Mulford CA, Osborn MJ (1983) An intermediate step in translocation of lipopolysaccharide to the outer membrane of *Salmonella typhimurium*. Proc Natl Acad Sci USA 80: 1159–1163
14. Abeyrathne P, Daniels C, Poon KK, Matewish MJ, Lam J (2005) Functional characterization of WaaL, a ligase associated with linking O-antigen polysaccharide to the core of *Pseudomonas aeruginosa* lipopolysaccharide. J Bacteriol 187:3002–3012
15. Heinrichs DE, Monteiro MA, Perry MB, Whitfield C (1998) The assembly system for the lipopolysaccharide R2 core-type of *Escherichia coli* is a hybrid of those found in *Escherichia coli* K-12 and *Salmonella enterica*. Structure and function of the R2 WaaK and WaaL homologs. J Biol Chem 273: 8849–8859
16. Heinrichs DE, Yethon JA, Amor PA, Whitfield C (1998) The assembly system for the outer core portion of R1- and R4-type lipopolysaccharides of *Escherichia coli*. The R1 core-specific β-glucosyltransferase provides a novel attachment site for O-polysaccharides. J Biol Chem 273:29497–29505
17. Schild S, Lamprecht AK, Reidl J (2005) Molecular and functional characterization of O antigen transfer in *Vibrio cholerae*. J Biol Chem 280:25936–25947
18. Kaniuk NA, Vinogradov E, Whitfield C (2004) Investigation of the structural requirements in the lipopolysaccharide core acceptor for ligation of O antigens in the genus *Salmonella*: WaaL "ligase" is not the sole determinant of

acceptor specificity. J Biol Chem 279:36470–36480
19. Olsthoorn MM, Petersen BO, Schlecht S, Haverkamp J, Bock K, Thomas-Oates JE, Holst O (1998) Identification of a novel core type in *Salmonella* lipopolysaccharide. Complete structural analysis of the core region of the lipopolysaccharide from *Salmonella enterica* sv. Arizonae O62. J Biol Chem 273:3817–3829
20. Heinrichs DE, Valvano MA, Whitfield C (1999) Biosynthesis and genetics of lipopolysaccharide core. In: Brade H, Morrison DC, Vogel S, Opal S (eds) Endotoxin in health and disease. Marcel Dekker, New York, pH 305–330
21. Meredith TC, Mamat U, Kaczynski Z, Lindner B, Holst O, Woodard RW (2007) Modification of lipopolysaccharide with colanic acid (M-antigen) repeats in *Escherichia coli*. J Biol Chem 282:7790–7798
22. Sperandeo P, Lau FK, Carpentieri A, De Castro C, Molinaro A, Dehò G, Silhavy TJ, Polissi A (2008) Functional analysis of the protein machinery required for transport of lipopolysaccharide to the outer membrane of *Escherichia coli*. J Bacteriol 190:4460–4469
23. Barr K, Klena J, Rick PD (1999) The modality of enterobacterial common antigen polysaccharide chain lengths is regulated by *o349* of the *wec* gene cluster of *Escherichia coli* K-12. J Bacteriol 181:6564–6568
24. Pérez JM, McGarry MA, Marolda CL, Valvano MA (2008) Functional analysis of the large periplasmic loop of the *Escherichia coli* K-12 WaaL O-antigen ligase. Mol Microbiol 70:1424–1440
25. Islam ST, Taylor VL, Qi M, Lam JS (2010) Membrane tpology mapping of the O-antigen flippase (Wzx), polymerase (Wzy), and ligase (WaaL) from *Pseudomonas aeruginosa* PAO1 reveals novel domain architectures. mBio 1:e00189-00110–e00189-00119
26. Pan Y, Ruan X, Valvano MA, Konermann L (2012) Validation of protein topology models by oxidative labeling and mass spectrometry. J Am Soc Mass Spectrom 23:889–898
27. Ruan X, Pérez JM, Marolda CL, Valvano MA (2012) The WaaL O-antigen lipopolysaccharide ligase has features in common with metal ion-independent inverting glycosyltransferases. Glycobiology 22:288–299
28. Han W, Wu B, Li L, Zhao G, Woodward R, Pettit N, Cai L, Thon V, Wang PG (2012) Defining the function of lipopolysaccharide O-antigen ligase WaaL using chemoenzymatically synthesized substrates. J Biol Chem 287:5357–5365
29. Hug I, Couturier MR, Rooker MM, Taylor DE, Stein M, Feldman MF (2010) *Helicobacter pylori* lipopolysaccharide is synthesized via a novel pathway with an evolutionary connection to protein *N*-glycosylation. PLoS Pathog 6:e1000819
30. Neuhard J, Thomassen E (1976) Altered deoxyribonucleic pools in P2 eductants of *Escherichia coli* K-12 due to deletion of the *dcd* gene. J Bacteriol 126:999–1001
31. Valvano MA, Crosa JH (1989) Molecular cloning and expression in *Escherichia coli* K-12 of chromosomal genes determining the O7 lipopolysaccharide antigen of a human invasive strain of *E. coli* O7:K1. Infect Immun 57:937–943
32. Marolda CL, Lahiry P, Vinés E, Saldías S, Valvano MA (2006) Micromethods for the characterization of lipid A-core and O-antigen lipopolysaccharide. Methods Mol Biol 347:237–252
33. Lee CH, Tsai CM (1999) Quantification of bacterial lipopolysaccharides by the purpald assay: measuring formaldehyde generated from 2-keto-3-deoxyoctonate and heptose at the inner core by periodate oxidation. Anal Biochem 267:161–168
34. Pan Y, Konermann L (2010) Membrane protein structural insights from chemical labeling and mass spectrometry. Analyst 135: 1191–2000

Chapter 16

Functional Identification of Bacterial Glucosyltransferase WbdN

Yin Gao, Anna Vinnikova, and Inka Brockhausen

Abstract

The outer membrane of gram-negative bacteria is stabilized by lipopolysaccharides (LPS). The O-antigenic polysaccharides of LPS are composed of repeating units that are exposed to and can interact with the environment. The glycosyltransferases that assemble these repeating units are encoded by the O-antigen gene cluster and utilize undecaprenol-phosphate-linked intermediates as natural acceptor substrates, and nucleotide sugars as donor substrates on the cytoplasmic face of the inner membrane. Many of the glycosyltransferase genes are known but the enzymatic functions of most of them remain to be identified. We describe here how the function of a recombinant glucosyltransferase WbdN from *Escherichia coli* O157 can be determined by NMR analysis of the enzyme product, using a synthetic acceptor substrate analog. A fluorescent acceptor substrate analog can be used in highly sensitive enzyme assays that allow the characterization of enzyme activity without the use of radioactive nucleotide sugar donor substrates.

Key words Glucosyltransferase, WbdN, NMR, Linkage analysis, Fluorescent acceptor substrate

1 Introduction

The outer membrane of gram-negative bacteria has a thick coat of lipopolysaccharides (LPS) that consist of polymerized oligosaccharide repeating units (O-antigen) attached to the outer oligosaccharide core which is attached to lipid A [1]. The O-antigen is exposed to the external environment and plays a role not only in antigenic variability but also in host defenses [2, 3]. O-antigen structures vary according to the corresponding glycosyltransferase genes in the O-antigen gene cluster. Many of these genes are known but the enzymatic activities and functions of very few enzymes have been identified. The characterization of these putative glycosyltransferases remains a major challenge.

The synthesis of the first sugar of the O-antigen repeating unit starts with the transfer of sugar-phosphate from nucleotide sugar to undecaprenol-phosphate at the cytoplasmic side of the inner

Inka Brockhausen (ed.), *Glycosyltransferases: Methods and Protocols*, Methods in Molecular Biology, vol. 1022, DOI 10.1007/978-1-62703-465-4_16, © Springer Science+Business Media New York 2013

membrane to form sugar-diphosphate-undecaprenol (*see* Chapter 14) [4]. The second step is usually strain-specific. We have characterized a number of glycosyltransferases that transfer the second sugar residue to GlcNAcα-diphosphate-phenoxyundecyl as the acceptor analog of sugar-diphosphate-undecaprenol [5–7]. These glycosyltransferases require the diphosphate group in the acceptor for full activity [8]. The availability of a synthetic acceptor substrate GlcNAcα-PO_3-PO_3-phenoxyundecyl [9] has made it possible to characterize these enzymes and determine the specific linkage formed.

The O157:H7 strain of *Escherichia coli* (*E. coli*) is responsible for outbreaks of severe diarrhea and hemolytic-uremic syndrome due to the production of Shiga-like toxin. The O157-antigen has the repeating unit structure [-2-D-Rha4NAcα1-3-L-Fucα1-4-D-Glcβ1-3GalNAcα1-].

The second step of the assembly pathway involves β1,3-Glucosyltransferase WbdN [10]. In this protocol, we describe the biochemical identification of the β1,3-Glucosyltransferase activity from *E. coli* O157 encoded by the *wbdN* gene in the O157-antigen gene cluster. In order to prove the biochemical function of the enzyme, the structure of the enzyme product is determined using GalNAcα-diphosphate-phenoxyundecyl as the acceptor substrate. The linkage of the newly added Glc residue in the WbdN enzyme product Glcβ1-3GalNAcα-diphosphate-phenoxyundecyl can be identified by NMR analysis of the enzyme product and establishes the specific enzyme activity. In addition, a new fluorescent acceptor substrate analog was used in highly sensitive glycosyltransferase assays. This analog can serve to eliminate the use of radioactive nucleotide sugar donor substrate in characterization studies of the enzyme. The nonradioactive enzyme product can serve as an acceptor substrate for enzymes that act subsequently in the O-antigen synthesis pathway.

2 Materials

2.1 Preparation of Bacterial Enzyme Homogenate

1. *E. coli* BL21 bacteria harboring the pET28a expression plasmid containing *wbdN* (Gao et al. [10]) (*see* **Note 1**).
2. Luria–Bertani (LB) broth: 0.01 g of tryptone/mL (Fisher Scientific, Toronto, Ontario, Canada), 0.05 g of yeast extract/mL (Fluka Biochemika-Sigma-Aldrich, Oakville, Ontario, Canada), 0.01 g of NaCl, adjusted to pH 7.5 with NaOH.
3. Special additives: Kanamycin, isopropyl-thio-galactoside (IPTG) (Sigma).
4. Shaker water bath at 37 °C, autoclave, sterile culture tubes, flasks, and pipettes.
5. Centrifuge J2-21M, JA-17 rotor (Beckman, Mississauga, ON, Canada) with 50 mL sterile centrifuge tubes.

6. Microcentrifuge.
7. Sonicator (Sonic Dismembrator Model 100, Fisher Scientific).
8. Sonication buffer: 10 mM Na_2HPO_4, 300 mM NaCl, 2 mM $MgCl_2$, 0.1 mM ethylene glycol tetraacetic acid (EGTA), protease inhibitor cocktail (Sigma), 0.2 mM ATP, 5 mM 2-mercaptoethanol, adjusted to pH 8.0 with NaOH.
9. Washing solution: phosphate-buffered saline (PBS, pH 7.4).
10. Storage solution: PBS with 10 % glycerol.
11. Protein dye (Bio-Rad, Mississauga, Ontario) and 4 mg/mL of bovine serum albumin (BSA, Sigma) as a protein standard.

2.2 Glucosyltransferase Assay

1. Acceptor substrate: 2 mM GalNAcα-PO_3-PO_3-$(CH_2)_{11\text{-O-Ph}}$ (**640**) is prepared according to Utkina et al. [11] (*see* **Note 2**).
2. Donor substrate: 5 mM UDP-Glc (Sigma).
3. Buffer: 0.5 M 2-(*N*-morpholino) ethanesulfonic acid (MES) buffer (pH 7).
4. Cofactor: 5 mM $MnCl_2$.
5. Enzyme homogenate containing 50–80 μg protein (*see* **Note 3**).

2.3 Purification of Enzyme Product

1. C18 Sep-Pak cartridge (Waters Corporation, Mississauga, ON, Canada) activated with 4 mL methanol, followed by 10 mL of deionized water.
2. High-performance liquid chromatography (HPLC) system: two pumps, controller, manual injector, ultraviolet absorbance monitor (Waters), recorder, fraction collector, C18 column, 250 × 4.60 mm 10 μm (Phenomenex, Torrance, CA, USA).
3. Organic solvents: methanol, acetonitrile (Caledon Laboratories, Georgetown, ON, Canada).
4. Rotary evaporator.
5. Lyophilizer.

2.4 Analysis of Enzyme Product by Mass Spectrometry

1. Mass spectrometer: Qstar XL QqTOF mass spectrometer with matrix assisted laser desorption ionization (MALDI) source. Electrospray Ionization source run in the negative ion mode.
2. Matrix: DHB (2,5-dihydroxybenzoic acid).

2.5 Analysis of Enzyme Product by NMR

1. D_2O (Sigma).
2. NMR spectrometer: Bruker Avance 600-MHz for ^1H, ^{13}C, and ^{31}P NMR spectroscopy.

2.6 Glucosyltransferase Assay Using Fluorescent Substrate

1. Assay reagents as described above in Subheading 2.2.
2. Fluorescent acceptor substrate replacing substrate **640**: 2.5 mM GalNAcα-PO_3-PO_3-$(CH_2)_{11}$-O-CH_2-9-anthracenyl **724**, synthesized as described in Vinnikova et al. [12].

3. Fluorescence spectrometer.
4. Protein assay components: Bradford dye reagent (Bio-Rad, Hercules, CA, USA), Bovine serum albumin protein standard (Sigma).

3 Methods

3.1 Preparation of Bacterial Enzyme Homogenate

1. Prepare LB broth in sterile culture tubes and flasks with specific antibiotic (50 μg/mL Kanamycin).
2. Start an overnight culture by transferring bacteria into a sterile tube containing 5 mL LB broth. Incubate in a shaker water bath at 37 °C. Gently shake at 150 rpm (*see* **Note 4**).
3. Transfer the 5 mL overnight culture into a flask with 125 mL pre-warmed LB broth. Cover the flask with a cotton plug and two layers of aluminum foil. Incubate in the shaker water bath at 37 °C for about 90 min or until the absorbance at 600 nm reaches 0.8.
4. Add IPTG to the bacterial culture to obtain a final concentration of 1 mM. Incubate for 4 h at 37 °C to induce protein expression (*see* **Note 5**).
5. Harvest bacteria by centrifugation in capped 50 mL tubes for 15 min at 6,000 × *g*. Discard supernatant. Resuspend the bacterial pellet in phosphate-buffered saline (PBS), pH 7.4, and centrifuge again.
6. Resuspend bacterial pellet in 10 mL PBS containing 10 % glycerol and store in 1 mL aliquots at −20 °C for enzyme activity assays (*see* **Note 6**).
7. Thaw bacterial suspension, centrifuge in microcentrifuge at 10,000 × *g* for 5 min, and wash with 1 mL PBS.
8. Suspend bacteria in 0.5 mL cold sonication buffer and sonicate for 15 s two times at setting 3, with 2 min intervals. Keep the bacterial suspension on ice at all times (*see* **Note 7**).
9. Determine the protein concentration of the bacterial homogenate with the Bio-Rad (Bradford) protein assay using bovine serum albumin (0–80 μg protein) to establish the calibration curve (*see* **Note 8**).

3.2 Large Scale Production of WbdN Product Using Nonradioactive UDP-Glc

The glucosyltransferase assay conditions should be adjusted according to prior enzyme characterization to achieve the maximal yield of enzyme product and optimal conversion of precious acceptor substrate to product without major degradation and side reactions.

1. The total volume of the standard assay mixtures is 40 mL (1,000 times the volume of a small scale radioactive assay) [10].

The incubation mixture contains: 0.025 mM acceptor substrate **640** (2 μmol), 5 mM $MnCl_2$, 0.125 M MES buffer, pH 7, 10 mL enzyme homogenate, 0.46 mM UDP-Glc (*see* **Note 9**).

2. The mixture is incubated for 30 min at 37 °C, and reactions quenched by addition of 20 mL of ice-cold water (*see* **Note 10**).

3.3 Purification of Enzyme Product

1. The reaction product is isolated using 60 Sep-Pak C18 columns. 1 mL of the mixture is added to each column. Water-soluble components of the assay are eluted with 5 mL water. Enzyme product is then eluted with 3 mL methanol (*see* **Note 11**).
2. Methanol elution fractions containing acceptor and product are combined, dried by rotary evaporation and resuspended in a minimal volume of water (2 mL).
3. A small aliquot (10 μL) of the methanol elution fractions is submitted to mass spectrometry analysis to ensure the presence of enzyme product (*see* **Note 12**).
4. Further purification of the enzyme product is achieved by reversed-phase HPLC using a C18 column [10] with an acetonitrile–water mixture (28/72 %) as the mobile phase at a flow rate of 1 mL/min. Aliquots of 150 μL are injected. The retention time of the acceptor substrate is about 50 min, and that of the enzyme product is 35 min which is sufficient for separation. Elution is monitored by measuring absorbance at 195 nm and using standard compounds, for example acceptor substrate and radioactive enzyme products (*see* **Note 13**).
5. Fractions of 2 mL are collected and the product peaks are pooled after comparison of the elution patterns. Pooled fractions that contain the enzyme product are dried by rotary evaporation to remove acetonitrile, followed by lyophilization.

3.4 Analysis of Enzyme Product by MALDI-TOF

1. A small amount of dried enzyme product is dissolved in 20 μL of water/methanol (1/1). One μL of this sample is mixed with 1 μL matrix DHB (2,5-dihydroxybenzoic acid). A sample of the mixture (0.5 μL) is loaded directly onto the MALDI target plate, followed by drying at room temperature (*see* **Note 14**).
2. Enzyme product is analyzed by matrix-assisted laser desorption ionization (MALDI) MS in the negative-ion mode. Spectra are obtained by submitting each spot to pulsed nitrogen laser shots over a range of m/z ratio from 400 to 2,000. The laser spot should be moved to different areas of the target during acquisition, and spectra should be averaged until a satisfactory signal-to-noise ratio is obtained. A large peak at m/z of 788 (for substrate plus 162) indicates that one glucose (hexose) residue has been added to the acceptor substrate to form the product.

3.5 Analysis of Enzyme Product by NMR

Mass spectrometry can determine the molecular weight of the enzyme product, the number of sugar residues added, and gives a measure of its purity. To establish the type of anomeric linkage synthesized by the enzyme, NMR remains the primary tool for analysis.

1. Dissolve 1 μmol of the sample in 0.6 mL D_2O in an NMR tube (*see* **Note 15**).
2. Experiment 1: ^1H NMR spectroscopy with 16 scans is performed first and saved as Experiment 1. This provides the information on the numbers of different hydrogens and neighboring hydrogens present in the enzyme product (Table 1, Fig. 1).
3. Experiment 2: After determining the chemical shift of the anomeric protons, COSY with four scans is performed to assign ^1H resonances with the effect of strong coupling in 2D spectra (*see* **Note 16**). Figure 2 shows the two routes for the determination of the resonances in Glc and GalNAc.
4. Experiment 3: HSQC (16 scans) experiment is required for the further determination of carbohydrate structures (Fig. 3). It provides sensitivity of ^{13}C chemical shifts with proton correlation.

Table 1
^1H and ^{13}C NMR parameters of WbdN enzyme product Glcβ1-3GalNAcα-O-PO_3-PO_3-$(CH_2)_{11}$-O-Ph and substrate GalNAcα-O-PO_3-PO_3-$(CH_2)_{11}$-O-Ph for comparison

	^1H (ppm)	^{13}C (ppm)	^1H (ppm)	^{13}C (ppm)
Residue	Substrate		Product	
Glc-1			4.48 $J_{1,2}=7.8$	104.0
Glc-2			3.20	72.9
Glc-3			3.42	75.4
Glc-4			3.33	75.4
Glc-5			3.34	69.2
Glc-6			3.67, 3.75	60.3
GalNAc-1	5.39	94.6	5.53	94.6
GalNAc-2	4.13	49.8	4.31	48.3
GalNAc-3	3.78–3.88	66.9	3.98	77.4
GalNAc-4	3.90–3.94	67.3	4.19	68.6
GalNAc-5	4.07	72.1	4.13	71.6
GalNAc-6	3.58–3.69	61.1	3.65, 3.70	60.9

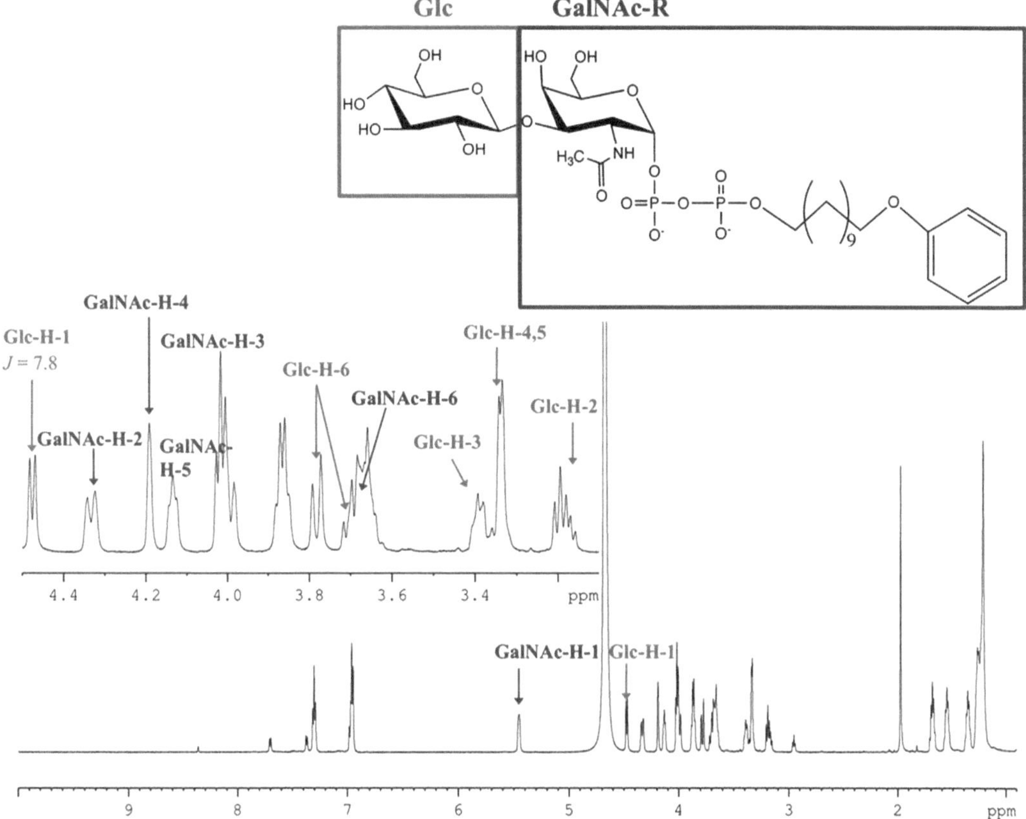

Fig. 1 One-dimensional ^{1}H-NMR spectrum of WbdN product Glcβ1-3GalNAcα-PO_3-PO_3-phenoxyundecyl, and assignments based on two-dimensional NMR analyses

Assigned ^{13}C chemical shifts are used for the carbohydrate linkage analysis in HMBC.

5. Experiment 4: High resolution (*see* **Note 17**) of heteronuclear multiple bond correlation (HMBC) and through-space correlation experiments are used to confirm the assigned ^{1}H and ^{13}C chemical shifts. The long range coupling between Glc and GalNAc provides the linkage information (Fig. 4).
6. Experiment 5: NOESY (eight scans) experiment is performed to confirm the linkage between Glc and GalNAc (Fig. 5). The exact 90° pulse is determined before performing each experiment. The selected mixing time for the sample is 0.7 s (*see* **Note 18**).

3.6 Glucosyltransferase Assay Using Fluorescent Acceptor Substrate

The fluorescent acceptor substrate **724**, P^1-11-(9-anthracenylmethoxy)undecyl, P^2-2-acetamido-2-deoxy-α-D-galactopyranosyl diphosphate sodium salt (GalNAc-PP-AntrU), has the same GalNAcα-PO_3-PO_3-undecyl- structure as substrate **640** but has an additional fluorescent 9-anthracenyl moiety. The WbdN reaction using **724** is shown in Fig. 6.

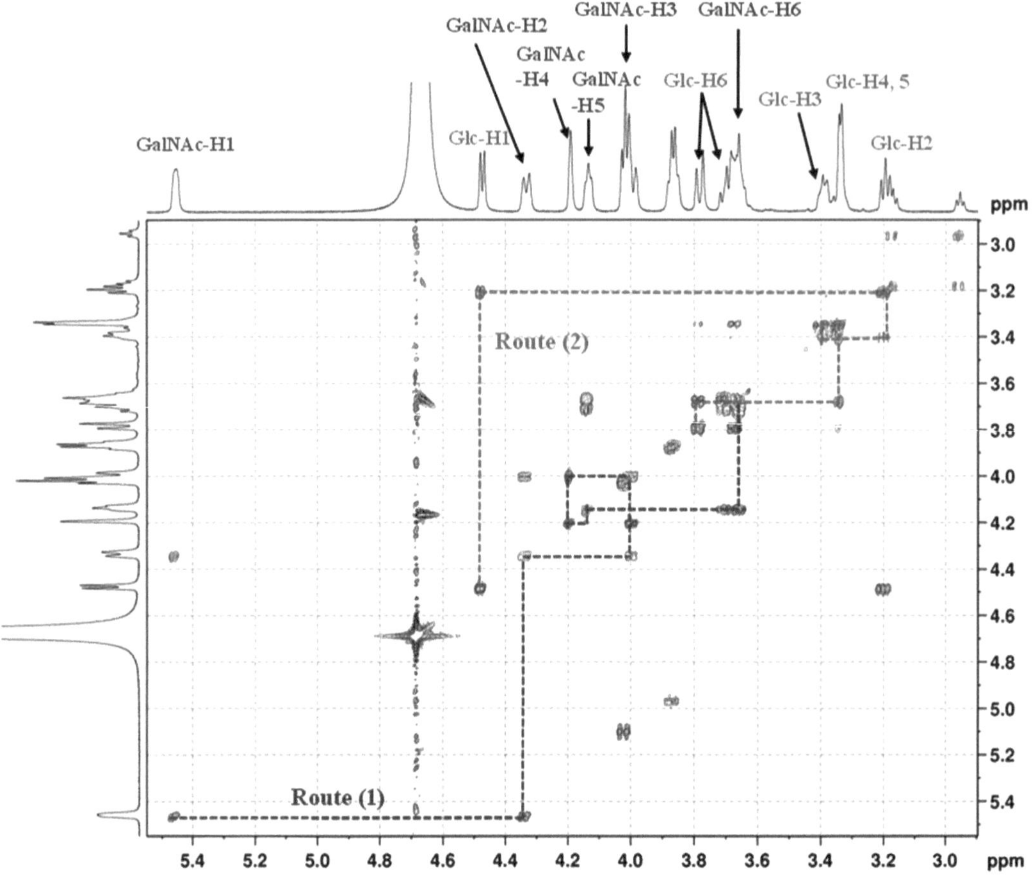

Fig. 2 COSY spectrum of WbdN enzyme product Glcβ1-3GalNAcα-PO_3-PO_3-phenoxyundecyl. The *dashed lines* indicate the protons that are spin–spin coupled. Once the anomeric protons are assigned, the indicated routes allow the determination of the adjacent protons

1. Prepare the assay mixture in a 40 μL total volume containing 0.25 mM of the sodium salt of the synthetic acceptor **724**, GalNAc-PP-AntrU in aqueous solution, 12.5 mM $MnCl_2$, 0.125 M of MES buffer, pH 7.0, 0.46 mM UDP-Glc (*see* **Note 19**) and 10 μL of enzyme homogenate. All assays are done in duplicate. As a background control two assays lack acceptor substrate.
2. Mixtures are incubated for 10 min (*see* **Note 20**). The reactions are stopped by the addition of 0.7 mL of ice-cold water and freezing.
3. The incubation mixture is applied to a recycled C18 Sep-Pak column at flow rate of not more than 2 mL/min. The column is washed with 5 mL of water to elute the hydrophilic compounds in 1 mL fractions (*see* **Note 21**). The hydrophobic product and substrate are eluted with 3 mL methanol, collected in 1 mL fractions.

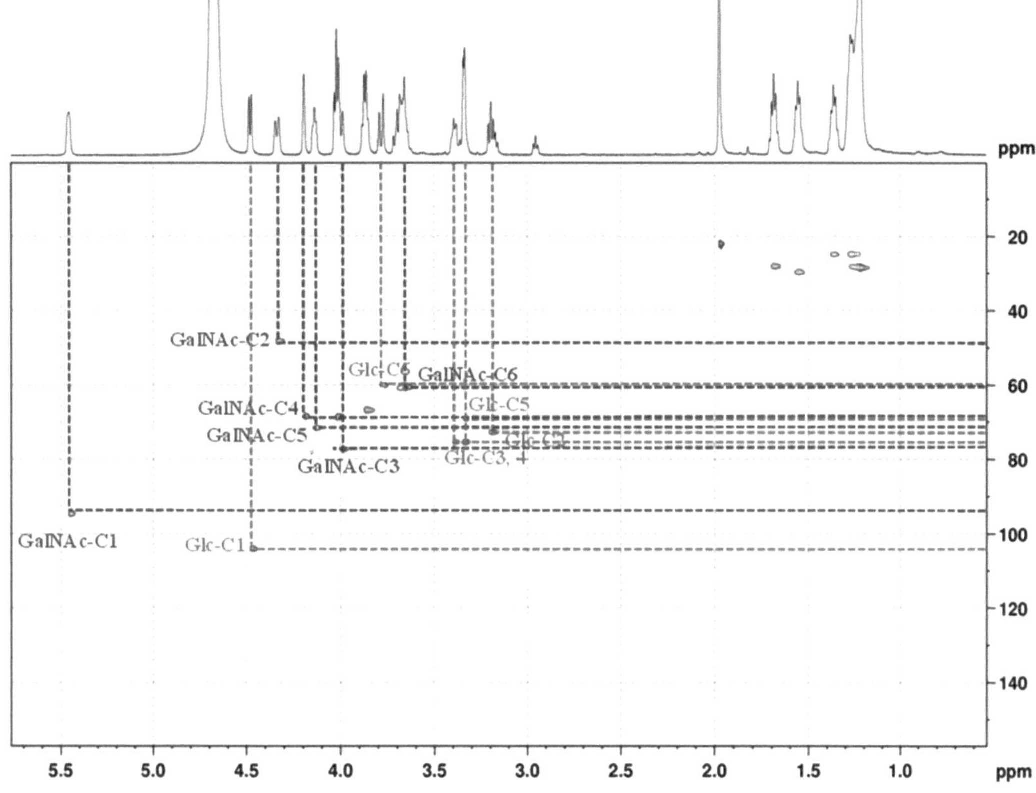

Fig. 3 HSQC experiment of WbdN enzyme product Glcβ1-3GalNAcα-PO$_3$-PO$_3$-phenoxyundecyl. The *dashed lines* indicate signals of the aliphatic carbons and the corresponding protons attached

4. After use, the C18 Sep-Pak columns are regenerated by washing them with 6 mL of methanol and then with 10 mL of water at flow rate 2 mL/min. The columns can be reused at least ten times.
5. Fractions eluted with methanol from Sep-Pak columns containing enzymatic product are concentrated by drying (*see* **Note 22**).
6. Methanol elution fractions are taken up in 120 μL water for HPLC analysis as described in Subheading 3.3. The mobile phase is acetonitrile–water 35/65. To monitor elution the absorbance at 247 nm is measured. The GalNAc-PP-AntrU acceptor substrate elutes at 38 min (Fig. 7, *see* **Note 23**).
7. Fractions of 1 mL are collected and analyzed with a fluorescence spectrometer. The fluorescence intensity for every fraction is measured (excitation wavelength 250 nm, emission wavelength 415 nm). Absorbance and fluorescence intensity are recorded and the nmol of enzyme product present in each fraction calculated based on the standard curve obtained for substrate alone in the same solvent (Fig. 7, *see* **Note 24**).

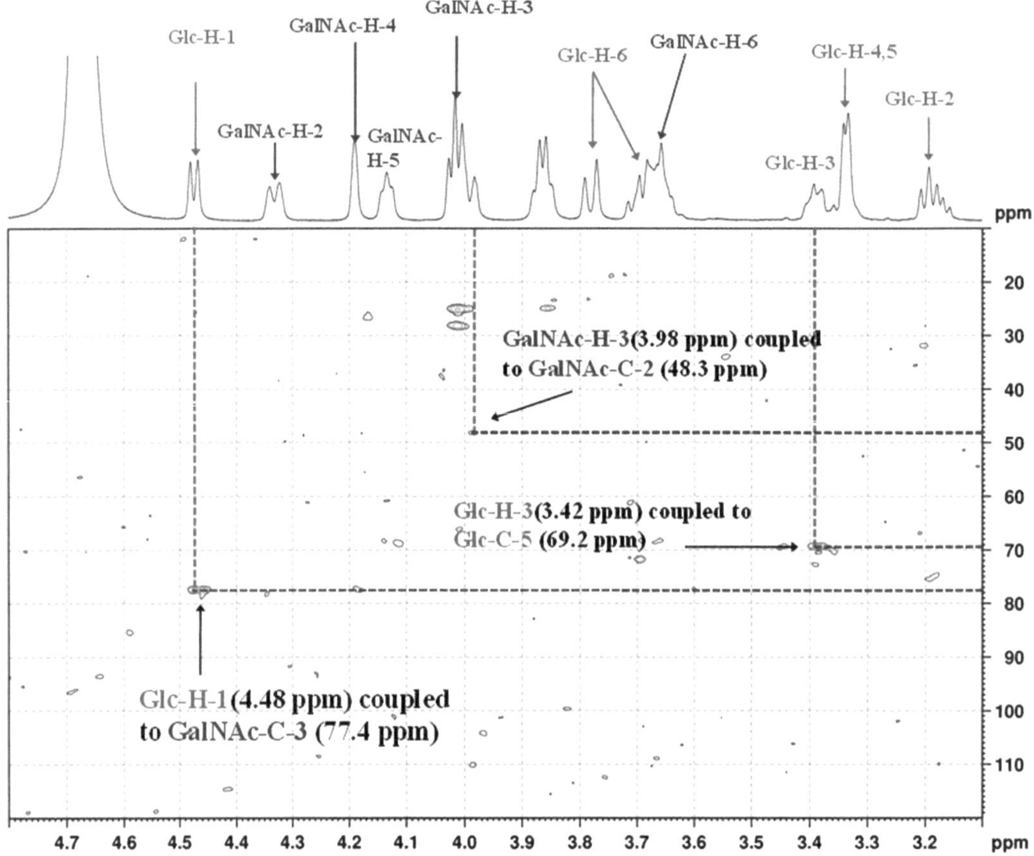

Fig. 4 HMBC experiment of WbdN enzyme product Glcβ1-3GalNAcα-PO_3-PO_3-phenoxyundecyl. The *dashed lines* show proton–carbon J-coupling due to protons three bonds away from the carbon

8. Determine the protein concentration in a 1:5 dilution of the bacterial homogenate (or purified protein) using a standard curve established with 1–80 g bovine serum albumin. The enzyme activity is then calculated as nmol/min incubation/mg protein in the assay.

4 Notes

1. The enzyme has a His_6 tag at the C-terminus and can be purified using a Nickel-Nitrilo-triacetate (Ni-NTA) column, eluted with imidazole.
2. The original synthesis of the acceptor analog GlcNAc-PO_3-PO3-phenoxyundecyl was described in Montoya et al. [9]. Modifications of this method have been made to produce both the GlcNAc- and the GalNAc-derivatives. Depending on the type of phosphate salt, the compound may be water-soluble or

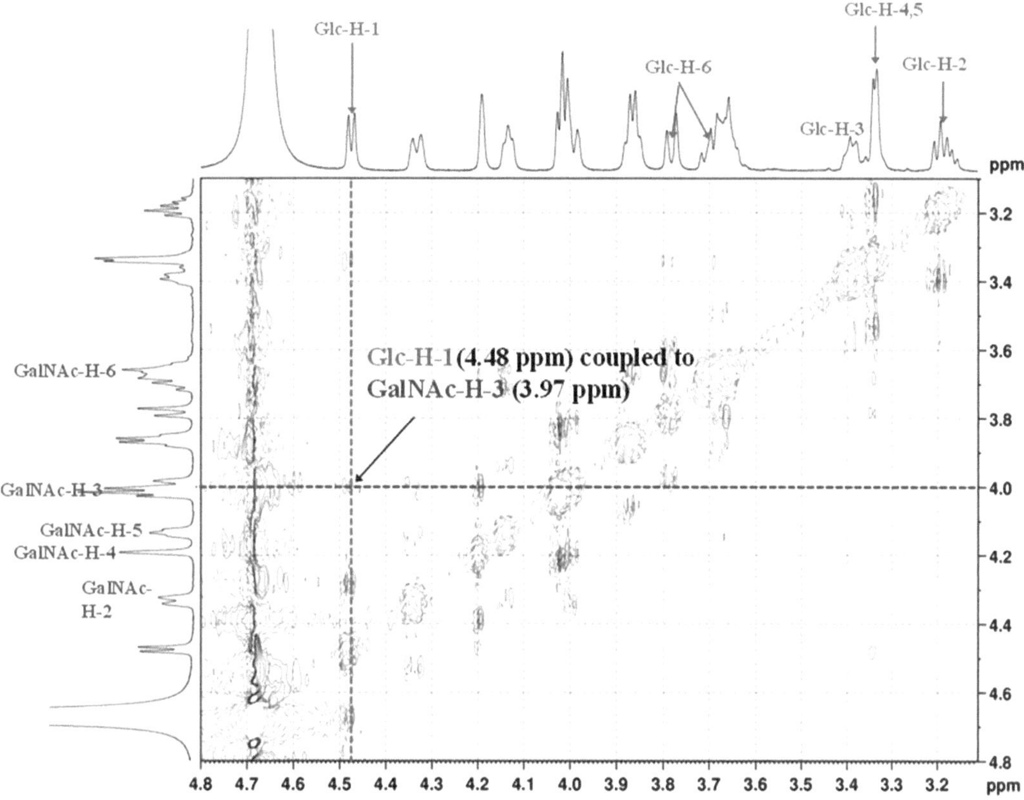

Fig. 5 NOESY experiment of WbdN enzyme product Glcβ1-3GalNAcα-PO_3-PO_3-phenoxyundecyl. *Dashed lines* indicate the proton–proton coupling through-space interactions via nuclear Overhauser effect

Fig. 6 Glucosyltransferase (Glc-T) WbdN reaction using the synthetic fluorescent acceptor substrate analog P^1-11-(9-anthracenylmethoxy)undecyl, P^2-2-acetamido-2-deoxy-α-D-galactopyranosyl diphosphate disodium salt and radioactive UDP-[^{3}H]Glc

may require methanol to be dissolved. The same concentration of methanol should then be present in all assays. 10 % methanol in the assay is usually well tolerated by these glycosyltransferases.

3. The enzyme homogenate is useful for making large amounts of enzyme product. However, purified enzyme may also be used for this purpose and this may make the product purification easier.

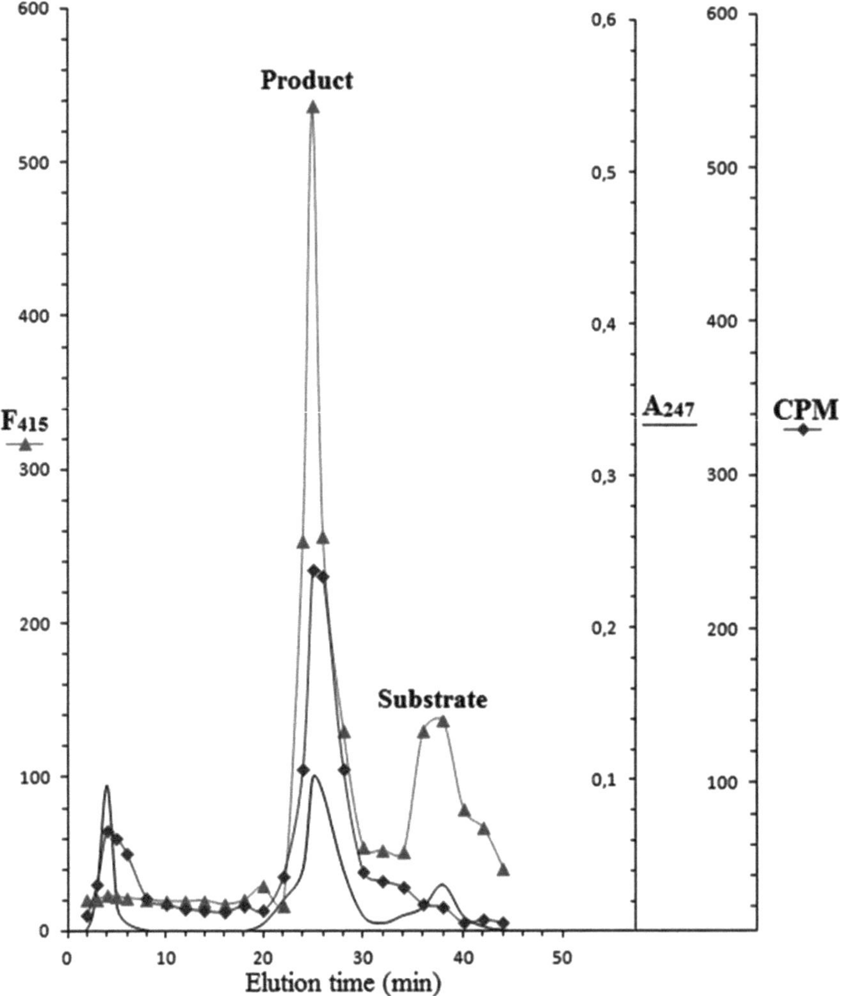

Fig. 7 HPLC separation after WbdN assays using the fluorescent acceptor substrate. The *graph* shows the absorbance at 247 nm, and radioactivity (CPM) and fluorescence intensity at 415 nm (F415) in HPLC fractions. Substrate was added as an internal standard to the sample to visualize its elution

4. The temperature and time of bacterial growth may vary. Bacteria can be taken from −80 °C bacterial stock, from bacteria in agar gels, or from bacteria in LPB broth, soaked on filter paper. A piece of the filter paper can be placed in the sterile tube for bacterial growth. If contaminations occur, growth on agar plates to isolate pure colonies may be helpful.
5. The concentration of IPTG, the time and temperature may be varied according to the expression levels of the protein that can be determined by SDS-PAGE and Western blots (*see* Chapters 3 and 12) using anti-His-tag antibodies.
6. Bacteria suspended in PBS/10 % glycerol yield active enzyme after storage for at least 6 months at −20 °C. About 1 g bacteria pellet per 15 mL will give a suitable enzyme concentration.

7. A number of different sonication buffers can be used. Sucrose (0.05 M) is also suitable to keep the enzyme active after homogenization. The time of sonication can be increased to solubilize more enzyme.
8. Usually, a 1:5 dilution of the homogenate (in water) yields protein concentrations that lie within the initial part of the calibration curve.
9. The concentration of the acceptor can be increased to yield more product, but the conversion of substrate to product may be lower. Thus, when only small amounts of acceptor are available, it is wise to keep the concentration low. The optimal incubation time and conditions should be determined first in small scale radioactive assays.
10. The amount of water added can be increased to ensure solubility of substrate and product.
11. The substrate is often not detected after HPLC separation.
12. Since the conversion of substrate to product is always <100 %, both, substrate (m/z 626) and product (m/z 788) are present. The type of phosphate salt cannot be seen from the MS data.
13. The standards for HPLC are GlcNAc (25 nmol) to ensure all parts of the system are working, substrate to establish a convenient retention time, and radioactive enzyme product to ensure a good separation between product and other assay components and to confirm the identity of the nonradioactive product peak. Radioactive enzyme product can be prepared by using UDP-[^{3}H]Glc (American Radiolabeled Chemicals, St. Louis, MO, USA) which is diluted with nonradioactive UDP-Glc to a specific activity of 2,000–4,000 cpm/nmol. Standards can also be added to the sample (as internal standards) to confirm the retention time which may be slightly different in complex samples, compared to pure compounds.
14. If the signal is too low, load another 0.5 μL sample of the mixture to enhance the signal.
15. The concentration of the sample in D_2O is critical. Since the nature of this enzyme product allows it to form micelles in aqueous solution that lead to a line broadening effect, less concentrated sample is recommended (Fig. 8). Also, it is important that no insoluble materials are present that cause poor homogeneity of the magnetic field. One-dimensional 600 MHz ^{1}H NMR spectrum and the two-dimensional $^{1H\text{-}1H}$ correlation spectroscopy (COSY), ^{1}H-^{1}H total correlated spectroscopy (TOCSY), and ^{1}H-^{13}C heteronuclear single quantum coherence (HSQC) NMR experiments are required for distinguishing and assigning the ^{1}H resonances in the enzyme product. The linkage information of enzyme product can be determined through two-dimensional heteronuclear multiple bond correlation (HMBC) and nuclear Overhauser effect spectroscopy

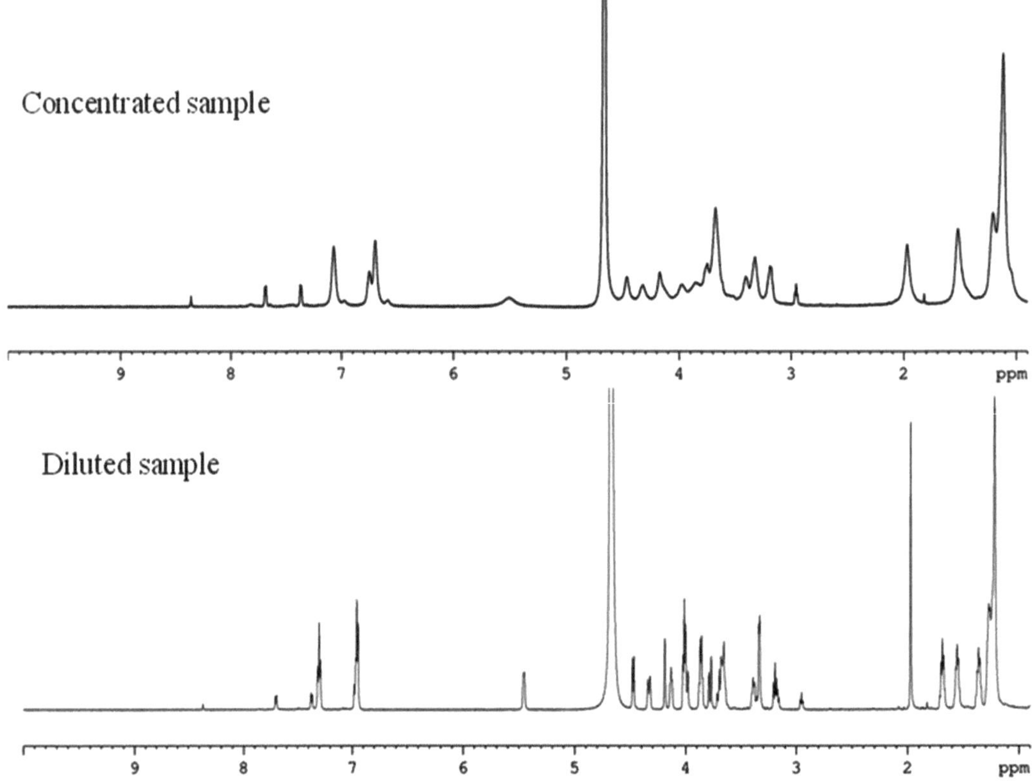

Fig. 8 The proton NMR spectrum of WbdN enzyme product shows broad signals in the concentrated D_2O solution but sharper signals upon dilution

(NOESY) experiments. Most of the ^1H carbohydrate resonances are clustered in the region of 3.0–6 ppm. The signals of anomeric protons are usually located between 4 ppm and 6 ppm with the beta configuration at 4–4.7 ppm and alpha configuration at 4.9–6 ppm. The importance of determining the chemical shift and integrated intensity of the anomeric protons is that it can reveal the number of monosaccharide residues in the sample and can indicate whether the sample contains a mixture of different carbohydrates. The anomeric configuration can be determined based on the magnitude of coupling constant ($J_{1,2}$). The value of 7–9 Hz is associated with a β-configuration and the value of 2–4 Hz indicates an α-configuration. The spectra of the enzyme product Glcβ1-3GalNAcα-PO_3-PO_3-$(CH_2)_{11}$-O-Ph are shown in Figs. 1–5, Table 1. It is generally helpful to compare spectra with those of related carbohydrate structures.

16. The COSY experiment is performed with a special window set up to focus the region of sugar ^1H resonances, usually between 3 and 6 ppm.

17. Since ^{13}C NMR is less sensitive than ^{1}H NMR, HMBC experiment is performed with 32 scans. Higher scan numbers are recommended for less concentrated samples.
18. The mixing time for small molecules should be selected in the range of the relaxation time, whereas a smaller mixing time is recommended to avoid spin diffusion for a large molecule.
19. The concentration of the donor substrate can be increased to yield more enzyme product but should be kept constant when comparing enzyme properties. UDP-Glc can be radioactive [10] or nonradioactive.
20. The linearity of initial enzyme rates is important for the characterization of enzyme. 10 min is usually the maximal time ensuring linearity. The incubation time can be increased to 30 min for near-maximal production of enzyme product.
21. It is important to load on samples slowly (<1 mL per min). The capacity of the Sep-Pak columns depends also on the number of regeneration cycles.
22. Fractions can be concentrated by rotary evaporation, SpeedVac centrifuge, or drying under a stream of nitrogen, and lyophilization. A small amount of water is always present in the first methanol elution fraction which may require longer drying periods.
23. The elution time of substrate and product depends on the age and regeneration of the C18 column used. Often, substrate is not clearly visible after the assay, and therefore internal standard is added for HPLC analysis.
24. Any fluorescence spectrometer can be used. A standard curve to assess to amount of fluorescent compound is constructed using the substrate **724**. The fluorescence intensity of substrate and product at 415 nm is identical.

References

1. Raetz CRH, Whitfield C (2002) Lipopolysaccharide endotoxins. Annu Rev Biochem 71:635–700
2. Duerr CU, Zenk SF, Chassin C, Pott J, Gutle D, Hensel M, Hornef MW (2009) O-antigen delays lipopolysaccharide recognition and impairs antibacterial host defense inmurine intestinal epithelial cells. PLoS Pathog 5:e1000567
3. Sheng H, Lim JY, Watkins MK, Minnich SA, Hovde CJ (2008) Characterization of an *Escherichia coli* O157:H7 O-antigen deletion mutant and effect of the deletion on bacterial persistence in the mouse intestine and colonization at the bovine terminal rectal mucosa. Appl Environ Microbiol 74:5015–5022
4. Whitfield C (1995) Biosynthesis of lipopolysaccharide O antigens. Trends Microbiol 3:178–185
5. Riley JG, Menggad M, Montoya-Peleaz P, Szarek WA, Marolda CL, Valvano MA, Schutzbach JS, Brockhausen I (2005) The wbbD gene of E. coli strain VW187 (O7:K1) encodes a UDP-Gal: GlcNAc(alpha)-pyrophosphate-R (beta)1,3-galactosyltransferase involved in the biosynthesis of O7-specific lipopolysaccharide. Glycobiology 15:605–613
6. Brockhausen I, Hu B, Liu B, Lau K, Szarek WA, Wang L, Feng L (2008) Characterization of two UDP-GlcNAc: β1,3-glucosyltransferases from the *Escherichia coli* serotypes O56 and O152. J Bacteriol 190:4922–4932

7. Xu C, Liu B, Hu B, Han Y, Feng L, Allingham JS, Szarek WA, Wang L, Brockhausen I (2011) Biochemical characterization of UDP-Gal:GlcNAc-pyrophosphate-lipid β-1,4-Galactosyltransferase WfeD, a new enzyme from Shigella boydii typH 14 that catalyzes the second step in O-antigen repeating-unit synthesis. J Bacteriol 193:449–459
8. Riley JG, Xu C, Brockhausen I (2010) Synthesis of acceptor substrate analogs for the study of glycosyltransferases involved in the second step of the biosynthesis of O antigen repeating units. Carbohydr Res 345: 586–597
9. Montoya-Peleaz PJ, Riley JG, Szarek WA, Valvano MA, Schutzbach JS, Brockhausen I (2005) Identification of a UDP-Gal: GlcNAc-R galactosyltransferase activity in *Escherichia coli* VW187. Bioorg Med Chem Lett 15:1205–1211
10. Gao Y, Liu B, Strum S, Schutzbach JS, Druzhinina TN, Utkina NS, Torgov VI, Danilov LL, Veselovsky VV, Vlahakis JZ, Szarek WA, Wang L, Brockhausen I (2012) Biochemical characterization of WbdN, a β1,3-glucosyltransferase involved in O-antigen synthesis in enterohemorrhagic Escherichia coli O157. Glycobiology 22:1092–1102
11. Utkina NS, Danilov LL, Druzhinina TN, Veselovsky VV (2010) Simple synthesis of P1-(11-phenoxyundecyl)-P2-(2-acetamido-2-deoxy-α-D-galactopyranosyl)diphosphate. Russian J Bioorg Chem 36:783–785
12. Vinnikova AN, Druzhinina TN, Danilov LL, Utkina NS, Torgov VI, Veselovsky VV, Wang S, Liu B, Wang L, Brockhausen I (2013) Synthesis of a fluorescent acceptor substrate for glycosyltransferases involved in the assembly of O-antigens of enterohemorrhagic Escherichia coli O157 and O5. Carbohydr Res 366:17–24

Chapter 17

Methods for the *Pasteurella* Glycosaminoglycan Synthases: Enzymes that Polymerize Hyaluronan, Chondroitin, or Heparosan Chains

Paul L. DeAngelis

Abstract

Multiple glycosyltransferases (GTases) that produce glycosaminoglycans (GAGs) have been identified; several distinct putative architectures and catalytic abilities have been noted from microbes and vertebrates. Here the preparation and use of a class of enzymes (Class II) from the gram-negative bacterium, *Pasteurella multocida*, with utility in chemoenzymatic synthesis of GAGs is described. The analyses of sugar products include thin layer chromatography (TLC), matrix-assisted laser desorption ionization mass spectroscopy (MALDI-ToF MS), and polyacrylamide gel electrophoresis (PAGE). The related Chapter 18 is focused on larger molecular weight GAGs analyses as well as the Class I membrane streptococcal and mammalian HA synthases.

Key words Thin layer chromatography, Polyacrylamide gel electrophoresis, Mass spectroscopy, Oligosaccharide, Polysaccharide

1 Introduction

The prototypical *Pasteurella* GAG synthases that form the hyaluronan, chondroitin, or heparosan polysaccharides [1–3] share some similarities, but possess crucial differences as well (*see* Table 1). These catalysts have proven to be useful tools for constructing both oligosaccharides (from ~2 to 25 monosaccharide units) [4] and polysaccharides (from >25 to ~10,000 monosaccharide units or ~2,500 kDa) [5] in vitro. GAG synthases from microbes are more robust than their vertebrate counterparts in general. The various *Pasteurella* enzymes are well expressed in (~100 mg/L culture) in *Escherichia coli* and may be isolated in a relatively pure state due to chromatography based on fusion protein tags (e.g., maltose-binding protein or histidine$_6$). The enzyme activity may be monitored by many methods, but here we describe the analyses with focus on smaller molecular weight (MW) chains; the companion

Inka Brockhausen (ed.), *Glycosyltransferases: Methods and Protocols*, Methods in Molecular Biology, vol. 1022, DOI 10.1007/978-1-62703-465-4_17,

Table 1
GAG structures and *Pasteurella* enzymes

GAG	Enzyme	Backbone repeat structures	Size, *n*	Notes[a]
Hyaluronan, Hyaluronic Acid, HA	PmHAS	[-4-GlcUAβ1-3-GlcNAc-β1-]$_n$	~10^{2-4}	none
Chondroitin	PmCS	[-4-GlcUAβ-1-3-GalNAc-β1-]$_n$	~$0.5–8 \times 10^2$	*a, c*
Heparosan (unmodified Heparin)	PmHS1 PmHS2	[-4-GlcUAβ1-4-GlcNAc-α1-]$_n$	~$0.5–8 \times 10^2$	*a, b, c*

GlcUA glucuronic acid, *GlcNAc N*-acetylglucosamine, *GalNAc N*-acetylgalactosamine

[a]No modifications are known to occur during biosynthesis in bacteria, but in animals, sulfation can occur at the *O*- or the *N*-positions of the sugar ring as well as C-5 epimerization of the GlcUA to iduronic acid (*a*, *b*, or *c*, respectively)

Chapter 18 of Weigel et al. describes the methods for higher MW GAG chains. In addition to the *Pasteurella* enzyme, guidelines useful for preparing or testing novel enzymes are also presented.

2 Materials

In general, standard chemical, microbiological, and biochemical guidelines should be followed unless special care is noted. The plasmids used for preparing *Escherichia coli*-derived PmHAS, PmCS, and PmHS1 and two enzymes are described in refs. [2–4] (*see* **Notes 1** and **2**).

2.1 Recombinant Bacterial Culture and Enzyme Preparation Components

1. Luria Broth or Superior Broth (Athena Enzyme Systems, Baltimore, MD): Autoclaved as recommended for the volume of media. Typically, 0.3–0.4 L cultures in 2-L flasks are employed. After media has cooled and before use, drugs are added (here ampicillin and carbenicillin from 100 mg/mL stocks in dimethylformamide (DMF)).
2. Inducers: If using phage-lysin-containing *E. coli* hosts (*see* **Note 2**), prepare 1 M L-ARABINOSE in water stock and filter-sterilize (Sigma-Aldrich, St. Louis, MO, USA; make sure *not* to use D-arabinose because this isomer will not work!). For the referenced *Pasteurella* synthase methods, to induce target enzyme at mid-log phase growth, isopropyl-thio-galactoside (IPTG; filter sterilized 0.1 M stock in water; Sigma) is employed.
3. Lysis buffer: 50 mM Tris, pH 7.2, plus protease inhibitor cocktail (for PmHAS and PmCS, also add 1 M ethylene glycol; enzyme grade, Fisher, Fairlawn, NJ, USA).
4. Purification Buffers: For heparosan synthases, maltose-binding protein (MBP) tag system components are made according to

guidelines of manufactures (New England Biolabs, Ipswich, MA, USA); typically 20 mM maltose in 50 mM Tris, pH 7.2, is used for elution.

2.2 Chemoenzymatic Synthesis Components

1. UDP-sugar stocks: Prepare 10–100 mM solutions (note the actual assay content in manufacturer specifications and adjust mass used to compensate; Sigma) in 10–20 mM Tris or HEPES, pH 7–8. Store at −80 °C in small aliquots so that any one tube is not freeze-thawed more than 2–4 times. Thaw and mix thoroughly, then store on ice until used.
2. Divalent cation stocks: Prepare 1 M $MgCl_2$ or $MnCl_2$ stocks in water. Can be stored for 1–3 months at room temperature. Working stocks at lower concentrations (e.g., 10–50 mM) should be made fresh daily.
3. Enzyme stocks: *See* **step** 7 of Subheading 3.1 for details.
4. Water, Buffers, and Stabilizers: Use high quality reagents free of metals and microbes to prepare 1 M HEPES or Tris buffers (pH 7.2 is useful for all for *Pasteurella* GAG synthases) and other reagents. Use enzyme grade (or better) ethylene glycol.
5. Acceptors: HA oligosaccharide acceptors are available from Hyalose, LLC (Oklahoma City, OK, USA). HA or chondroitin may be prepared as in ref. [4], but heparosan acceptors may be prepared as described in ref. [2].

2.3 Normal Phase Thin Layer Chromatography Analyses Components

1. Silica TLC plates (0.25 mm aluminum-backed plates; Whatman AL SIL G 20×20 cm; Maidstone, England): These plates can be cut with sharp scissors (*caution*: avoid silica dust formation and inhalation) to desired size depending on sample number and resolution requirements; typically, (a) for length, use 1 cm/sample plus 2 cm for end and (b) for height, ~7–10 cm is useful for routine separations. If the UDP or UDP-sugars need to be tracked nondestructively before staining, then use fluorescent indicator plates ("shadow" the absorbing spots using 254 nm light in a darkened room; *caution*: wear eye protection as the white plate reflects significant ultraviolet light). Take care to keep the plates sealed (e.g., in zip-lock bags) from the atmosphere as the silica will adsorb water and organic compounds from air. If a problem with staining background, then plates can be pre-run in methylene chloride and dried before use.

   For better resolution, use high performance TLC plates with a preload zone (Whatman Partisil LHPKDF); these glass-backed plates must be used intact or carefully cut with a glass-cutter.
2. TLC Solvent: A good resolving system for various GAG oligosaccharides that is compatible with many sample types is based on *n*-butanol–glacial acetic acid–water. Use HPLC grade solvents and water to avoid solvent background staining effects.

The solvent ratio is tuned for the size and the sulfation state of polymer that is being analyzed. For example, use 1.5:1:1 (v/v/v) for unsulfated GAG tetrasaccharide and hexasaccharide separations. Use 1:1:1 or 1:1.5:1 for oligosaccharides up to 10–20 monosaccharides units or when using sulfated GAGs. If using very small acceptors or molecules with hydrophobic aglycones (e.g., fluorescein), then 2:1:1–3:1:1 is used. When mixing, the order of addition should be butanol, acetic acid, and then water.

3. Napthoresorcinol (NR) Stain:

   Stock Reagent—Prepare 0.2 % NR (Sigma) in 95 % ethanol. Store in dark (wrap tube in foil); good for ~6–12 months.

   Working Reagent—Add 0.4 mL concentrated sulfuric acid to 9.6 mL of Stock Reagent in a 50 mL tube, mix carefully, and store in dark. Make fresh daily for best results.

4. Scintillant: If radioactive samples used, EnHANCE spray (Perkin Elmer, Waltham, MA, USA).
5. Chromatography Tank: Either purpose-built glass TLC tank or a 250–1,000-mL glass beaker capped with aluminum foil.
6. Fume hood: Safety item to contain solvents during running and processing of plates.

2.4 Matrix-Assisted Laser Desorption Ionization Time of Flight Mass Spectroscopy (MALDI-ToF MS) Components

1. Matrix solutions: (a) 5 mg/mL THAP (2′,4′,6′-Trihydroxyacetophenone monohydrate; Aldrich, Milwaukee, WI, USA) or (b) ATT (6-aza-2-thiothymine; Aldrich) in 50 % Acetonitrile–49.9 % water–0.1 % trifluoroacetic acid. Make fresh daily for best performance.
2. MALDI-ToF Mass Spectrometer and accompanying sample plate.

2.5 Polyacrylamide Gel Components

1. Stock acrylamide solution: 30:1 monomer/bis acrylamide mixture in water (ready-made stock or prepare from solid stocks and filter).
2. Running buffer: standard 1× TBE (0.1 M Tris-borate, 1 mM Na_2EDTA, pH 8.3) formulation [6].
3. 5× sample buffer: 2 M sucrose in 1× TBE buffer with Orange G dye to track (Sigma; dye content is variable so adjust for a strong color when diluted to 1×).
4. Staining and Destaining Solutions:
   (a) Alcian Blue Stain: 0.05 % dye (w/v) (Sigma) in 2 % acetic acid; destain with water.
   (b) Silver Stain: Components of the Bio-Rad staining kit (Hercules, CA, USA; after Alcian Blue staining, use the same reagents as for SDS-PAGE gels from the oxidizer solution onwards).

(c) Acridine Orange: 0.1 % dye (w/v) in water; destain with water.

(d) Stains-all Stain: 0.005 % dye (w/v) (Sigma) in 50 % ethanol; destain with water (*caution*: light-sensitive, store in dark or wrap bottle with foil).

5. Gel Apparatus: The Bio-Rad mini-Protean System is a useful and commonly available setup (used by many laboratories for SDS-PAGE of proteins, etc), but any small format gel system will suffice.

3 Methods

3.1 Expression and Purification Guidelines

1. The night before growing large-scale cultures, using a colony from a streaked out plate, inoculate small starter culture with media containing selection drug. Grow at 30 °C, a temperature with a good compromise between optimal growth and avoiding deletion or mis-folding issues common to glycosyltransferases. However, if needed, then room temperature growth will also work provided culturing times are extended. The goal is to have heavy growth, but not to stationary phase; a rough guide is 20–50 mL culture (in a 250–500-mL flask) per colony with ~16 h growth.
2. The next day, media with drug in the larger flasks is pre-warmed to 30 °C. For phage-lysin strains, also add L-arabinose to 3.3 mM final.
3. Inoculate large cultures with 1/20–1/100 volume of starter culture. Grow with high level of aeration.
4. Typically, induction with 0.1–1 mM isopropyl-thio-galactoside is performed when the OD_{600} is ~0.3–0.5 and then growth is continued for 4–16 h. The culture is then chilled on ice for 10–20 min before centrifugation.
5. Cells are collected centrifugation (4–10,000 × *g* for 10–25 min at 4 °C), resuspended in cold 50 mM Tris, pH 7.2, and recentrifuged. The pellet is then suspended as a 10–20 % cell suspension in the same Tris buffer with protease inhibitors (aminoethylsulfonyl fluoride, pepstatin, benzamidine, *N*-[*N*-(L-3-trans-carboxyoxirane-2-carbonyl)-L-leucyl]-agmatine, leupeptin, etc.) (*see* **Note 3**). If PmHAS or PmCS are used, then also include 1 M ethylene glycol (but *not* for PmHS1 or 2).
6. *E. coli* Xja cells lyse upon 2 or 3 cycles of freeze-thawing (e.g., cycle between −80 °C freezer and room temperature water). Alternatively, if using other hosts, then sonicate the suspension on ice as typical for releasing recombinant proteins from the cytoplasm. The lysate is clarified by centrifugation (20,000 × *g* for 25 min at 4 °C); if needed, then also filter through 0.45 μm membrane.

(a) PmHAS and PmCS are purified by pseudo-affinity chromatography on Red resin (Tosoh) with salt gradient elution (50 mM HEPES, pH 7.2, 1 M ethylene glycol with 0–1.5 M NaCl gradient in 1 h; [4]).

(b) PmHS1 and PmHS2 are purified via their fusion tag partner on amylose resin (NEB), maltose-binding protein with maltose elution as in ref. [2] and NEB guidelines (*see* **Note 4**).

7. Fractions are held on ice until protein assays (e.g., Bradford; Bio-Rad) are performed to assist pooling of the main peak; SDS-PAGE verification of the target protein is also recommended. The pooled fractions are concentrated and then transferred 50 mM Tris or HEPES, pH 7.2, (with 1 M ethylene glycol as needed) buffer using ultrafiltration devices (Amicon units; Millipore, Billerica, MA, USA) at 4 °C; typically a final concentration of 10–30 mg/mL protein is useful. The enzyme stocks may be stored as small aliquots at −80 °C for months. Minimize the number of freeze-thaw cycles, but usually one or two is not overly detrimental with the synthase constructs described here.

3.2 Chemoenzymatic GAG Synthesis

1. Thaw all frozen components and keep on ice directly before using; any concentrated stocks should also be well-mixed after melting. Reactions are typically performed in polypropylene tubes ("microfuge" tubes work well).

2. Typically the order of addition and recommended final concentration is: (a) water and stabilizer (e.g., 1 M ethylene glycol for PmHAS, PmCS) as needed, (b) 20–50 mM Tris or HEPES, pH 7.2, (c) 1–15 mM divalent metal ions (Mg or Mn) (d) 1–50 mM UDP-sugars (relevant precursors), (e) any acceptor, if needed, and (f) finally, ~0.1–5 mg/mL enzyme solution (*see* **Note 5**). Mix thoroughly at each step, but avoid generating foam and bubbles. When performing radioactive assays destined for TLC analysis, the use of small reactions volumes (5–10 μL) with lower range concentrations of buffer and metal ions are helpful.

3. Incubate in water bath (but if small volume, then incubate in a PCR block or an air incubator otherwise reactions may dry out prematurely) at 22–30 °C for desired time (1 min to overnight range, as needed).

4. Quench reaction, if needed, with (a) sufficient EDTA to counter the divalent metal ion, or (b) 1 % sodium dodecyl sulfate (SDS), or (c) boiling 1–2 min followed by clarification via centrifugation at ~20,000 × *g* for ~7–10 min; the choice of termination method will affect downstream processing, and therefore, plan ahead. In general, termination by (a) or (c) is the best for all three analytical methods described here, but (b)

is useful for paper chromatography described in Chapter 18. Placing the reactions simply on ice will slow the reaction, but it may still proceed over time; therefore, caution must be observed if performing kinetic tests.

5. Intermediate cleanup of GAG polymers can be done with anion exchange or gel filtration chromatography, ultrafiltration or dialysis, or ethanol precipitation (note the latter method does not completely remove UDP-sugars and UDP from the GAG polymer).

3.3 TLC of GAG Oligosaccharides

Caution: Preparing the solvents as well as running and staining the TLC plate should be done in a fume hood. Use gloves.

1. Add filter paper on wall of tank/beaker to facilitate vapor equilibration.
2. Prepare chromatography buffer (*see* **Note 6**) and add to tank or beaker so that fluid level is about ~5 mm high; cover the chamber tightly. Wait ~5–30 min to equilibrate the chamber with vapor. While the tank is equilibrating, proceed to **step 3**.
3. (a) If using high performance plates, then simply spot the sample (load ~3–5 μg of each sugar target if using NR reagent) using a pipettor in the middle third of the pre-load zone (make sure sample is always higher than ~8 mm above the future solvent level). Try to use the same sample volume and location in the zone to spot for improved reproducibility.

   (b) If using regular TLC plates, gently mark the plate with a series of dots ~10 mm above the predicted fluid layer using pencil (graphite, *not* ink!); each sample spot should be ~5–10 mm apart as well as add an extra ~10 mm on each side of the plate (the edges do not migrate identically to the middle of the plate).

   Directly spotting 1–3 μl of the reaction mixture/sample (load ~10 μg of each sugar target if using NR reagent) by pipetting is typically acceptable. Take care to avoid gouging/damaging the silica layer; if the liquid is dispensed just above the plate surface, then the silica will "grab" the drop thus no direct touching of the surface with the pipette tip occurs. Standards such as HA oligosaccharides and UDP-sugars also need to be spotted on spots adjacent to the samples. For adding large volumes, use the "bulls-eye target" method; e.g., to test a 6-μL sample, try to spot 3 μL, dry with room temperature air (e.g., hair drier without heat setting), spot 2 μL, dry, spot 1 μL, and dry. Also, the best test for identity between two molecules is to mix the samples before spotting and compare to each individual sample alone (e.g., load "A", "A+ B", and "B" in three adjacent lanes).

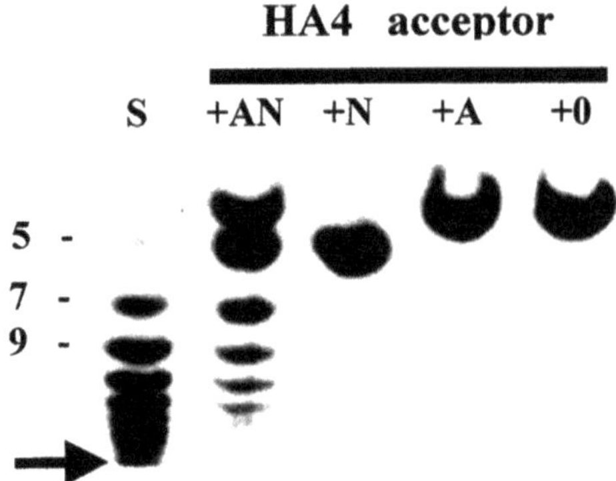

Fig. 1 Thin layer chromatography of HA oligosaccharides. This fluorograph of a silica high performance TLC plate (developing solvent 1.5:1:1 *n*-butanol–glacial acetic acid–water) depicts the separation of radioactive HA oligosaccharides. Four reactions of recombinant PmHAS enzyme acting on a radioactive HA tetrasaccharide ("HA4" = GlcUA-GlcNAc-GlcUA-GlcNAc-reducing end) acceptor are shown: +AN = both UDP-sugars added thus polymerization (new bands for HA 5,7,9,11, and 13 in addition to some residual HA4) is observed; +N = only UDP-GlcNAc added thus a single sugar is added to make HA5 (GlcNAc-GlcUA-GlcNAc-GlcUA-GlcNAc-reducing end); +A = only UDP-GlcUA added thus no extension observed (due to incorrect nonreducing end terminal residue on HA4); +0 = no UDP-sugar added thus HA4 remains with the same mobility. For resolving longer or shorter oligosaccharides, the relative amount of water in the development solvent is increased or decreased, respectively. Similarly, sulfated chains require more water in the solvent while hydrophobic reducing end groups (aglycones) require less water. Lane S = ladder of HA oligosaccharides prepared by digestion of HA polysaccharide with hyaluronidase; numbers on *left side* (*5-*, *7-*, *9-*) indicate the number of monosaccharide units. *Arrow* indicates the start of the silica layer of the plate

If using radiochemistry, then applying ~5,000–50,000 dpm of ^{3}H or ^{14}C of any expected target will yield an autographic signal within about a week of exposure upon fluorography.

4. After all liquid has been spotted, dry ~1–3 min with room temperature air (hair drier without heat setting) until the spot "disappears" and looks like the silica alone (but if much salt in reaction, then a "whiter" spot may remain) (*see* **Note** 7). If the reaction mixture contains DMSO or DMF solvents, then the plate must be dried under strong vacuum (a lyophilizer works well) for ~10–20 min otherwise smearing will result.
5. Put plate in tank and develop for ~20–60 min or until ~1–2 cm from top of plate. Remove plate (but keep in the hood), mark the solvent front with a pencil (not pen).
6. Air-dry with heat until plate is dry. Check with ultraviolet light (254 nm) for shadowing UDP or UPD-sugars as needed. If radioactive, then use EnHANCE spray or similar and expose to film (but to image standards using NR on the same TLC plate,

it is convenient to cut the TLC plate to separate the radioactive portion from the lanes with standards and stain it as in **step 7**) (*see* Fig. 1).

7. If nonradioactive, dip quickly or flood surface of plate with napthoresorcinol (NR) Working Reagent (take ~5 s to evenly distribute). Air-dry with heat briefly.
8. For NR staining, incubate on 100–150 °C hot plate in a fume hood (or use oven) for ~1–5 min as needed until blue or dark purple colored spots corresponding to GAGs, UDP-sugars, and UDP appear. Keep protected from light to avoid fading. Plates may be stored in plastic bags or wraps. As the stained plate is somewhat corrosive, use care in handling.

3.4 MALDI-ToF MS of GAG Oligosaccharides

1. Reaction mixtures may need dilution into water 1:5, 1:10, etc. and should have low salt concentrations (<25–50 mM total, if possible) by synthesis strategy, gel filtration or ultrafiltration.
2. Spot 0.5–1 μL matrix solution on plate, then quickly add equivalent volume of sample (load ~0.1–1 μg of each sugar target; avoid evaporation until mixed) to the same drop. Reaction mixtures without enzyme, or matrix alone, should also be spotted as negative controls. The inclusion of a mass calibrant, such as HA_4 (Hyalose), is needed in a different spot to obtain the proper masses.
3. Dry plate under vacuum or gentle air stream. Examine the spots with a magnifying lens or loupe. Crystal formation in the spot is a very good sign. On the other hand, totally amorphous or "wet" spots are not promising and indicate that better sample clean-up or sample dilution may be required.
4. Proceed with MS in negative mode as directed by spectrometer manufacturer (*see* **Note 8**); either linear or reflector mode are suitable (*see* Fig. 2). Assistance by an expert on the instrument is recommended in operation and interpretation. For the Bruker Daltonics Ultraflex II with a SMART beam laser mass spectrometer (Billerica, MA, USA), settings to start GAG analyses in reflector mode are: Ion Source 1, 20 kV; Ion Source 2, 17.4 kV; Lens, 7.00; Reflector 1, 21 kV; Reflector 2, 11.1 kV; Detector Gain, 4–5×; Laser power, 50–70 %; Low mass gate, 200–500 Da; High mass gate, set as needed for target sugar.

3.5 PAGE of GAG Oligosaccharides and Polysaccharides

1. Typical PAGE protocols [6] for pouring gels are done (with TEMED/ammonium persulfate catalysis), but the final concentration of acrylamide is tuned depending on the size of the GAG target polymer. For example, for oligosaccharides in the 8–20mer range, use 20 % gels (*see* Fig. 3) while for polysaccharides in the 5–200 kDa range, use 4 % or 6 % gels (*see* **Note 9**).

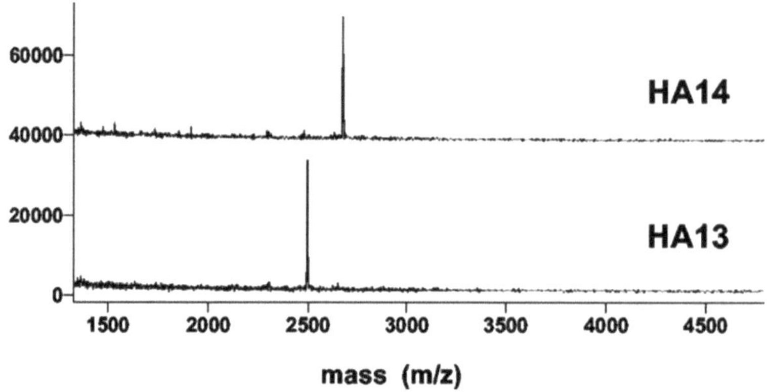

Fig. 2 Mass spectroscopy of HA oligosaccharides. These MALDI-ToF MS spectra (ATT matrix; reflector mode) depict the analysis of HA oligosaccharides with 13 or 14 sugar units (*HA13* or *HA14*) formed by stepwise extension of chains by recombinant PmHAS mutant enzymes. In some samples (but not these materials), minor peaks at multiples of +22 Da are also noted; these are the sodiated versions (23 Da for Na^+ instead of the +1 for a H^+ on the GlcUA residues). Molecules with masses greater than ~600 are observable above the matrix fragment background, but controls with matrix solution alone (or with sample buffer but no sugar analyte) should always be run in parallel when searching the low mass range. With optimization, unsulfated GAG molecules with up to ~20–30 sugar units are readily observable. The testing of heavily sulfated GAGs requires different protocols than presented here due to poor ionization and sulfate losses and/or migration

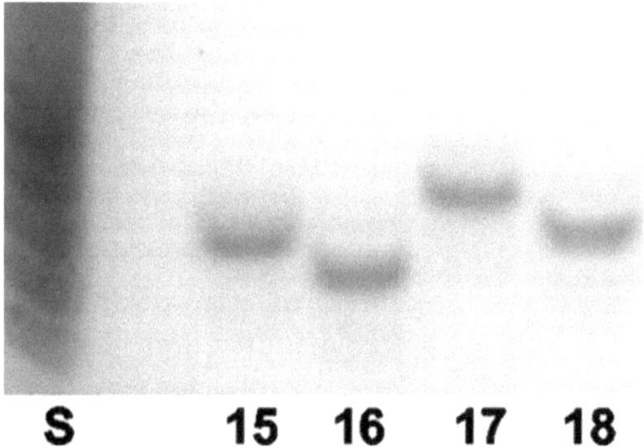

Fig. 3 Gel analysis of HA oligosaccharides. This PAGE gel (20 % acrylamide; Acridine Orange stain) depicts the analysis of HA oligosaccharides with 15–18 sugar units prepared by chemoenzymatic synthesis (Lane S = ladder of HA oligosaccharides prepared by digestion with hyaluronidase). The different charge–mass ratio for odd- and even-numbered oligosaccharides causes the pattern observed here (e.g., the HA16 has an additional GlcUA residue thus one extra negative charge compared to HA15 and migrates faster even if "smaller" size). Note that gel detection methods for small oligosaccharides (less than 10 or 12 sugar units) are somewhat problematic; such molecules are typically resolved on the gel, but diffuse out during processing and/or have low color yield thus TLC may be the better method for small molecules. If resolving larger unsulfated GAGs up to ~100 kDa, then use lower percentage gels (e.g., 4–6 % for polymers of 20–100 kDa or 100–500 sugar units). If resolving unsulfated GAGs beyond ~100 kDa (and up to ~8,000 kDa), then use agarose gels as described in Chapter 18. If resolving sulfated GAGs, then the increased charge to mass ratio of these polymers will necessitate denser PAGE gels

Table 2
Staining sensitivity for GAGs on PAGE gels

Stain	Monodisperse (μg/lane)	Polydisperse (μg/lane)
Alcian blue or acridine orange	0.5–2	1–25
Alcian-silver	0.1–0.5	0.5–2
Stains-all	0.05–0.5	0.5–2

Suggested values for mini-gel format (~8–10 cm × 8–10 cm dimensions) gels are listed. Monodisperse polymers yield tight narrow size-distribution bands while polydisperse polymers yield broad size-distribution smears, and therefore, the latter preparations require more total mass to be visualized

2. Samples (e.g., chemoenzymatic reaction mixtures, GAG digest samples in the digestion buffer, or purified oligosaccharides in water) are mixed with a one-fifth volume of sucrose sample buffer and 2–4 μL of the mixtures are applied directly to the gel (*see* **Notes 10** and **11**). Loading depends on staining method and GAG product polydispersity (*see* Table 2).
3. The gels are run at 250 V until Orange G tracking dye is ~1 cm from the bottom of the gel.
4. Stain gel with Acridine Orange or Alcian Blue solutions with gentle rocking for 5–30 min as needed, and then destain with water. For combined Alcian Blue/Silver staining, after destaining with water, proceed starting at the "oxidizer" step as directed. For Stains-all [7], stain for 30–120 min with gentle rocking, then destain for 1–16 h as needed (*caution*: dye adducts are light sensitive so keep gel trays wrapped in foil). For ^{125}I materials, gels can be dried directly and subjected to autoradiography. For ^{3}H, ^{14}C, ^{35}S, use fluorography (i.e., impregnate with scintillant, dry, then expose to film) utilizing guidelines by the scintillant manufacturer; staining with Alcian Blue or Acridine Orange should help fix GAGs but consider that more processing can result in losses.

4 Notes

1. *E. coli* production strains with a goal of enzyme preparation should not possess an active UDP-glucose dehydrogenase unless polysaccharide production in vivo is required.
2. To facilitate cell disruption after harvesting, the phage-lysin containing strains are very useful; *E. coli* XJa (Zymogenetics) is very useful especially if multiple enzymes or mutants need to be analyzed in parallel.
3. The selection of protease inhibitors should be made after screening for undesirable inhibitory effects; while not expected,

some glycosyltransferases may be inactivated. Similarly, prescreening of agents typically used to enhance protein stability (e.g., glycerol, DMSO, ethylene glycol, etc.) should be done before settling on a protocol.

4. Synthases may also be purified using poly-histidine tag purification systems (Qiagen), but the histidine or imidazole elution conditions should be prescreened. Typically, imidazole or histidine elution (if pH is controlled) is useful, but the use of pH shift (e.g., reduce pH from neutrality to pH 4.5) is often detrimental to enzyme activity.
5. An important control for chemoenzymatic synthesis of GAG is to check the effect of having zero, one, or both UDP-sugars present; these reactions represent the negative control, a single sugar extension, or polymerization, respectively. In the case of a repeating GAG disaccharide unit, the choice of which UDP-sugar to employ depends on the identity of the sugar unit at the nonreducing end of the acceptor.
6. The TLC solvent should be made daily to avoid butyl acetate formation over time; the creation of variable amounts of this product could change solvent polarity and thus shift analyte mobility.
7. For samples with much salt, buffer, etc., before running the TLC plate in the normal solvent, use the same developing solvent without water to pre-run the plate (allow solvent to rise to go ~5 cm once or twice depending on contaminant load); most small molecules (Tris, Mn^{2+}, etc), but not underivatized sugars (however, use caution with hydrophobic aglycones as they may also migrate in the wash step), will migrate out of the origin of application. The plate is then air-dried without heat, water is added to the tank buffer in the required amount (with vigorous shaking to mix!), and the plate is run as normal.
8. Mass spectroscopy is an art with respect to both forming crystals (i.e., drying rate and drying method) and the actual locations on the spot that yield the highest signal (e.g., test the edge, middle, or central regions of a spot for every sample). Try to acquire the best signal by surveying the spot before settling on final data collection.
9. In general, note that oligosaccharides less than 5–15 sugar units long may not stain efficiently as the longer polymers using these staining methods. If important to monitor very small sugar chains and the use of TLC is not an option, then use radioactive precursors or acceptors.
10. For optimal analyses, MS and PAGE require that samples should have low salt concentrations or in the worst case, all samples, standards, and blanks should have the equivalent amount of salt (or buffer) to avoid disparate effects.

11. The use of quasi-monodisperse GAG size standards, as exemplified by SelectHA (Hyalose, LLC), is preferred to polydisperse natural GAGs. Unsulfated chondroitin and heparosan migrate in a similar, but nonidentical, fashion to hyaluronan thus the same HA standards are useful for initial studies. Please note that sulfated and non-sulfated GAGs cannot be directly compared due to the differences in charge to mass ratio.

Acknowledgment

Thanks to Dixy E. Green for gathering technical details.

References

1. DeAngelis PL (2002) Microbial glycosaminoglycan glycosyltransferases. Glycobiology 12:9R–16R
2. Sismey-Ragatz AE, Green DE, Otto NJ, Rejzek M, Field RA, DeAngelis PL (2007) Chemoenzymatic synthesis with distinct *Pasteurella* heparosan synthases: monodisperse polymers and unnatural structures. J Biol Chem 282:28321–28327
3. Jing W, DeAngelis PL (2003) Analysis of the two active sites of the hyaluronan synthase and the chondroitin synthase of *Pasteurella multocida*. Glycobiology 13:661–671
4. DeAngelis PL, Oatman LC, Gay DF (2003) Rapid chemoenzymatic synthesis of monodisperse hyaluronan oligosaccharides with immobilized enzyme reactors. J Biol Chem 278: 35199–35203
5. Jing W, DeAngelis PL (2004) Synchronized chemoenzymatic synthesis of monodisperse hyaluronan polymers. J Biol Chem 279:42345–42349
6. Ikegami-Kawai M, Takahashi T (2002) Microanalysis of hyaluronan oligosaccharides by polyacrylamide gel electrophoresis and its application to assay of hyaluronidase activity. Anal Biochem 311:157–165
7. Lee HG, Cowman MK (1994) An agarose gel electrophoretic method for analysis of hyaluronan molecular weight distribution. Anal Biochem 219:278–287

Chapter 18

Methods for Measuring Class I Membrane-Bound Hyaluronan Synthase Activity

Paul H. Weigel, Amy J. Padgett-McCue, and Bruce A. Baggenstoss

Abstract

Detecting and quantifying hyaluronan (HA) made by Class I HA synthase (HAS) and determining the level of activity of these membrane-bound enzymes is critical in studies to understand the normal biology of HA and how changes in HAS activity and HA levels or size are important in inflammatory and other diseases, tumorigenesis, and metastasis. Unlike the products made by the vast majority of glycosyltransferases, HA products are more complicated since they are made as a heterogeneous population of sizes spanning a broad mass range. Three radioactive and nonradioactive assay methods are described that can give the amount of HA made with or without information about the distribution of product sizes.

Key words Synthesis, Size, Mass, Paper chromatography, Agarose gel electrophoresis, SEC-MALLS, Radioactive assay, Nonradioactive assay

1 Introduction

Progress in many HA fields has been aided by quantitative assays to determine the HA content [1] and sometimes the HA size distribution [2, 3] in samples (e.g., cells, tissues, serum or conditioned medium) or the activity of membrane-bound [4–6] or purified HAS [7, 8] (e.g., from mouse or *Streptococcus equisimilis*; *SeHAS*). The accompanying chapter 17 by Paul DeAngelis describes three techniques (thin layer chromato graphy, mass spectroscopy, and polyacrylamide gel electrophoresis) to assay soluble GAG polymerases or mutants (e.g., Class II HAS from *Pasteurella multocida*) that transfer single sugars and are useful for chemoenzymatic GAG synthesis [9]. The Class I HASs, in contrast, are lipid-dependent enzymes with 6–8 membrane domains that are difficult to solubilize and purify in active form [10]. Although the streptococcal HASs can be readily purified and studied at the biochemical and mechanistic levels, mouse HAS 1 is the only mammalian HAS to be successfully purified [8]. Thus, essentially all studies to monitor

Inka Brockhausen (ed.), *Glycosyltransferases: Methods and Protocols*, Methods in Molecular Biology, vol. 1022, DOI 10.1007/978-1-62703-465-4_18, © Springer Science+Business Media New York 2013

mouse or human HAS activity are performed using membranes isolated from established or recombinant cultured cells. There are multiple methods to prepare membranes and multiple techniques for measuring HAS activity and detecting HA. Here we describe three techniques (ranging in complexity of equipment and cost) that can be used to quantify HAS activity: paper chromatography (PChr), agarose gel electrophoresis (AGE), and size exclusion chromatography coupled to multiangle laser light scattering (SEC-MALLS). PChr involves radioactive labeling of HA, MALLS does not use radioactivity, and AGE can be performed using either unlabeled or radiolabeled HA. The oldest, and lowest "tech," PChr method uses solid–liquid phase partitioning (adsorption and desorption) to separate insoluble HA from soluble UDP-sugars [11]. AGE and MALLS [12] use SEC as a sample moves through a gel or beads to fractionate HA by size throughout the gel [13] or eluted column fractions [14]. For AGE, detection and analysis of HA amount and size is based on staining and densitometry. For MALLS, detection and analysis of HA amount and size are performed simultaneously and data output requires higher "tech" software and equipment [15, 16].

2 Materials

All solutions are made using analytical grade reagents from Sigma-Aldrich (St. Louis, MO, USA) unless noted otherwise, and purified water prepared by reverse osmosis and then deionization, to a resistance of 18 MΩ cm at room temperature (RT). Unlabeled UDP-sugars were from Sigma and UDP-[^{14}C]GlcUA (380 mCi/mmol) was from Perkin-Elmer New England Nuclear Radiochemicals (Waltham, MA, USA). Salt solutions are stored at RT unless noted otherwise; biochemical solutions are stored at 4 °C for short-term (<1 week) or longer-term at −20 °C. Final volumes (V_f) are indicated for solutions. Radioactive monitoring, record keeping, and waste disposal procedures are per the appropriate institution, state and Federal regulations. Unless noted otherwise, sample mixes and pipetting are in an ice bucket, and sodium azide is not present in reagents.

2.1 Master Stock Solution Components Common to the Three Assay Reaction Mixes

1. 500 mM sodium potassium phosphate, pH 7.4 (at 25 mM): combine 90 mL 0.5 M Na_2HPO_4 and 15 mL 0.5 M KH_2PO_4 (6:1 ratio), filter (250 mL 0.22 μm unit), and store in sterile bottle.
2. 5.0 M NaCl: dissolve 14.61 g of NaCl to 50 mL V_f and filter (0.22 μm) into a sterile 50 mL Falcon tube.
3. 80 % (v/v) glycerol: mix 40 mL glycerol and 10 mL water. Store at RT in covered 50 mL Falcon tube to minimize light exposure.

4. 100 mM DTT: dissolve 15.42 mg dithiothreitol to 1.0 mL V_f. Store at −20 or −80 °C and use for only a few months.
5. 50 mM UDP-GlcNAc: dissolve 65.13 mg uridine 5′-diphospho-*N*-acetylglucosamine, disodium salt to 2.0 mL V_f. Store at −20 °C.
6. 50 mM UDP-GlcUA: dissolve 64.63 mg uridine 5′-diphosphoglucuronic acid, trisodium salt to 2.0 mL V_f. Store at −20 °C.
7. 400 mM $MgCl_2$: dissolve 8.13 g $MgCl_2 \cdot 6H_2O$ to 100 mL V_f and filter (0.22 μm) into sterile 50 mL conical Falcon tubes.
8. AP (1 U/μL): dilute calf intestinal alkaline phosphatase (AP) to a final activity of 1 U/μL in 5 mM Tris, pH 7.0, 5 mM $MgCl_2$, 100 μM $ZnCl_2$, and 50 % glycerol. Store at 4 °C.
9. 500 mM EDTA: dissolve 2.081 g ethylenediaminetetraacetic acid tetrasodium salt·$2H_2O$ to 10 mL V_f. Also make a 50 mM EDTA stock by diluting 1.0 mL of the 500 mM EDTA with 9.0 mL of water. The 500 mM stock is for quenching reactions and the 50 mM stock is for assay reaction mixes.

2.2 Reaction Mix 4× Stock Solution Common to the Three Assays

Final (1×) concentrations in individual sample assay reactions will be 25 mM sodium-potassium phosphate, pH 7.4, 50 mM NaCl, 1.0 mM DTT, 0.1 mM EDTA, 0.2 mM UDP-GlcNAc, and 0.2 mM UDP-GlcUA. Additional UDP-sugars are added for some assay methods, UDP-[^{14}C]GlcUA and glycerol content vary depending on the method, and AP is added later to 0.02 U/mL.

1. General Reaction Mix buffers are first made as partially complete 4× solutions using the Master Stock solutions. For ten assays (V_f = 1.0 mL; *see* **Note 1**) mix the following volumes (in parentheses) of each stock solution in a clean sterile 15 mL conical Falcon tube to give 0.25 mL (4×), and the desired final concentration [in brackets] at V_f: 500 mM Na_2KPO_4, (50 μL), [25 mM]; 5.0 M NaCl, (10.0 μL), [50 mM]; 100 mM DTT, (10 μL), [1.0 mM]; 50 mM EDTA, (2.0 μL), [0.1 mM]; 50 mM UDP-GlcUA, (4.0 μL), [0.2 mM]; 50 mM UDP-GlcNAc, (4.0 μL), [0.2 mM]; and water (170 μL) to give 250 μL.
2. If all assays will be under identical conditions, then subsequent additions (700 μL total) to complete the ten assay reaction mix (except for $MgCl_2$ addition) will include fixed volumes of AP, UDP-[^{14}C]GlcUA (if required), glycerol, and HAS-containing membranes. The reaction mix (0.950 mL) is then dispensed into nine sample tubes (95 μL each) and (5.0 μL) of 0.4 M $MgCl_2$ [20 mM] is added to initiate synthesis.
3. In many cases, however, individual samples will be set up using variable volumes of UDP-[^{14}C]GlcUA (if required), glycerol, and HAS-containing membranes before addition of $MgCl_2$ (*see* **Note 2**).

2.3 Membranes

The experience of those in the field, including us, is that the specific conditions and technique used to prepare eukaryotic cell membranes that retain HAS activity is dependent on the particular cell type. For *E. coli* cells expressing recombinant streptococcal HASs, active membranes can be prepared by a relatively standard membrane procedure using sonication of lysozyme/EDTA-treated cells [17]. For mammalian cultured cells, the methods of Spicer [5] and Yoshida et al. [8] are good starting points for developing a suitable protocol for the cell type of interest (*see* **Note 3**).

2.3.1 Equipment

Dounce homogenizer (5 mL or larger) with a type B pestle suitable for cell disruption, low-speed and medium-speed refrigerated centrifuges, and a tissue culture hood with aspirator.

2.3.2 Reagents

1. Protease inhibitors: aprotinin, leupeptin and phenyl methyl sulfonyl fluoride (PMSF).
2. Hypotonic Lysis Buffer: 10 mM Tris–HCl, pH 7.4 10 mM KCl, 1.5 mM $MgCl_2$ containing 0.7 μg/mL aprotinin, and 0.5 μg/mL leupeptin. Store at 4 °C. Just before use, PMSF is dissolved in water and added to 46 μg/mL final concentration.
3. Phosphate buffered saline (PBS).
4. Radioactive assay buffer (*see* Subheading 2.2) with 0.2 mM UDP-sugars and UDP-[^{14}C]GlcUA at ~100,000 dpm per assay (0.1 mL). For our typical commercial preparations, this corresponds to ~2 μL of UDP-[^{14}C]GlcUA per assay.
5. PC3M-LN4 [18] or other cells to be tested. PC3M-LN4 is a human prostate tumor cell line.

2.4 Paper Chromatography Assay Components

1. Radioactive substrate: UDP-[^{14}C]GlcUA (150–300 MBq/mmol).
2. Paper chromatography: a cylindrical or rectangular glass tank (e.g., 61 cm H × 30.5 cm W × 30.5 cm D) with a ground glass top, glass cover, stainless steel internal frame that holds multiple upper solvent trays for descending chromatography, and thick glass rods (bent in a broad U to fit in tray) to hold the paper strips in place; Whatman 3 mm Chr Chromatography Paper (46 × 57 cm) and; chromatography buffer, 1 M ammonium acetate, pH 5.5, and 95 % ethanol (7:13).
3. Preparation of PCr strips (cut in advance and store until needed). Draw four pencil (not ink) lines parallel to the 57 cm length of the chromatography paper, at 2.5, 6, 8, and 10 cm from the top of the paper. Cut the paper sheet (perpendicular to the lines), using a paper cutter, along the 46 cm width to make 2 cm wide strips (2 × 46 cm). Pencil lines can be drawn as a cutting guide.

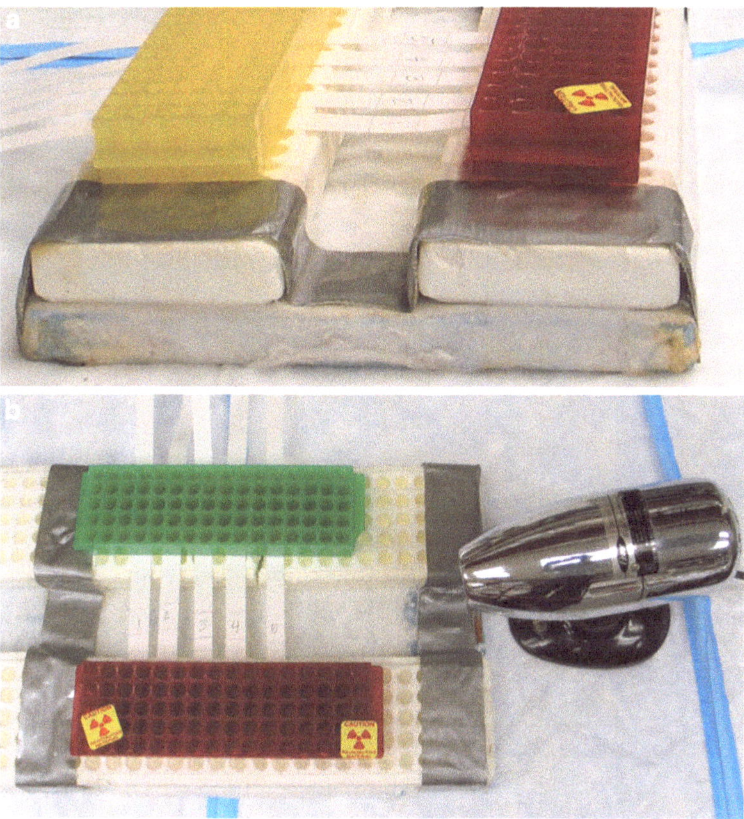

Fig. 1 A paper chromatography spotting station. (**a**) An end-on view of a lab-made unit shows five paper strips (labeled 1–5) elevated over a "trough" that allows the samples to be spotted, while the origin zone remains isolated so that sample liquid does not wick away to another surface. (**b**) The *view from above* shows the paper strips held in place by flat tube holders (for weight) and a blower positioned to flow air (not heated) over and under the origins to dry the spotted samples

4. Paper spotting support station: serves to hold many paper strips (to be spotted and dried) in place and suspended slightly so the origins (between 8 and 10 cm) do not touch another surface. There is no standard device. The goal is to elevate and isolate the origin and prevent the strip from moving during spotting or blowing away during drying. A trough is created to suspend strips and to help channel blowing air under/over the strips for more efficient drying. We use a 90 × 25 cm station made from styrofoam pieces held together with duct tape to create a parallel ~5 cm gap between them (Fig. 1a).
5. Liquid scintillation: Scintillation fluid suitable for ^{14}C, screw cap scintillation vials (e.g., 16 × 57 mm) with screw cap, and a liquid scintillation counter.

2.5 SEC-MALLS Analysis Components

1. Equipment: PL aquagel –OH 60, and –OH 40 size exclusion columns (Agilent Technologies, Santa Clara, CA, USA) in series. DAWN DSP Laser Photometer and Optilab DSP Interferometric Refractometer (Wyatt Technology, Santa Barbara, CA, USA) or equivalent for multiangle static light scattering measurements. An HPLC and a speed vacuum centrifuge are required; an autosampler is useful.
2. MALLS Assay Mix (2×): Per 1.0 mL of 2× MALLS assay mix to be made, use 500 μL 4× Master Reaction Mix stock, 25 μL 80 % glycerol, 32 μL 50 mM UDP-GlcNAc, and 32 μL 50 mM UDP-GlcUA (to give 1 mM final concentrations of each UDP-sugar), 40 μL AP (to give 0.02 U/μL; AP also contributes 1 % glycerol at V_f) and 371 μL water.
3. UDP (120 mM): dissolve 56.4 mg uridine 5′-diphosphate (trisodium salt) to 1.0 mL V_f in water.
4. Chloroform–methanol (2:1); mix 20 mL chloroform and 10 mL methanol in a fume hood, and store in a dark glass bottle with Teflon lined screw cap or sintered glass stopper.
5. MALLS Running buffer (5×): (150 mM NaCl, 50 mM Na_3PO_4, pH 7, and 0.05 % sodium azide); dissolve 175.32 g NaCl, 46.92 g NaH_2PO_4, 93.69 g Na_2HPO_4 in 3.9 L water, bring the pH to 6.64 with NaOH. Add 2.0 mL 10 % (w/v) sodium azide. Bring to 4 L V_f with water.

2.6 Agarose Gel Electrophoresis Assay

1. Equipment: An AGE apparatus with a gel running cell of about 57×59×7 cm (*W*×*L*×*H*) and a gel tray size of about 15×10×2.5 cm (*W*×*L*×*H*) can accommodate 20–30 samples, depending on the comb used. Our lab uses a Bio-Rad Wide Mini-Sub Cell system (Bio-Rad Life Science research, Hercules, CA, USA).
2. AGE buffer (20× stock): dissolve 387.52 g Tris, 32.8 g sodium acetate (anhydrous), and 29.76 g disodium EDTA in water to ~3.8 L, pH to 7.6 with glacial acetic acid (~100 mL) and bring to 4 L V_f.
3. Agarose gel: several hours before use, dissolve an amount of agarose to give the desired final weight-percent (%). For a 1 % gel, add 1.0 g agarose to 100 mL of 1× AGE buffer in a 250 mL flask with swirling to disperse the solid and minimize clumps. Dissolve by heating in a microwave oven until it starts to bubble. Carefully remove and after bubbling stops, swirl for ~15 s and then reheat briefly (e.g., 15 s). Continue swirling and reheating until all gel solids are dissolved. Cool to ~55 °C by running tap water on the outside of flask and pour slowly into the gel casting tray.
4. Stains-All (0.005 %): dissolve 0.1 g in 2 L 50 % ethanol in a dark bottle and store dark at RT. The dye is sensitive to photobleaching and fades upon light exposure.

5. AGE Sample buffer (6×): for V_f of 10.0 mL, dissolve 5 mg bromophenol blue (0.05 % final concentration, w/v) in 3.0 mL 20× AGE buffer and 2.9 mL water, and then add 4.1 mL 80 % glycerol. Store at 4 °C.

2.7 AGE Radioactive Assay

All components are identical to those in Subheading 2.6, **steps 1–5** above, except that UDP-[^{14}C]GlcUA is added in the final assay buffer. Radiolabeled HA is visualized by autoradiography of the dried gel [19], which also requires a gel drying apparatus and X-ray film suitable for autoradiography and fluorography.

3 Methods

3.1 Eukaryotic Cell Membrane Preparation

1. Grow the eukaryotic cells in their appropriate medium and serum to near confluence (~80–90 %) in a number of T-75 flasks corresponding to half the number of replicate assays desired.
2. Remove the flasks from the incubator and chill on ice.
3. Remove the old media by aspiration. Set the flask on its end and tilt so that the side with the cells is up and aspirate from a corner with no cells. Do not touch the cell layer with the aspirating pipette.
4. Hold the flask upside down and gently pipette 3 mL of cold PBS into a corner. Recap and slowly tilt the flask back and forth until the PBS coats the entire cell layer. Aspirate the PBS and repeat, aspirating the second wash.
5. Add 3 mL of Hypotonic Lysis buffer (*see* section 2.3.2, **step 2**) and incubate the flask on ice for 10 min.
6. Perform **steps 3–5** for all flasks.
7. Bring each flask to the tissue culture hood one at a time to harvest cells. Aspirate the Lysis buffer from the tissue culture flasks as described in **step 3**. Add 3 mL of fresh cold Lysis buffer (without PMSF).
8. Use a prechilled sterile cell scraper to remove the layer of cells by gently scraping back and forth across the surface of the tissue culture flask.
9. Transfer the harvested cell suspension to a prechilled 15 mL or 50 mL Falcon tube on ice.
10. Repeat **steps 7–9** with all T-75 tissue culture flasks.
11. Pool the suspensions into same Falcon tube (e.g., 50 mL) on ice and gently add and mix freshly prepared PMSF.
12. Using a clean Dounce homogenizer and a type B pestle, disrupt the cells on ice. Place the type B pestle at the bottom of the Dounce homogenizer and add 1.5–2.0 mL of cell suspension

into the homogenizer (or more if a >5 mL Dounce is used) and slowly raise and lower the pestle once allowing the suspension to flow past the pestle. Slowly raise the pestle one more time and with continuous rotation through the liquid; remove the pestle. Collect and place the homogenate into a new, cold 50 mL Falcon tube.

13. Repeat with each portion of cell suspension until all have been homogenized and pooled.
14. Aliquot the total homogenate equally into 2 mL microfuge tubes (~1.5 mL each); larger tubes can be used if a larger volume is being processed.
15. Centrifuge at 4,000 × *g* for 5 min in a refrigerated centrifuge to pellet nuclei and large cell debris.
16. Transfer the supernatants to another set of prechilled tubes and centrifuge at 20,000 × *g* for 20 min in a refrigerated centrifuge to pellet membranes (*see* **Note 4**).
17. Discard supernatants and resuspend the membrane pellets (e.g. by vortex mixing and pipetting) in one-tenth the original volume of General Reaction Mix buffer (without UDP-sugars or AP) with protease inhibitors.
18. Pool the resuspended membranes and divide for use in the three different assays or to assess different incubation times or membrane protein concentrations. The radioactivity PCr assay is recommended as the first choice (being the most sensitive) until HAS activity or HA content is known.

3.2 PCr Radioactive Assay

The radioactive assay mix for PCr is prepared for nine reactions plus 0.1 mL extra.

1. Set up reaction samples in triplicate in clean conical microfuge tubes (0.6–2.0 mL) numbered A1-3, B1-3, etc.
2. For membranes with high activity (e.g., streptococcal) add UDP-[^{14}C]GlcUA (150–300 MBq/mmol) to the 4× reaction stock to give ~50,000 DPM per sample; depending on the specific radioactivity, this is a final concentration of ~0.7 μM labeled nucleotide. Unlabeled UDP-sugar concentrations can also be increased from 200 μM to 1.0 mM to ensure more optimal kinetics [20]; *see* **Note 5**.
3. For eukaryotic cell membranes (e.g., PC3M-LN4) with lower or unknown activity, add UDP-[^{14}C]GlcUA to the 4× reaction stock to give ~100,000 DPM per 100 μL final sample and do not add additional unlabeled UDP-sugars (*see* **Note 6**).
4. Mix 250 μL of 4× Reaction Mix, 188 μL of 80 % glycerol master stock, 20 μL AP, and 42 μL water so that the reaction mix volume is 500 μL; the reaction mix is now 2×.

5. Add 50 µL of the 2× assay mix to each labeled sample tube and the desired amount of membrane protein suspension. Different concentrations of membrane protein (e.g., 50–400 µg) should be used initially to assess the detection sensitivity and linearity of synthesis (*see* **Note 7**). Alternatively, if all samples will have the same membrane content, then the membrane suspension and water can be added to the 2× mix to a volume of 0.95 mL.
6. Add water to give a volume of 95 µL in each sample tube and gently agitate all tubes on a mixer (e.g., MicroMixer E-36 from Taitec; Bio Nexus Inc, Oakland, CA) at 30–37 °C (*see* **Note 8**) for 10–15 min to prewarm the samples.
7. Start reactions by adding 5 µL of 0.4 M $MgCl_2$ to give 20 mM final concentration in 100 µL. Mix well, place tubes back on the mixer, and incubate at the same temperature for 1–4 h.
8. Stop reactions (and inhibit endogenous hyaluronidases) by adding 20 µL of 12 % (w/v) SDS and mixing (final concentration of 2 % SDS). The samples can be stored at 4 °C if analyzed within a few days or at −20 °C for longer storage.

3.3 Paper Chromatography

1. For spotting the PCr strips: On a clean surface (e.g., lined absorbent pads) lay out the same number of strips as sample tubes and label (pencil) a strip with each tube code between the lines drawn at 8 and 10 cm (the origin) from the top of the strip.
2. Align strips on the spotting station so the labels are easily read, they are ~0.5 cm apart and their origins are centered and suspended over the gap between the parallel supports. A 90 cm station should allow up to 36 strips to be spotted at a time. Plastic tube holders or other clean flat lab objects are put on top of the strips on both sides of the station to weigh down and hold in place the strips spanning the gap (Fig. 1b).
3. Set up a hair dryer at the end of the station to blow air (cool only) over and under the spotted origins. Before spotting, briefly run and adjust the hair dryer to confirm that all strips are held in place firmly and do not touch, flap too much, or shift position.
4. Pipette half of each reaction sample (~55 µL) onto each appropriate strip at the center of the origin square between the 8 and 10 cm lines (*see* **Note 9**). Be certain all liquid is well absorbed into the paper before turning on the hair dryer (to avoid SDS bubbles blowing between strips, causing cross-contamination).
5. After the first spots have dried (~20–30 min), add the second half of each reaction sample over the first spot and then air-dry the strips as above.
6. Once all strips are dry, fold each one forward (up) at the first top line and backwards (down) at the second line. The first

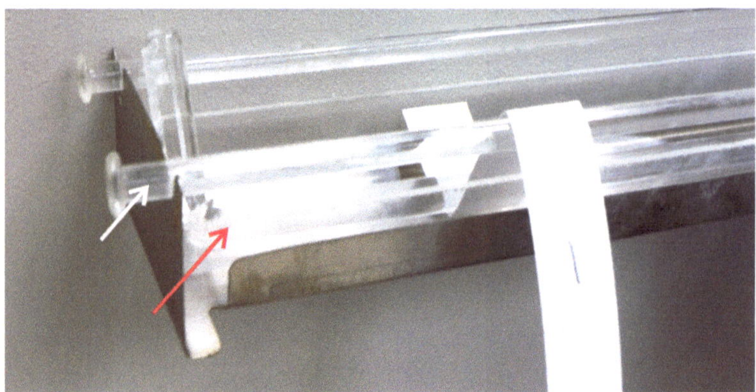

Fig. 2 Paper orientation during chromatography. A portion of an upper buffer tray shows a paper strip held in position at the 2.5 cm line (up-fold) by a U shaped glass rod (*red arrow*). The 6 cm line (down-fold) hangs over a glass rod at the tray top (*white arrow*) to position the origin (between the 8 and 10 cm lines) for descending solvent flow through sample #1

two lines help placement of the strips in the buffer reservoir tray and the next two lines define the origin. Add chromatography buffer to the bottom of the tank to help keep the tank atmosphere equilibrated and prevent too much solvent evaporation from the reservoir trays. This is particularly important if using only one tray. Place the strips carefully in the top (dry) trays in the chromatography tank by aligning the folded sections of the strip to fit (like a V) into the tray and hold in place with the heavy glass U rod (red arrow; Fig. 2). The strips should not touch each other and if the folded 6 cm lines are on the supporting glass rod above the top edge of the buffer tray (white arrow; Fig. 2), then the origin areas are aligned.

7. Carefully add chromatography buffer to ~2/3 fill each tray without splashing on the strips.
8. Cover top of the tank, which should be sealed tightly enough to minimize evaporative loss of solvent. Use stopcock grease on the ground glass and a distributed weight on the cover to help make a good seal.
9. Processing the PCr strips for scintillation counting: Fig. 3 shows the radioactive profile along a typical PCr strip from origin (at 8–10 cm) to near the end of the strip. It illustrates the ^{14}C-HA signal obtained at the origin with PC3M-LN4 cell membranes and the difference when incubated with and without the second unlabeled substrate, UDP-GlcNAc (Fig. 3 inset). HA at the origin is well separated from other smaller, more soluble, ^{14}C-material by the overnight chromatography.
10. The next morning, carefully remove and dry the strips well (e.g., lay on absorbent paper) in a fume hood, cut out the labeled origins between the 8 and 10 cm lines (*see* **Note 10**).

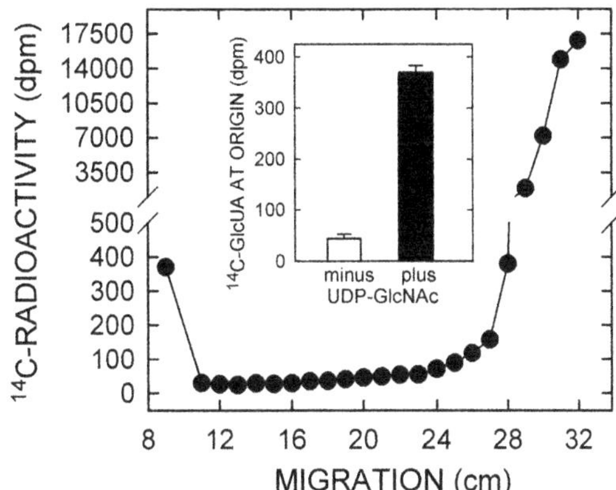

Fig. 3 Paper chromatography separates UDP-[^{14}C]GlcUA from ^{14}C -HA made by PC3M-LN4 cell membranes. Most of the chromatography strip is shown for a sample with unlabeled UDP-GlcNAc; the origin between 8 and 10 cm contains labeled HA at the *left* and the large peak at the *right* is well separated labeled UDP-[^{14}C]GlcUA near the end of the strip. Membranes were prepared and assayed as described in Subheading 3.1 with each 0.1 mL assay reaction containing ~1/2 of the total membrane preparation from one T-75 flask, ~95,000 DPM of UDP-[^{14}C]GlcUA, and 200 μM total UDP-GlcUA. The ^{14}C incorporation remaining at the origin in the absence of UDP-GlcNAc (*inset, white bar*) is about 10 % of the incorporation obtained with UDP-GlcNAc (*black bar, inset*)

Using tweezers, place each 2 × 2 cm paper square into the corresponding labeled scintillation vial with the pencil label facing the vial center, not the wall.

11. Add 1 mL of water to each scintillation tube in a holder that allows them to be rocked gently for 1 h to dissolve and elute the radiolabeled HA (this increases counting efficiency by decreasing quenching). Position the holder so that the tubes are at a 45° angle while rocking, so that the water more fully covers the paper square.
12. Add the appropriate volume of liquid scintillation cocktail for 1 mL water to each tube, cap and mix well, being certain that the paper remains submerged and is not above the scintillation fluid.
13. Place the scintillation vials in a scintillation counter rack, load into the counter and run a suitable program for detection of ^{14}C with quenching correction and output as DPM as per the manufacturer's instructions.

3.4 SEC-MALLS Assay

3.4.1 Setting Up SEC-MALLS Assays

1. Set up reaction samples (0.1 mL per sample) in triplicate using clean labeled conical microfuge tubes (0.6–2.0 mL).
2. Add 50 μL of 2× MALLS assay mix to each tube and the desired amount of membrane suspension.

3. Add water to each tube to give a volume of 95 μL and gently agitate at 30–37 °C for 10–15 min (*see* **Note 8**).
4. Start reactions by adding 5 μL of 0.4 M $MgCl_2$. Mix well, place tubes back on the mixer, and incubate at the same temperature for 1–4 h (*see* **Note 7**).
5. To stop reactions add 10 μL of 500 mM EDTA and 10 μL of 120 mM UDP (to give final concentrations of 42 and 10 mM, respectively), mix well, and chill on ice for 20 min. Heat samples at 100 °C for 1 min and store at 4 °C if analyzed within a few days or at −20 °C for longer storage.

3.4.2 MALLS Sample Preparation

1. Centrifuge quenched reaction samples at 20,000 × *g* for 30 min (4 °C) to pellet membranes.
2. Quantitatively remove a measured volume (e.g., 115 μL) of supernatant to a clean microfuge tube to perform a Folch extraction [21] to remove lipids. Add three volumes of chloroform–methanol (2:1), agitate vigorously to mix phases (e.g., vortex mixer) at RT for 15 min, and centrifuge at 6,800 × *g* for 5 min to separate the organic and aqueous layers.
3. Transfer the top aqueous layer quantitatively to a new microfuge tube, remove residual organic by speed vacuum centrifugation (e.g., 2 h at RT) and add MALLS running buffer to reconstitute to the original volume (*see* **Note 11**).

3.4.3 MALLS Sample Analysis

1. Chromatographic SEC separation of samples is achieved at 22 °C at a flow rate of 0.4–0.5 mL/min in degassed MALLS running buffer using one PL aquagel-OH60 column in series with either another PL aquagel-OH60 (for >1 MDa HA) or a PL aquagel-OH40 column (for <1 MDa HA or when a broad range of HA sizes is present and better separation of smaller HA from residual protein and other contaminants is required). A 0.1 μm or smaller in-line solvent filter is critical to minimize light scattering noise (*see* **Note 12**).
2. Briefly heat samples if needed and centrifuge to collect liquid at the bottom. Remove 100 μL to inject manually or load into vials for automatic injection using an autosampler. Run times are typically 1–2 h per sample, to allow time to return to baseline (*see* **Note 13**).
3. Analyze data using Astra (or other suitable) software, using a dn/dc value of 0.153 [22], an A2 value of 0.0023, and either first-order Zimm or second-order Berry fits (*see* **Note 14**). Results are typically presented as one continuous plot of refractive index versus elution volume overlaid with a discontinuous plot of molar mass values, with symbols representing data from discrete elution volumes (Fig. 4). Depending on the HA size range detected, there may be co-migration of other large species (e.g., proteoglycans) that contribute to the light scattering signal (*see* **Note 15**).

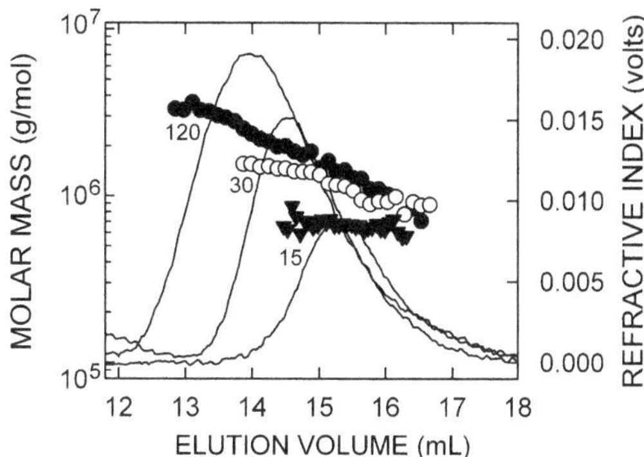

Fig. 4 SEC-MALLS data shows the kinetics of HA synthesis and increasing HA size. The refractive index tracings (*black lines*; *right scale*) show the amount of HA at 15 (*inverted triangle*), 30 (*open circle*), and 120 (*filled circle*) min of synthesis (using *E. coli* membranes expressing recombinant *Streptococcus equisimilis* HAS) at 30 °C. With increasing time, the size (area) of each peak increases, indicating the amount of HA, and the position of each peak shifts to the *left* (indicating larger size in the SEC fractionation). The individual symbols show values of molar mass (*left scale*) calculated for 4 s volume increments (not all datum points are shown). At 15, 30, and 120 min the Mw (weight-average molar mass) values were 0.65, 1.33, and 2.24 MDa and the HA contents were 19, 34, and 59 μg/mL, respectively

3.5 Nonradioactive AGE Assay

1. Assay mix and membrane sample incubations are identical to that described for PCr assays (Subheading 3.2), except that UDP-[^{14}C]GlcUA is omitted.

3.6 Agarose Gel Electrophoresis

1. Centrifuge quenched reaction samples at 20,000 × *g* for 30 min (4 °C) to remove membranes (*see* **Note 16**).
2. Agarose gel electrophoresis [13] is performed using 0.5–1.5 % (w/v) gels, depending on the size of the HA to be fractionated (*see* **Note 17**). After it cools, transfer the cast gel to the flat cell and overlay and cover the gel with AGE buffer.
3. Mix five volumes of sample supernatant with one volume of 6× AGE Sample buffer.
4. Slowly load an amount of sample per well (by underlaying below the AGE buffer) that contains 1–5 μg HA keeping all loading volumes the same (~2 μg generally gives a good signal). For a 100 mL gel with a 20 well comb, 30–40 μL can be loaded per well; 55–60 μL can be loaded per well using a 150 mL gel (*see* **Note 18**).
5. Electrophoresis is at 80–100 V for 2–3 h until the dye front is near the bottom of the gel. When running at these voltages, the gel will become too hot; keep the apparatus cool either by having it in an ice bath tray or in a cold room.

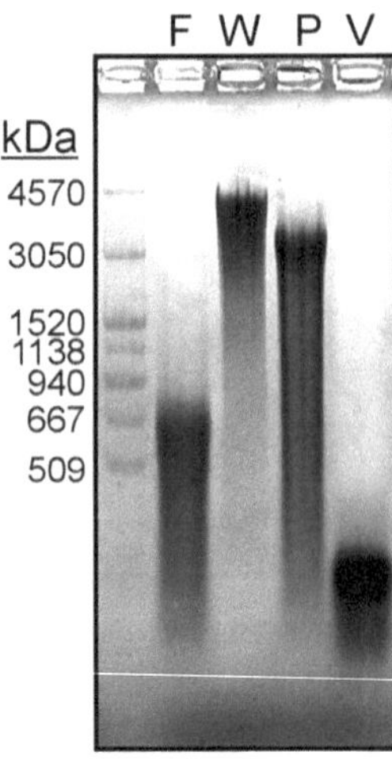

Fig. 5 Agarose gel of HA made by various SeHAS mutants. Membranes from *E. coli* Sure cells expressing wildtype (*lane* 3, W), or K48F (*lane* 2, F), K48P (*lane* 4, P), and K48V (*lane* 5, V) variants of SeHAS were incubated for 2 h at 30 °C and then reaction mixes were quenched and centrifuged as described in the text. Sample supernatants (50 μL) were mixed with 10 μL 6× AGE sample buffer and 55 μL was loaded into lanes of a 0.8 % (150 mL) agarose gel. HiLadder and MegaLadder HA standards (5 μL each from Hyalose, LLC) were mixed with sample buffer and loaded into *lane* 1. Electrophoresis was at 100 V for 2 h in an ice-filled tray and the gel was stained and destained as noted in the text; a digital image was captured as a JPEG file using a FluorChem FC imaging system from Alpha Innotech. The gel illustrates the broad range of sizes that can be separated and visualized by AGE [13]

6. Decant the buffer from the AGE cell and slide the gel carefully into a flat container (tray) with a lid. Immerse in Stains-All solution and incubate overnight, with rocking, at RT; cover to keep out light.
7. Gels are de-stained by decanting the Stains-All solution, washing with water and then incubating in water while exposing to light until the background is acceptable and HA staining is robust enough to capture. Using a digital imaging station, save an image file in JPEG or other suitable format (Fig. 5).

3.7 Radioactive AGE Assay

1. Assay mix and membrane sample incubations are identical to that described for PCr assays (Subheading 3.2); UDP-[^{14}C] GlcUA is included.
2. The AGE protocol is identical to **steps 1–6** in Subheading 3.6 above (*see* **Note 19**). However, since the gels will not be stained, there is no visual reference to identify the top left (or right) lane and thus the order of sample lanes. Cut a diagonal piece from one corner, avoiding any sample lanes or wells, to define the orientation of the gel in subsequent steps.

3. Wash the gel in a clean tray with water to remove remaining UDP-[^{14}C]GlcUA (*see* **Note 20**) and underlay the gel with a clean piece of filter paper (cut to appropriate size) to help support the gel as it is transferred and dried.
4. Blot excess liquid and transfer to a gel dryer (paper down), cover with plastic wrap to minimize contamination and then cover with a heavy plastic sheet to help keep the gel flat under uniform pressure and minimize air bubbles. Dry the gel using mechanical vacuum at <50 °C to avoid melting the agarose.
5. Perform autoradiography with the dried gel against a suitable blue-sensitive X-ray film (e.g., X-Omat AR, Kodak) at −80 °C, until a desired exposure level is attained (*see* **Note 21**).

4 Notes

1. After all additions, the 4× reaction stock for V_f of 1.0 mL, gives nine sample reactions (e.g., three sets of triplicates) and an extra 0.1 mL. A slightly larger volume (at least one additional reaction portion of 0.1 mL) should be made to allow for sample pipetting differences and for taking samples (e.g., 10 and 20 μL) to determine total radioactivity in the final mix (from which percent of substrate incorporation into HA can be calculated). As the 4× reaction volume is increased to handle more samples, a higher fraction of total volume can be used for samples. Thus, for a V_f of 10 mL reaction mix, 96–98 samples could be assayed.
2. We prefer to increase specific radioactivity to optimize signal strength, in a cost effective way, by increasing the expensive UDP-[^{14}C]GlcUA (twofold to fourfold) and by lowering the concentration of unlabeled UDP-GlcUA. The trade-off is that enzyme velocity will be slower at the lower substrate concentration. In our experience, this strategy generally works well—keeping the substrate concentrations at 100–200 μM, which is near the K_m values for most HASs [4, 10].
3. Since HASs seem particularly sensitive to inactivation by proteases released during homogenization and protease profiles may vary with cell type, one should include a selection of protease inhibitors with different targets and mechanisms of action. The number of inhibitors used and their concentrations likely need to be optimized, especially if it is known that the cells being studied make and secrete HA into the medium. This can be confirmed using an HA test kit that uses an ELISA-like assay with a specific HA-binding protein (e.g., from Echelon or Corgenics).
4. Centrifugation time or g-force can be varied to assess if increasing yield of HAS-containing membranes can be obtained. Larger membrane fragments, such as sheets, may be generated

by Dounce homogenization and recovered at the relatively low speed centrifugations used here. Recovery of smaller microsomal membranes, which may be more favored by sonic disruption procedures, would be increased at higher g-force or time.

5. Avoid using different total concentrations of each UDP-sugar, since an excess of one can inhibit HAS activity by competing for the UDP binding site of the second substrate [23].
6. Uniformly radiolabeled GlcUA usually has the highest specific radioactivity. We do not use radiolabeled UDP-GlcNAc because much higher background levels result, due to the large number of endogenous GTases that will transfer GlcNAc to many different acceptors. An additional advantage of using UDP-[^{14}C]GlcUA is the ability to do an independent control for HAS-dependent incorporation into HA by omitting the UDP-GlcNAc.
7. Some reactions should proceed for different times to assess linearity of synthesis with enzyme and to determine if weak signals can only be detected after longer time. Individual assay reactions will often have different amounts of membrane protein to assess linearity of HAS activity (HA synthesis), variable glycerol concentrations to assess effects on stabilizing HAS activity, or different amounts of UDP-[^{14}C]GlcUA to increase specific radioactivity and thus sensitivity of detection. Accordingly, 25 μL of the 4× reaction stock will be pipetted into each sample tube and different volumes of the Master Stock solutions of glycerol, membranes, UDP-[^{14}C]GlcUA, UDP-sugars, and water may then be added to a volume of 95.0 μL per tube. Reactions are then initiated by adding 5 μL of 0.4 M $MgCl_2$ to each reaction tube.
8. If initial trials show little or no HAS activity at 37 °C (e.g., due to enzyme instability), then test different temperatures ranging down to 30 °C, or even 25 °C. Lower temperatures might enable HAS to stay active longer, although at reduced synthetic rates as the temperature is lowered; compensate for this by performing assays over longer time periods.
9. Spot the sample gently and slowly (stabilizing your arm or hand on a surface helps), allowing the liquid to wick into the paper and not to pool or flow over the paper. Avoid puncturing the wet paper. It is important that the paper not touch anything or part of the sample can be lost by transfer (wicking) to a supporting material (Fig. 1). The spots will spread in a circular pattern and occupy most of the 4 cm^2 area.
10. HA larger than about 8–12 sugars remains at the origin, while soluble UDP-[^{14}C]GlcUA migrates far down the strip (sometimes dripping onto and contaminating the tank bottom), resulting in a low background at the origin (typically <50 cpm). This should be verified by comparing an unused

(no radioactivity) paper piece with a control strip run with radioactive reaction buffer without HAS-containing membranes. The fact that HA fragments >8–12 sugars can stay at the origin also means that controls to verify that HA is the true product cannot reliably use hyaluronidase digestion. The single UDP-[^{14}C]GlcUA control without UDP-GlcNAc (*see* **Note 6**) is a good alternative control (Fig 3).

11. Failure to remove lipid by extracting with chloroform–methanol can cause artifacts due to co-eluting large MDa material; this results in apparently larger HA molecular mass values. The SEC columns will also soon malfunction, due to accumulation of lipids. This artifact can be eliminated or minimized by briefly heating samples just prior to injection, in a 100 °C water or sand bath for 1 min/100 μL volume [24]. Regular column washes with 0.1 % SDS, 20 % methanol, 0.05 % sodium azide in water generally restores and maintains column performance and HA fractionation; wash out the phosphate buffer first with 0.05 % azide in water prior to the SDS/methanol wash to avoid salt precipitation.
12. MALLS analysis simultaneously determines the weight-average (Mw) and number average (Mn) molar mass, the range of masses present (size distributions or polydispersity) and the amount of HA in the sample—without the need for calibration standards. Data are collected continuously as the fractionated column eluate passes through a laser photometer and then a refractometer [15].
13. Higher HA concentrations (>0.1 mg/mL of HA >200–300 kDa) should be avoided with SEC–MALLS because artificially high mass values may occur. An optimal HA concentration for a 100 μL sample using a 2-column system is ~0.05 mg/mL (~5 μg total HA). Injecting more HA often results in poor separation and skewed peaks [24].
14. In our experience, first-order Zimm fits have less uncertainty in analysis of data for HA of <2 MDa, whereas second-order Berry analysis of data for HA >2 MDa gives better fits [24].
15. An important control is to verify that any signal seen is HA, by its elimination when samples are treated with the *Streptomyces hyalurolyticus* lyase that is specific for HA. Testicular hyaluronidase should not be used, since it will also degrade other large glycosaminoglycans (and proteoglycans) that may be present and could contribute to a light scattering signal not associated with HA.
16. If membrane protein is >20–25 μg/mL, higher g-force or longer times may be needed to clarify samples. If gels run aberrantly, then include the chloroform–methanol extraction steps to remove lipid as described for samples undergoing SEC-MALLS analysis (*see* Subheading 3.4.2).

17. The best resolution and results are obtained if sample HA does not migrate near the top or near the bottom of the gel. If samples migrate too slowly (close to the top wells), then decrease the % agarose or if samples migrate too fast (too far), then increase the % agarose.
18. Trials may be necessary to determine appropriate sample volumes to load. The amount of HA is estimated by comparison to the staining intensity of known amounts of purified HA samples used as "standards" in the same gel (e.g., depending on the HA size range produced by HASs in the sample membranes, a range of unique very narrow-sized HA or broad-sized HA can be obtained, respectively, from Hyalose or Lifecore). If HA content is too low, samples can be concentrated by speed-vacuum centrifugation or lyophilization and redissolving in a smaller volume (e.g., 30 μL). Speed-vacuum concentration is at RT or heating to only 30–35 °C. It may take 2–4 h to get back to the original volume prior to Folch extraction; then concentrate further to ~1/3 of initial volume. If samples are to be lyophilized following Folch extraction, they should first have residual organic material removed by speed-vacuum centrifugation to minimize bumping and potential loss of sample. Care should be taken to measure the volumes of samples before and after concentration, to perform appropriate book-keeping for recovery and HA content.
19. Since radioactivity is used, the equipment and buffers can be contaminated and all appropriate safety, monitoring and accounting procedures for radioactive materials should be followed.
20. Unless the HA made is very small, it will remain in the gel. The cut-off for molecules diffusing out of a gel during washing will be inversely related to the % agarose. The reason to remove unused ^{14}C substrate is that only a very small percent of the total radioactivity will be in HA; if too much UDP- [^{14}C] GlcUA remains, the resulting autoradiogram will be compromised by the broad band of substrate and other small labeled molecules that migrate near the front; higher background throughout the film due to high decay events that travel parallel to the film. If this happens, cut and remove this section of the gel with the front and expose the gel again on fresh film.
21. If bands are weak even after 2–4 weeks of exposure, sensitivity can be increased in several ways: (1) by increasing specific radioactivity in the assay mix, (2) by pooling and concentrating (using speed vacuum centrifugation or lyophilization) several replicates into one final sample; (3) by performing fluorography—impregnating the washed gel with scintillants (e.g., Enlightening, Perkin-Elmer) that enhance β-emission energy capture and transfer to light; or (4) by using intensifying screens designed to achieve higher sensitivity without fluorophore impregnation.

References

1. Fosang AJ, Hey NJ, Carney SL, Hardingham TE (1990) An ELISA plate based assay for hyaluronan using biotinylated proteoglycan G1 domain (HA-binding region). Matrix 10: 306–313
2. Itano N, Sawai T, Yoshida M, Lenas P, Yamada Y, Imagawa M, Shinomura T, Hamaguchi M, Yoshida Y, Ohnuki Y, Miyauchi S, Spicer AP, McDonald JA, Kimata K (1999) Three isoforms of mammalian hyaluronan synthases have distinct enzymatic properties. J Biol Chem 274:25085–25092
3. Brinck J, Heldin P (1999) Expression of recombinant hyaluronan synthase (HAS) isoforms in CHO cells reduces cell migration and cell surface CD44. Exp Cell Res 252: 342–351
4. Itano N, Kimata K (2002) Mammalian hyaluronan synthases. IUBMB Life 54:195–199
5. Spicer AP (2001) In vitro assays for hyaluronan synthase. Methods Mol Biol 171: 373–382
6. Weigel PH (2002) Functional characteristics and catalytic mechanisms of the bacterial hyaluronan synthases. IUBMB Life 54:201–210
7. Tlapak-Simmons VL, Baggenstoss BA, Clyne T, Weigel PH (1999) Purification and lipid dependence of the recombinant hyaluronan synthases from *Streptococcus pyogenes* and *Streptococcus equisimilis*. J Biol Chem 274: 4239–4245
8. Yoshida M, Itano N, Yamada Y, Kimata K (2000) In vitro synthesis of hyaluronan by a single protein derived from mouse HAS1 gene and characterization of amino acid residues essential for the activity. J Biol Chem 275:497–506
9. DeAngelis PL, Oatman LC, Gay DF (2003) Rapid chemoenzymatic synthesis of monodisperse hyaluronan oligosaccharides with immobilized enzyme reactors. J Biol Chem 278: 35199–35203
10. Weigel PH, DeAngelis PL (2007) Hyaluronan synthases: a decade-plus of novel glycosyltransferases. J Biol Chem 282:36777–36781
11. Prehm P (1983) Synthesis of hyaluronate in differentiated teratocarcinoma cells. Characterization of the synthase. Biochem J 211:181–189
12. Sheehan JK, Arundel C, Phelps CF (1983) Effect of the cations sodium, potassium, and calcium on the interactions of hyaluronate chains: a light scattering and viscometric study. Int J Biol Macromol 5:222–228
13. Lee HG, Cowman MK (1994) An agarose gel electrophoretic method for analysis of hyaluronan molecular weight distribution. Anal Biochem 219:278–287
14. Prehm P (1983) Synthesis of hyaluronate in differentiated teratocarcinoma cells. Mechanism of chain growth. Biochem J 211: 191–198
15. Wyatt PJ (1997) Multiangle light scattering combined with HPLC. LCGC 15:160–168
16. Wyatt PJ (1993) Light scattering and the absolute characterization of macromolecules. Anal Chim Acta 272:1–40
17. Tlapak-Simmons VL, Kempner ES, Baggenstoss BA, Weigel PH (1998) The active streptococcal hyaluronan synthases (HASs) contain a single HAS monomer and multiple cardiolipin molecules. J Biol Chem 273: 26100–26109
18. Pettaway CA, Pathak S, Greene G, Ramirez E, Wilson MR, Killion JJ, Fidler IJ (1996) Selection of highly metastatic variants of different human prostatic carcinomas using orthotopic implantation in nude mice. Clin Cancer Res 2:1627–1636
19. Kumari K, Tlapak-Simmons VL, Baggenstoss BA, Weigel PH (2002) The streptococcal hyaluronan synthases are inhibited by sulfhydryl modifying reagents but conserved cysteine residues are not essential for enzyme function. J Biol Chem 277:13943–13951
20. Tlapak-Simmons VL, Baggenstoss BA, Kumari K, Heldermon C, Weigel PH (1999) Kinetic characterization of the recombinant hyaluronan synthases from *Streptococcus pyogenes* and *Streptococcus equisimilis*. J Biol Chem 274:4246–4253
21. Folch J, Lees M, Sloane Stanley GH (1957) A simple method for the isolation and purification of total lipides from animal tissues. J Biol Chem 226:497–509
22. Ghosh S, Khobal I, Zanette D, Reed WF (1993) Conformational contraction and hydrolysis of hyaluronate in sodium hydroxide solutions. Macromolecules 26:4684–4691
23. Tlapak-Simmons VL, Baron CA, Weigel PH (2004) Characterization of the purified hyaluronan synthase from *Streptococcus equisimilis*. Biochemistry 43:9234–9242
24. Baggenstoss BA, Weigel PH (2006) Size exclusion chromatography-multiangle laser light scattering analysis of hyaluronan size distributions made by membrane-bound hyaluronan synthase. Anal Biochem 352:243–251

Chapter 19

Creation and Characterization of Glycosyltransferase Mutants of *Trypanosoma brucei*

Luis Izquierdo, M. Lucia S. Güther, and Michael A.J. Ferguson

Abstract

The survival strategies of protozoan parasites frequently involve the participation of glycoconjugates. *Trypanosoma brucei* expresses complex glycoproteins throughout its life cycle and a review of its repertoire of glycosidic linkages suggests a minimum of 38 glycosyltransferase activities. Here we describe a functional characterization workflow in which we create glycosyltransferase null or conditional null mutants in both the bloodstream and procyclic life-cycle forms of the parasite. Subsequently, we characterize the biochemical phenotype of the mutant strains generated and assign precise functions to the genes involved in glycoconjugate biosynthesis and processing in *T. brucei*. In this way, a comprehensive picture of *T. brucei* glycosylation associated genes, their specificities and their relationship to similar genes in other organisms can be obtained.

Key words Glycosyltransferase, Galactosyltransferase, *N*-acetylglucosaminyltransferase, Mass spectrometry, Gene replacement, Trypanosoma

1 Introduction

The African trypanosomes are protozoan parasites that cause human African sleeping sickness and Nagana in cattle. The parasite undergoes a complex life cycle between the mammalian host and the blood-feeding tsetse fly vector (*Glossina* sp.). Throughout this life cycle, *T. brucei* is coated by glycosylphosphatidylinositol (GPI) anchored proteins. The bloodstream form of the parasite in the mammalian host is covered by a coat of 5×10^6 variant surface glycoprotein (VSGs) homodimers and it evades the immune system by replacing one VSG coat by another, in a process known as antigenic variation [1–4]. The VSG GPI anchors contain side chains of 0-6 Gal residues, depending on the VSG variant [5, 6] and between 1 and 3 N-linked glycans. The latter can be of the conventional oligomannose or paucimannose or complex types [7–11]. When bloodstream-form parasites are ingested by the tsetse fly, they differentiate to the procyclic form in the insect midgut. During this process the

Inka Brockhausen (ed.), *Glycosyltransferases: Methods and Protocols*, Methods in Molecular Biology, vol. 1022, DOI 10.1007/978-1-62703-465-4_19, © Springer Science+Business Media New York 2013

VSG coat is replaced by a mix of a non-GPI surface coat [12] and about 3×10^6 GPI-anchored procyclin glycoproteins. Procyclins are characterized by internal dipeptide (EP) or pentapeptide (GPEET) repeats which confer a rod-like structure to the protein [13, 14]. Most expressed EP procyclins contain a single N-glycosylation site, occupied exclusively by a conventional $Man_5GlcNAc_2$ oligosaccharide, at the N-terminal side of the EP repeat domain [14, 15]. The GPEET procyclins are not N-glycosylated but are phosphorylated on six of seven Thr residues [16, 17]. Both, GPEET and EP procyclins, contain similar GPI membrane anchors. These are the largest and most complex anchors known and they are characterized by the presence of large poly-disperse branched *N*-acetyllactosamine (Galβ1-4GlcNAc) and lacto-*N*-biose (Galβ1-3GlcNAc)-containing side chains that can be capped with α2–3-linked sialic acid residues [14, 18]. Sialic acid is transferred from serum sialoglycoconjugates to terminal β-galactose residues by the action of a cell surface trans-sialidase enzyme [19–22] and trans-sialylation of surface components plays a role in the successful colonization of the tsetse fly [23]. In addition, it has been postulated that the branched side chains of the anchor form a dense glycocalyx that contributes to the protective function of the coat against digestive enzymes in the fly midgut [24]. The procyclins exist alongside free GPI glycolipids of similar structure to the procyclin anchors, except that they contain, on average, 4 Gal-GlcNAc repeats as opposed to the ten of the procyclin anchors [23]. These GPI-anchored structures, in turn, sit alongside a network of glycosylated transmembrane transporters [25]. As the parasite infection develops in the tsetse vector, they differentiate into epimastigote forms that colonize the tsetse salivary glands and that express GPI-anchored BARP glycoproteins [26]. These finally differentiate into VSG-coated nondividing metacyclic trypomastigotes that are ready to infect a mammalian host upon tsetse bite. As well as the aforementioned major GPI-anchored surface molecules, *T. brucei* expresses numerous other GPI-anchored and transmembrane glycoproteins at the cell surface, in the flagellar pocket and in the intracellular endosomal/lysosomal system, some of which are life-cycle-stage-specific or display life-cycle-stage-specific glycosylation differences. For example, the transmembrane invariant surface glycoproteins ISG65 and ISG75 [27] and the GPI-anchored flagellar pocket ESAG6/ESAG7 heterodimeric transferrin receptors [28–30] are specific to the bloodstream life-cycle stage while the major lysosomal glycoprotein p67 is common to bloodstream and procyclic stages but contains complex N-glycans only in the bloodstream stage [31]. This control of stage-specific glycosylation probably resides at the level of oligosaccharyltransferase (OST) expression [32]. Thus, in bloodstream-form *T. brucei* it is known that both the *TbSTT3A* and *TbSTT3B* genes are expressed and it appears that TbSTT3A co-translationally scans for glycosylation sequons in relatively acidic local environments, transferring exclusively $Man_5GlcNAc_2$ that is destined to be processed to paucimannose or complex N-glycans, while TbSTT3B post-translationally modifies any remaining sequons with $Man_9GlcNAc_2$ that is destined to be processed no further than $Man_5GlcNAc_2$ in the conventional oligomannose series. Conversion from the oligomannose series to the complex series by the conventional mammalian-type route cannot occur because the parasite lacks a Golgi alpha-mannosidase II gene [11]. In procyclic form *T. brucei* the expression of *TbSTT3A* is highly repressed [32], strongly favoring the transfer of $Man_9GlcNAc_2$ and the predominant expression of the conventional $Man_9GlcNAc_2$—$Man_5GlcNAc_2$ series [33].

The survival strategies of protozoan parasites frequently involve the participation of glycoconjugates. *T. brucei* expresses many glycoproteins containing Gal and GlcNAc, including glycoproteins with novel bloodstream-form-specific giant poly-*N*-acetyllactosamine (poly-LacNAc) containing N-linked glycans [9, 34]. The creation of UDP-Glucose 4′-epimerase (*TbGalE*) conditional null mutants showed that this gene, and hence UDP-Gal, is essential for the survival of the parasite in both the bloodstream and procyclic-form life-stages [35–37]. Similarly, the creation of UDP-GlcNAc pyrophosphorylase (*TbUAP*) and glucosamine 6-phosphate *N*-acetyltransferase (*TbGNA1*) conditional null mutants has shown that UDP-GlcNAc is also essential for bloodstream-form *T. brucei* [38, 39]. From these experiments, it is possible to conclude that one or more of the UDP-Gal- and UDP-GlcNAc-dependent glycosylation pathways are essential to the parasite. This has provided the impetus to identify and characterize the UDP-Gal and UDP-GlcNAc-dependent glycosyltransferase (GT) genes in the parasite. In a recent study, we mined the *T. brucei* genome for GTs and found a family of 21 genes with predicted amino acid sequences consistent with being UDP-sugar-dependent GTs. All 21 putative *T. brucei* GT amino acid sequences are similar to those of the mammalian β3GT family [40]. The mammalian β3GT family includes Gal, Glc, GlcA, GlcNAc, and GalNAc β1,3-transferases and its members contain N-terminal transmembrane domains followed by three conserved motifs of (I/L)R*XX*WG, (F/Y)(V/L/M)*XXX*D*X*D, and (E/D) D(A/V)(Y/F)*X*G*X*(C/S). The comparable motifs in the *T. brucei* genes are slightly different: WG, Y(I,V,F)*X*K*X*DDD, and ED(A/V/I/L/M)(M/L)*X*(G/A) but are, nevertheless, identifiable [41]. The first of these genes to be studied (*TbGT8*) is indeed encoding a β1,3-GlcNAc-transferase that modifies the complex GPI anchor side chains of the procyclins [41]. However, it is conceivable that other members of the *T. brucei* β3GT family might encode the β1,4 and β1,6 GTs needed to synthesize the Galβ1-4GlcNAc lactosamine units and GlcNAcβ1-6Gal inter-lactosamine linkages that are abundant in both life-cycle stages, i.e., in the complex N-linked glycans of the bloodstream form and the procyclin GPI side chains of the procyclic form. Certainly, BLAST searches with mammalian β4Gal-T sequences and β6Gn-T sequences fail to return any convincing candidates. Thus, we have postulated that there are either additional classes of currently unidentifiable UDP-Gal/GlcNAc dependent GT genes to be found in the *T. brucei* genome and/or this organism has adapted the β1,3-glycosyltransferase family to catalyze a variety of different linkages. In this article, we describe the methodology we use to biochemically characterize null and conditional null mutants of trypanosome GTs.

2 Materials

2.1 Generation of Glycosyltransferase (GT) Null-Mutant and Conditional Null-Mutant Cell Lines

2.1.1 Generation of the Constructs

1. Standard vector for cloning DNA fragments in *Escherichia coli* (pGEM-5Zf(+)).
2. pLew100 tetracycline-inducible based vector [42].
3. Oligonucleotide primers.
4. High-Fidelity DNA polymerase (Life Technologies, Carlsbad, CA, USA).
5. PCR purification/gel extraction kit and miniprep plasmid DNA purification kit (QIAGEN, Valencia, CA, USA).
6. T4 DNA ligase (New England Biolabs, Ipswich, MA, USA).
7. Large-scale plasmid DNA purification kit (QIAGEN).
8. Restriction enzymes (New England Biolabs).

2.1.2 Transfection of T. brucei Cells by Electroporation

1. Bloodstream-form *T. brucei* cells strain 427 (variant 221), known as single marker cells because they express T7-RNA polymerase and tetracycline repressor (TETR) [43] protein under continuous geneticin (G418) selection [42].
2. For *T. brucei* bloodstream-form culture, Medium HMI-9T, which is HMI-9 [44] containing 56 μM monothioglycerol instead of 2-mercaptoethanol, 10 % (v/v) serum Plus (SAFC, Sigma, St. Louis, MO, USA) and supplemented either with 10 % heat-inactivated fetal bovine serum (FBS) or Tetracycline (Tet)-system-approved FBS.
3. Procyclic-form *T. brucei* cells, cell line 29.13.6 (derived from *T. brucei* strain 427) containing T7-RNA polymerase and TETR genes respectively under G418 and hygromycin control [42].
4. Medium for *T. brucei* procyclic-form culture: Medium SDM-79 [45] containing 7.5 mg/L hemin and 2 g/L sodium bicarbonate, pH adjusted to 7.3, filter-sterilized and further supplemented with 15 % heat-inactivated FBS or Tet-system-approved FBS, and 2 mM sterile fresh Glutamax I (Life Technologies).
5. Drug concentrations: *see* Table 1.
6. Cytomix: *see* Table 2.
7. Electroporation cuvette (0.4 cm gap).
8. BTX ECM-830 Square Porator with BTX 630B safety stand.

Table 1
Drug concentrations

Drug	Concentration bloodstream-form culture (µg/mL)	Concentration procyclic-form culture (µg/mL)
Blasticidin[a]	2.5	10
G418	2.5	15
Hygromycin	4	50
Puromycin	0.1	1
Phleomycin	2.5	2.5
Tetracycline[b]	0.5	0.5

[a]Blasticidin 10 mg/mL stock is prepared in water, sterile filtered, aliquoted, and stored frozen at −20 °C. Fresh stocks should be prepared every 2 months due to chemical instability. If kept at 4 °C or in medium at 37 °C should be discarded after 2 weeks. This is particular important during the selection process and fresh addition of antibiotics is required every 2 weeks
[b]Tetracycline is labile in water. A fresh stock solution (10 mg/ml in 70% ethanol in water) should be prepared every week at and stored at -20 C. A fresh aliquot should be added to the cultures every 48 h

Table 2
Cytomix

Component	Final concentration
EGTA	2 mM
KCl	120 mM
$CaCl_2$	0.15 mM
K_2HPO_4/KH_2PO_4 pH 7.6[a]	10 mM
HEPES	25 mM
$MgCl_2 \cdot 6H_2O$	5 mM
Glucose	0.5 % (w/v)
BSA (defatted)	100 µg/mL
Hypoxanthine[b]	1 mM

[a]Prepare by mixing 8.66 mL 1 M K_2HPO_4 with 1.34 mL 1 M KH_2PO_4 to 90 mL H_2O
[b]Hypoxanthine 0.1 M stock prepared in 0.1 M NaOH

9. Cell culture conditions—procyclic-form: cells are cultivated at 28 °C without CO_2 using non-treated plastic culture flasks, fully capped (Beckton-Dickson, BD Biosciences, San Jose, CA, USA), while bloodstream-form cells are cultivated in filter capped culture flasks or otherwise partially open caps at 37 °C with 5 % CO_2.

2.1.3 Determination of T. brucei Chromosomal Background

1. DNAzol, for *T. brucei* genomic DNA isolation (Life Technologies, Carlsbad, CA, USA).
2. Materials required for Southern blotting transfer of DNA.
3. Random prime labelling system and [α-^{32}P]-dCTP (GE Healthcare, Little Chalfont, UK).
4. DNA labelled-probe(s).

2.2 Analysis of the N-glycan and GPI Anchor of the Variant Surface Glycoprotein

1. Phosphate buffered saline (PBS): 10 mM phosphate buffer, 2.7 mM KCl and 137 mM NaCl, pH 8.0.
2. Protease inhibitors: 0.1 mM 1-chloro-3-tosylamido-7-amino-2-heptone (TLCK), 1 μg/mL leupeptin and 1 μg/mL aprotinin.
3. DE52 anion exchange resin (Whatman, GE Healthcare, Little Chalfont, UK).
4. 10 mM sodium phosphate buffer (NaH_2PO_4), pH 8.0 adjusted with NaOH.
5. Microcon YM-10 concentrator (Amicon, Merck Millipore, Billerica, MA, USA).
6. Methanol. Analytical grade is preferable for further MS applications.
7. Formic acid. Analytical grade, to be used in MS applications.
8. Nanotips (Waters type F) for mass spectrometry.
9. Electrospray mass spectrometry system (ES-MS) (with MS/MS capability, *see* **step 9** of protocol in Subheading 3.2) with software converting spectra into mass data (e.g., ABI-Sciex Bayesian protein/peptide reconstruct or Waters maximum entropy program).
10. Ammonium bicarbonate.
11. Pronase (Sigma).
12. Calcium acetate.
13. Supelclean ENVI-Carb cartridge (Supelco, Sigma).
14. Omix C4 pipette tips, 10–100 μL (Agilent Technologies, Santa Clara, CA, USA).

2.3 Lectin Blotting of Bloodstream-Form Cell Lysates

1. Trypanosome dilution buffer (TDB) 1×: *see* Table 3.
2. Protease inhibitors: 0.1 mM 1-chloro-3-tosylamido-7-amino-2-heptone (TLCK), 1 μg/mL leupeptin and 1 μg/mL apronitin.
3. SDS-PAGE system and Bis-Tris 4–12 % gradient polyacrylamide gels (Life Technologies).
4. Western blotting transfer system.
5. Biotin-conjugated lectins (other detection systems or even HRP-conjugated lectins can be found from different suppliers).

Table 3
Trypanosome dilution buffer (TDB) 1×

Component	Final concentration (mM)
KCl	5
NaCl	80
$MgSO_4 \cdot 7H_2O$	1
Na_2HPO_4	20[a]
$NaH_2PO_4 \cdot 2H_2O$	2[a]
Glucose	20

[a]pH should be 7.7 without adjustment; stock can be prepared 5× concentrated and kept at −20 °C

6. PBS.
7. Lectin inhibitors, depending on the lectin specificity (e.g., D(+) galactose and α-lactose for ricin and chitin hydrolysate for wheat germ agglutinin and tomato lectin).
8. Horseradish peroxidase (HRP)-labelled ExtrAvidin (Sigma).
9. Enhanced chemiluminescence (ECL) detection system (GE Healthcare).
10. ECL Film (GE Healthcare).

2.4 Procyclin Characterization by SDS-PAGE

1. PBS.
2. Freeze drying equipment.
3. Glass or chloroform-resistant plasticware.
4. Chloroform, methanol, and butan-1-ol (VWR, West Chester, PA, USA). Analytical grade is preferable, especially if the purified procyclins are to be used in subsequent protocols in Subheadings 3.5 and 3.6.
5. Reagents for periodate-Schiff stain: acetic acid, methanol, sodium periodate, Schiff's fuchsin-sulfite reagent (Sigma), and sodium metabisulfite.

2.5 Procyclin N-glycan Analysis

1. Acid washed 2 mL Reacti-Vials (Pierce, Thermo Fisher Scientific, Waltham, MA, USA) and other laboratory glassware.
2. Peptide N-glycosidase F (PNGase F) (Roche, Penzberg, Germany).
3. 0.25 M sodium phosphate buffer (NaH_2PO_4), adjusted to pH 7.5 with NaOH.
4. SpeedVac concentrator.
5. Dimethyl sulfoxide (DMSO).
6. Sodium hydroxide.
7. Iodomethane (Sigma).

8. Sodium thiosulfate (VWR).
9. Acetonitrile. Analytical grade is preferable for MS applications.
10. Sodium acetate. Analytical grade, to be used in MS applications.
11. Nanotips (Waters type F) for mass spectrometry.
12. Electrospray mass spectrometry system (ES-MS) (with MS/MS capability).

2.6 Procyclin GPI-Anchor Characterization by ES-MS[3]

1. Reagents for glycan permethylation (1 and 4–10, in previous Subheading 2.5).
2. 2,5-dihydroxybenzoic acid (Sigma).
3. MALDI-TOF mass spectrometer.
4. LTQ Orbitrap mass spectrometer.

3 Methods

3.1 Generation of GT Null-Mutant and Conditional Null-Mutant Cell Lines

3.1.1 Primer Design and Generation of the Constructs

1. As in other organisms, *T. brucei* homologous recombination is strictly dependent on substrate length and it is impeded by base mismatches [46]. Thus, the gene replacement of our target GT gene requires the cloning of a drug resistance marker between the gene 5′ and 3′-untranslated region (UTR) sequences of at least 200 bp in length, but preferably 400–500 bp (*see* **Note 1**). To achieve this, two different sets of primers are designed (*see* Fig. 1a): one set is used to amplify the 5′-UTR region, with the tailed-reverse primer (primer 2) annealing right before the initiator ATG; the second set will amplify the 3′-UTR region with the tailed-forward primer (primer 3) annealing right after the stop codon (*see* **Note 2**). This will avoid altering the functionality of the open reading frame (ORF) promoter after the recombination event. The tailed-primers include complementary sequences at their 5′ ends containing, for instance, *Hin*dIII and *Bam*HI cloning sites to introduce the resistance gene (*see* **Note 3**). 5′ and 3′-UTR regions are PCR-amplified from genomic DNA using a high-fidelity DNA polymerase and the two PCR products generated are used together in a further PCR reaction (using primers 1 and 4, *see* Fig. 1a) to yield a product containing the 5′-UTR linked to the 3′-UTR by a short *Hin*dIII and *Bam*HI cloning site (*see* **Note 4**).
2. The fragment generated, containing both UTRs, is cloned into any standard cloning vector (such as pGEM-5Zf(+)). At least three different clones should be sequenced and compared to the *T. brucei* genome sequence to rule out any possible mutation introduced by the DNA polymerase that could affect the

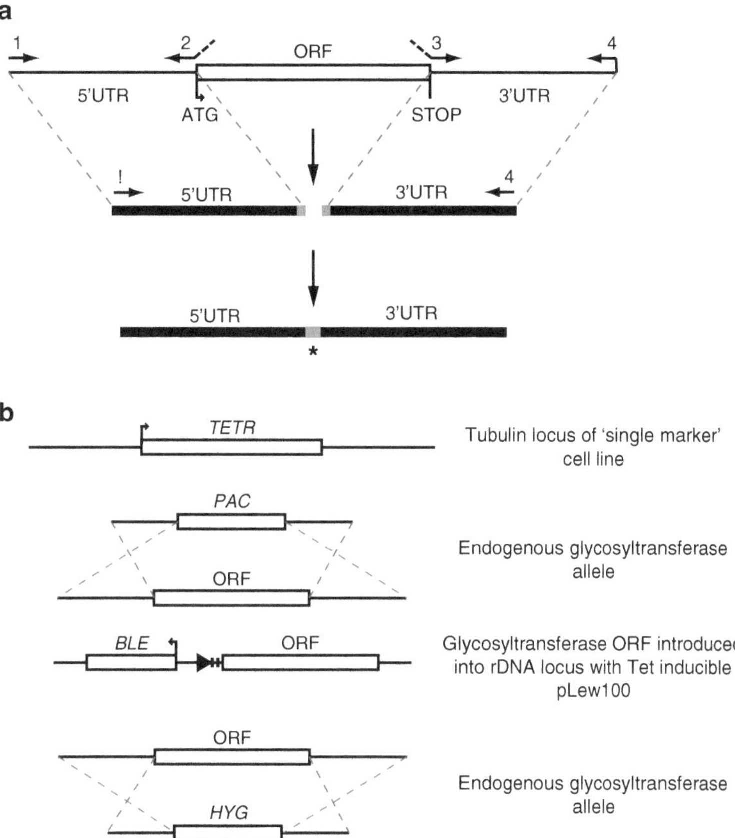

Fig. 1 Construction of *T. brucei* conditional null-mutants. (**a**) Creation of the constructs. The *top line* represents the region of the chromosome where the GT is located (*open square*, ORF) and the annealing position of the primers, indicated by numbers 1–4. The *middle line* shows the two PCRs used to generate fragments 5′UTR and 3′ UTR. The *bottom line* shows a further PCR reaction product obtained using 5′UTR and 3′UTR fragments as template and primers 1 and 4. This product will contain *Hin*dIII and *Bam*HI restriction sites instead of the ORF (marked with an *asterisk*). The PCR primers 2 and 3 are complementary over 11–20 nucleotides (represented by *dashed lines*) containing the restriction recognition sequences. Primers 1 and 4 may have different restriction sites in their 5′ ends so that the fusion product can be cloned into pGEM 5Zf(+). (**b**) Schematic representation showing the location of the *TETR* gene in *T. brucei* bloodstream-form "single marker" cells, the targeted replacement of one GT ORF with *PAC*, the introduction of an ectopic tetracycline-inducible copy of the GT into the rDNA locus, and the replacement of the second allele with *HYG*. *Small arrows* indicate direction of T7-constitutive promoters; *black solid arrow* shows the pLew100 procyclin promoter; *black solid bars* indicate the position of tetracycline operator sequences

expression of the marker or the efficiency of the homologous recombination event. The new vector generated will be used to introduce the desired resistance gene in the *Hin*dIII and *Bam*HI cloning site. Generally, puromycin *N*-acetyltransferase

(*PAC*) and hygromycin phosphotransferase (*HYG*) genes are used to generate a null-mutant cell line in single-marker bloodstream-form cells resistant to G418, co-expressing a T7-RNA polymerase and a Tet repressor (derived from *T. brucei brucei* strain 427) [43]. In the case of procyclic-form parasites, the cell line 29.13.6 (derived, as well, from *T. brucei brucei* strain 427) contains a T7-RNA polymerase and a tetracycline repressor (TETR), and it is resistant to G418 and hygromycin [43]. Thus, the resistance genes used to generate a null-mutant in this procyclic line are *PAC* and blasticidin S deaminase (*BSD*).

3. In order to generate a conditional null-mutant cell line either in bloodstream or in procyclic-form parasites, the GT ORF is amplified from *T. brucei* genome and cloned into the pLew100 tetracycline-inducible expression vector [42], carrying the bleomycin resistance (*BLE*) gene that confers resistance to phleomycin. This expression vector allows the integration of an ectopic copy of the ORF into the trypanosome rDNA locus, driven by a trypanosome promoter regulated by two tetracycline operator sequences. Thus, the addition of tetracycline to the media relieves binding of the TETR protein to the operators and allows the expression of the GT gene.
4. The plasmids generated are extracted and purified with a large-scale preparation (maxi-preparation) to be used for the electroporation of the parasites. The purity of the plasmid preparation is relevant and should have a ratio 260/280 nm above 1.7.

3.1.2 Transfection of T. brucei Cells by Electroporation

1. Digest vector (10 μg/cuvette) to release the knockout cassette (*see* **Note 5**). It is recommended to analyze a small aliquot of the digestion by agarose gel. Ethanol-precipitate and wash it twice with cold 70 % ethanol. Leave tube(s) open in the laminar flow hood to dry in sterile conditions.
2. In the hood, resuspend DNA at 1 μg/μL in sterile water. 10 μL of DNA will be typically used per cuvette.
3. Bloodstream and procyclic-form cultures should be between 0.9 and 2×10^6/mL and 0.9–1.3×10^7 cells/mL, respectively.

3.1.3 Bloodstream-Form Transfection

All the steps will be carried out at room temperature (RT) with minimal delay. Bloodstream-form cells are grown in HMI-9T media described above. Each transfection will require 3×10^7 cells/cuvette and 10 μg of DNA/cuvette (*see* **Note 6**).

1. Pellet the cells for 10 min at 800 × g and wash three times with 10 mL of cytomix, at room temperature.
2. During the cytomix washes, cells can be counted in the hemocytometer to adjust the numbers.
3. Resuspend bloodstream-form cells at 6×10^7 cells/mL in cytomix.

4. Aliquot 500 μL (3×10^7 cells) into each electroporation cuvette (allowing one without DNA as a control).
5. Zap in BTX830 squareporator delivering around 1,600 V (device set between 1,700 and 1,900 V) in three pulses, with a pulse length of 100–200 μs of interval (*see* **Note 7**).
6. Cells are promptly transferred to 24 mL of HMI-9T media (they can be pooled together, keeping the control without DNA separately) and plated at 1 mL/well in 12-well culture plates.
7. Cells should recover for at least 8 h at the required growth conditions, meaning HMI-9T at 37 °C plus 5 % CO_2. Note that at this stage only antibiotics to which the cells were already resistant to should be used.
8. Add 1 mL/well of HMI-9T with 2× the new antibiotic selection, containing the usual amounts of the other antibiotics present since the recovery period.
9. Cells should start to die in 3–7 days (*see* **Note 8**), depending on the antibiotic selection used. The no DNA control will guide for cell death.
10. Transfectants should be subcultured into 10 mL flasks of HMI-9T when they reach a density around $1–2 \times 10^6$ cells/mL.

3.1.4 Procyclic-Form Transfection

All the steps will be carried out at room temperature (RT) with minimal delay. Procyclic-form cells are grown in SDM-79 media [45], with supplements described above (*see* Subheading 2.1.2, **item 3**). Each transfection will require 1×10^7 cells/cuvette and 10 μg of DNA/cuvette (*see* **Note 9**).

1. Pellet the cells for 10 min at 600 × g and wash three times with 10 mL of cytomix.
2. During the cytomix washes, cells can be counted in the hemo cytometer to adjust the numbers.
3. Resuspend procyclic-form cells at 2×10^7 cells/mL in cytomix.
4. Aliquot 500 μL (1×10^7 cells) into each electroporation cuvette (allowing one without DNA as a control).
5. Zap in BTX830 squareporator delivering around 1,600 V (the device should be set between 1,700 and 1,900 V) in three pulses, with a pulse length of 100–200 μs of interval.
6. Procyclic-form cells should be transferred to 10 mL of SDM-79 media in non-treated plastic 25 cm² culture flasks and recovered overnight at 28 °C without CO_2. Note that at this stage only antibiotics to which the cells were already resistant to should be used.
7. After recovery add equal volume of media containing 2× the new antibiotic selection and distribute 1 mL/well in 24-well culture plates (non-treated plastic).

8. Cells should start to die in 3–7 days (*see* **Note 8**), depending on the antibiotic selection used. The no DNA control will guide for cell death.
9. Transfectants should be cultured in plates and not fed until the culture density is very high. Then careful feeding will be required to avoid cell death. This can be achieved by daily drop additions of the media. When many dividing cells are obtained, the cells can be tested diluted at 1:2 and 1:3, inside of 24-well plates. When these cultures grow, transfer into 25 cm^2 non-treated culture flasks and avoid dilutions below 1×10^6 cells/mL.

3.1.5 Determination of *T. brucei* Chromosomal Background

1. Southern blot analysis is the method of choice to confirm the replacement of both chromosomal GT alleles with antibiotic resistance genes and to assess the insertion of the tetracycline inducible ectopic copy. To validate the appropriate replacement of the chromosomal alleles (*see* Fig. 1b) it is recommendable to use probes against the 3′ and/or 5′ UTRs of chromosomic DNA digested with an assorted panel of restriction enzymes. It is advisable as well to do southern blot using probes against the resistance genes (*PAC*, *HYG* or *BSD*) to determine if there is non-homologous recombination or duplication events of the selectable markers.
2. To test whether the GT gene is essential, the conditional null mutant cells are washed three times in medium without tetracycline supplemented with Tet-system-approved fetal bovine serum, and cultures (with and without 0.5 μg/mL of tetracycline) are inoculated at various cell densities.
3. Cells should be counted daily and the cultures split when densities approach 3×10^6 cells/mL in bloodstream-form and 4×10^7 cells/mL in procyclic form.

3.2 Analysis of the N-glycan and GPI Anchor of the Variant Surface Glycoprotein

The highly abundant and easily purified VSG of bloodstream-form *T. brucei* is a convenient reporter of the effects of gene replacement and/or chemical treatment on both GPI anchor and N-glycan structure [10, 11, 32, 37, 38, 41, 47]. Soluble form VSG (sVSG), released from the cell surface by the action of endogenous GPI-specific phospholipase C in response to osmotic lysis of the parasite at 37 °C [48], is purified by ion exchange chromatography and analyzed by electrospray mass spectrometry (ES-MS), **steps 1–5** below. The deconvolved spectra of the sVSG preparations from both wild type and the GT mutant parasites quickly indicate whether the GT knock out has affected the attachment and/or processing of the GPI anchor or the N-linked glycans (*see* Fig. 2a, b). No change in VSG glycoform pattern directs further analysis of the total glycoprotein repertoire of the parasite by lectin blotting (*see* Subheading 3.3) whereas a change in glycoform pattern directs further analysis of the VSG by analysis of its Pronase glycopeptides

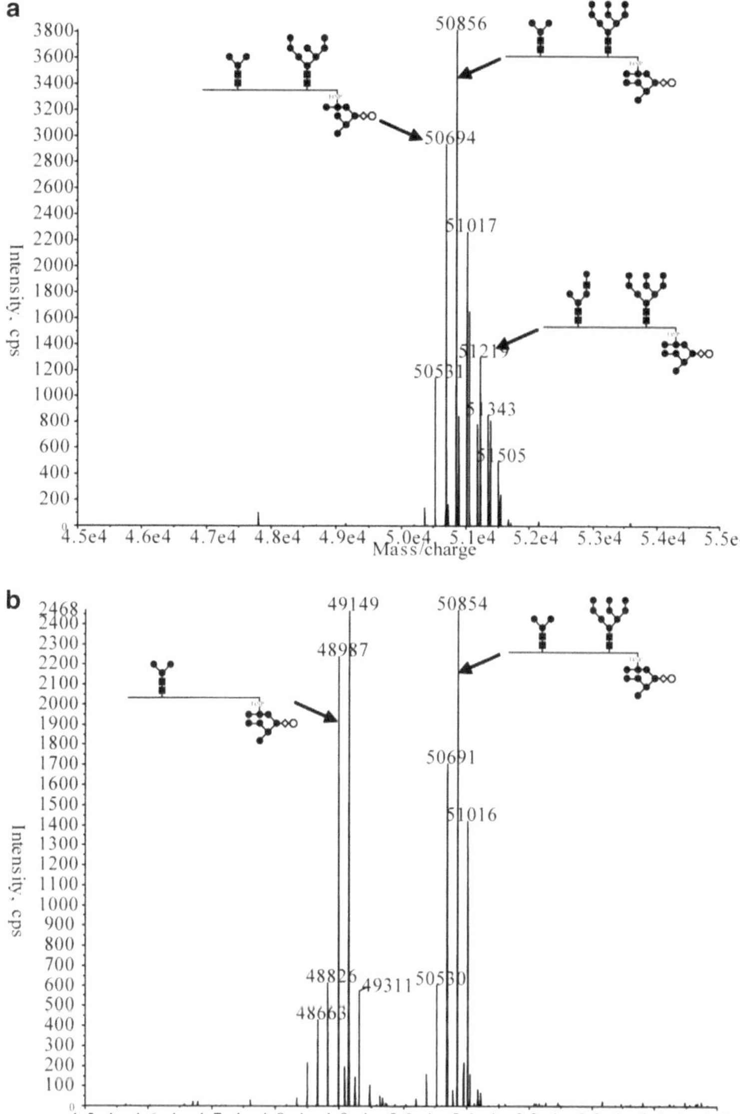

Fig. 2 Analysis of sVSG in conditional mutant trypanosomes. (**a**) Deconvolved ES-MS spectrum of intact sVSG from a conditional null mutant under permissive conditions (the same as wild type sVSG). (**b**) Deconvolved ES-MS spectrum of intact sVSG from a conditional null mutant under non-permissive conditions, showing a change in sVSG glycoforms. (**c**) Deconvolved ES-MS spectrum of the Pronase glycopeptides of the mutant sVSG shown in panel **b**. The identities of the glycopeptides are inferred by their accurate mass and their MS/MS spectra (*see* panel **d**). (**d**) Representative MS/MS spectrum of one of the ions giving rise to data in panel **c**. In this case, an $[M + 2H]^{2+}$ ion at *m/z* 874.8 (that gives rise to the 1744.55 mass in panel **c**) was subjected to collision induced dissociation. The inset shows the interpretation of this GPI anchor glycopeptide. The data in panels **a**–**c** were originally published in reference [38]. The synthesis of UDP-*N*-acetylglucosamine is essential for bloodstream-form *Trypanosoma brucei* in vitro and in vivo and UDP-*N*-acetylglucosamine starvation reveals a hierarchy in parasite protein glycosylation [38]

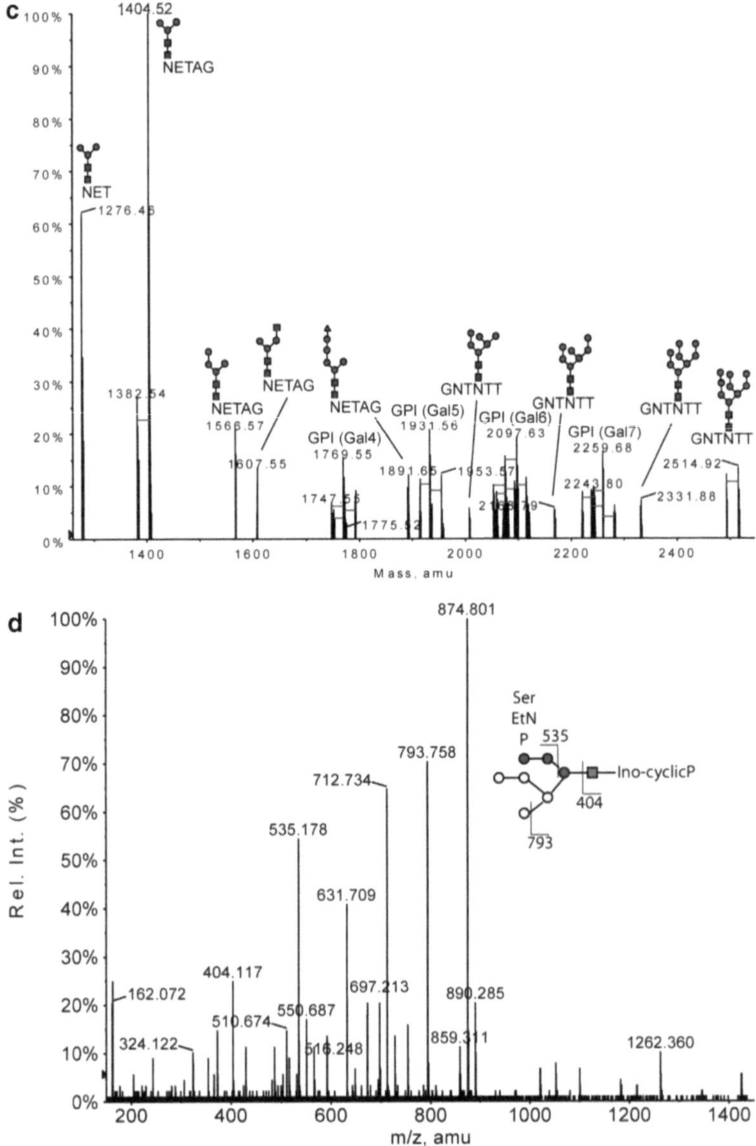

Fig. 2 (continued)

by ES-MS (*see* Fig. 2c) and ES-MS/MS (*see* Fig. 2d) to determine which of the GPI anchor or the two N-glycosylation sites has been affected, **steps 6–9**, below.

1. Approximately 2×10^8 bloodstream-form *T. brucei* cells expressing VSG variant 221 (from about 100 mL of culture) are chilled on ice for 10 min, centrifuged at $2,500 \times g$ for 10 min, resuspended in 10 mL of ice-cold PBS and centrifuged again.
2. The cell pellet is resuspended in 300 μL of 10 mM sodium phosphate buffer, pH 8.0, containing 0.1 mM TLCK, 1 μg/

mL leupeptin and 1 μg/mL aprotinin. After 5 min at 37 °C, the mixture is cooled on ice and centrifuged (14,000 × *g*, 5 min).

3. The supernatant is applied to a small (0.2 mL) DE52 anion exchange column equilibrated in 10 mM sodium phosphate buffer, pH 8.0 (*see* **Note 10**) and eluted with 0.8 mL of the same buffer.
4. The column eluate (containing about 100 μg sVSG) is concentrated to about 20 μL and diafiltered five times with 0.5 mL water on a Microcon YM-10 concentrator and recovered in 100 μL of water.
5. Samples of sVSG are diluted to 0.05 μg/μL in 50 % methanol 1 % formic acid and loaded into nanotips (or directly infused) into an appropriate ES-MS system with software capable of converting the primary positive ion spectrum into mass data (e.g., using the ABI-Sciex Bayesian protein reconstruct program or the Waters maximum entropy program).
6. Aliquots of sVSG (approximately 50 μg in 50 μL water) are mixed with 5 μL of 1 M ammonium bicarbonate and 10 μL of 1 mg/mL Pronase dissolved in 5 mM calcium acetate and incubated at 37 °C for 36 h.
7. Envicarb graphitized carbon micro-columns are prepared as follows: The contents of an Envicarb cartridge are suspended in methanol, and a bed of approximately 20 μL of graphitized carbon is packed into to a 100 μL C4 OMIX pipette tip. The micro-columns are prepared by attaching them to a Gilson pipette, set at 100 μL, and pipetting up and down ten times with each of 80 % methanol, 1 % formic acid; 60 % methanol, 1 % formic acid; and 1 % methanol, 1% formic acid.
8. The Pronase glycopeptides are purified as follows: The Pronase digest is adjusted to 1 % methanol, 1 % formic acid and applied to the micro-column by pipetting up and down 20 times. The micro-columns are washed by pipetting up and down 20 times with 1 % methanol, 1 % formic acid. The pipette is reset to 50 μL and the glycopeptides eluted by pipetting up and down 20 times with 50 μL of 60 % methanol, 1 % formic acid.
9. Aliquots of these samples are loaded into nanotips and analyzed by ES-MS in positive ion mode on an appropriate instrument and ES-MS/MS spectra are collected for the glycopeptide ions.
10. ES-MS spectra are converted into mass graphs using appropriate software, e.g., the Bayesian peptide reconstruction program in ABI-Sciex Analyst software, and assigned based on their mass and the corresponding MS/MS spectra.

3.3 Lectin Blotting of Bloodstream-Form Lysates

The high-molecular-weight branched poly LacNAc-containing N-glycans of bloodstream-form *T. brucei* are attached to glycoproteins other than VSG [34]. Therefore, the next step in the biochemical phenotyping of GT-null-mutants (or conditional null-mutants) is the SDS/urea extraction of total glycoproteins from trypanosome ghosts, depleted of VSG, and Western blotting analysis with different lectins.

Lectins recognize glycans with a high degree of specificity. However, for most of them the natural ligands are complex glycoconjugates and their binding properties can be extremely complex and not easily defined. Thus, although the use of lectins cannot replace detailed structural analysis of glycans they are excellent tools to study differential glycosylation and to deduce many aspects of glycan structure. Bloodstream-form trypanosomes are usually analyzed with tomato lectin, whose primary ligands are structures containing β1-3-linear poly *N*-acetyl-lactosamine repeats [49], and ricin, which binds structures bearing terminal galactose residues [50]. Nonetheless, other lectins such as wheat germ agglutinin, succinylated wheat germ agglutinin, *Datura stramonium* lectin, *Erythrina crystagalli* lectin can be of use. An example using tomato lectin, wheat germ agglutinin, and ricin is shown in Fig. 3.

1. Bloodstream-form cells from 100 to 150 mL of cell culture are washed twice with chilled TDB, spinning them for 10 min at $800 \times g$. Wild type and null mutant (or conditional null-mutants grown in the absence or presence of tetracycline in the media) strains should be used for comparison purposes.
2. Cells are hypotonically lysed with prewarmed 10 mM sodium phosphate buffer (pH 8.0) for 5 min at 37 °C (*see* **Note 11**) in the presence of protease inhibitors (0.1 mM TLCK, 1 μg/mL leupeptin, and 1 μg/mL apronitin).
3. Cell ghosts are centrifuged at $13{,}000 \times g$ for 10 min and solubilized by boiling for 15 min in 2 % SDS and 4 M urea (at 1×10^9 cell equivalents/mL).
4. The extracts are subjected to SDS-PAGE at 2×10^7 cells/lane on bis-Tris 4–12 % gradient acrylamide gel and transferred to nitrocellulose membranes in a semidry transfer apparatus at 45 mA for 1 h (*see* **Note 12**).
5. Membranes are blocked with 3 % BSA in PBS for 1 h (at this point, membranes can be blocked as well over night at 4 °C, in this case supplement with 0.05 % sodium azide).
6. After blocking, the membranes are incubated for 1 h with biotin-conjugated lectin with or without sugar inhibitors (*see* **Note 13**).
7. Membranes are washed three times with the appropriate buffer (for instance, PBS plus 0.1 % v/v Tween 20) before incubation with HRP-labelled ExtrAvidin (0.2 μg/mL) for 1 h.
8. Membranes are washed three times and developed with ECL reagent.

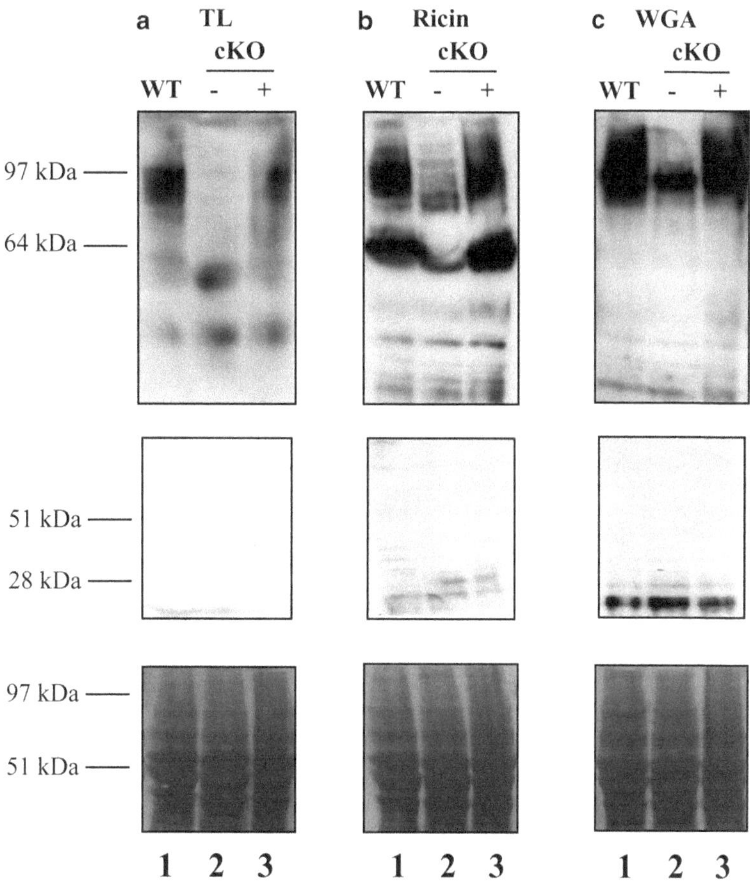

Fig. 3 Lectin blotting of bloodstream-form Trypanosoma ghost lysates. (**a**) Trypanosome ghost lysates (depleted of VSG) from wild type (*lane 1*) and a GT conditional null mutant under non-permissive (*lane 2*) and permissive (*lane 3*) conditions were separated by SDS-PAGE, transferred to nitrocellulose, stained with Ponceau S (loading control, lower panel) and blotted with tomato lectin without sugar inhibitors (TL, upper panel) and with inhibitors (specificity control, middle panel). (**b**) As panel **a** but using ricin lectin. (**c**) As panel (**a**) but using wheat germ agglutinin. The data in panels (**a**)–(**c**) were originally published in reference [41]

3.4 Procyclin Characterization by SDS-PAGE

Bloodstream-form trypanosome cells differentiate to the procyclic-form in the tsetse fly midgut, replacing their VSG coat by a new one that includes the GPI-anchored procyclins. Procyclins are characterized by internal dipeptide (EP) or pentapeptide (GPEET) repeats which confer a rod-like structure to the protein [13, 14]. Most EP-procyclins contain a single N-glycosylation site, occupied exclusively by a conventional triantennary $Man_5GlcNAc_2$ oligomannose oligosaccharide at the N-terminal side of the EP-repeat domain [15]. Procyclins can be purified by organic extraction, providing a sample with a degree of purity adequate to determine differences in their glycosylation state.

1. Procyclic-form cells from 100 mL of cell culture are washed twice in chilled PBS, spinning them for 10 min at $600 \times g$ (*see* **Note 14**). Wild type and null mutant (or conditional null-mutants grown in the absence or presence of tetracycline in the media) strains should be used for comparison purposes.
2. Freeze-dry the cell pellet. It can be stored at −80 °C until the organic extraction for several months.
3. Homogenize the freeze-dried pellet and extract the powder with chloroform–methanol (2:1, v/v; at 1×10^9 cell equivalents/mL) with thorough mixing and sonication (*see* **Note 15**).
4. Spin the sample and extract the pellet with chloroform–methanol (1:2, v/v; at 1 x 10^9 cell equivalents/mL). Mix thoroughly and sonicate.
5. Spin and extract the pellet three times with chloroform–methanol–water (10:10:3 v/v/v; at 1×10^9 cell equivalents/mL) as in previous steps.
6. Dry the pellet almost to completion under a nitrogen line and extract it three times with 9 % butan-1-ol in water (at 1×10^9 cell equivalents/mL), pooling the extracts together.
7. Freeze-dry the 9 % butan-1-ol extracts containing the procyclins (*see* **Note 16**).
8. From this point procyclins are dissolved in 40 % propan-1-ol. The appropriate amount (*see* **Note 14**) is dried in a SpeedVac (or freeze-dried) and subjected to SDS-PAGE bis-Tris 4–12 % gradient acrylamide gels (*see* **Note 17**) followed by periodate-Schiff staining (**steps 9–16**).
9. After the electrophoresis has finished, remove the gel and rinse it briefly with water. Then incubate with Fixing Solution (acetic acid–methanol–water, 10:35:25, v/v/v) for 15–25 min.
10. Wash with water for 5 min, three times.
11. Incubate for 30 min with Oxidation solution (1 % sodium periodate in 3 % acetic acid).
12. Repeat washes.
13. Incubate in Schiff reagent for 1 h (or longer if needed).
14. Pour off Schiff reagent and incubate with Reducing solution (1 % sodium metabisulfite) for 30 min.
15. Wash again three times to stop the reaction.
16. Procyclins will show a typical polydisperse pattern of migration with an apparent molecular mass of 20–35 kDa for GPEET-procyclins and 37–47 kDa for EP-procyclins.

3.5 Procyclin N-glycan Analysis

To analyze potential effects of genetic modifications on procylin N-glycosylation, procyclin can be treated with PNGase F to release the N-glycans and these can be permethylated and analyzed by MALDI-TOF and/or ES-MS [11].

1. Purified procyclin samples, corresponding to about 2×10^8 cells (*see* Subheading 3.4), are dried in a 2 mL Reacti-Vial (*see* **Note 18**) and deglycosylated with 15 U of peptide N-glycosidase F (PNGase F) in 10 μL of 0.25 M sodium phosphate buffer (pH 7.5), 24 h, 37 °C.
2. The samples are dried in a SpeedVac concentrator and resuspended in 50 μL DMSO.
3. A suspension of 120 mg/mL NaOH in DMSO is prepared by grinding NaOH pellets in a clean glass pestle and mortar (approx. two pellets in 1 mL gives a suitable concentration). Add 50 μL of this slurry to each sample and leave for 20 min, tapping the vials occasionally (*see* **Note 19**).
4. In a fume hood, add 10 μL iodomethane, leave for 10 min.
5. Add another 10 μL iodomethane and leave for 10 min.
6. Add 20 μL iodomethane and leave for 20 min.
7. Add 250 μL chloroform and 1 mL 100 mg/mL sodium thiosulfate and vortex for 2 min. Separate the phases by low-speed centrifugation and remove the upper aqueous phase.
8. Add 1 mL water, vortex for 2 min, separate the phases by low-speed centrifugation and remove the upper aqueous phase. Repeat this process a further two times.
9. Dry the lower chloroform phase under a stream of nitrogen in a fume hood.
10. Immediately prior to analysis by ES-MS, redissolve the permethylated N-glycans in 50 μL 80 % acetonitrile and mix aliquots of 5 μL with an equal volume of 80 % acetonitrile containing 1 mM sodium acetate.
11. Load the samples into nanotips or introduce into the ES-MS system by direct infusion and collect positive ion spectra. The permethylated N-glycans typically appear as $[M+2Na]^{2+}$ ions, the identities of which can be confirmed by MS/MS.

3.6 Procyclin GPI-Anchor Characterization by ES-MS³

Procyclins, both GPEET and EP, contain similar GPI membrane anchors. These are the largest and most complex anchors known and they are characterized by the presence of large poly disperse branched *N*-acetyllactosamine (Galβ1-4GlcNAc)- and lacto-N-biose (Galβ1-3GlcNAc)-containing side-chains (with an average of about 8–12 repeats, depending on the preparation) that can be capped with α2-3-linked sialic acid residues [18, 41]. The characterization of these complex structures requires the permethylation of the GPI-anchors and their analysis by MS^3.

1. More than 4 μg of procyclins from wild type and null mutant backgrounds are treated with 50 μL of ice-cold 50 % aqueous hydrogen fluoride for 48 h at 0 °C to cleave the GPI anchor ethanolamine-phosphate bond (*see* **Note 20**).

2. Samples are freeze-dried and then re-dried twice from 50 μL of water to remove traces of aqueous hydrogen fluoride.
3. The sample, containing the released GPI glycans, is finally dissolved in 50 μL of water and transferred into a 2 mL Reacti-Vial (*see* **Note 18**).
4. Samples are dried then re-dried from 50 μL methanol in a SpeedVac concentrator to ensure complete removal of water.
5. Samples are then resuspended in 50 μL of DMSO, left for 20 min at room temperature and deprotonated by adding a solution of 120 mg/mL sodium hydroxide in dimethyl sulfoxide for 20 min (*see* **Note 19** and **steps 2–11** of Subheading 3.5).
6. Once ready for permethylation, iodomethane is added in three intervals over a total of 30 min (10 μL, 10 min incubation; another 10 μL, 10 min incubation, then 20 μL, 10 min incubation), and the permethylation reaction is continued for another 30 min.
7. The permethylated glycans are partitioned between 250 μL chloroform and 1 mL of 100 mg/mL sodium thiosulfate in water. The upper aqueous phase is removed and the chloroform phase re-extracted at least five times with 1 mL of water each.
8. The permethylated glycans recovered in the lower chloroform phase are transferred into a new glass vial and dried under N_2 (*see* **Note 21**).
9. The samples are redissolved in 20 μL methanol and one aliquot (1 μL) loaded onto a MALDI plate after mixing with 2,5-dihydroxybenzoic acid matrix and analyzed in positive ion mode using an ABI Voyager DE-STR MALDI-TOF mass spectrometer (or similar) (*see* **Note 22**). This analysis should show the presence of $[M]^+$ molecular ions correspondent to core structures of $Hex_5HexN(Me_3)^+Ino$ and $Hex_6HexN(Me_3)^+Ino$ substituted by several HexHexNAc units (449.2 Da mass differences) and with or without sialic acids (361.2 Da mass differences) generating more complex or simple spectra depending on the branching level of the structures (*see* **Note 23**).
10. After MALDI analysis, samples are dried and redissolved in 50 μL of 80 % acetonitrile. Aliquots (5–10 μL) are dried again and recovered in 80 % acetonitrile, 0.5 mM sodium acetate before loading into nanotips for positive ion ES-MS, ES-MS^2, and ES-MS^3 on a LTQ Orbitrap mass spectrometer (*see* **Note 24**).
11. Ions of the "selected species" (*see* **Note 23**) can be observed triply ($[M+2Na]^{3+}$) or doubly ($[M+Na]^{2+}$) charged by ES-MS. The ion is fragmented (MS^2) resulting in the simultaneous non-neutral loss of the permethylated D-*myo*-inositol moiety and quaternary ammonium group of the glucosamine [41, 51]. This dominant non-neutral loss produces a mass transition for the permethylated GPI glycan from the triply charged $[M+2Na]^{3+}$ precursor ion to the doubly charged

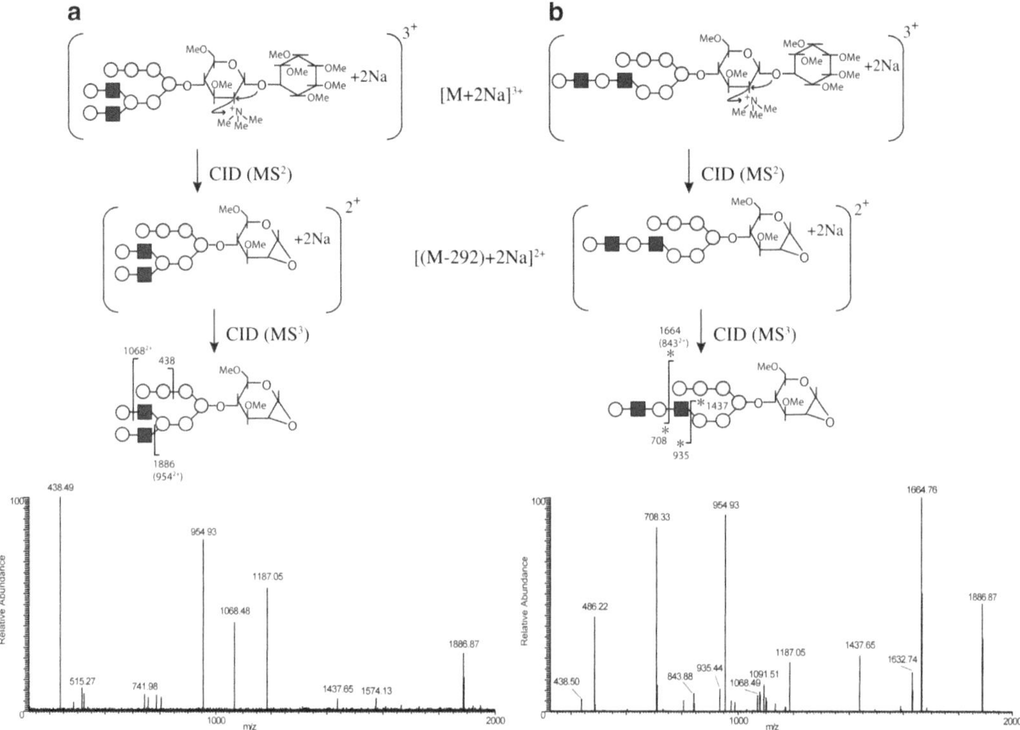

Fig. 4 Proposed MS^2 and MS^3 fragmentation patterns of permethylated GPI glycans with different side chain substitutions. Isobaric ions of permethylated GPI glycans isolated from the procyclins of wild type (**a**) and a GT mutant (**b**) procyclic trypanosomes (*top cartoons*) give rise to a major isobaric collision induced dissociation (CID) MS^2 fragment ions through the elimination of inositol and the trimethylated amine group (*middle cartoons*). Further fragmentation of these ions give rise to the MS^3 mass spectra shown. The proposed and MS^3 fragmentation pathways, diagnostic of the proposed structures, are shown (*bottom cartoons*). Data are adapted from those originally published [41]

product ion $[(M\text{-}292)+2Na]^{2+}$, or from the doubly charged $[M+Na]^{2+}$ precursor ion to the singly charged product ion $[(M\text{-}292)+Na]^{+}$ (*see* Fig. 4).

12. Spectra arising from further fragmentation of the previous product ions (MS^3) will show diagnostic ions and will allow, by comparison between null mutant and wild type permethylated GPI glycan structures, the characterization of the specificity of the GT (see an example of comparison in Fig. 4).

4 Notes

1. If the GT is not expressed in one particular life stage we cannot expect the selection marker to be expressed by the GT native promoter in that stage. Therefore, an analysis of the developmental regulation of the GT expression during the *T. brucei* life stages is recommended to address this problem.

2. Usually, 18–20 bp length primers are optimal. However, given the high AT-content of the UTR regions, sometimes increasing the length of the primers can enhance the PCR results, specially considering that it is better to avoid altering the position of reverse primer 2 and forward primer 3 (*see* Fig. 1a).
3. Restriction sites should be chosen carefully, since they should not be present in the UTRs amplified or in the drug resistance gene.
4. It is useful to test various PCR-reactions with different concentrations of the 5′ and 3′-UTR templates, within a range from 25 to 200 ng (which can be adjusted according to the size of the amplified regions).
5. In the case of the pLew100 tetracycline-inducible expression vector usually NotI linearization is fine (unless your ORF contains a NotI restriction site).
6. Generally transfecting 5 cuvettes (plus one control without DNA) in the case of the knockout cassettes and 2–3 cuvettes (plus control) in the case of the pLew100 tetracycline-inducible expression vector is enough. This should be taken into account in order to calculate the volume of $1–3 \times 10^6$ cells/mL culture needed.
7. Excellent transformation efficiencies can be obtained as well with the AMAXA Nucleofector II apparatus, using program X-100 and Human T-cell Nucleofector solution. Generally, two or three cuvettes with 1×10^7 cells and 1 μg DNA per cuvette are enough.
8. When using phleomycin as a selection drug, usually when transfecting pLew100 tetracycline-inducible expression vector, cells need on average 7 days or more to die. In this case it is of use to monitor the cells and if they are starting to overgrow dilute them again and continue selecting.
9. Transfect 2 cuvettes with the BTX electroporator (plus one control without DNA) which is generally enough. This should be taken into account in order to calculate the volume of $0.9–1 \times 10^7$ cells/mL culture needed.
10. DE52 resin must be pre-cycled by sequentially washing with 5 volumes each of 0.5 M HCl, water, 0.5 M NaOH and water. Subsequently, add phosphoric acid until the pH of the supernatant is 8.0 and store over a small volume of chloroform. Start equilibrating the DE52 resin using 1 volume of cold 100 mM sodium phosphate buffer (pH 8.0). Then use at least 5 volumes of cold 10 mM sodium phosphate buffer (pH 8.0). Measurements of buffer pH with pH test strips before and after passing through the resin can be used to assess that the DE52 resin is properly equilibrated.
11. For the hypotonic lysis to take place in optimal conditions, cells should be diluted in sodium phosphate buffer approximately at

1×10^9 cells/mL. Conditional-null-mutant cells should be grown in presence or absence of tetracycline in the media and, if the mutated GT is essential, cells should be harvested about 12 h before cell death is initiated.

12. It is convenient to "duplicate" the number of lanes/condition (i.e., wild type and null-mutant or wild type, conditional null-mutant minus tetracycline and plus tetracycline) in order to assess the specificity of the lectin blot procedure with specific inhibitors. Thus, one piece of the nitrocellulose membrane will be incubated with the lectin as normal, whereas the other will be incubated with the lectin plus the inhibitor, according to the manufacturer's specifications. Due to the high molecular mass of the extracted glycoproteins it is of use to test for adequate protein transference, for instance staining the membranes with Ponceau S. Images of the Ponceau S red-stained membranes can be used to demonstrate equal lane-loading.
13. For most lectins 0.5–2 μg/mL concentrations freshly prepared in the manufacturer specified buffer or in phosphate buffered saline (PBS, pH 7.4) are optimal. Please note any lectin requirement for divalent cations, if so you should include them. The same buffer should be used for washes and horseradish peroxidase labelled Extravidin dilution/incubation.
14. Generally $1–1.5 \times 10^8$ cell equivalents are detected well by SDS-PAGE followed by periodic acid-Schiff staining. For Western blot procedures, usually 2×10^7 cell equivalents allow the detection of EP- and GPEET-procyclins, using anti-EP-procyclin mouse monoclonal antibody 247 [52] and antiphosphorylated-GPEET-procyclin mouse monoclonal antibody 5H3 [53], respectively (at 1: 2,500 dilution). Increased transference times to nitrocellulose membranes may enhance the results, since procyclins are proteins highly negatively charged (due to their glutamic acid content and sialic acid residues) [14]. For detailed structural analysis of glycans procyclin purification from at least 1×10^{11} cells and further octyl-Sepharose chromatography is recommended.
15. For homogenization of the freeze-dried pellet a razor blade or a scalpel are optimal. From this step use glass or chloroform-resistant plasticware. Stored chloroform will react with air over time to form phosgene. Using chloroform in good condition is key to the delipidation/organic extraction procedure. For the sonication steps, tubes can be immersed in an ultrasound bath.
16. Further purification of procyclins for structural analysis can be achieved by octyl-Sepharose chromatography. The pooled 9 % butan-1-ol extract is redissolved in 1 mL of buffer A (5 % propan-1-ol in 100 mM ammonium acetate) and applied to an octyl-Sepharose column (130×10 mm) at 6 mL/h. Binding of molecules to the column may be enhanced by keeping the sample in the resin overnight (18 h). The column is washed

with 20 mL of buffer A and eluted with a linear gradient from buffer A to 60 % propan-1-ol over 40 mL. Fractions are dried, dissolved in 50 μL 40 % propan-1-ol and aliquots (2 × 1 μL) of each two fractions are sequentially applied directly to an HPTLC plate and carbohydrate content estimated by staining the plate with orcinol reagent. The myo-inositol contents of aliquots of the fractions in the orcinol-positive peak region are determined by GC-MS [54]. Inositol-containing fractions are pooled and dried by rotary evaporation. These samples will have the purity needed for mass spectrometry analyses.

17. Procyclins are best resolved in SDS-PAGE bis-Tris 4–12 % gradient acrylamide gels, although good results can be obtained as well in 10 % polyacrylamide gels. For better resolution, allow the gel to run 5 min after the bromophenol blue front (~3–5 kDa) reaches the end of the separating gel. Alternatively, you can use a pre-stained molecular weight marker and allow the gel to run out until the 15 kDa band reaches the gel front.
18. Acid washed laboratory glassware should be used beyond this point. Acid washing with 3 M nitric acid for at least 24 h assures that all potentially contaminating ions are removed from the surfaces of the glass.
19. Analytical grade reagents should be used throughout this protocol since very sensitive instruments will be used for the analysis of the samples.
20. Aqueous hydrogen fluoride dephosphorylation releases GPI anchor glycans from the procyclin peptide and from the lysophosphatidic acid component of the phosphatidylinositol lipid. This procedure also removes some sialic acid residues capping *N*-acetyllactosamine—containing side-chains.
21. The permethylation procedure removes the fatty acid from the inositol ring of the GPI glycans, methylates all free hydroxyl groups, and converts the free amine of the single non-N-acetylated glucosamine residue to a quaternary amine carrying a fixed positive charge. Permethylation greatly enhances the mass spectrometric detection of dephosphorylated GPI glycans thanks in part to this quaternary amine feature [30, 51, 55].
22. The accelerating voltage is 25 000 V, and grid voltage is set at 94 % with an extraction time delay of 700 ns. Data is collected manually at 500 shots/spectrum with the laser intensity set at 2,500.
23. It is ideal to use the MALDI spectra as a guide to find singly charged ions correspondent to core structures with 1–3 HexHexNAc repeats (or even with terminal GlcNAc residues not substituted with Gal). These species will be used for fragmentation analysis (MS^2 and MS^3) by positive ion ESI-MS to determine the structure of the GPI-anchor side chain.

24. Source and capillary voltages are 0.63 kV and 48 V, respectively. Collision energy is set between 17 and 25 %.

Acknowledgments

We thank Angela Mehlert, Isabelle Nett, Sujatha Manthri, Deuan Jones, and Alvaro Acosta-Serrano who all contributed to the development of the analyses described herein. This work was supported by a programme grant (085622) and a strategic award (083481) from the Wellcome Trust.

References

1. Cross GA (1996) Antigenic variation in trypanosomes: secrets surface slowly. Bioessays 18(4):283–291
2. Pays E, Vanhamme L, Perez-Morga D (2004) Antigenic variation in Trypanosoma brucei: facts, challenges and mysteries. Curr Opin Microbiol 7(4):369–374. doi:10.1016/j.mib.2004.05.001, S1369527404000645 [pii]
3. Horn D (2004) The molecular control of antigenic variation in Trypanosoma brucei. Curr Mol Med 4(6):563–576
4. Stockdale C, Swiderski MR, Barry JD, McCulloch R (2008) Antigenic variation in Trypanosoma brucei: joining the DOTs. PLoS Biol 6(7):e185. doi:08-PLBI-P-2322 [pii], 10.1371/journal.pbio.0060185
5. Ferguson MA, Homans SW, Dwek RA, Rademacher TW (1988) Glycosyl-phosphatidylinositol moiety that anchors Trypanosoma brucei variant surface glycoprotein to the membrane. Science 239(4841 Pt 1):753–759
6. Mehlert A, Richardson JM, Ferguson MA (1998) Structure of the glycosylphosphatidylinositol membrane anchor glycan of a class-2 variant surface glycoprotein from Trypanosoma brucei. J Mol Biol 277(2):379–392
7. Zamze SE, Ashford DA, Wooten EW, Rademacher TW, Dwek RA (1991) Structural characterization of the asparagine-linked oligosaccharides from Trypanosoma brucei type II and type III variant surface glycoproteins. J Biol Chem 266(30):20244–20261
8. Zamze SE, Wooten EW, Ashford DA, Ferguson MA, Dwek RA, Rademacher TW (1990) Characterisation of the asparagine-linked oligosaccharides from Trypanosoma brucei type-I variant surface glycoproteins. Eur J Biochem 187(3):657–663
9. Mehlert A, Zitzmann N, Richardson JM, Treumann A, Ferguson MA (1998) The glycosylation of the variant surface glycoproteins and procyclic acidic repetitive proteins of Trypanosoma brucei. Mol Biochem Parasitol 91(1):145–152
10. Jones DC, Mehlert A, Guther ML, Ferguson MA (2005) Deletion of the glucosidase II gene in Trypanosoma brucei reveals novel N-glycosylation mechanisms in the biosynthesis of variant surface glycoprotein. J Biol Chem 280(43):35929–35942
11. Manthri S, Guther ML, Izquierdo L, Acosta-Serrano A, Ferguson MA (2008) Deletion of the TbALG3 gene demonstrates site-specific N-glycosylation and N-glycan processing in Trypanosoma brucei. Glycobiology 18(5):367–383
12. Guther ML, Lee S, Tetley L, Acosta-Serrano A, Ferguson MA (2006) GPI-anchored proteins and free GPI glycolipids of procyclic form Trypanosoma brucei are nonessential for growth, are required for colonization of the tsetse fly, and are not the only components of the surface coat. Mol Biol Cell 17(12):5265–5274
13. Roditi I, Schwarz H, Pearson TW, Beecroft RP, Liu MK, Richardson JP, Bühring HJ, Pleiss J, Bülow R, Williams RO, Overath P (1989) Procyclin gene expression and loss of the variant surface glycoprotein during differentiation of Trypanosoma brucei. J Cell Biol 108(2):737–746
14. Treumann A, Zitzmann N, Hulsmeier A, Prescott AR, Almond A, Sheehan J, Ferguson MA (1997) Structural characterisation of two forms of procyclic acidic repetitive protein expressed by procyclic forms of Trypanosoma brucei. J Mol Biol 269(4):529–547
15. Acosta-Serrano A, Cole RN, Mehlert A, Lee MG, Ferguson MA, Englund PT (1999) The procyclin repertoire of Trypanosoma brucei. Identification and structural characterization of the Glu-Pro-rich polypeptides. J Biol Chem 274(42):29763–29771

16. Mehlert A, Treumann A, Ferguson MA (1999) Trypanosoma brucei GPEET-PARP is phosphorylated on six out of seven threonine residues. Mol Biochem Parasitol 98(2):291–296
17. Schlaeppi AC, Malherbe T, Butikofer P (2003) Coordinate expression of GPEET procyclin and its membrane-associated kinase in Trypanosoma brucei procyclic forms. J Biol Chem 278(50):49980–49987
18. Ferguson MA, Murray P, Rutherford H, McConville MJ (1993) A simple purification of procyclic acidic repetitive protein and demonstration of a sialylated glycosylphosphatidylinositol membrane anchor. Biochem J 291(Pt 1):51–55
19. Engstler M, Reuter G, Schauer R (1993) The developmentally regulated trans-sialidase from Trypanosoma brucei sialylates the procyclic acidic repetitive protein. Mol Biochem Parasitol 61(1):1–13
20. Pontes de Carvalho LC, Tomlinson S, Vandekerckhove F, Bienen EJ, Clarkson AB, Jiang MS, Hart GW, Nussenzweig V (1993) Characterization of a novel trans-sialidase of Trypanosoma brucei procyclic trypomastigotes and identification of procyclin as the main sialic acid acceptor. J Exp Med 177(2):465–474
21. Montagna G, Cremona ML, Paris G, Amaya MF, Buschiazzo A, Alzari PM, Frasch AC (2002) The trans-sialidase from the african trypanosome Trypanosoma brucei. Eur J Biochem 269(12):2941–2950
22. Montagna GN, Donelson JE, Frasch AC (2006) Procyclic Trypanosoma brucei expresses separate sialidase and trans-sialidase enzymes on its surface membrane. J Biol Chem 281(45):33949–33958
23. Nagamune K, Acosta-Serrano A, Uemura H, Brun R, Kunz-Renggli C, Maeda Y, Ferguson MA, Kinoshita T (2004) Surface sialic acids taken from the host allow trypanosome survival in tsetse fly vectors. J Exp Med 199(10):1445–1450
24. Acosta-Serrano A, Vassella E, Liniger M, Kunz Renggli C, Brun R, Roditi I, Englund PT (2001) The surface coat of procyclic Trypanosoma brucei: programmed expression and proteolytic cleavage of procyclin in the tsetse fly. Proc Natl Acad Sci USA 98(4):1513–1518
25. Guther ML, Beattie K, Lamont DJ, James J, Prescott AR, Ferguson MA (2009) Fate of glycosylphosphatidylinositol (GPI)-less procyclin and characterization of sialylated non-GPI-anchored surface coat molecules of procyclic-form Trypanosoma brucei. Eukaryot Cell 8(9):1407–1417. doi:EC.00178-09 [pii], 10.1128/EC.00178-09
26. Urwyler S, Studer E, Renggli CK, Roditi I (2007) A family of stage-specific alanine-rich proteins on the surface of epimastigote forms of Trypanosoma brucei. Mol Microbiol 63(1):218–228
27. Ziegelbauer K, Overath P (1992) Identification of invariant surface glycoproteins in the bloodstream stage of Trypanosoma brucei. J Biol Chem 267(15):10791–10796
28. Steverding D (2000) The transferrin receptor of Trypanosoma brucei. Parasitol Int 48(3):191–198. doi:S1383-5769(99)00018-5 [pii]
29. Steverding D, Stierhof YD, Fuchs H, Tauber R, Overath P (1995) Transferrin-binding protein complex is the receptor for transferrin uptake in Trypanosoma brucei. J Cell Biol 131(5):1173–1182
30. Mehlert A, Ferguson MA (2007) Structure of the glycosylphosphatidylinositol anchor of the Trypanosoma brucei transferrin receptor. Mol Biochem Parasitol 151(2):220–223
31. Kelley RJ, Brickman MJ, Balber AE (1995) Processing and transport of a lysosomal membrane glycoprotein is developmentally regulated in African trypanosomes. Mol Biochem Parasitol 74(2):167–178
32. Izquierdo L, Schulz BL, Rodrigues JA, Guther ML, Procter JB, Barton GJ, Aebi M, Ferguson MA (2009) Distinct donor and acceptor specificities of Trypanosoma brucei oligosaccharyltransferases. EMBO J 28(17):2650–2661. doi:emboj2009203 [pii], 10.1038/emboj.2009.203
33. Acosta-Serrano A, O'Rear J, Quellhorst G, Lee SH, Hwa KY, Krag SS, Englund PT (2004) Defects in the N-linked oligosaccharide biosynthetic pathway in a Trypanosoma brucei glycosylation mutant. Eukaryot Cell 3(2):255–263
34. Atrih A, Richardson JM, Prescott AR, Ferguson MA (2005) Trypanosoma brucei glycoproteins contain novel giant poly-N-acetyllactosamine carbohydrate chains. J Biol Chem 280(2):865–871
35. Roper JR, Guther ML, Milne KG, Ferguson MA (2002) Galactose metabolism is essential for the African sleeping sickness parasite Trypanosoma brucei. Proc Natl Acad Sci U S A 99(9):5884–5889
36. Roper JR, Guther ML, Macrae JI, Prescott AR, Hallyburton I, Acosta-Serrano A, Ferguson MA (2005) The suppression of galactose metabolism in procylic form Trypanosoma brucei causes cessation of cell growth and alters procyclin glycoprotein structure and copy number. J Biol Chem 280(20):19728–19736
37. Urbaniak MD, Turnock DC, Ferguson MA (2006) Galactose starvation in a bloodstream form Trypanosoma brucei UDP-glucose

4′-epimerase conditional null mutant. Eukaryot Cell 5(11):1906–1913
38. Stokes MJ, Guther ML, Turnock DC, Prescott AR, Martin KL, Alphey MS, Ferguson MA (2008) The synthesis of UDP-N-acetylglucosamine is essential for bloodstream form trypanosoma brucei in vitro and in vivo and UDP-N-acetylglucosamine starvation reveals a hierarchy in parasite protein glycosylation. J Biol Chem 283(23):16147–16161
39. Marino K, Guther ML, Wernimont AK, Qiu W, Hui R, Ferguson MA (2011) Characterization, localization, essentiality, and high-resolution crystal structure of glucosamine 6-phosphate N-acetyltransferase from Trypanosoma brucei. Eukaryot Cell 10(7):985–997. doi:EC.05025-11 [pii], 10.1128/EC.05025-11
40. Narimatsu H (2006) Human glycogene cloning: focus on beta 3-glycosyltransferase and beta 4-glycosyltransferase families. Curr Opin Struct Biol 16(5):567–575. doi:S0959-440X(06)00148-5 [pii], 10.1016/j.sbi.2006.09.001
41. Izquierdo L, Nakanishi M, Mehlert A, Machray G, Barton GJ, Ferguson MA (2009) Identification of a glycosylphosphatidylinositol anchor-modifying beta1-3 N-acetylglucosaminyl transferase in Trypanosoma brucei. Mol Microbiol 71(2):478–491
42. Wirtz E, Leal S, Ochatt C, Cross GA (1999) A tightly regulated inducible expression system for conditional gene knock-outs and dominant-negative genetics in Trypanosoma brucei. Mol Biochem Parasitol 99(1):89–101
43. Wirtz E, Clayton C (1995) Inducible gene expression in trypanosomes mediated by a prokaryotic repressor. Science 268(5214):1179–1183
44. Hirumi H, Hirumi K (1989) Continuous cultivation of Trypanosoma brucei blood stream forms in a medium containing a low concentration of serum protein without feeder cell layers. J Parasitol 75(6):985–989
45. Brun R, Schonenberger M (1979) Cultivation and in vitro cloning or procyclic culture forms of Trypanosoma brucei in a semi-defined medium. Short communication. Acta Trop 36(3):289–292
46. Barnes RL, McCulloch R (2007) Trypanosoma brucei homologous recombination is dependent on substrate length and homology, though displays a differential dependence on mismatch repair as substrate length decreases. Nucleic Acids Res 35(10):3478–3493. doi:gkm249 [pii], 10.1093/nar/gkm249
47. Izquierdo L, Atrih A, Rodrigues JA, Jones DC, Ferguson MA (2009) Trypanosoma brucei UDP-glucose:glycoprotein glucosyltransferase has unusual substrate specificity and protects the parasite from stress. Eukaryot Cell 8(2):230–240
48. Ferguson MA, Duszenko M, Lamont GS, Overath P, Cross GA (1986) Biosynthesis of Trypanosoma brucei variant surface glycoproteins. N-glycosylation and addition of a phosphatidylinositol membrane anchor. J Biol Chem 261(1):356–362
49. Merkle RK, Cummings RD (1987) Relationship of the terminal sequences to the length of poly-N-acetyllactosamine chains in asparagine-linked oligosaccharides from the mouse lymphoma cell line BW5147. Immobilized tomato lectin interacts with high affinity with glycopeptides containing long poly-N-acetyllactosamine chains. J Biol Chem 262(17):8179–8189
50. Baenziger JU, Fiete D (1979) Structural determinants of Ricinus communis agglutinin and toxin specificity for oligosaccharides. J Biol Chem 254(19):9795–9799
51. Nett IR, Mehlert A, Lamont D, Ferguson MA (2010) Application of electrospray mass spectrometry to the structural determination of glycosylphosphatidylinositol membrane anchors. Glycobiology 20(5):576–585. doi:cwq007 [pii], 10.1093/glycob/cwq007
52. Richardson JP, Beecroft RP, Tolson DL, Liu MK, Pearson TW (1988) Procyclin: an unusual immunodominant glycoprotein surface antigen from the procyclic stage of African trypanosomes. Mol Biochem Parasitol 31(3):203–216
53. Richardson JP, Jenni L, Beecroft RP, Pearson TW (1986) Procyclic tsetse fly midgut forms and culture forms of African trypanosomes share stage- and species-specific surface antigens identified by monoclonal antibodies. J Immunol 136(6):2259–2264
54. Ferguson MAJ (1994) Glycobiology: a practical approach. GPI membrane anchors: isolation and analysis, vol Chapter 8. Fukuda, M. and Kobata, A. (eds.) IRL Press at Oxford University Press, Oxford
55. Baldwin MA (2005) Analysis of glycosylphosphatidylinositol protein anchors: the prion protein. Methods Enzymol 405:172–187

Chapter 20

Mannose-6-Phosphate: A Regulator of LLO Destruction

Ningguo Gao and Mark A. Lehrman

Abstract

Oligosaccharyltransferase (OT) catalyzes the signature reaction of the asparagine-linked glycosylation pathway, namely, the transfer of preformed glycans from the lipid-linked oligosaccharide $Glc_3Man_9GlcNAc_2$-P-P-Dolichol ($G_3M_9Gn_2$-LLO) to appropriate asparaginyl residues on acceptor polypeptides. We have identified a reaction, possibly catalyzed by OT, that results in the hydrolysis or "transfer to water" of host LLOs in response to viral infection with release of a free $G_3M_9Gn_2$ glycan. The loss of LLO ostensibly hinders *N*-glycosylation of viral polypeptides. This response is achieved by a novel stress-activated signaling pathway in which free mannose-6-phosphate (M6P) acts as a second-messenger. Here, we describe methods with permeabilized mammalian cells for activation of the M6P-regulated LLO hydrolysis, or transfer of glycan to water, in vitro.

Key words Mannose-6-phosphate, $Glc_3Man_9GlcNAc_2$-P-P-Dolichol, Dolichol, Lipid-linked oligosaccharide, Free glycan, Streptolysin-O

1 Introduction

The Type I Congenital Disorders of Glycosylation are genetic diseases characterized by multisystem abnormalities, and diagnosed by the presence of hypoglycosylated serum transferrin [1] Seventeen distinct Type 1 genotypes have been identified to date, all of which are thought to involve reduction of the synthesis of $G_3M_9Gn_2$-LLO and thus result in underglycosylation of glycoproteins, leading to the characteristic pathology of the diseases. The most common is CDG-Ia, with a deficiency of phosphomannomutase (PMM) encoded by the *PMM2* locus. CDG-Ia patients always have at least one allele with partial activity, since a total absence of PMM function is probably lethal at the early embryonic stage. PMM converts M6P to M1P, the immediate precursor of GDP-mannose. Thus, it is thought that CDG-Ia etiology may involve suppressed levels of M1P and GDP-mannose, leading to lower production of $G_3M_9Gn_2$-LLO, and/or accumulation of abnormal LLOs that are weak substrates for OT. According to this hypothesis, the disease may also

Inka Brockhausen (ed.), *Glycosyltransferases: Methods and Protocols*, Methods in Molecular Biology, vol. 1022, DOI 10.1007/978-1-62703-465-4_20, © Springer Science+Business Media New York 2013

involve the reduction of other glycoconjugates that use GDP-mannose as a precursor, i.e., glycosylphosphatidylinositol lipids and anchors, O-mannose, and C-mannose.

We proposed a second hypothesis, not necessarily in conflict with the first, based upon the likelihood that PMM deficiency would cause accumulation of M6P as well as suppression of M1P [2]. According to this hypothesis, the accumulated M6P may also have detrimental effects on LLO synthesis, usage, or stability. In this way, underglycosylation in CDG-Ia could be due to both elevated M6P and diminished M1P, each impairing the LLO supply by its own mechanism. To demonstrate the plausibility of M6P impacting the CDG-Ia LLO pool, we showed that M6P added to permeabilized mammalian cells caused specific cleavage of $G_3M_9Gn_2$-LLO with release of free $G_3M_9Gn_2$ [2]. This cleavage reaction was highly specific for M6P, because related sugar phosphates such as G6P and M1P were ineffective, and LLOs shorter than $G_3M_9Gn_2$-LLO were left uncleaved.

The M6P-dependent cleavage reaction has been observed with all normal permeabilized mammalian cell types tested to date, i.e., human dermal fibroblasts, mouse embryonic fibroblasts, rat hepatocytes, and Chinese hamster ovary cells [2]. However, it is not detected in permeabilized Lec35 cells, which synthesize the truncated LLO M_5Gn_2-LLO rather than $G_3M_9Gn_2$-LLO, although the reaction is detected with these cells after genetic rescue to restore $G_3M_9Gn_2$-LLO production [2].

Recently, we have identified the M6P-dependent cleavage reaction as the terminus of a novel signaling pathway [3]. This pathway is activated by infection with an enveloped virus, herpes simplex virus-1, which causes ER stress. This results in activation of the ER stress kinase IRE1-alpha, which in turn promotes breakdown of glycogen by glycogen phosphorylase. The released G1P is converted to M6P, increasing its concentration twofold to fourfold. This M6P appears to co-opt OT, to direct it for destruction of LLOs that would otherwise be used by the virus to synthesize glycosylated components of its envelope. Thus, the ability of M6P to activate the "transfer to water" reaction of OT may be a form of anti-viral host defense. Here, we list the supplies and present the methods used to activate this M6P-dependent cleavage of $G_3M_9Gn_2$-LLO in permeabilized cells.

2 Materials and Cell Cultures

2.1 Cell Cultures

Adherent mammalian cells should be grown under normal conditions with physiological glucose, i.e., at least 5 mM. This is important because we have found that with normal dermal fibroblasts, glucose concentrations below 2 mM can artificially abrogate LLO synthesis, resulting in accumulation of truncated LLO intermediates. Cells are typically grown to approximately 80–90 % of confluence, and refed 16–24 h before use. Although we have only used adherent cells, it may also be possible to use non-adherent cells as these should also be susceptible to permeabilizing treatment (see below).

2.2 Cells with Plasma Membranes Gently Permeabilized with Streptolysin-O (SLO)

To promote M6P-dependent hydrolysis of LLOs, it is necessary for M6P to gain access to the cytosolic components of the ER membrane. Since M6P is membrane-impermeant, some means of disrupting the plasma membrane is necessary. However, we found that the M6P-responsive entity is inactivated by robust physical perturbation. Thus, conventional microsomal membrane preparations are not suitable for this reaction, because of the strong forces necessary to break open cells. However, the M6P-responsive entity is stable with SLO treatment. SLO gently and selectively introduces large pores in the plasma membranes of mammalian cells, capable of allowing diffusion of proteins up to ~1,000 kDa, without permeabilizing or otherwise perturbing intracellular membranes. Thus, it is an ideal reagent for introducing enzyme substrates directly into the cytosolic compartment. Soluble activities that might degrade the substrates are removed by simply washing the permeabilized cells.

2.3 Commercial SLO Preparations

Although many commercial preparations of SLO are suitable as antigens for screening anti-SLO antibodies, it appears that most are not suitable for specific permeabilization of the plasma membrane. It is strongly recommended that investigators obtain verification from the supplier that the preparation they are considering has been validated for cell permeabilization. Active preparations are typically expressed in terms of International Units (I.U.), which should not be confused with other unit measurements by which poorly active preparations are often sold. The SLO preparations used in our prior studies are no longer commercially available. However, what appears to be a comparable preparation is manufactured by BioAcademia and is offered by each of three different suppliers: Abcam (www.abcam.com), Cosmo Bio (www.cosmobio.co.jp), and Sceti K. K. (www.sceti.co.jp) (*see* **Note 1**). We have not tested other methods of selectively permeabilizing the plasma membrane, such as with other pore-forming toxins like perfringens-O, or with the detergent digitonin, but these may be useful alternatives to SLO.

2.4 Other Supplies

1. GDP-[2-^{3}H]mannose (20 Ci/mmol): American Radiolabeled Chemicals (www.arc-inc.com).
2. Mannose-6-phosphate (sodium salt): Sigma-Aldrich (www.sigmaaldrich.com).

All other supplies (nonradioactive nucleotide sugars, solvents, etc.) should be obtained from any reputable company.

3. Nucleotide sugars: GDP-mannose, UDP-GlcNAc, UDP-glucose.
4. Castanospermine.
5. AMP.
6. Solvents.
7. Salts: NaCl, Na-Phosphate, KCl, $MgCl_2$.
8. Buffer: Potassium-HEPES.
9. Trypan Blue.

3 Methods

M6P-dependent destruction of LLOs can be demonstrated by incubating SLO-permeabilized cells with 50–100 μM M6P at 37 °C for periods up to 1 h. The readout for this reaction may be either a loss of intact $G_3M_9Gn_2$-LLO, or an increase of free $G_3M_9Gn_2$ released from the LLO. This decision depends partly upon the presence or absence of nucleotide sugars that allow the resynthesis of LLO from Dol-P, generated by recycling of Dol-P-P released during the LLO cleavage reaction. When the concentrations of the three nucleotide sugars used to make LLOs (i.e., UDP-GlcNAc, GDP-Man, and UDP-Glc) are not limiting, and in the presence of M6P, LLO levels do not change because they are continuously regenerated, but free glycan concentrations continue to rise over time. Thus, free glycans should be measured to assay the M6P effect. In contrast, when one of the nucleotide sugars is limiting (for example, GDP-mannose below 200 nM) in the presence of M6P, LLO resynthesis rates fall below the rate of cleavage, causing a measurable decline in LLO levels as well as increasing free glycan concentrations. When nonradioactive nucleotide sugars are used, the products can be detected by fluorophore-assisted carbohydrate electrophoresis (FACE) [4] or any other method not requiring isotopes [5]. If radioactive nucleotide sugars are included, products can be detected by HPLC or other methods allowing radioisotopic detection.

3.1 Permeabilization of Cells with SLO

1. PBS: A solution containing 150 mM sodium chloride and 15 mM sodium phosphate (pH 7.4).
2. SLO: According to the supplier's instruction, prepare a 2 U/mL solution in PBS just before use, and place on ice.
3. Transport buffer: A solution containing 78 mM potassium chloride, 4 mM magnesium chloride, and 50 mM potassium-HEPES (pH 7.2) (*see* **Note 2**).

4. Refeed cells in a 10 cm dish the day before the experiment. On the day of the experiment cells should be 75–90 % confluent.
5. Place dishes on ice and wash cells twice with ice-cold PBS.
6. Add 2 ml 2 U/mL SLO (ice-cold) per dish on ice. Be sure all cells are covered and periodically swirl the dishes.
7. After 4 min, wash twice with ice-cold PBS keeping the dishes on ice.
8. Add 10 mL transport buffer, prewarmed to 37 °C. Quickly transfer the dish to a 37 °C incubator for 4 min, then place on ice for 10 min. Keep the dish on ice until the next step.

3.2 Testing Permeability with Trypan Blue

For the initial SLO treatment, and at subsequent times to confirm the efficacy of the SLO, cell permeability can be verified by uptake of trypan blue.

1. Trypan blue suspension: Prepare a 0.2 % (w/v) trypan blue suspension in transport buffer. Store at 4 °C.
2. Add the trypan blue suspension to cells, before or after SLO treatment.
3. Incubate 15 min, room temperature. Replace the trypan blue suspension with transport buffer lacking trypan blue.
4. Examine cells under a conventional light microscope. Estimate the percentage of blue cells, which are permeabilized.

3.3 Incubation with M6P

1. Prepare reaction buffer by supplementing transport buffer with (*all final reaction concentrations*) 100 μM UDP-GlcNAc, 500 μM UDP-glucose, 2 mM 5' AMP, 100 μg/mL castanospermine, and either (A) 200 nM radioactive GDP-mannose (20 nM GDP-[^{3}H]mannose, 180 nM unlabeled GDP-mannose) to generate a mixture of truncated and mature LLOs, (B) 400 nM radioactive GDP-mannose (20 nM GDP-[^{3}H]mannose, 380 nM unlabeled GDP-mannose) to generate primarily mature LLOs, or (C) 20 μM nonradioactive GDP-mannose for a pool of mature, unlabeled LLOs.
2. Remove the transport buffer from SLO-permeabilized cells kept on ice, prepared above.
3. Add 3 mL of reaction buffer A, B, or C and the desired final concentration of M6P (typically 100 μM). Incubate at 37 °C for 60 min.
4. Remove the reaction buffer, transfer the dish to ice, and wash cells twice with ice-cold PBS.
5. While the dish is still on ice, add 5 mL room temperature methanol. Transfer the dish to a room temperature surface, and collect the buffer and cellular material with a plastic scraper or rubber policeman.

6. Transfer the suspension to a 10 mL glass tube and dry the entire contents under a stream of N_2 gas, or by centrifugal evaporation under vacuum.
7. The dried pellet is now suitable for analytical determination of LLOs and free glycans by any preferred method (*see* **Note 3**).

4 Notes

1. Catalog numbers for SLO are as follows: Abcam (#ab63978), Cosmo Bio (#BAM-01-531-EX), and Sceti K. K. (#01-531).
2. Since SLO is subject to oxidation in solution, the transport buffer may be supplemented with a reductant such as 2 mM dithiothreitol. However, we have not found this addition to be necessary in our own experiments, and intentionally omit it to avoid altering disulfide bonding of luminal ER proteins.
3. If the methanol-denatured material is not suitable for the analytical method of choice, **steps 4–6** can be altered accordingly.

Acknowledgments

This work was supported by generous funding from the National Institutes of General Medical Sciences (NIH, grant GM38545) and the Robert Welch Foundation (grant I-1168).

References

1. Jaeken J, Matthijs G (2001) Congenital disorders of glycosylation. Annu Rev Genomics Hum Genet 2:129–151
2. Gao N, Shang J, Lehrman MA (2005) Analysis of glycosylation in CDG-Ia fibroblasts by fluorophore-assisted carbohydrate electrophoresis: implications for extracellular glucose and intracellular mannose-6-phosphate. J Biol Chem 280:17901–17909
3. Gao N, Shang J, Huynh D, Manthati VL, Arias C, Harding HP, Kaufman RJ, Mohr I, Ron D, Falck JR, Lehrman MA (2011) Mannose-6-phosphate regulates destruction of lipid-linked oligosaccharides. Mol Biol Cell 22: 2994–3009
4. Gao N, Lehrman MA (2002) Analyses of dolichol pyrophosphate-linked oligosaccharides in cell cultures and tissues by fluorophore-assisted carbohydrate electrophoresis. Glycobiology 12:353–360
5. Lehrman MA (2007) Teaching dolichol-linked oligosaccharides more tricks with alternatives to metabolic radiolabeling. Glycobiology 17(8): 75r–85r

Chapter 21

N-Acetylglucosaminyltransferase (GnT) Assays Using Fluorescent Oligosaccharide Acceptor Substrates: GnT-III, IV, V, and IX (GnT-Vb)

Shinji Takamatsu, Hiroaki Korekane, Kazuaki Ohtsubo, Suguru Oguri, Jong Yi Park, Akio Matsumoto, and Naoyuki Taniguchi

Abstract

Determining glycosyltransferase activities gives a clue for better understanding an underlying mechanism for glycomic alterations of carrier molecules. *N*-glycan branch formation is concertedly regulated by cooperative and competitive activities of *N*-acetylglucosaminyltransferases (GnTs). Here, we describe methods for large scale preparation of the oligosaccharide acceptor substrate, fluorescence-labeling of oligosaccharides by pyridylamination, quality control, and reversed phase HPLC-based measurement of GnT activities including GnT-III, IV, V, and IX.

Key words *N*-acetylglucosaminyltransferase, GnT-III, GnT-IV, GnT-V, GnT-IX, GnT-Vb, PA-labeled oligosaccharide, Reversed phase HPLC

1 Introduction

The branching of the core of *N*-glycans has been implicated in many physiological and pathological events including cell adhesion, signal transduction, cancer metastasis and invasion [1, 2]. In vertebrates, seven different *N*-acetylglucosaminyltransferases (GnTs) I through VI, and IX (also designated as Vb) [3–5], have been known to be involved in the initiation of the branching of the *N*-glycan core structure. As indicated in Fig. 1, each GnT catalyzes the transfer of *N*-acetylglucosamine (GlcNAc) from the common donor substrate UDP-GlcNAc to a specific position on the core of *N*-glycans via a specific glycosidic linkage. GnT-III catalyzes the transfer of GlcNAc to the core β-mannose (Man) in *N*-glycans via a β1,4-linkage (Fig. 2a) forming a unique structure called

Authors Shinji Takamatsu and Hiroaki Korekane have contributed equally to this work.

Inka Brockhausen (ed.), *Glycosyltransferases: Methods and Protocols*, Methods in Molecular Biology, vol. 1022, DOI 10.1007/978-1-62703-465-4_21, © Springer Science+Business Media New York 2013

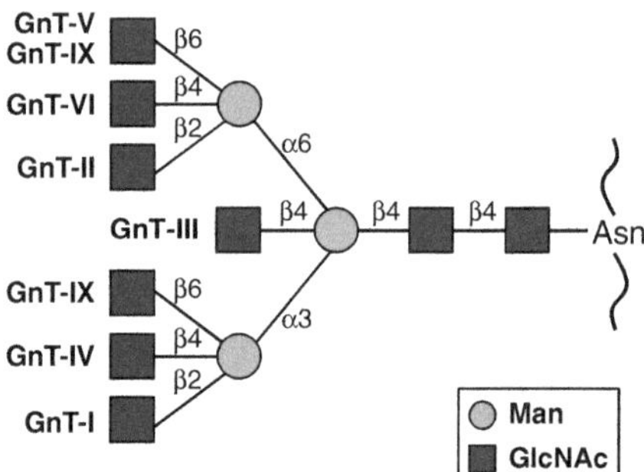

Fig. 1 *N*-Acetylglucosaminyltransferases (GnTs) involved in the branch formation on the core of complex type *N*-glycans. Seven different GnTs (I through VI, and IX) are known to be present. The GnT-IX is also designated as GnT-Vb. Two additional GnT activities (GnT-VII and VIII) were also demonstrated in CHO mutant cells, but their corresponding genes have not yet been identified [21]

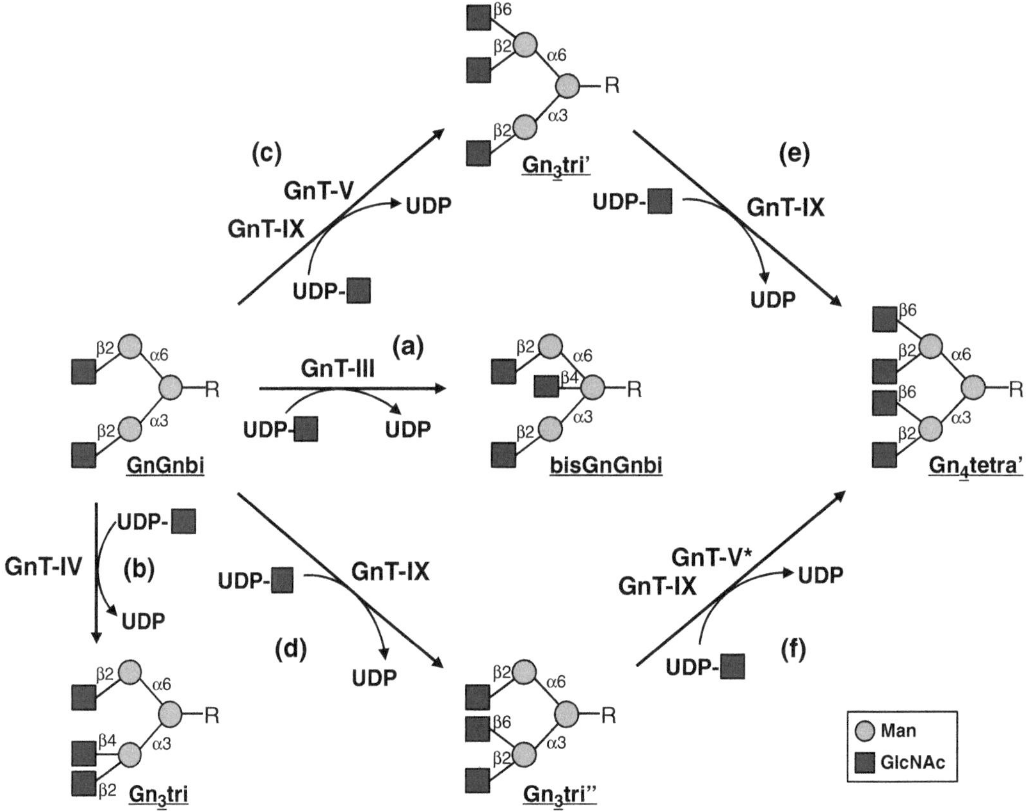

Fig. 2 Reactions catalyzed by GnT-III (*a*), IV (*b*), V (*c*,*f*), and IX (*c*, *d*, *e*, *f*), as well as the structures of substrates and enzyme products. R, β1-4GlcNAcβ1-4GlcNAcβ1-Asn or β1-4GlcNAcβ1-4GlcNAc-PA. *Our unpublished observation

"bisecting GlcNAc." The introduction of a bisecting GlcNAc is known to inhibit further branch formation catalyzed by GnT-IV or GnT-V because the bisected *N*-glycan cannot be their acceptor substrate [6, 7]. GnT-IV has two isoenzymes, a and b, bearing identical substrate specificity [8]. GnT-IV is responsible for the formation of multi-antennary complex type *N*-glycans, which are important for regulating protein functions and are associated with maintenance of cellular functions [9, 10]. The enzyme catalyzes the transfer of GlcNAc to α3-Man residue on the core of *N*-glycans via a β1,4-linkage (Fig. 2b). GnT-V transfers GlcNAc to the α6-Man residue on the core of *N*-glycans via a β1,6-linkage (Fig. 2c, f), and the production of the β1,6-GlcNAc branch is strongly implicated in cancer metastasis [1, 2]. GnT-IX is the most recently identified GnT that is responsible for the branching of the core of *N*-glycans [4, 5]. This enzyme is a closely related homolog of GnT-V, being mainly expressed in brain and mildly in testis. GnT-IX has broader substrate specificity than GnT-V and transfers GlcNAc to both α3- and α6-Man residues on the core of *N*-glycans via a β1,6-linkage (Fig. 2c, d, e, f), and also reported to be involved in the formation of a β1,6-branched structure in brain-type *O*-mannosyl glycan [11, 12]. The GnT activities used to be measured by detecting the incorporated radioactivity in the glycan products of enzymatic reactions in conjunction with separation by column chromatography, gel filtration, or paper electrophoresis that required special work place in a radiation controlled area and are tedious and time-consuming procedures [3, 5]. However, the development of techniques for the fluorescent labeling of oligosaccharide acceptor substrate with 2-aminopyridine and the separation of oligosaccharides using reversed phase high-performance liquid chromatography (HPLC) with a fluorescence detector [13, 14] has enabled us to detect and to determine specific glycosyltransferase activities sensitively and conveniently by quantifying the specific fluorescent peak in the elution charts [6, 14–17]. For setting up glycosyltransferase activity assays, highly purified oligosaccharide acceptor substrate is required. The bi-antennary *N*-glycan structure, intrinsically encompassing the "GnGnbi-Asn" structure (Fig. 2), abundantly exists in Apo-transferrin, so that the acceptor substrate, GnGnbi-Asn, can be obtained by enzymatic cleavage of *N*-glycans from glycoproteins using peptide-N4-(*N-acetyl-βglucosaminyl*) asparagine amidase (PNGase) F, followed by enzymatic trimming of terminal sialic acid and galactose residues [17]. GnGnbi-Asn is a good substrate for GnT-IV isoenzymes as well as for GnT-III and V, so that it is a suitable acceptor substrate for profiling the portfolio of GnT activities in sample enzyme extracts [15, 18]. We can assay GnT-IX activity also by using Gn_3tri′-Asn structure (Fig. 2) as an acceptor substrate.

2 Materials

Prepare all solutions using ultrapure water prepared with Milli-Q system (Millipore, Billerica, MA, USA) and analytical grade reagents.

2.1 Preparation of Fluorescent Oligosaccharide Acceptor Substrates

1. Human apo-transferrin (Sigma-Aldrich, St. Louis, MO, USA).
2. PNGase F (Roche, Basel, Switzerland).
3. Biogel P-100 (Bio-Rad, Hercules, CA, USA). Pack the resin into a glass column (2.5 × 70 cm).
4. 5 % (w/w) Phenol: Weigh 5 g phenol to a dark glass bottle. Add 95 g water to the bottle. Shake it vigorously and store at 4 °C (*see* **Note 1**).
5. Sulfuric acid.
6. PA reagent: Dissolve 27 g of 2-aminopyridine (PA, fluorescence-labeling grade) in 13 mL of glacial acetic acid.
7. Reducing reagent: Dissolve 57 g of borane-dimethylamine complex (Wako, Osaka, Japan) and 23 mL of glacial acetic acid in 14 mL of water.
8. Toyopearl HW-40 F (Tosoh, Tokyo, Japan). Pack the resin into a glass column (5 × 25 cm).
9. *Arthrobactor ureafaciens* sialidase (Nakalai Tesque, Kyoto Japan).
10. *Aspergillus sp.* β-galactosidase (TOYOBO, Osaka, Japan).

2.2 GnT Reactions

1. 0.8 M 2-(*N*-morpholino)ethanesulfonic acid (MES), pH 6.25: Weigh 8.53 g MES monohydrate (MES·H_2O) and transfer to a glass beaker containing ca 40 mL water. Mix and adjust pH with NaOH solution, then make up to 50 mL with water and store at 4 °C.
2. 0.8 M *N*-2-hydroxyethylpiperazine-*N*-ethanesulfonic acid (HEPES), pH 7.5: Weigh 9.53 g HEPES and transfer to a glass beaker containing ca 40 mL water. Mix and adjust pH with NaOH solution, then make up to 50 mL with water and store at 4 °C.
3. 1 M 3-(*N*-Morpholino)propanesulfonic acid (MOPS), pH 6.8: Weigh 20.9 g of MOPS and transfer to a 200 mL beaker containing about 60 mL of water and dissolve by stirring on the magnetic stirrer. After dissolving adjust pH to 6.8 with KOH and make up to 100 mL with water. Wrap the bottle with aluminum foil and store at room temperature.
4. 1 M $MnCl_2$ solution: Weigh 1.258 g of $MnCl_2$ (anhydrous) and transfer to a 15 mL conical tube containing about 9 mL of water. Dissolve by vortex and make up to 10 mL. No need to make pH adjustments (*see* **Note 2**).

5. 0.5 M EDTA solution: Weigh 10.31 g Ethylenediamine-*N*,*N*,*N*′,*N*′-tetraacetic acid trisodium salt trihydrate (EDTA·3Na·3H_2O, DOJINDO, Kumamoto, Japan) to a glass beaker containing ca 40 mL water. Mix and then make up to 50 mL with water. No need to make pH adjustments.
6. 10 % (w/v) Triton X-100 solution: Weigh 5 g Triton X-100 to a glass beaker containing ca 40 mL water. Mix and then make up to 50 mL with water.
7. 10 mg/mL bovine serum albumin (BSA): Weigh 100 mg BSA to a glass beaker containing ca 8 mL water. Mix and then make up to 10 mL with water. Store at −20 °C.
8. Stock GnT-III reaction buffer (2×): 0.25 M MES (pH 6.25), 20 mM $MnCl_2$, 0.4 M GlcNAc, 1 % Triton X-100, 2 mg/mL BSA (*see* **Note 3**). Leave one aliquot at 4 °C for routine use and store remaining aliquots at −20 °C.
9. Stock GnT-IV reaction buffer (2×) (minus UDP-GlcNAc): 0.25 M MOPS (pH 7.3), 20 % glycerol, 1 % Triton X-100, 0.4 M GlcNAc, 10 mg/ml BSA, 15 mM $MnCl_2$ (*see* **Note 4**). Store at −20 °C.
10. Stock GnT-IV reaction buffer (2×) (+UDP-GlcNAc): GnT-IV reaction buffer (2×) (minus UDP-GlcNAc), 40 mM UDP-GlcNAc (*see* **Note 5**). Store at −20 °C.
11. Stock GnT-V reaction buffer (2×): 0.25 M MES (pH 6.25), 20 mM EDTA, 400 mM GlcNAc, 1 % Triton X-100, 2 mg/mL BSA (*see* **Note 6**). Leave one aliquot at 4 °C for routine use and store remaining aliquots at −20 °C.
12. Stock GnT-IX reaction buffer (2×): 0.1 M HEPES (pH 7.5), 20 mM $MnCl_2$, 1 % Triton X-100, 2 mg/mL BSA (*see* **Note 7**). Leave one aliquot at 4 °C for routine use and store remaining aliquots at −20 °C.
13. 400 mM Uridine 5′-diphospho-*N*-acetylglucosamine (UDP-GlcNAc). Weigh 782 mg UDP-GlcNAc sodium salt (Sigma-Aldrich) to a 5 mL graduated cylinder. Mix and then make up to 3 mL with water. Store at −20 °C.
14. 200 mM UDP-GlcNAc. Dilute 400 mM UDP-GlcNAc two-fold with water. Store at −20 °C.
15. 4 mM GnGnbi-PA. Adjust concentration of the PA-oligosaccharide (*see* Subheading 3) with water (*see* **Note 8**). Store at −20 °C.
16. 100 μM GnGnbi-PA. Adjust concentration of the PA-oligosaccharide (*see* Subheading 3) with water (*see* **Note 8**). Store at −20 °C.
17. 100 μM Gn_3tri′-PA. Adjust concentration of the PA-oligosaccharide (*see* Subheading 3) with water (*see* **Note 8**). Store at −20 °C.
18. Bioruptor sonicator (Cosmo Bio, Tokyo, Japan).

2.3 Reversed Phase HPLC

1. HPLC stock buffer: 0.1 M Ammonium acetate, pH 4.0. Weigh 18.02 g (17.17 mL) acetic acid (HPLC grade) to a 3 L glass beaker containing 2.7 L water. Mix and adjust pH with 10 % ammonia solution, then make up to 3 L with water. Store at room temperature.
2. Eluent A: 20 mM Ammonium acetate, pH 4.0. Dilute HPLC stock buffer fivefold with water (*see* **Note 9**). De-gass the solvent in a sonic bath for 10 min before use.
3. Eluent B: 20 mM Ammonium acetate containing 1 % 1-butanol (HPLC grade), pH 4.0. Add 5 mL 1-butanol to 500 mL of eluent A. De-gass the solvent in a sonic bath for 10 min before use.
4. Eluent C: 50 mM Ammonium acetate, pH 4.0. Dilute HPLC stock buffer twofold with water (*see* **Note 9**). De-gass the solvent in a sonic bath for 10 min before use.
5. Eluent D: 50 mM Ammonium acetate containing 1 % 1-butanol (HPLC grade), pH 4.0. Add 5 mL 1-butanol to 500 mL of eluent C. De-gass the solvent in a sonic bath for 10 min before use.
6. HPLC system. A standard system equipped with a fluorescence detector.
7. TSKgel ODS-80TM (*see* **Note 10**). Purchase two sizes of the columns (4.6 × 150 mm and 4.6 × 250 mm) from Tosoh.
8. Vydac 218TP152010 (10 × 250 mm, Grace, Deerfield, IL, USA).

3 Methods

3.1 Preparation of Oligosaccharide Acceptor Substrate [15, 17]

1. Dissolve 2.5 g of human apo-transferrin in 100 mL of 50 mM Tris–HCl (pH 7.2) containing 1.0 % (v/v) β-mercaptoethanol and 2.5 % (w/v) SDS.
2. Mix gently with stirring on a magnetic stirrer and incubate at 60 °C for 2 h.
3. After cooling down to room temperature, add 4.0 % (w/v) of Nonidet P-40 and 75 U of *N*-glycosidase F. To prevent bacterial growth, add 100 μL of toluene and incubate at 37 °C overnight.
4. To achieve the complete digestion, add another 75 U of *N*-glycosidase F and further incubate at 37 °C overnight.
5. Analyze the aliquot of sample by SDS-PAGE.
6. Divide into 20 mL aliquots in 50 mL conical tubes. To extract free glycans from the reaction solution, add an equal volume of TE-Buffer saturated phenol–chloroform solution (phenol–chloroform–isoamylalcohol = 25:24:1) and mix vigorously with a vortex mixer.

7. Centrifuge at >3,000×g for 20 min at room temperature (*see* **Note 11**).
8. Transfer the clear aqueous phase to conical tubes and let stand at −80 °C until frozen.
9. Lyophilize the sample.
10. Dissolve the lyophilized powder in 20 mL of 50 mM ammonium acetate buffer (pH 7.2).
11. Apply the sample to a Biogel P-100 gel filtration column equilibrated with 50 mM ammonium acetate buffer for desalting.
12. Collect the eluate at 2 mL per fraction.
13. Detect the sugar chains by phenol–H_2SO_4 colorimetric method for hexoses [19] (*see* **Note 12**). The sugar chains would be eluted between 130 and 170 fractions.
14. Transfer the carbohydrate-positive fractions to a 500 mL glass bottle and lyophilize.

3.2 Pyridylamination of Oligosaccharide [17, 20]

1. Add the PA reagent to the lyophilized sugar chains and stir on a magnetic stirrer until completely dissolved. Wrap the bottle with aluminum foil and incubate at 90 °C for 1 h in a bath.
2. Add the reducing reagent to the PA-reacted solutions in the bottle and stir on a magnetic stirrer. Further incubate at 80 °C for 1 h in a bath inside a fume hood.
3. After cooling down to room temperature, add about 100 mL of water and finally adjust the total volume to 250 mL.
4. Transfer the sample to a separating funnel. To remove the excess 2-aminopyridine, extract the sample twice with equal volumes of TE-saturated phenol–chloroform, and once with chloroform.
5. After chloroform extraction, let the separating funnel stand for a while and recover the aqueous phase containing the PA-labeled sugar chains (*N*-glycans).
6. Concentrate the sample to about 40 mL by using a rotary evaporator.
7. Apply the concentrated sample to the Toyopearl HW-40 F gel filtration column equilibrated with 10 mM ammonium acetate buffer (pH 6.0). Collect the eluate at 5 mL per fraction.
8. Detect sugar chains by phenol–H_2SO_4 colorimetric method. The absorption of 2-aminopyridine can be monitored at 300 nm. The PA-labeled sugar chains will be eluted between 40 and 70 fractions. Collect the carbohydrate positive fractions and lyophilize.
9. Analyze the yield of PA-labeled oligosaccharides by the phenol–H_2SO_4 colorimetric method and HPLC.

3.3 Exoglycosidase Digestions of PA-Oligosaccharide [17]

1. Dissolve the lyophilized PA-oligosaccharides in 10 mL of 0.4 M sodium acetate buffer (pH 5.0) containing 4 mM $MgCl_2$.
2. Add 10 U of sialidase derived from *Arthrobactor ureafaciens* and 10 mg of β-galactosidase derived from *Aspergillus sp.* in a PA-oligosaccharides solution. Add one drop of toluene to the solution for prevention of bacterial growth. Incubate at 37 °C overnight.
3. To determine the effectiveness of digestion, analyze a part of the sample by reversed phase HPLC.
4. Purify the digested product, PA-labeled sugar chain (GnGnbi-PA), by reversed phase HPLC under the following conditions: [column, Vydac 218TP152010; eluent, 100 mM ammonium acetate (pH 4.0) containing 0.15 % 1-butanol (*see* **Note 13**); flow rate, 2.5 mL/min; column temperature, 50 °C; detection, 300 nm].
5. Perform this purification by repeating 1 mL of sample injection ten times (*see* **Note 14**).
6. Pool the fractions containing the PA-labeled sugar chain (GnGnbi-PA) and concentrate by lyophilization.
7. To prepare an acceptor substrate for GnT-IX, that is, Gn_3tri′-PA (Fig. 2), transfer an aliquot of the prepared GnGnbi-PA (500 nmol) to a 1.5 mL tube and carry out the reaction catalyzed by GnT-V with 40 mM UDP-GlcNAc in a final volume of 1 mL at 37 °C for 24 h (*see* **Note 15**). Purify and collect the formed Gn_3tri′-PA by reversed phase HPLC under the same conditions as the above step (*see* **Note 14**).

3.4 Confirmation of the Purity of the PA-Oligosaccharides [17]

1. Check the purity of PA-labeled oligosaccharide acceptor substrates (GnGnbi-PA or Gn_3tri′-PA) by reversed phase HPLC under the following conditions: [column, TSKgel ODS-80TM (4.6 × 150 mm); eluent, 50 mM ammonium acetate (pH 4.0) containing 0.15 % 1-butanol; flow rate, 1.2 mL/min; column temperature, 50 °C; fluorescence detection, ex. 320 nm, em. 400 nm].
2. Analyze the homogeneity and molecular weight of the samples by MALDI-TOF-MS [matrix: 2,5-dihydroxybenzoic acid].
3. Approximately 25–30 μmol GnGnbi-PA and ca 350 nmol Gn_3tri′-PA can be prepared from 2.5 g of human apo-transferrin (yield, 43 %) and 500 nmol GnGnbi-PA, respectively (*see* **Note 16**).

3.5 Preparation of Crude Enzyme Extracts

1. In the case of cell lines, wash cultured cells (5×10^6–1×10^7 cells) with ice-chilled PBS (−) twice and then harvest them with a cell scraper into a 1.5-mL tube. Centrifuge at 600 × *g* at 4 °C for 5 min and discard the resulting supernatant.

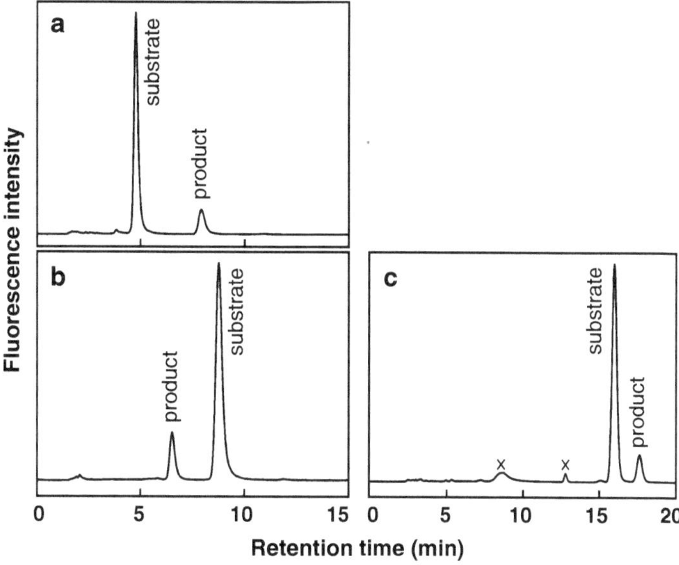

Fig. 3 Typical chromatograms for GnT-III, V, and IX assays. (**a**) GnT-III assay for a crude enzyme extract from GnT-III-transfected COS-1 cells. (**b**) GnT-V assay for a crude enzyme extract from GnT-V-transfected COS-1 cells. (**c**) GnT-IX assay for a crude enzyme extract from GnT-IX-transfected HEK293 cells. Xs in *panel* (**c**) were contaminants that existed in the enzyme sample

2. Resuspend the cell pellet in 100–200 μL ice-chilled PBS (−) containing protease inhibitors (*see* **Note 17**) and then sonicate for 10 min in an ice-chilled water bath with the Bioruptor sonicator. Use the resulting homogenate as crude enzyme extracts for GnT assays (*see* **Note 18**).

3.6 GnT-III Assay

1. Mix the following components in a final volume of 10 μL in a small plastic tube: 5 μL of stock GnT-III reaction buffer, 1 μL of 200 mM UDP-GlcNAc, 1 μL of 100 μM GnGnbi-PA, and 3 μL enzyme sample (*see* **Note 19**).
2. Incubate the mixture at 37 °C for 1–4 h.
3. Stop the reaction by boiling for 2 min. After cooling, add 40 μL water and centrifuge at 20,000 × *g* for 5 min (*see* **Note 20**).
4. Inject 10 μL of the resulting supernatant to an HPLC system equipped with a TSKgel ODS-80TM column (4.6 × 150 mm) to separate and quantitate the product. Carry out the elution isocratically at a flow rate of 1 mL/min at 55 °C using 20 mM ammonium acetate buffer (pH 4.0) containing 0.25 % 1-butanol as the eluent (Eluent A/B = 75/25) (*see* **Note 21**). Monitor fluorescence using excitation and emission wavelengths of 320 and 400 nm, respectively (Fig. 3a).

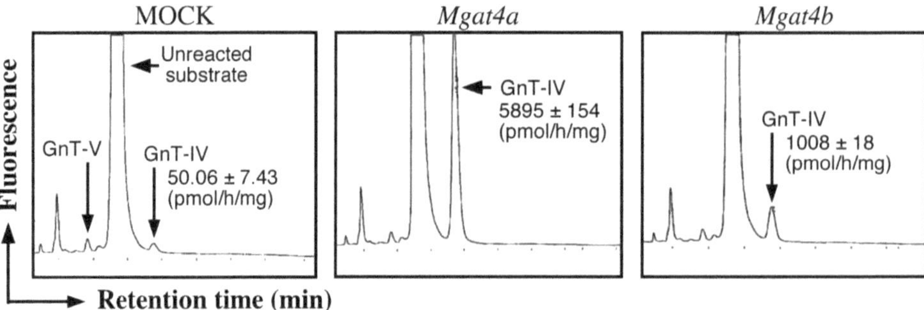

Fig. 4 GnT-IV enzymatic activity of GnT-IV isoenzymes. Either empty vector or expression vector of *Mgat4a* or *Mgat4b* was introduced into COS-7 cells, and the resultant GnT-IV enzymatic activities were assayed [18]. Each HPLC chart indicates the elution of the fractions containing the substrate (retention time = about 14 min) and the GnT-IV reaction products (retention time = about 15 min). The fractions containing the GnT-V reaction products were simultaneously detected. The indicated activities are obtained from the chart and normalized according to their introduced each gene's transcript levels. These transcript levels were estimated by using quantitative RT-PCR analysis

5. Estimate the amount of the product from the fluorescence intensity (*see* **Note 22**).

3.7 GnT-IV Assay

1. Mix the following components in a final volume of 10 μL in a small plastic tube: 5 μL of stock GnT-IV reaction buffer (+UDP-GlcNAc), 2 μL of 4 mM GnGnbi-PA, and 3 μL enzyme sample (*see* **Note 23**).
2. Mix gently by pipetting, and incubate at 37 °C for 2–4 h (*see* **Note 24**).
3. Stop the reaction by adding 30 μL of 10 mM EDTA (pH 7.0) and boil for 2 min.
4. Transfer the reaction mixture to a 1.5 mL microtube and centrifuge at 20,000 × *g* for 15 min at 4 °C (*see* **Note 20**).
5. Inject 10 μL of the resulting supernatant to an HPLC system equipped with a TSKgel ODS-80TM column (4.6 × 150 mm) to separate and quantitate the product (*see* **Note 25**). Carry out the elution isocratically at a flow rate of 1.2 mL/min at 50 °C using 50 mM ammonium acetate buffer (pH 4.0) containing 0.12 % 1-butanol as the eluent (Eluent C/D = 88/12). Monitor fluorescence using excitation and emission wavelengths of 320 and 400 nm, respectively (Fig. 4) (*see* **Note 22**).

3.8 GnT-V Assay

1. Mix the following components in a final volume of 10 μL in a small plastic tube: 5 μL of stock GnT-V reaction buffer, 1 μL of 400 mM UDP-GlcNAc, 1 μL of 100 μM GnGnbi-PA, and 3 μL enzyme sample (*see* **Note 26**).
2. Incubate the mixture at 37 °C for 2–4 h.

3. Stop the reaction by boiling for 2 min. After cooling, add 40 μL water and centrifuge at 20,000 × *g* for 5 min (*see* **Note 20**).
4. Inject 10 μL of the resulting supernatant to an HPLC system equipped with a TSKgel ODS-80TM column (4.6 × 150 mm) to separate and quantitate the product. Carry out the elution isocratically at a flow rate of 1 mL/min at 55 °C using 20 mM ammonium acetate buffer (pH 4.0) containing 0.1 % 1-butanol as the eluent (Eluent A/B = 90/10). Monitor fluorescence using excitation and emission wavelengths of 320 and 400 nm, respectively (Fig. 3b).
5. Estimate the amount of the product from the fluorescence intensity (*see* **Note 22**).

3.9 GnT-IX Assay

1. Mix the following components in a final volume of 10 μL in a small plastic tube: 5 μL of stock GnT-IX reaction buffer, 1 μL of 200 mM UDP-GlcNAc, 1 μL of 100 μM Gn_3tri′-PA, and 3 μL enzyme sample (*see* **Note 27**).
2. Incubate the mixture at 37 °C for 4 h.
3. Stop the reaction by boiling for 2 min. After cooling, add 40 μL water and centrifuge at 20,000 × *g* for 5 min (*see* **Note 20**).
4. Inject 10 μL of the resulting supernatant to an HPLC system equipped with a TSKgel ODS-80TM column (4.6 × 250 mm) to separate and quantitate the product. Carry out the elution isocratically at a flow rate of 1 mL/min at 55 °C using 20 mM ammonium acetate buffer (pH 4.0) containing 0.1 % 1-butanol as the eluent (Eluent A/B = 90/10). Monitor fluorescence using excitation and emission wavelengths of 320 and 400 nm, respectively (Fig. 3c).
5. Estimate the amount of the product from the fluorescence intensity (*see* **Note 28**).

4 Notes

1. Phenol is highly corrosive and can cause severe burns. Wear gloves when handling it.
2. Long time storage of manganese chloride solution is not recommended. It should be freshly prepared for making 2× enzyme assay reaction premixture.
3. To prepare the stock buffer, mix the following components in a final volume of 5 mL: 1.56 mL of 0.8 M MES buffer (pH 6.25), 0.1 mL of 1 M $MnCl_2$, 0.44 g GlcNAc, 0.5 mL of 10 % Triton X-100, 1 mL of 10 mg/mL BSA, and appropriate

volume of water. GnT-III requires divalent cations for its enzyme activity. Since precipitation of metal ions in the stock buffer may arise when it passes for a long time, mix well before use. GlcNAc is added as an inhibitor for β-*N*-acetylhexosaminidase which is likely to be present in the crude enzyme extract.

4. Prepare the stock buffer as follows. Add 25 mL of 1 M MOPS buffer, 20 mL of glycerol, 10 mL of 10 % (w/v) Triton X-100, 8.85 g of GlcNAc, and 1 g of BSA to a 200 mL glass beaker containing about 30 mL of water. Dissolve by stirring on the magnetic stirrer and adjust pH to 7.3 with KOH. Additionally add 1.5 mL of 1 M $MnCl_2$ and make up to 100 mL with water.
5. Prepare the stock buffer as follows. Weigh 104 mg of UDP-GlcNAc sodium salt (Sigma-Aldrich) and transfer to an appropriate tube. Add 4 mL of stock GnT-IV reaction buffer (minus UDP-GlcNAc) and dissolve by gentle pipetting. Dispense to microtubes and store at −20 °C.
6. To prepare the stock buffer, mix the following components in a final volume of 5 mL: 1.56 mL of 0.8 M MES buffer (pH 6.25), 0.2 mL of 0.5 M EDTA, 0.44 g GlcNAc, 0.5 mL of 10 % Triton X-100, 1 mL of 10 mg/mL BSA, and appropriate volume of water. GnT-V does not require divalent cations for its enzyme activity. The addition of EDTA is effective in inhibiting other GnTs including GnT-III and IV which act divalent cation-dependently on the common acceptor substrate, GnGnbi-PA.
7. To prepare the stock buffer, mix the following components in a final volume of 5 mL: 0.625 mL of 0.8 M HEPES buffer (pH 7.5), 0.1 mL of 1 M $MnCl_2$, 0.5 mL of 10 % Triton X-100, 1 mL of 10 mg/mL BSA, and appropriate volume of water. Although GnT-IX is active in the presence of EDTA, the enzyme activity is clearly promoted twofold to threefold in the presence of divalent cations such as Mn^{2+} or Mg^{2+} [5, our unpublished observation]. Thus, we are assaying GnT-IX activity in the presence of divalent cations; however, when GnT-III and IV coexist with GnT-IX, GnT-III and IV also act divalent cation-dependently on the acceptor substrate Gn_3tri′-PA under these conditions. GlcNAc is not supplied to the reaction buffer, because we found that GlcNAc inhibited GnT-IX by itself at 0.2 M of final concentration (our unpublished observation).
8. We are using a commercially available GnGnbi-PA (PA-Sugar Chain 012, TaKaRa Bio, Shiga, Japan) as standard for quantification of PA-oligosaccharides. At present, many kinds of standard PA-oligosaccharides are commercially available.
9. This diluted aqueous solvent should not be stored for more than 3–5 days especially in the summer season because of the

risk of microbial growth. A general rule in reversed phase HPLC is to use freshly prepared mobile phases.

10. Separation of PA-oligosaccharides on reversed phase HPLC changes depending on the kind of reversed phase column. In the case of using a different column from the TSKgel ODS-80TM, optimization of reversed phase HPLC conditions would be necessary before analysis.

11. Centrifuge within the tolerable centrifugal force of the equipped centrifuge and conical tubes.

12. We are routinely carrying out the phenol–sulfuric acid method using a 96-well microplate as follows. Mix 10 μL sample and 10 μL of 5 % phenol in a well on the plate. Add rapidly 100 μL concentrated sulfuric acid and then measure absorbance at 490 nm with a microplate reader. The standard curve should be made by using mannose (10–80 μg/mL).

13. Use stock HPLC buffer without any dilution after the addition of 0.15 % (v/v) 1-butanol (HPLC grade). De-gass the solvent in a sonic bath for 10 min before use.

14. Check the retention time of GnGnbi-PA or Gn_3tri′-PA in advance on reversed phase HPLC using the standard PA-oligosaccharides.

15. We are routinely using GnT-V-transfected COS-1 cells as the enzyme source. Carry out the GnT-V reaction as follows. Lyophilize GnGnbi-PA (500 nmol) in a 1.5 mL tube and redissolve it in 100 μL water. Add the following components to this tube: 500 μL stock GnT-V reaction buffer, 100 μL of 400 mM UDP-GlcNAc, and 300 μL of the GnT-V-transfected cell lysate. Mix gently and start the reaction to synthesize Gn_3tri′-PA enzymatically.

16. Expected yield of GnGnbi-PA from apo-transferrin is approximately from 25 to 40 % when the protein is fully glycosylated [17].

17. Any commercially available protease inhibitor cocktail is applicable. In most cases of cultured cells, 1 mM PMSF (phenylmethylsulfonyl fluoride) could be effective enough without further addition of other protease inhibitors.

18. In the case of tissues, add 4× volume of ice-chilled 10 mM Tris–HCl buffer (pH 7.4) containing 0.25 M sucrose and protease inhibitors (Complete EDTA-free, Roche) to the tissue. Homogenize it with a Polytron homogenizer or a Potter-Elvehjem homogenizer on ice and centrifuge at 900 × *g* at 4 °C for 5 min. Use the resulting supernatant as tissue homogenate for GnT assays. Alternatively, we may use cell lysate as an enzyme source. Preparation of cell lysate is carried out as follows. Suspend the cell pellet in 100–200 μL of ice-chilled 10 mM HEPES buffer (pH 7.5) containing 0.15 M NaCl, 1 %

Triton X-100, and protease inhibitors. Sonicate the sample for 5 min in an ice-chilled water bath with the Bioruptor sonicator and then centrifuge at 20,000 × *g* at 4 °C for 10 min. Use the resulting supernatant as cell lysate for GnT assays.

19. Final concentration of each component in GnT-III reaction buffer is as follows: 125 mM MES (pH 6.25), 10 mM $MnCl_2$, 200 mM GlcNAc, 0.5 % Triton X-100, 1 mg/mL BSA, 20 mM UDP-GlcNAc, and 10 μM GnGnbi-PA. When GnT-IV and V coexist with GnT-III in assay samples, note that these three enzymes react simultaneously with a common acceptor substrate, GnGnbi-PA under these conditions.
20. In some cases the method of deproteinization by boiling and centrifugation becomes insufficient; if so, carry out the following alternative procedure. At the end point of the enzyme reaction, add 40 μL water and 50 μL water-saturated phenol–chloroform (1:1). Immediately vortex the sample vigorously for 10 s and centrifuge at 20,000 × *g* for 1 min. Collect 40 μL of the resulting upper layer (aqueous phase) in a new tube. Add 40 μL chloroform and vortex vigorously for 10 s. Centrifuge at 20,000 × *g* for 1 min. Analyze 10 μL of the resulting upper layer (aqueous phase) by reversed phase HPLC to separate and quantitate the product. This step can be substituted by filtration of the heat denatured reaction mixture through a Millipore filter (0.22 μm), and then the filtrate can be directly subjected to HPLC.
21. Under the elution conditions, the GnT-IV and V products, Gn_3tri-PA and Gn_3tri′-PA, respectively (Fig. 2), are co-eluted with the unreacted acceptor substrate, GnGnbi-PA (Fig. 3a).
22. We are routinely calculating the conversion ratio to the enzymatic product of the acceptor substrate from each peak area. It is better that the consumption of substrate by the enzyme is kept below 20 % to ensure accurate initial velocity measurement. The specific activity of the enzyme is expressed as picomoles of GlcNAc transferred/h/mg of protein.
23. Final concentration of each component in GnT-IV reaction buffer is as follows: 125 mM MOPS (pH 7.3), 7.5 mM $MnCl_2$, 200 mM GlcNAc, 0.5 % Triton X-100, 5 mg/mL BSA, 10 % glycerol, 20 mM UDP-GlcNAc, and 0.8 mM GnGnbi-PA.
24. In the case of GnT-III assays, the enzyme activity was proportional to the enzyme protein concentration in the range of 0–200 μg per incubation mixture (50 μL). The reaction rate should be constant in the time window to 90 min [15]. However, 2–4 h incubation is required for GnT-IV assays, because the GnT-IV activity is lower than that of GnT-III. This GnT-IV enzyme assay is based on our previous method [15]

with some optimization. Therefore, GnT-III and V activities can be simultaneously assayed by this method. However, GnT-V does not require Mn^{2+} for activity and the optimal pH for GnT-III and V is different from that for GnT-IV.

25. Sample injection volume is depending on the enzyme activity and the dynamic range of fluorescence detector. If the enzyme activity is out of range, dilute the enzyme extract or change injection volume. Also applied amount of protein and/or the incubation time of the enzyme reaction may need to be optimized. The concentration of 1-butanol affects the retention time. If the peaks of the unreacted acceptor substrate and the GnT-IV product appear too closely, the use of slightly lower concentration of 1-butanol will improve the separation. GnT-IVa and GnT-IVb have identical substrate specificity, so that this assay cannot distinguish between their activities [8].
26. Final concentration of each component in GnT-V reaction buffer is as follows: 125 mM MES (pH 6.25), 10 mM EDTA, 200 mM GlcNAc, 0.5 % Triton X-100, 1 mg/mL BSA, 40 mM UDP-GlcNAc, and 10 μM GnGnbi-PA.
27. Final concentration of each component in GnT-IX reaction buffer is as follows: 50 mM HEPES (pH 7.5), 10 mM $MnCl_2$, 0.5 % Triton X-100, 1 mg/mL BSA, 20 mM UDP-GlcNAc, and 10 μM Gn_3tri′-PA.
28. When we carry out GnT-IX assay using a sample that contains substantial GnT-III and IV activities, GnT-III and IV also act on the acceptor substrate for GnT-IX as described in **Note 7**. In such a case, calculate the conversion ratio to the GnTs-IX, III, and IV products of acceptor substrate from each peak area. Products of GnT-III and IV are eluted at around 32 min and 24 min, respectively, under our experimental conditions. It would be better to check in advance the elution time of GnT-III and IV products on reversed phase HPLC using standard PA-oligosaccharides.

Acknowledgments

This work was mainly supported by Grant-in-Aid for Scientific Research (A), No. 20249018, and the Global COE Program of Osaka University funded by the Ministry of Education, Culture, Sports, Science, and Technology of Japan, the Naito foundation, and the National Institute of Biomedical Innovation (NIBIO) in Japan. We also thank Ms. Fumi Ota for supporting in the preparation of this manuscript.

References

1. Taniguchi N, Miyoshi E, Gu J, Honke K, Matsumoto A (2006) Decoding sugar functions by identifying target glycoproteins. Curr Opin Struct Biol 16:561–566
2. Taniguchi N, Korekane H (2011) Branched *N*-glycans and their implications for cell adhesion, signaling and clinical applications for cancer biomarkers and in therapeutics. BMB Rep 44:772–781
3. Brockhausen I, Narasimhan S, Schachter H (1988) The biosynthesis of highly branched *N*-glycans: studies on the sequential pathway and functional role of *N*-acetylglucosaminyltransferases I, II, III, IV, V and VI. Biochimie 70:1521–1533
4. Inamori K, Endo T, Ide Y, Fujii S, Gu J, Honke K, Taniguchi N (2003) Molecular cloning and characterization of human GnT-IX, a novel β1,6-N-acetylglucosaminyltransferase that is specifically expressed in the brain. J Biol Chem 278:43102–43109
5. Kaneko M, Alvarez-Manilla G, Kamar M, Lee I, Lee JK, Troupe K, Zhang WJ, Osawa M, Pierce M (2003) A novel β(1,6)-*N-acetylglucosaminyltransferase* V (GnT-VB). FEBS Lett 554:515–519
6. Oguri S, Minowa MT, Ihara Y, Taniguchi N, Ikenaga H, Takeuchi M (1997) Purification and characterization of UDP-N-acetylglucosamine: alpha1,3-D-mannoside beta 1,4-*N*-acetylglucosaminyltransferase (*N*-acetylglucosaminyltransferase-IV) from bovine small intestine. J Biol Chem 272:22721–22727
7. Gu J, Nishikawa A, Tsuruoka N, Ohno M, Yamaguchi N, Kangawa K, Taniguchi N (1993) Purification and characterization of UDP-N-acetylglucosamine: alpha-6-D-mannoside beta 1-6 *N*-acetylglucosaminyltransferase (*N*-acetylglucosaminyltransferase V) from a human lung cancer cell line. J Biochem 113:614–619
8. Oguri S, Yoshida A, Minowa MT, Takeuchi M (2006) Kinetic properties and substrate specificities of two recombinant human *N*-acetylglucosaminyltransferase-IV isozymes. Glycoconj J 23:473–480
9. Ohtsubo K, Takamatsu S, Minowa MT, Yoshida A, Takeuchi M, Marth JD (2005) Dietary and genetic control of glucose transporter-2 glycosylation promotes insulin secretion in suppressing diabetes. Cell 123: 1307–1321
10. Ohtsubo K, Chen MZ, Olefsky JM, Marth JD (2011) Pathway to diabetes through attenuation of pancreatic beta cell glycosylation and glucose transport. Nat Med 17:1067–1075
11. Inamori K, Endo T, Gu J, Matsuo I, Ito Y, Fujii S, Iwasaki H, Narimatsu H, Miyoshi E, Honke K, Taniguchi N (2004) *N*-acetylglucosaminyltransferase IX acts on the GlcNAcβ1,2-Manα1-Ser/Thr moiety, forming a 2,6-branched structure in brain *O*-mannosyl glycan. J Biol Chem 279:2337–2340
12. Abbott KL, Matthews RT, Pierce M (2008) Receptor tyrosine phosphatase β (RPTPβ) activity and signaling are attenuated by glycosylation and subsequent cell surface galectin-1 binding. J Biol Chem 283: 33026–33035
13. Hase S, Ikenaka T, Matsushima Y (1978) Structure analyses of oligosaccharides by tagging of the reducing end sugars with a fluorescent compound. Biochem Biophys Res Commun 85:257–263
14. Hase S, Ibuki T, Ikenaka T (1984) Reexamination of the pyridylamination used for fluorescence labeling of oligosaccharides and its application to glycoproteins. J Biochem 95:197–203
15. Taniguchi N, Nishikawa A, Fujii S, Gu JG (1989) Glycosyltransferase assays using pyridylaminated acceptors: *N*-acetylglucosaminyltransferase III, IV, and V. Methods Enzymol 179:397–408
16. Taguchi T, Ogawa T, Kitajima K, Inoue S, Inoue Y, Ihara Y, Sakamoto Y, Nagai K, Taniguchi N (1998) A method for determination of UDP-GlcNAc:GlcNAcβ1-6(GlcNAcβ1-2)Manα1-R [GlcNAc to Man] β1-4 *N*-acetylglucosaminyltransferase VI activity using a pyridylaminated tetraantennary oligosaccharide as an acceptor substrate. Anal Biochem 255:155–157
17. Tokugawa K, Oguri S, Takeuchi M (1996) Large scale preparation of PA-oligosaccharides from glycoproteins using an improved extraction method. Glycoconj J 13:53–56
18. Takamatsu S, Antonopoulos A, Ohtsubo K, Ditto D, Chiba Y, Le DT, Morris HR, Haslam SM, Dell A, Marth JD, Taniguchi N (2010) Physiological and glycomic characterization of *N*-acetylglucosaminyltransferase -IVa and -IVb double deficient mice. Glycobiology 20:485–497
19. Dubois M, Gilles K, Hamilton JK, Rebers PA, Smith F (1956) Colorimetric method for determination of sugars and related substances. Anal Chem 28:350–356
20. Natsuka S, Hase S (1998) Analysis of *N*- and *O*-glycans by pyridylamination. Methods Mol Biol 76:101–113
21. Raju TS, Stanley P (1998) Gain-of-function Chinese hamster ovary mutants Lec18 and Lec14 each express a novel *N*-acetylglucosaminyltransferase activity. J Biol Chem 273:14090–14098

Chapter 22

A Method for Determination of UDP-GlcNAc: GlcNAcβ1-6(GlcNAcβ1-2)Manα1-R [GlcNAc to Man] β1-4*N*-Acetylglucosaminyltransferase VI Activity

Tomohiko Taguchi and Naoyuki Taniguchi

Abstract

To characterize and purify glycosyltransferases, it is essential to establish a simple and sensitive assay method. Here, we describe a method for determination of the activity of GnT VI (UDP-GlcNAc: GlcNAcβ1-6(GlcNAcβ1-2)Manα1-R [GlcNAc to Man] β1-4*N*-acetylglucosaminyltransferase VI) using a fluorescently labeled oligosaccharide.

Key words Pentaantennary *N*-linked glycan, *N*-Acetylglucosaminyltransferase, Pyridylamination, Fluorescent labeling, HPLC

1 Introduction

Until now, a pentaantennary *N*-linked core structure has been the most branched structure identified in *N*-linked glycoproteins [1, 2] (Fig. 1). Six different *N*-acetylglucosaminyltransferases (GnTs) I through VI are involved in the GlcNAc branching of the core [3]. These enzymes were purified and the corresponding genes have been cloned [4]. As evident from a number of recent publications, knockdown/knockout or overexpression of these genes has shown the importance of multiantennary *N*-linked glycoproteins in many aspects of biology, such as cell adhesion, signaling, and development [5]. Moreover, dysregulation of these genes is often linked to carcinogenesis. It is important to note that functional glycotopes, such as the sialyl Le^x structure, are known to be expressed more preferentially on multiantennary *N*-linked glycans with poly-*N*-acetyllactosamines rather than on biantennary *N*-linked glycans. Human GnT VI has recently been shown to contribute to the surface expression of CD133 through the processing of multiantennary *N*-linked glycans on CD133 [6].

Inka Brockhausen (ed.), *Glycosyltransferases: Methods and Protocols*, Methods in Molecular Biology, vol. 1022, DOI 10.1007/978-1-62703-465-4_22, © Springer Science+Business Media New York 2013

```
GnT V    GlcNAcβ1 \
                   6
GnT VI   GlcNAcβ1- 4Manα1
                   2        \
GnT II   GlcNAcβ1 /          6
GnT III          GlcNAcβ1-  4Manβ1-4GlcNAcβ1-4GlcNAcβ1-Asn
                             3
GnT IV   GlcNAcβ1 \         /
                   4Manα1
                   2
GnT I    GlcNAcβ1 /
```

Fig. 1 Structure of a pentaantennary *N*-linked oligosaccharide with the bisecting GlcNAc residue (GnT III product) and GlcNAc transferases (GnT I through VI) involved in the incorporation of six nonreducing terminal GlcNAc residues shown in *italic* font

Here, we describe a highly sensitive method for determination of the activity of GnT VI using a fluorescently labeled tetraantennary oligosaccharide as an acceptor substrate [7]. A tetraantennary oligosaccharide was chosen to make this assay specific to GnT VI, since GnT-I, II, IV, and V cannot use it as an acceptor substrate [8, 9].

2 Materials

Prepare all solutions using ultrapure water and analytical grade reagents.

2.1 Preparation of the GnT VI Substrate

1. Human α_1-acid glycoprotein (Sigma, St. Louis, MO, USA).
2. Anhydrous hydrazine (Nacalai Tesque, Kyoto, Japan).
3. 2-aminopyridine (Takara Bio, Inc., Shiga, Japan).
4. Jack bean β-galactosidase (Seikagaku Corp., Tokyo, Japan).
5. H_2SO_4.
6. Toluene.
7. $NaHCO_3$.
8. Acetic anhydride.
9. HCl.
10. Citric acid.
11. Na_2HPO_4.
12. Ammonium acetate.
13. *n*-Butanol.
14. Dowex-50 (Sigma, St. Louis, MO, USA).
15. Bio-Gel P-4 (Bio-Rad, Hercules, CA, USA).
16. COSMOSIL 10C18 high performance liquid chromatography (HPLC) column (10 × 250 mm, Nacalai Tesque).
17. Glass test tube with screw cap.
18. Lyophilizer.

19. Glass vacuum desiccator.
20. Dry block heating system.
21. Glass column.
22. Rotary evaporator.
23. A column jacket for water circulation.
24. PALSTATION® (Takara Bio, Inc., Shiga, Japan).
25. HPLC system with a fluorescent detector (excitation 320 nm, emission 400 nm).

2.2 GnT VI Assay

1. HEPES (4-(2-hydroxyethyl)-1-piperazineethanesulfonic acid).
2. $MnCl_2$.
3. GlcNAc.
4. Triton X-100.
5. Bovine serum albumin.
6. UDP-GlcNAc (Sigma).
7. TSK-gel ODS-80TM (4.6×75 mm, TOSOH, Tokyo, Japan).

2.3 Preparation of Tissue Homogenate as Enzyme Source

1. Tissue of interest.
2. Sucrose.
3. Tris-HCl.
4. Protease inhibitor tablet "COMPLETE, Mini, EDTA free" (Roche, Switzerland).
5. PBS.
6. Dounce homogenizer.

3 Methods

3.1 Preparation of the GnT VI Substrate

1. Dissolve human α_1-acid glycoprotein (about 400 mg) into 4 mL of H_2O and make 10 aliquots of 0.4 mL in glass test tubes with screw cap.
2. Lyophilize them overnight.
3. Add 2 mL of anhydrous hydrazine into individual tubes.
4. Heat them at 100 °C for 10 h on a dry block heating system.
5. The reaction mixture is evaporated to dryness under reduced pressure over concentrated H_2SO_4. The residue is repeatedly evaporated with toluene to remove completely the remaining hydrazine.
6. Recover the material into 20 mL of saturated $NaHCO_3$ solution.
7. Free primary amino groups are then acetylated with 2 mL of acetic anhydride for 20 min at room temperature. Repeat the

acetylation reaction with further addition of 20 mL of saturated $NaHCO_3$ solution and 2 mL of acetic anhydride.

8. Pass the reaction mixture through a column of Dowex-50 (H^+ form, about 400 mL). The column is further washed with five bed volumes of water.
9. The flow through fraction is evaporated to dryness.
10. Dissolve the material in 40 mL of 0.15 M HCl.
11. Heat the solution for desialylation (80 °C, 1 h).
12. The solution is evaporated to dryness, and dissolved in 6 mL of H_2O.
13. Load the material to a Bio-Gel P-4 column with a jacket (2.5 × 90 cm, equilibrated with H_2O, 55 °C, 0.2 mL/min). The elution profile is monitored by measuring absorption at 230 nm.
14. The peak that corresponds to asialo-tetraantennary glycan is pooled and evaporated (*see* **Note 1**).
15. The material is reacted with 2-aminopyridine by a commercial pyridylamination apparatus, PALSTATION®.
16. The pyridylaminated (PA-) asialo-tetraantennary glycan is digested with 1 U of Jack bean β-galactosidase for 48 h at 37 °C in 4 mL of 50 mM citrate phosphate buffer (pH 4.0).
17. The agalacto-tetraantennary PA-oligosaccharide is purified with an HPLC system (COSMOSIL column) in an isocratic mode (20 mM ammonium acetate/0.1 % *n*-butanol, pH 4.0, 2.0 mL/min, 40 °C) with the detection of fluorescence (excitation 320 nm, emission 400 nm).
18. The purified agalacto-tetraantennary PA-oligosaccharide is used as a substrate for GnT VI.

3.2 Preparation of Tissue Homogenate

1. All the following processes are carried out at 4 °C. The buffers used need to be prechilled.
2. Tissue of interest is washed twice with PBS.
3. Tissue (typically 2–3 g) is homogenized with Dounce homogenizer with 10 mL of buffer: 0.25 M sucrose, 10 mM Tris–HCl (pH 7.5), protease inhibitors.
4. Spin the homogenate at 2,000 × *g* for 10 min to remove cell debris.
5. Recover the supernatant and use it as crude enzyme source (*see* **Note 2**).

3.3 GnT VI Assay

1. The standard incubation mixture contains the following components in a total volume of 10 μL: 150 mM HEPES (pH 8.0), 10 mM UDP-GlcNAc, 100 mM GlcNAc, 30 mM $MnCl_2$,

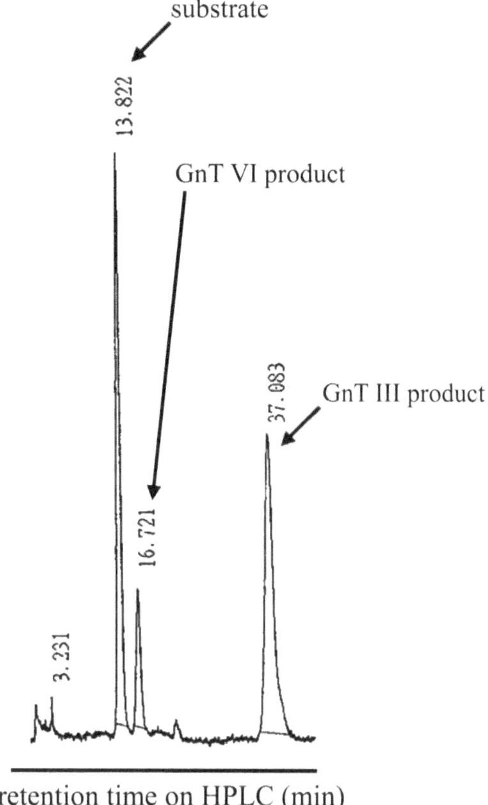

Fig. 2 A typical elution pattern of the reaction mixture on a TSK-gel ODS-80TM column (4.6 × 75 mm, TOSOH) at 55 °C with 20 mM ammonium acetate/0.02 % *n*-butanol, pH 4.0, at a flow rate of 1.6 mL/min. Fluorescence was detected with excitation and emission wavelengths of 320 and 400 nm, respectively. Two GnT products (GnT VI product, agalacto-pentaantennary PA-oligosaccharide; GnT III product, agalacto-tetraantennary PA-oligosaccharide with a bisecting GlcNAc residue) are detected

0.5 % Triton X-100, 2 mg/mL bovine serum albumin, 20 pmol of substrate (agalacto-tetraantennary PA-oligosaccharide), and 2 μL of tissue homogenate. Incubate the mixture at 37 °C for 4 h (*see* **Notes 3** and **4**).

2. Stop the reaction by adding 40 μL of H_2O and boil for 1 min.
3. After centrifugation at 13,000 rpm (16,000 × *g*) for 5 min, 10 μL of the supernatant from the reaction mixture is applied to a TSK-Gel ODS-80TM column (4.6 × 75 mm; TOSOH, Tokyo, Japan) (*see* **Note 5**). Elution is performed at 55 °C with 20 mM ammonium acetate (pH 4.0) at a flow rate of 1.6 mL/min. Fluorescence is monitored with excitation and emission wavelengths of 320 and 400 nm, respectively. A typical elution pattern of the reaction mixture is shown in Fig. 2.

4 Notes

1. Human α_1-acid glycoprotein contains mainly tetraantennary and triantennary *N*-linked glycans. One expects two major peaks on Bio-Gel P4 column chromatography with absorption at 230 nm. The first peak corresponds to asialo-tetraantennary glycan.
2. It is ideal to do the enzyme assay immediately after you prepare tissue homogenates. However, you can freeze the tissue homogenates at −80 °C for longer storage.
3. GnT VI stringently requires β1,6-*N*-acetylglucosaminylation on the Manα1-6 arm (GnT V product) as an acceptor substrate. Agalacto-triantennary oligosaccharide (GnT V product) can also be used as an alternate substrate. However, HPLC conditions to separate GnT IV product (agalacto-tetraantennary oligosaccharide) from GnT VI product needs to be considered.
4. GnT VI activity in hen oviduct is divalent cation-dependent [9]. The activity is maximal with Mn^{2+}. Co^{2+} and Mg^{2+} could partially substitute for Mn^{2+}.
5. Before loading the reaction sample to HPLC, a brief boil and spin to remove protein aggregates is essential for the column maintenance, particularly when the enzyme source is crude.

Acknowledgment

We are grateful to Takara Bio, Inc. (Shiga, Japan) for the guidance in the preparation of GnT VI substrate.

References

1. Yamashita K, Kamerling JP, Kobata A (1982) Structural study of the carbohydrate moiety of hen ovomucoid. Occurrence of a series of pentaantennary complex-type asparagine-linked sugar chains. J Biol Chem 257:12809–12814
2. Taguchi T, Seko A, Kitajima K, Muto Y, Inoue S, Khoo KH, Morris HR, Dell A, Inoue Y (1994) Structural studies of a novel type of pentaantennary large glycan unit in the fertilization-associated carbohydrate-rich glycopeptide isolated from the fertilized eggs of Oryzias latipes. J Biol Chem 269:8762–8771
3. Brockhausen I, Hull E, Hindsgaul O, Schachter H, Shah RN, Michnick SW, Carver JP (1989) Control of glycoprotein synthesis. Detection and characterization of a novel branching enzyme from hen oviduct, UDP-N-acetylglucosamine:GlcNAc beta 1–6 (GlcNAc beta 1–2)Man alpha-R (GlcNAc to Man) beta-4-N-acetylglucosaminyltransferase VI. J Biol Chem 264:11211–11221
4. Stanley P, Schachter H, Taniguchi N (2009) N-Glycans. In: Essentials of Glycobiology, 2nd edn. Cold Spring Harbor Laboratory Press, New York, pH 101–114
5. Taniguchi N, Korekane H (2011) Branched N-glycans and their implications for cell adhesion, signaling and clinical applications for cancer biomarkers and in therapeutics. BMB Rep 44:772–781
6. Mak AB, Blakely KM, Williams RA, Penttila PA, Shukalyuk AI, Osman KT, Kasimer D, Ketela T, Moffat J (2011) CD133 protein N-glycosylation processing contributes to cell surface recognition

of the primitive cell marker AC133 epitope. J Biol Chem 286:41046–41056

7. Taguchi T, Ogawa T, Kitajima K, Inoue S, Inoue Y, Ihara Y, Sakamoto Y, Nagai K, Taniguchi N (1998) A method for determination of UDP-GlcNAc: GlcNAc beta 1-6(GlcNAc beta 1–2) Man alpha 1-R [GlcNAc to Man]beta 1-4 N-acetylglucosaminyltransferase VI activity using a pyridylaminated tetraantennary oligosaccharide as an acceptor substrate. Anal Biochem 255:155–157
8. Sakamoto Y, Taguchi T, Honke K, Korekane H, Watanabe H, Tano Y, Dohmae N, Takio K, Horii A, Taniguchi N (2000) Molecular cloning and expression of cDNA encoding chicken UDP-N-acetyl-D-glucosamine (GlcNAc): GlcNAcbeta 1-6(GlcNAcbeta 1–2)- manalpha 1-R[GlcNAc to man]beta 1,4 N-acetylglucosaminyltransferase VI. J Biol Chem 275: 36029–36034
9. Taguchi T, Ogawa T, Inoue S, Inoue Y, Sakamoto Y, Korekane H, Taniguchi N (2000) Purification and characterization of UDP-GlcNAc: GlcNAcbeta 1-6(GlcNAcbeta 1–2)Manalpha 1-R [GlcNAc to Man]-beta 1, 4-N-acetylglucosaminyltransferase VI from hen oviduct. J Biol Chem 275: 32598–32602

Chapter 23

In Vitro Assays of Orphan Glycosyltransferases and Their Application to Identify Notch Xylosyltransferases

Maya K. Sethi, Falk F.R. Buettner, Angel Ashikov, and Hans Bakker

Abstract

Here we describe a systematic approach to determine the activity of putative glycosyltransferases with a focus on orphan members of the glycosyltransferase 8 family. An assay that measures the hydrolysis activity of glycoslytransferases can indicate the donor nucleotide sugar specificity without previous knowledge about the acceptor. Knowing the donor specificity, the acceptor specificity can subsequently be determined using synthetic acceptors. Three putative glycosyltransferases, now renamed *GXYLT1*, *GXYLT2*, and *XXYLT1*, have been identified this way as xylosyltransferases and in addition have been shown to act on *O*-glucosylated EGF repeats of Notch.

Key words Glycosyltransferase assay, Xylosyltransferase, Nucleotide sugar, UDP-xylose, Notch, Hydrolysis, Epidermal growth factor (EGF) repeat

1 Introduction

Typical eukaryotic glycosyltransferases of the endoplasmic reticulum and Golgi apparatus use nucleotide-sugars (*see* **Note 1**) as donors to build the growing glycan structures covering the cell surface. Glycosyltransferase assays have traditionally been used to link structures and enzymatic activities [1], and, after the cloning of glycosyltransferase genes, to determine the relationship between genes and enzymatic activities. Putative glycosyltransferase genes have been identified in genetic screens in *C. elegans* and *Drosophila* and especially the sequencing of whole genomes has uncovered many glycosyltransferase-like genes with unknown function [2]. This provided the challenging opportunity to link potential glycosyltransferases with a known enzymatic activity or structure. Here, we describe the in vitro assays that we have used to determine the enzymatic function of human members of the glycosyltransferase 8 family. Three genes of this family, formerly named *GLT8D3*, *GLT8D4*, and *C3ORF21*, have been determined to

Inka Brockhausen (ed.), *Glycosyltransferases: Methods and Protocols*, Methods in Molecular Biology, vol. 1022, DOI 10.1007/978-1-62703-465-4_23,

encode xylosyltransferases that extend *O*-glucose on epidermal growth factor (EGF)-like repeats of Notch to form the trisaccharide Xyl-Xyl-Glc-Ser and are now named *GXYLT1* (glucoside xylosyltransferase), *GXYLT2*, and *XXYLT1* (xyloside xylosyltransferase) according to their function [3, 4].

Although there are reports of successful expression of mammalian glycosyltransferases in yeast, bacteria or plants, the most compatible expression systems are mammalian or insect cells. The type II transmembrane glycosyltransferases can often be expressed as secreted enzymes by replacing the N-terminal transmembrane domain with a cleavable signal sequence (*see* **Note 2**). In addition, a protein tag can be added to allow convenient purification of the enzyme. We here describe a protocol in which the enzymes are expressed in Sf9 (*Spodotera frugiperda*) insect cells with an N-terminal protein A tag.

We furthermore describe a strategy to determine the enzymatic activity of orphan glycosyltransferases. This consists of determining the donor substrate by a hydrolysis assay, and subsequently determining the acceptor specificity using synthetic carbohydrate acceptors with a hydrophobic aglycon. Finally, an assay using the natural substrates, in this case EGF repeats of Notch, shows the probable biological functions of the enzymes.

2 Materials

2.1 Expression and Purification of Glycosyltransferases

1. Plasmids containing glycosyltransferase clones and Notch fragments were used as described [3, 4].
2. Baculoviral expression system in Sf9 (ATCC, Manassas, VA, USA) insect cells using the vector pFastBac (Life Technologies, Carlsbad, CA, USA). A baculoviral stock (passage (P) 3) is prepared according to the protocol provided by Life Technologies. Insect-XPRESS protein-free insect cell medium (Lonza, Basel, Switzerland) for growth of Sf9 cells.
3. IgG Sepharose 6 Fast Flow (GE Healthcare Europe, Freiburg, Germany).
4. Millex-GP Filter, 0.22 μm, PES 33 mm (Millipore, Billerica, MA, USA).
5. MultiScreen-BV Plates and Amicon ultracentrifuge filters with a cutoff of 10 kDa (Millipore).
6. 1 mL Ni-Sepharose™ High Performance HisTrap (GE Healthcare).
7. Poly-Prep column (Bio-Rad, Hercules, CA, USA).
8. Buffers and solutions: TBST (50 mM Tris–HCl pH 7.6, 150 mM NaCl, 0.05 % Tween-20)/2 mM $MnCl_2$, 5 mM NH_4Ac pH 5.2/2 mM $MnCl_2$, 100 mM MOPS pH 7/2 mM $MnCl_2$, glycerol.

2.2 Determination of Donor Sugar Specificity by Hydrolysis Assays

1. AG1-X8 resin (Bio-Rad).
2. Buffers: 100 mM NaAc pH 5.8, 10 mM MOPS pH 7.
3. Radiolabeled nucleotide sugars: UDP-[^{3}H]glucose, UDP-[^{3}H]*N*-acetylgalactosamine, and UDP-[^{14}C]xylose (American Radiolabeled Chemicals, Saint Louis, MO, USA).
4. UDP-[^{3}H]galactose, UDP-[^{3}H]*N*-acetylglucosamine, and UDP-[^{14}C]glucuronic acid (PerkinElmer, Waltham, MA, USA, *see* **Note 3**).
5. UDP-xylose (Complex Carbohydrate Research Center, Athens, GA, USA).
6. Nonradioactive nucleotide sugars (Sigma, St Louis, MO, USA, *see* **Note 4**).
7. Glycosyltransferase assay buffer: 100 mM MOPS pH 7 (*see* **Note 5**), 10 mM $MnCl_2$ (*see* **Note 6**), 10 mM ATP (*see* **Note 7**), 5 μM UDP-[^{3}H or ^{14}C]sugar (*see* **Note 8**).

2.3 Glycosyltransferase Assays and Identification of Reaction Products by Mass Spectrometry

1. Assay reagents listed in Subheading 2.2.
2. C18 3 mL 200 mg vacuum cartridges (Macherey-Nagel, Düren, Germany).
3. Buffers and solutions: loading buffer A (for FPLC, 20 mM Tris–HCl pH 8, 150 mM NaCl), elution buffer B (loading buffer with 500 mM imidazole), Methanol.
4. Equipment: ÄKTA prime FPLC (GE Healthcare), liquid chromatography electrospray mass spectrometry (LC-ESI-MS); nanoACQUITY UPLC device equipped with an analytical column (BEH130 C18, 100 μm × 100 mm, 1.7-μm particle size) coupled online to an ESI Q-TOF Ultima mass spectrometer, MassLynx V4.1 Software (all Waters, Milford, MA, USA). SDS-PAGE system (Biometra, Göttingen, Germany).
5. Scintillation fluid. scintillation cocktail Lumasafe plus (Perkin Elmer).

3 Methods

3.1 Expression and Purification of Secreted Glycosyltransferases in Sf9 Insect Cells

A classical way to express glycosyltransferases is as secreted protein fused to protein A (*see* **Note 9**). This method has been used first in mammalian cells [5] but has also been adapted to insect cells [6]. The system that we describe here is based on baculoviral expression in Sf9 insect cells using the vector pFastBac. For details on generation of baculoviruses we would like to refer to the handbook of the Bac-to-Bac baculoviral expression system found on the homepage of Life Technologies. Using standard cloning methods, the honeybee melittin signal sequence from pMelBac, the protein A sequence from pPROTA, and the C-terminal luminal part of glycosyltransferases, containing the stem and catalytic domain, starting just after the transmembrane domain (*see* **Note 10**), was cloned in pFastBac. The resulting protein is targeted to the ER where the signal sequence is cleaved off. The soluble fusion protein, having an N-terminal protein A tag, is then secreted and can be purified using IgG beads.

1. A baculoviral stock (passage (P) 3) is prepared according to the protocol provided by Life Technologies.
2. From this stock, 300 mL of an Sf9 culture in a 1 L Erlenmeyer flask with a density of 2×10^6 is infected, usually in a ratio of 1:1,000 from P3 (the optimal infection rate can be experimentally determined on a small scale).
3. The culture is incubated in a shaker (80 rpm) at 26 °C for 72 h.
4. The cell suspension is cooled down and centrifuged at $800 \times g$ for 5 min at 4 °C (in 50 mL Falcons or bigger tubes if available). The supernatant is collected and filtered through Millex-GP Filter to remove remaining cells or cell fragments.
5. The cleared medium is then incubated with 50 μL of IgG Sepharose (*see* **Note 11**) per 50 mL (in a Falcon tube) of culture supernatant for 12 h at 4 °C on a tube roller.
6. The Sepharose beads are spun down at $800 \times g$ for 5 min and the supernatant is discarded.
7. Beads (combined from all tubes) are transferred to an empty Poly-Prep drip column. The beads are washed five times with two bed volumes TBST/2 mM $MnCl_2$, with two bed volumes of 5 mM NH_4Ac pH 5.2/2 mM $MnCl_2$ (*see* **Note 12**) followed by two bed volumes of 100 mM MOPS pH 7/2 mM $MnCl_2$.
8. Beads are then transferred to an Eppendorf tube (wash out the column with 100 mM MOPS pH 7/2 mM $MnCl_2$). Let the beads settle down (or centrifuge for 5 min at $800 \times g$), remove the supernatant and take up the beads in 100 mM MOPS pH 7/2 mM $MnCl_2$ with 50 % glycerol (*see* **Note 13**). This enzyme stock can be aliquoted and stored at −20 °C.
9. For assays, 20 μL of IgG beads (50 % beads in 50 % glycerol) are transferred (with a cut yellow tip) to a 1.5 mL Eppendorf tube. To remove the glycerol, 180 μL of MOPS buffer is added, resuspended gently and spun down at $800 \times g$ for 5 min. The supernatant is carefully removed and the washing step repeated. Reagents for glycosyltransferase assays can be added directly to the Sepharose beads (*see* **Note 14**).

3.2 Determination of Donor Sugar Specificity by Hydrolysis Assays

The separation of the reaction product of a glycosyltransferase from unreacted nucleotide sugar can be achieved by anion exchange chromatography [7]. The nucleotide sugar is retained on the column while the product runs through (provided that the transferred

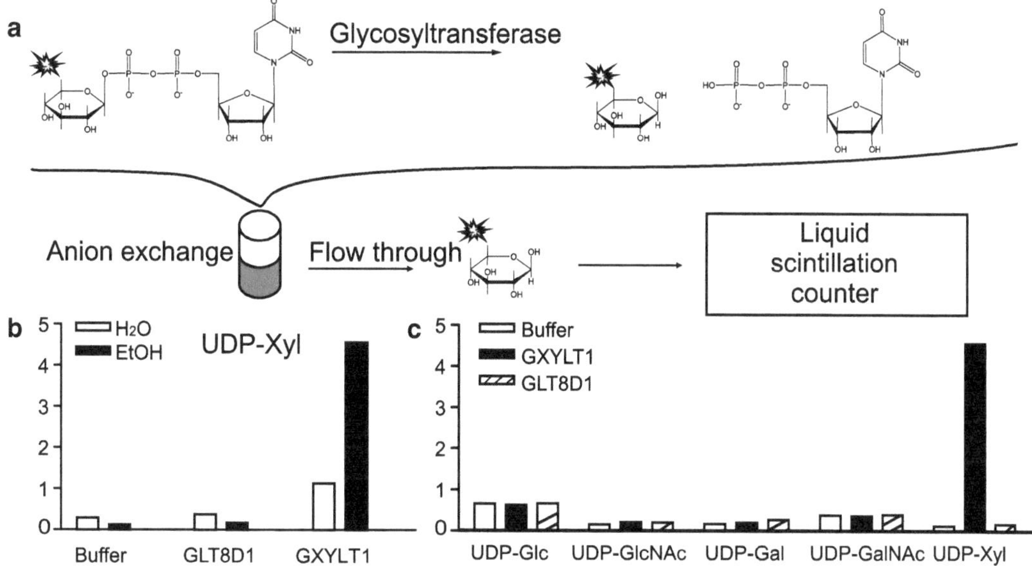

Fig. 1 Strategy to determine the donor specificity of glycosyltransferases with unknown activity. (**a**) A side reaction of glycosyltransferases is the hydrolysis of the donor substrate. The produced monosaccharide (if not charged) can be separated from the nucleotide sugar by anion exchange. (**b**) Ethanol enhanced the observed activity, probably by functioning as acceptor (*see* **Note 21**), for xylosyltransferase GXYLT1. (**c**) The activity, here in presence of ethanol, is donor sugar specific, which gave the first indication that the enzyme GXYLT1 was a xylosyltransferase. Activity is presented in pmol/h/assay

sugar is not charged). Transfer activity to a specific acceptor is classically determined by comparing the activity with and without acceptor. Although the activity without acceptor is considered as background, it is still partly enzyme dependent and represents transfer to other compounds present in the reaction mixture or hydrolysis, which can be seen as transfer to water. This reaction is low but specific and can be used to determine the donor sugar specificity of a purified glycosyltransferase, even when the acceptor is unknown. In addition, small molecules, having at least one OH group, like for example ethanol, can often function as acceptor molecule. Here, we have applied this method to members of the glycosyltransferase 8 family and could show that the enzyme GXYLT1 is a xylosyltransferase (Fig. 1). Other Notch xylosyltransferases (GXYLT2 and XXYLT1) show UDP-xylose specific hydrolysis activity as well (not shown). For another member of this family, GLT8D1, we were not able to determine the donor sugar using this method. The described method uses microtiter plates and requires a microtiter plate centrifuge. The method can also be adapted to small drip columns for separation.

1. Preparation of AG1-X8 resin: To 10 g AG1-X8 resin, 20 mL of 100 mM NaAc pH 5.8 is added in a 50 mL Falcon tube and gently swirled and spun down at 50 × *g* for 2 min. The supernatant

is removed and the washing step repeated. Finally the pellet is resuspended in 100 mM NaAc pH 5.8 (1.5 mL/g) and stored at 4 °C.

2. A glycosyltransferase/hydrolysis assay is carried out in 50 μL assay buffer in a 1.5 mL Eppendorf tube. To a 10 μL aliquot of washed enzyme-bound beads prepared as described in Subheading 3.1, all components are added from 10× concentrated stocks and water added to 45 μL. 5 μL radiolabeled nucleotide sugar is added last. The sample is incubated for 2 h at 37 °C in a shaker at about 200 rpm.
3. Meanwhile, a multi screen plate is placed on top of a 96 well round bottom plate. The round bottom plate is to collect the washing fractions.
4. The cover is removed and 200 μL of AG1-X8 suspension are loaded per well using a cut pipet tip.
5. The plate is spun down at 100×*g* for 1 min and the flow-through discarded.
6. Carefully, 100 μL 10 mM MOPS pH 7 are added without perturbing the beads bed.
7. The plate is spun down at 100×*g* for 1 min and the flow-through discarded.
8. The glycosyltransferase reaction mixture is added without perturbing the beads bed.
9. The plate is again spun at 100×*g* for 1 min.
10. The flow-through is transferred to a counting vial.
11. Beads are washed with 100 μL 10 mM MOPS pH 7.
12. The plate is spun at 100×*g* for 1 min.
13. The flow-through is combined with the first flow-through in the counting vial.
14. 3 mL of scintillation cocktail are added and vortexed. The vials are left at room temperature for 10–15 min and radioactivity is measured with a liquid scintillation counter.
15. An example of results obtained with this method is shown in Fig. 1b, c. Results can be expressed as measured cpm to obtain relative activities or, if necessary, calculated in molar amounts per mg or mole of enzyme.

3.3 Glycosyltransferase Assays Using Synthetic Glycans with Hydrophobic Aglycons as Acceptor

This assay is based on the protocol of Palcic et al. [8] often referred to as Sep-Pak assay in which synthetic hydrophobic acceptors (*see* **Note 15**) are used. This enables an easy separation of the reaction product from unreacted nucleotide sugar by solid phase extraction. The reaction product binds to C18 matrix, whereas nucleotide sugars do not. Traditionally the assay is carried out with C18 cartridges (Sep-Pak, Waters) to which the samples are applied with a syringe, but cartridges that can be used with a vacuum manifold can be used as well. We have adapted the latter to a system that can be used in a bench centrifuge (*see* **Note 16**). Using this protocol we have determined the acceptor specificity of the xylosyltransferases involved in Notch glycosylation [3, 4].

1. A standard glycosyltransferase assay is performed as described in Subheading 3.2, now with addition of 100 μM of a hydrophobic synthetic acceptor (*see* **Note 17**). Reactions can be set up with different acceptors, for example a series of different monosaccharides linked to *p*-nitrophenyl (pNP) and one control without acceptor.
2. The reaction is stopped by adding 950 μL of cold water and kept on ice until further processing.
3. A C18 cartridge is equilibrated with 3 mL methanol (*see* **Note 18**) and centrifuged at 100 × *g* for 1 min.
4. The cartridge is loaded with 3 mL of water and centrifuged again at 100 × *g* for 1 min.
5. **Step 4** is repeated and the combined flow-through discarded (non radioactive).
6. The reaction product is transferred to the cartridge and the tube washed out with another 1 mL of water that is loaded as well. Again the cartridge is spun at 100 × *g* for 1 min.
7. The cartridge is washed three times with 3 mL of water with in-between centrifugations.
8. A new Falcon tube is taken and a scintillation vial is placed inside. The cap with C18 cartridge is transferred from the centrifuged tube (these now contain radioactive waste) and placed on the new tubes. The tip of the C18 cartridge should hang inside the scintillation vial.
9. Elute the glycosyltransferase reaction product with two times 2 mL of methanol (*see* **Note 19**), by centrifugation at 100 × *g* for 1 min.
10. The collected methanol must be evaporated before the liquid scintillation cocktail can be added. This can be done under a stream of nitrogen or by placing the tubes in a heating block at 40 °C overnight in a fume hood.
11. 3 mL scintillation cocktail are added and vortexed. Vials are left at room temperature for 10–15 min and radioactivity measured with a liquid scintillation counter.

3.4 Xylosyltransferase Assays Using Notch EGF Domains as Acceptor and Analysis by Mass Spectrometry

If the natural acceptor of the glycosyltransferases is predicted, it can sometimes be produced and tested. Here, a protein fragment of mouse Notch containing EGF repeat 1–5 was produced in Sf9 cells, basically as described above in Subheading 3.1 for glycosyltransferases [3, 4]. The secreted protein fragment contained a C-terminal Myc-His (*see* **Note 20**) tag for purification using a

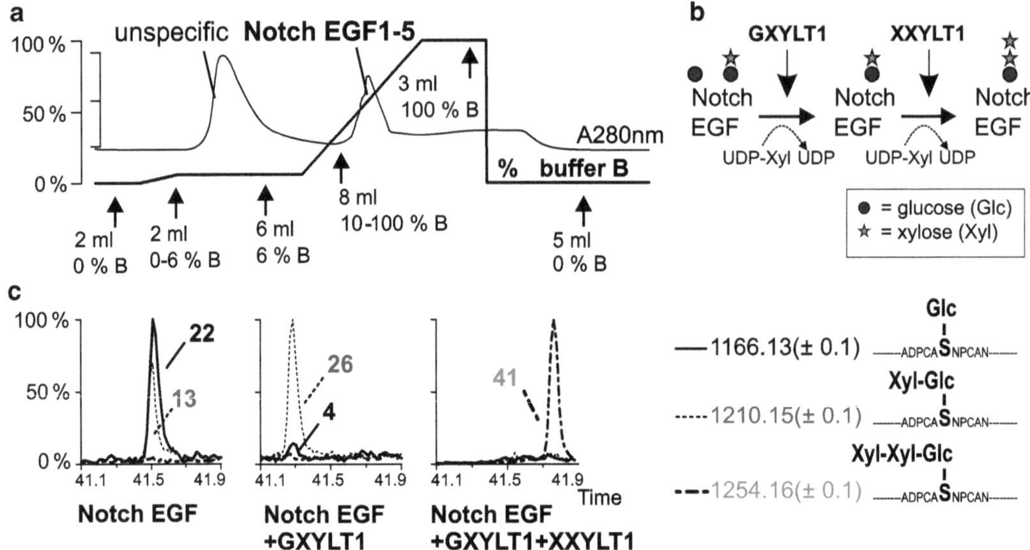

Fig. 2 Activity assays using a Notch fragment as acceptor. (**a**) Elution profile of Notch EGF repeat 1–5 on Ni-Sepharose. Buffer B consist of 500 mM imidazole. (**b**) Schematic reaction of GXYLT1 and XXYLT1 with the Notch fragment produced in insect cells. (**c**) Extracted ion chromatograms of the *O*-glucosylated Notch peptide, before and after incubation with xylosyltransferases. Mass spectrometry is used to quantify specific glycopeptides eluted from the LC column. The intensity of *m/z* values corresponding to the triple charged peptides with Glc, Xyl-Glc and Xyl-Xyl-Glc glycan modifications are plotted against the elution time from the LC column. By measurement of the relative peak areas (indicated in the *graphs*) the relative proportion of the differently glycosylated peptides can be determined. The exact elution time of the peptides might differ slightly from run to run

1 mL Ni-Sepharose column and was used as "natural" acceptor for xylosyltransferases.

1. Medium (100–300 mL) from a Sf9 expression culture is pumped over a 1 mL Ni-Sepharose column at about 1 mL/min overnight.
2. Before connection to the FPLC, the column is washed with about 10 mL loading buffer.
3. The 1 mL Ni-Sepharose column is connected to the FPLC and a step gradient as indicated in Fig. 2a is applied. Nonspecifically bound proteins are washed by applying 30 mM imidazole for 6 min, before a linear gradient to 500 mM imidazole over 8 min elutes the specifically bound protein. The Notch fragment elutes between 200 and 300 mM imidazole. Fractions of 0.5 mL are collected.
4. Analyze a sample (10 μL) of each fraction by western blotting to determine which fractions contain the Notch fragment.
5. Combined positive fractions are concentrated and buffer is exchanged to 100 mM MOPS pH 7/2 mM $MnCl_2$ on a centrifuge filter with a cutoff of 10 kDa.

6. 10 μL of purified Notch EGF1–5 are used as acceptor in a glycosyltransferase assay as described above for 4 h in the presence of 100 μM cold UDP-xylose as donor and IgG Sepharose beads-coupled enzyme (GXYLT or XXYLT).
7. The reaction product is cleared from enzyme beads by centrifugation and the supernatant applied to standard 15 % SDS-PAGE. The gel is stained with Coomassie and protein bands of Notch EGF1-5 are excised from gel, trypsinized, and peptides recovered as described [9].
8. Peptides are applied to a LC-ESI-MS and chromatography is run for 45 min at a flow rate of 0.3 μL/min. The chromatography starts with 99 % of eluent A (water with 0.1 % formic acid) and 1 % eluent B (acetonitrile with 0.1 % formic acid). After 0.33 min, eluent B is increased in a linear gradient from 1 to 35 % within 30 min, and from 35 to 85 % within 1 min. After 1 min at 85 % the system is re-equilibrated with 99 % eluent A for 13 min. Spectra are recorded in positive ion mode.
9. Spectra are analyzed using the MassLynx V4.1 Software (Waters). For quantitative determination of the amounts of differentially glycosylated peptides, extracted ion chromatograms (EICs) are generated by searching the theoretical masses of anticipated glycopeptides in chromatograms obtained in MS-mode applying the "Mass Chromatogram" function of MassLynx. The permitted mass error is set to 0.1 Da. Subsequently the peak area is obtained using the "Integrate Chromatogram" function of MassLynx setting the "Peak-to-peak amplitude" at 100. The identity of the MS peaks displayed as EICs should be confirmed by MS/MS analysis.

In the presented example, Notch EGF repeat 4 contains a site that is *O*-glucosylated. In Sf9 cells this site is partly xylosylated. The purified peptides containing glucose or glucose and xylose appear at m/z 1166.13 and 1210.15. GXYLT converts the glucosylated EGF repeat into the xylosylated form. Incubation with both a GXYLT and an XXYLT results in the complete conversion to the trisaccharide Xyl-Xyl-Glc (Fig. 2b). The extracted ion chromatogram enables quantification of specific (glyco)peptides eluted from the LC column based on their specific masses (Fig. 2c).

4 Notes

1. In vertebrates the following eight nucleotide sugars are used: UDP-Glc, UDP-Gal, UDP-GlcNAc, UDP-GalNAc, UDP-GlcA, UDP-Xyl, GDP-Fuc, and different variants of CMP-Sialic acid. GDP-Man is not directly used by luminal ER/Golgi enzymes but metabolized via Dolichol-P-Man.

2. Several glycolipid specific enzymes have been shown to require their transmembrane domain for activity [10–12].

3. American Radiolabeled Chemicals and Perkin Elmer, which acquired New England Nuclear, are global suppliers of radiolabeled nucleotide sugars. ^{3}H or ^{14}C labeled sugars can both be used. ^{3}H labeled compounds are usually less expensive; however, we use about five times more ^{3}H than ^{14}C (in Bq) in assays due to higher background and lower efficiency in counting. In the case of UDP-xylose, only ^{14}C is available.

4. Nucleotide sugars can be obtained from various companies, with the exception of UDP-xylose which is produced by CarboSource Services of the Complex Carbohydrate Research Center, Athens GA, USA and supported in part by NSF-RCN grant # 0090281.

5. pH 7 is a good starting point for glycosyltransferase assays. Traditionally, cacodylate buffer was often used [13], but is rather toxic. MOPS, MES, HEPES, or BIS-TRIS are usually good substitutes. TRIS is not recommended, as pH 7 is at the limit of the effective pH range and phosphate buffers might influence glycosyltransferase reactions.

6. Manganese ions, which sometimes can be replaced by magnesium ions, is required by most mammalian glycosyltransferases as a cofactor. Be careful not to use a too high concentration, as this has inhibitory effects. 2–10 mM is usually optimal.

7. ATP is not required in the reaction, but is often added as protective agent against enzymatic degradation of the nucleotide sugars. Various other protective agents can be added to the reaction mixture. These include protease inhibitors or BSA against proteases or monosaccharides as inhibitors of glycosidases that can degrade the reaction product or acceptor sugar. This is, however, usually not needed if purified enzyme is used. Triton-X-100 or another detergent is often included. This is certainly required when whole cell lysates or vesicle preparations are used, but might also improve solubility if hydrophobic acceptors are used. It is not included in this protocol.

8. Radiolabeled nucleotide sugars are sold with high specific activity. UDP-[^{14}C]Xyl for example has 8 kBq/nmol. We carry out standard assays with about 200 Bq/assay of 50 μL. This means that UDP-Xyl is present at 0.5 μM concentration if only radiolabeled material is used. It is desirable to increase the concentration by adding non labeled UDP-Xyl. The Km is mostly unknown, but for β4GalT, for example, was determined to be 25 μM. Increasing the concentration will protect against contaminating enzymes that degrade either the nucleotide sugar or the product. This is especially of concern with samples containing cell lysates, where the concentration can be further

increased. We use a donor nucleotide sugar concentration of 5 μM, which is still rather low. Labeled nucleotide sugars are usually stored in 70 % ethanol. The ethanol from a small aliquot is evaporated (best by a mild stream of nitrogen) and a solution of cold nucleotide sugar is added (50 μM) to make a 10× stock of which 5 μL is added to a 50 μL assay. Measure a small aliquot of this stock in a scintillation counter to determine the exact specific activity.

9. The first glycosyltransferase expressed in this way was cloned in the vector pPROTA, resulting in secretion of a hybrid protein between the *Staphylococcus aureus* protein A (a protein that binds to immunoglobulins) tag and the luminal domain of the glycosyltransferase [5]. This is still a widely used method and provides a safe expression system. The C-terminal luminal part of Type II glycosyltransferases contains a stem region linking the transmembrane domain with the catalytic domain. Leaving this stem region intact and placing the protein A tag at the N-terminus keeps the glycosyltransferase in its natural conformation. On the other hand, there are reports of successful expression of glycosyltransferases with C-terminal tags, e.g., in reference [14].
10. The single transmembrane domain of glycosyltransferases is usually well recognized by prediction programs. TMHHM (http://www.cbs.dtu.dk/services/TMHMM) was rated best in a comparative study [15].
11. IgG Sepharose beads can be pipetted with a cut-off pipet tip and should be washed two times with TBST before use to remove ethanol.
12. Be careful. Longer washing or incubation might elute the fusion protein. At lower pH (3.5) the IgG-protein A interaction is broken.
13. Consider that the beads volume already contains MOPS buffer. Adding two volumes of MOPS/$MnCl_2$/75 % glycerol will result in a final concentration of 50 % glycerol and 33 % Sepharose beads. Removing one volume of buffer results in a stock with 50 % beads in 50 % glycerol. In 50 % glycerol no ice crystals that can potentially damage the enzyme are formed at −20 °C.
14. The enzyme remains bound to the Sepharose beads in the glycosyltransferase assay and the preparation contains a high concentration of enzyme. This high enzyme concentration makes the preparation suitable for screening of unknown activities, for example with the herein described hydrolysis assay. Keeping the enzyme linked to the beads avoids the risky low pH elution step and easy separation of enzyme and reaction products is possible. This becomes especially important if proteins are used

as acceptor. Immobilized enzymes can also find applications in continuous, fixed-bed operation to produce large amounts of product. For kinetic studies, these enzyme preparations are less suitable. Immobilized enzyme is not homogeneous in solution and kinetic measurements require low enzyme concentration. The enzyme concentration can be determined by running a sample together with a serial dilution of protein A or BSA on SDS-PAGE and Coomassie staining.

15. Many different hydrophobic acceptors can be used. pNP- or para-methylumbelliferyl (MU)-conjugated monosacharides are sold by Sigma and many other companies as substrates for glycosidases but function often very well as acceptors for glycosyltransferases. Many disaccharides (or even longer saccharides) linked to pNP are also sold by Toronto Research Chemicals (Canada) or Carbosynth (Compton, UK). Acceptors containing acyl chains can be good acceptors as well. GlcNAc-β-$(CH_2)_8COOMe$ [8] or octyl-β-glucoside [16] function as acceptors for galactosyltransferases. In our xylosyltransferase assays we have successfully applied mono and disaccharides conjugated to $(CH_2)_3NH_2CO(CH_2)_6CH_3$ [3, 4, 17].

16. We use Sep-Pac vacuum cartridges 3cc 200 mg from Waters or the equivalent from Macherey-Nagel. To be able to use them in a centrifuge, we make a 10 mm hole in the cap of a conical 50 mL falcon tube to hold the cartridge. This allows collecting up to 15 mL of flow-through in the tube. In addition, it can exactly fit a liquid scintillation vial so that the tip of the cartridge hangs in the scintillation vial (Use a new falcon tube to avoid contamination of the outside of the scintillation vial). Depending on the capacity of the centrifuge, several assays can be done simultaneously.

17. We use an acceptor concentration of 100 μM, but for less precious compounds (for example commercially available pNP-sugars) a concentration of 1 mM is often used.

18. Always start and end with methanol. The C18 material does not become wet using water. After elution with methanol the cartridges can be dried and reused several times. If 3 mL of liquid does not go through in 1 min at 100 g it is time to replace the cartridge.

19. The second 2 mL normally contains little radioactivity but prevents cross-contamination if cartridges are reused.

20. The Myc-His tag is integrated in several expression vectors from life technologies. It allows separation by Ni-affinity and detection by an anti Myc (a ten amino acid epitope of the protein c-Myc) antibody.

21. This cannot be determined directly. It is also possible that ethanol stimulates hydrolysis. Ethanol may increase the reaction

rate without increasing background, resulting in higher sensitivity. This is observed for the xylosyltransferase GXYLT1, but not for all enzymes. The xylosyltransferase activity of LARGE2, for example, is not higher in presence of ethanol.

Acknowledgments

This work is supported by funding from the Deutsche Forschungsgemeinschaft (DFG, German Research Foundation) for the Cluster of Excellence REBIRTH (From Regenerative Biology to Reconstructive Therapy). The authors thank Françoise Routier for critical reading of the manuscript.

References

1. Beyer TA, Sadler JE, Rearick JI, Paulson JC, Hill RL (1981) Glycosyltransferases and their use in assessing oligosaccharide structure and structure-function relationships. Adv Enzymol Relat Areas Mol Biol 52:23–175
2. Henrissat B, Surolia A, Stanley P (2009) A genomic view of glycobiology (chapter 7). In: Varki A, Cummings RD, Esko JD, Freeze HH, Stanley P, Bertozzi CR, Hart GW, Etzler ME (eds) Essentials of glycobiology, 2nd edn. Cold Spring Harbor, NY
3. Sethi MK, Buettner FF, Krylov VB, Takeuchi H, Nifantiev NE, Haltiwanger RS, Gerardy-Schahn R, Bakker H (2010) Identification of glycosyltransferase 8 family members as xylosyltransferases acting on O-glucosylated notch epidermal growth factor repeats. J Biol Chem 285:1582–1586
4. Sethi MK, Buettner FF, Ashikov A, Krylov VB, Takeuchi H, Nifantiev NE, Haltiwanger RS, Gerardy-Schahn R, Bakker H (2012) Molecular cloning of a xylosyltransferase that transfers the second xylose to O-glucosylated epidermal growth factor repeats of notch. J Biol Chem 287:2739–2748
5. Larsen RD, Rajan VP, Ruff MM, Kukowska-Latallo J, Cummings RD, Lowe JB (1989) Isolation of a cDNA encoding a murine UDPgalactose: β-D-galactosyl- 1,4-N-acetyl-D-glucosaminide α-1,3-galactosyltransferase: expression cloning by gene transfer. Proc Natl Acad Sci USA 86:8227–8231
6. Shinkai A, Shinoda K, Sasaki K, Morishita Y, Nishi T, Matsuda Y, Takahashi I, Anazawa H (1997) High-level expression and purification of a recombinant human α-1, 3-fucosyltransferase in baculovirus-infected insect cells. Protein Expr Purif 10:379–385
7. Prieels JP, Monnom D, Dolmans M, Beyer TA, Hill RL (1981) Co-purification of the Lewis blood group N-acetylglucosaminide α1→4 fucosyltransferase and an N-acetylglucosaminide α1→3 fucosyltransferase from human milk. J Biol Chem 256:10456–10463
8. Palcic MM, Heerze LD, Pierce M, Hindsgaul O (1988) The use of hydrophobic synthetic glycosides as acceptors in glycosyltransferase assays. Glycoconj J 5:49–63
9. Shevchenko A, Wilm M, Vorm O, Mann M (1996) Mass spectrometric sequencing of proteins silver-stained polyacrylamide gels. Anal Chem 68:850–858
10. Johswich A, Kraft B, Wuhrer M, Berger M, Deelder AM, Hokke CH, Gerardy-Schahn R, Bakker H (2009) Golgi targeting of Drosophila melanogaster β4GalNAcTB requires a DHHC protein family-related protein as a pilot. J Cell Biol 184:173–183
11. Kraft B, Johswich A, Kauczor G, Scharenberg M, Gerardy-Schahn R, Bakker H (2011) "Add-on" domains of Drosophila β1,4-N-acetylgalactosaminyltransferase B in the stem region and its pilot protein. Cell Mol Life Sci 68:4091–4100
12. Schwientek T, Keck B, Levery SB, Jensen MA, Pedersen JW, Wandall HH, Stroud M, Cohen SM, Amado M, Clausen H (2002) The Drosophila gene brainiac encodes a glycosyltransferase putatively involved in glycosphingolipid synthesis. J Biol Chem 277: 32421–32429
13. Palcic MM (1994) Glycosyltransferases in glycobiology. Methods Enzymol 230:300–316

14. Haines N, Irvine KD (2005) Functional analysis of Drosophila β1,4-N-acetylgalactosaminyltransferases. Glycobiology 15:335–346
15. Moller S, Croning MD, Apweiler R (2001) Evaluation of methods for the prediction of membrane spanning regions. Bioinformatics 17:646–653
16. Senn HJ, Wagner M, Decker K (1983) Ganglioside biosynthesis in rat liver. Characterization of UDPgalactose-glucosylceramide galactosyltransferase and UDPgalactose-GM2 galactosyltransferase. Eur J Biochem 135:231–236
17. Krylov V, Ustyuzhanina N, Grachev A, Bakker H, Nifantiev N (2007) Stereoselective synthesis of the 3-aminopropyl glycosides of α-D-Xyl-(1 → 3)β-D-Glc and α-D-Xyl-(1 → 3)-α-D-Xyl-(1 → 3)-β-D-Glc and of their corresponding N-octanoyl derivatives. Synthesis 2007:3147–3154

Chapter 24

In Vitro Folding of β-1,4Galactosyltransferase and Polypeptide-α-*N*-Acetylgalactosaminyltransferase from the Inclusion Bodies

Boopathy Ramakrishnan and Pradman K. Qasba

Abstract

The aim of this article is to present a unique in vitro folding technique for glycosyltransferases to generate active proteins that can be used for X-ray crystallographic and bioconjugation protocols. Although a number of in vitro refolding methods are available, β1,4galactosyltransferases in large quantities can be made using the current protocol only. This technique is not only limited to glycosyltransferases alone but has been successfully used to refold single-chain antibodies and other proteins. Although this in vitro folding method is quite similar to other methods, it differs from them by the use of *S*-sulfonation of the inclusion bodies before setting up the in vitro refolding of the protein.

Key words Galactosyltransferases, Refolding, S-Sulfonation

1 Introduction

Recombinant protein expression systems using *E. coli* to express active mammalian proteins have been challenging [1]. Rather than a soluble active protein, often these proteins are expressed in *E. coli* in large quantities as insoluble and inactive inclusion bodies. This is due to many factors such as lack of chaperone molecules or lack of post-translational modifications such as glycosylation or inability to form correct disulfide bonds in the target proteins. To overcome these problems, several specialized expression cell lines have been developed where chaperone molecules that aid the folding of the target protein are co-expressed. However, even with all these efforts, the expression of these proteins may be limited to low yield, compared to their inclusion bodies. In humans, glycosyltransferases are a superfamily of enzymes; many of them reside in the Golgi apparatus of a cell and synthesize the oligosaccharide chains by transferring a monosaccharide moiety from an activated sugar donor to an acceptor molecule, forming a glycosidic bond [2].

Inka Brockhausen (ed.), *Glycosyltransferases: Methods and Protocols*, Methods in Molecular Biology, vol. 1022, DOI 10.1007/978-1-62703-465-4_24,

They have type II membrane protein topology, with a short *N*-terminal cytoplasmic domain, a membrane-spanning region, as well as a stem and a *C*-terminal catalytic domain that faces the Golgi lumen. Based on their structure and function studies, recently, novel glycosyltransferases with broader donor specificities have been designed, and these mutant transferases can transfer a sugar residue with a chemical handle to specific terminal moieties of glycoconjugates [3–6]. Site-specific labeling and conjugation of glyco and non-glycoproteins have been developed using these enzymes [7–9]. Therefore, an efficient way to make large quantities of recombinant proteins is necessary. Here we describe refolding of the catalytic domain of few glycosyltransferases and α-lactalbumin that are used for various applications.

1.1 S-Sulfonation Method in Refolding

Refolding of the inclusion bodies by many different in vitro folding methods has been described in the literature, e.g., by dilution, by dialysis, and on column [10]. In all these methods, the recombinant proteins that were expressed in *E. coli* as inclusion bodies are dissolved in either 5–6 M guanidine HCl or 8 M urea and used for the refolding. Often, prior to refolding, the inclusion bodies are treated with 50 mM DTT (dithioerythritol) at pH 8.0 in order to remove any undesired disulfide bonds (intermolecular and intramolecular) present among the protein molecules in the inclusion bodies. However, when Cys residues are present in the target protein molecule, it is possible to form a wrong or undesired disulfide bond while refolding the target protein molecule. In order to minimize the formation of such an undesired disulfide bond, free thiols in the protein molecules are *S*-sulfonated with a sulfite group, and while refolding, this modified sulfite group is removed by oxido-shuffling agents that facilitate the formation of the right disulfide bond between the thiol groups [11]. It has been shown that the *S*-sulfonation procedure causes a dramatic conformational change in the protein and results in a predominantly disordered ensemble of conformations. The treatment of the protein with oxido-shuffling agents generates reduced species that oxidize and refold to the native enzyme in a concerted process [11].

1.2 Refolding of β4Gal-T1

The catalytic domain of bovine β4Gal-T1 protein was first expressed as a soluble GST fusion protein in *E. coli* [12]. Although soluble and active β4Gal-T1 protein has been generated, most of the recombinant protein is expressed as inactive insoluble inclusion bodies. Only a small amount of soluble domain of human β4Gal-T1 is produced as a secreted recombinant protein. It was initially found that in order to generate active protein from the inclusion bodies one has to use oxido-shuffling agents during refolding [12]. Although some amount of soluble active protein can be generated by this method, the in vitro folding of the *S*-sulfonated protein is consistently able to produce 4 mg/L soluble active protein from inclusion bodies [13]. Furthermore, the mutation of the free Cys residue at position 342 of bovine β4Gal-T1 molecule to a Thr residue enhanced the refolding yield to 10–12 mg/L [14]. Later, it was found that the stem region of the β4Gal-T1 acts like a chaperone during the refolding and thereby enhanced in vitro folding yield to 40 mg/L [13]. The addition of other reagents such as arginine, polyethylene glycol, glycerol during the refolding of β4Gal-T1 also enhances the refolding efficiency [13].

The S-sulfonation method can be also used for in vitro folding of other glycosyltransferases. However, they may require the presence of additional reagents in the refolding solution. For example, while refolding the polypeptidyl-α-GalNAc-T2 it is essential to include 10 % glycerol and 0.5 M arginine–HCl in the refolding solution, while 5 mM lactose enhances the refolding yield [15]. Similarly, β4Gal-T1 ortholog protein β4GalNAc-T from *Drosophila* can be refolded; however, only as a soluble domain that includes the stem and the catalytic domain, while the catalytic domain alone could not be refolded [16]. Interestingly, the catalytic domain of *Drosophila* β4Gal-T7 molecule also has been refolded; however, its ortholog human β4Gal-T7 protein, either as a soluble domain or only as catalytic domain, could not be refolded [16].

1.3 Refolding of α-Lactalbumin

α-Lactalbumin is a calcium binding protein and has a strong affinity for Ca^{2+} ion [17]. Also, it has four disulfide bonds. Although it has high protein sequence and structure similarity with C-type lysozyme, it does not share its function. Lysozyme is a glycosidase enzyme that binds to and hydrolyzes glycans, while α-lactalbumin neither binds any glycan nor has any enzymatic activity. Rather, it modulates the acceptor specificity of β4Gal-T1 enzyme to glucose to synthesize lactose [18]. A number of in vitro refolding methods have been developed using lysozyme as the model protein, and the generation of the active protein is tested by its enzymatic activity. Although a similar in vitro folding method could be used to refold α-lactalbumin, it is essential to have a Ca^{2+} ion during the refolding in order to generate active protein [19]. The generation of the active α-lactalbumin is judged by its modulating ability where it alters the acceptor specificity of the β4Gal-T1 enzyme from GlcNAc to the Glc molecule to synthesize lactose. The *S*-sulfonation method used for in vitro folding of β4Gal-T1 can also be used for α-lactalbumin folding. Here we find that after *S*-sulfonation of α-lactalbumin inclusion bodies, the *S*-sulfonated protein remains soluble in water and can be directly used for refolding without any urea or guanidine HCl. The correctly refolded protein can be purified from the misfolded protein by ammonium sulfate precipitation, where 40 % ammonium sulfate precipitates all the misfolded proteins, while the correctly folded proteins can be precipitated only between 40 and 70 % ammonium sulfate concentrations. Using this *S*-sulfonation method, we could achieve

40–60 % refolding efficiency in refolding α-lactalbumin. It has been shown that when the calcium ion is removed, α-lactalbumin adopts a molten globule form. When the molten globule form of α-lactalbumin is complexed with oleic acid, it acquires a novel tumoricidal activity and it has been named as HAMLET [20]. We have been able to make this complex using in vitro refolded recombinant human α-lactalbumin and find that it also exhibits tumoricidal activity [21].

2 Materials

1. The recombinant glycosyltransferases, bovine β4Gal-T1, *Drosophila* β4Gal-T7 and human ppαGalNAc-T2, proteins are expressed as described previously [13, 15, 16]. Unless stated otherwise, all solutions are prepared in water that has a resistivity of 18.2 MΩ-cm at 25 °C. This is referred to as "water" in the text.

2.1 Preparation of NTSB for Protein Sulfonation

1. Sodium sulfide 0.3 M solution: 420 mg in 10 mL.
2. DTNB (5-5′-Dithio-bis(2-Nitrobenzoic acid)), Sigma-Aldrich, St. Louis, MO, USA.

2.2 Sulfonation Solution

1. 5 M guanidine-HCl and 0.3 M sodium sulfite solution: 420 mg in 100 mL.

2.3 Inclusion Body Purification

1. Suspension buffer: 1× PBS,pH 7.4 and 1 mM EDTA.
2. Washing buffer: 25 % sucrose, 1× PBS, pH 7.4, 1 mM EDTA, and 0.1 % Triton X100.

2.4 b4Gal-T1 Protein Refolding

1. Solubilizing solution: 5 M guanidine HCl, and 0.5 M arginine HCl.
2. Refolding solution: 100 mM Tris–HCl, 4 mM cysteamine, 2 mM cystamine, and 1 mM EDTA.

2.5 a-Lactalbumin Refolding

1. Refolding solution: 100 mM Tris–HCl, 4 mM cysteamine, 2 mM cystamine, and 5 mM $CaCl_2$.

2.6 ppαGalNAc-T2 Refolding Solution

1. Solubilizing solution: 5 M guanidine HCl, 0.5 M arginine HCl, and 5 mM lactose.
2. Refolding solution: 100 mM Tris–HCl, 10 % glycerol, 0.5 M arginine HCl, 4 mM cysteamine, 2 mM cystamine, 5 mM lactose, and 1 mM EDTA.

2.7 SDS-PAGE Analysis of Proteins

1. 14 % SDS-PAGE Tris-glycine gels (Invitrogen, Carlsbad, CA, USA).
2. Tris-glycine SDS running buffer (10×) (Invitrogen).
3. PowerEase500 protein gel electrophoresis apparatus (Invitrogen).

3 Methods

In order to *S*-sulfonate a protein, the sulfonating agent NTSB (2-nitro-5-thiosulfobenzoate) is prepared by reduced oxidation of DTNB (5, 5′-dithiobis (2-nitrobenzoic acid)) by sodium sulfite in the presence of oxygen [22] (Fig. 1a).

In this reaction, initially every 1 mole of DNTB is sulfonated by creating 1 mole each of NTSB and NTB (2-nitro-5-thiobenzoic acid). By bubbling air/oxygen into this solution, 2 moles of NTB are oxidized into 1 mole of DTNB and further converted to NTSB. This iterative process continues until all the NTB molecules in the solution are converted to NTSB molecules. In the beginning, with the creation of the NTB molecules, the color of the solution changes to bright red; and when all the NTB molecules are converted into NTSB, the solution color turns pale yellow. Similarly, during the *S*-sulfonation step, when NTSB is added to the dissolved inclusion bodies in solution, the solution color turns red; and when the *S*-sulfonation process is complete, it turns colorless or pale yellow (Fig. 2, *see* **Note 1**).

During the *S*-sulfonation of the protein solution, the inclusion bodies are dissolved in 5 M guanidine HCl solution with 0.3 M sodium sulfide, where every 1 mole of any disulfide bond present

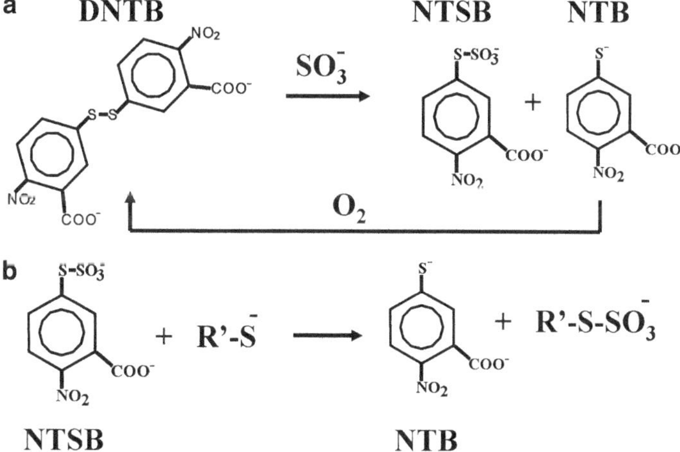

Fig. 1 Synthesis of NTSB and its use in S-sulfonation of proteins. (**a**) It shows the synthesis of NTSB from DTNB in the presence of sodium sulfite. As 1 mole of DTNB is treated with sodium sulfite, it creates 1 mole each of NTSB and NTB. In the presence of oxygen, NTB is oxidized into DTNB and goes through reaction again, creating more NTSB. This process goes on until all NTB molecules are converted into the NTSB molecule. (**b**) When the proteins carrying free thiol groups are treated with NTSB, every 1 mole of the free thiol group reacts with 1 mole of the NTSB molecule, creating an S-sulfonated protein and 1 mole of the NTB molecule. When a molar excess amount of NTSB is used, all the free thiol groups in the protein molecules are *S*-sulfonated

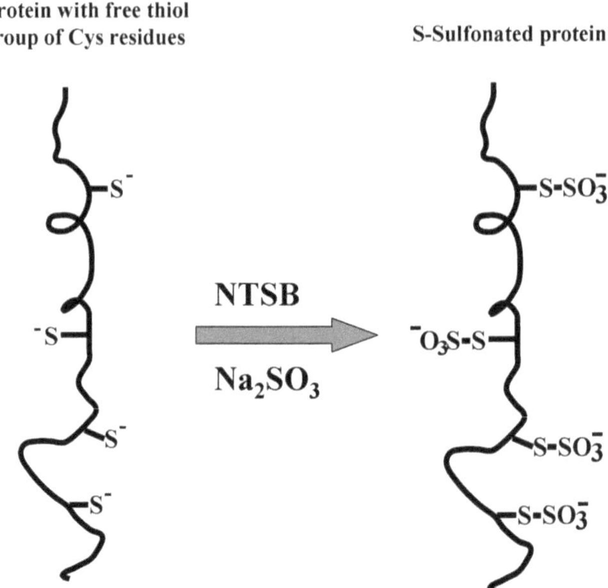

Fig. 2 The schematic diagram of *S*-sulfonation of a protein molecule. As explained in Fig. 1b, all the free thiol groups in the protein molecules are *S*-sulfonated by the NTSB molecule. This *S*-sulfonation causes a dramatic conformational change in the protein and results in a predominantly disordered ensemble of conformations. The treatment of the protein with oxido-shuffling agents generates reduced species that oxidize and refold to the native enzyme in a concerted process

among the protein molecules is converted into 1 mole of *S*-sulfonated thiol and 1 mole of the free thiol group (Fig. 3). Subsequently, when NTSB solution is added to the protein solution, every 1 mole of free thiol reacts with 1 mole of NTSB molecule to create 1 mole of *S*-sulfonated thiol and 1 mole of NTB molecule. Through the process of air oxidation, the NTB molecule is converted back into an NTSB molecule that changes the color of the solution from red to pale yellow or colorless (Fig. 1b). The extent of color change depends on the thiol concentration, which depends on the protein and its concentration in the solution. After the *S*-sulfonation is completed, the protein is purified by either precipitation by dilution of the protein solution or by dialysis to remove all the salts and NTSB molecules from the solution. In the case of most galactosyltransferases, the *S*-sulfonated protein precipitated out either by a simple dilution with water or dialysis against water. The *S*-sulfonated protein is collected by centrifugation and repeatedly washed with water to remove any residual NTSB molecules from it. The insoluble *S*-sulfonated protein is dissolved in 5 M guanidine HCl, and the protein concentration is estimated and used for in vitro folding. In the case of α-lactalbumin protein from most of the species, the *S*-sulfonated protein remains in soluble form; therefore, the salt and NTSB molecules are removed by

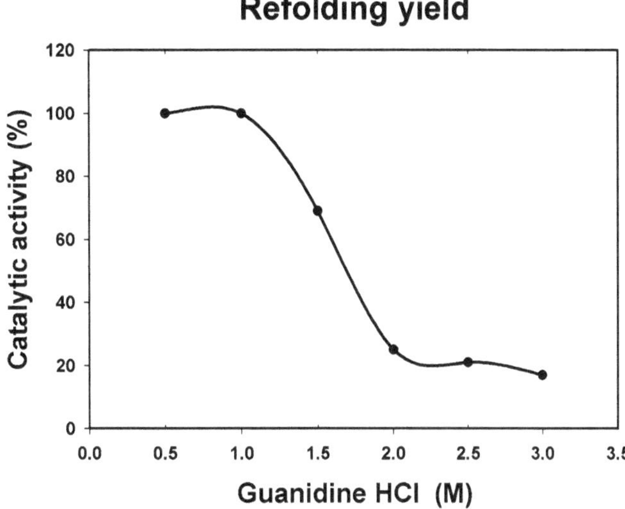

Fig. 3 Graph showing the catalytic activity of the catalytic domain of bovine β4Gal-T1 measured at different guanidine HCl concentrations in the refolding solution. The catalytic activity of the same amount of protein at different refolding conditions having different guanidine HCl measured under identical measurement conditions shows that at reduced guanidine HCl concentrations (<1 M), a maximum catalytic activity is observed. This shows that a high protein refolding yield is achieved at 0.5 M guanidine HCl refolding concentration and has been used in refolding these transferases

dialysis against water. At this stage, any protein precipitate is removed by centrifugation and the soluble protein in the supernatant is measured and used for in vitro folding (*see* **Note 2**).

After refolding is completed, the protein solution is dialyzed extensively against the dialysis buffer. During the dialysis, the misfolded proteins seem to precipitate out, while the active soluble protein remained soluble. After removing the precipitate by centrifugation, the protein solution can be concentrated using Centricon YM10 or using a pressure cell. The final protein concentration of the proteins is 0.5–1 mg/mL and can be frozen with 10 % glycerol at −20 °C for future use (*see* **Notes 3** and **4**).

3.1 NTSB Preparation

1. Dissolve 0.1 g of DTNB in 10 mL of 0.3 M sodium sulfite solution.
2. Set up the table top compressed air to bubble through the solution until the color of the solution turns to pale yellow.
3. Store the NTSB solution in 1 mL aliquots at −20 °C. The frozen NTSB solution can be used for at least 1 year.

3.2 Inclusion Bodies Purification

1. The bacterial pellet from the 1 L culture (ca. 2 g) was suspended in 10 mL suspension buffer.
2. Lyse the cells by through sonication (30 s for four times).

3. Dilute the suspension to 80 mL using the suspension buffer.
4. Centrifuge at 14,000 × *g* for 30 min.
5. Suspend the inclusion body pellet thoroughly in wash buffer and centrifuge at 14,000 × *g* for 30 min.
6. Repeat the washing step with the wash buffer for at least three times.
7. Finally, the protein pellet is washed one final time with suspension buffer by suspending and centrifugation.

3.3 S-Sulfonation of the Protein

1. In a 200-mL beaker, 100 mg of inclusion bodies (from a 2-L culture) is dissolved in 20 mL of sulfonation solution.
2. Add 1 mL of *S*-sulfonating agent, NTSB, to the protein solution, and stir vigorously on a magnetic stirrer at room temperature; wait until the color of the solution disappears or turns pale yellow.
3. The completion of the sulfonation is judged by the color change of the solution from red to pale yellow. This usually takes less than a few hours.

3.3.1 Purification of the S-Sulfonated Glycosyltransferases

1. Add 180 mL of water to the 20-mL sulfonation solution, in order to precipitate the *S*-sulfonated protein.
2. Collect the protein precipitate by centrifugation at 10,000 × *g* for 15 min.
3. Wash the protein pellet two times by resuspending in water, each time followed by centrifugation.

3.3.2 Purification of the S-Sulfonated α-Lactalbumin

1. Add 180 mL of water to the 20-mL sulfonation solution.
2. Dialyze the 200-mL protein solution for 12 h against 4 L of water at room temperature and change the water once.
3. Collect the soluble sulfonated α-lactalbumin protein in the supernatant by centrifugation at 10,000 × *g* for 15 min and discard the pellet.

3.4 Refolding of Glycosyltransferases

1. The sulfonated protein is dissolved in 30 mL of in 5 M guanidine HCl in the cold room at 4 °C.
2. The protein concentration is adjusted to 1 mg/mL which gives absorption of 1.8 at 280 nm.
3. A 100-mL protein solution is diluted tenfold (900 mL of dilution buffer at 4 °C), in nine portions (100 mL each portion).
4. The protein is allowed to fold for 48 h at 4 °C.
5. Dialyze 200 mL of protein solution against 3 × 4 L of water containing 10 mM Tris–HCl,pH 8.0, 4 mM cysteamine, and 2 mM cystamine at 4 °C.

6. The protein which precipitated during dialysis is removed by centrifugation.
7. Concentrate the supernatant either by Centricon YM10 or a stirred cell apparatus (Millipore) using aYM10 filter.
8. Analyze the protein by electrophoresis (*see* **Note 5**).

3.5 Refolding of α-Lactalbumin

1. The soluble sulfonated α-lactalbumin concentration is adjusted to 1 mg/mL, which gives absorption of 1.8 at 280 nm.
2. A 100-mL protein solution is diluted tenfold (900 mL of dilution buffer at room temperature), in nine portions (100 mL each portion).
3. The protein is left to fold for 48 h at room temperature.
4. Any protein that precipitates during refolding is removed by centrifugation.
5. The protein solution is concentrated either by Centricon YM10 or by a Amicon stir-cell apparatus (Millipore) using YM10 filter to a 1 mg/mL concentration.
6. Add 2.5 g of ammonium sulfate for every 10-mL protein solution and allow 30 min for protein precipitation to complete.
7. Centrifuge at 10,000 × *g* for 30 min to collect the supernatant protein solution.
8. Add the same amount of ammonium sulfate to the clear protein solution and allow 30 min for protein precipitation to complete.
9. Centrifuge at 10,000 × *g* for 30 min to collect the protein precipitate.
10. Dissolve the protein precipitate in water and adjust the protein concentration to 1 mg/mL which gives absorption of 2.0 at 280 nm.
11. Dialyze the protein solution against 4 L of water containing 10 mM Tris–HCl buffer (pH 8.0) and 1 mM $CaCl_2$ for 8 h.
12. Analyze the protein by electrophoresis (*see* **Note 6**).

3.6 SDS-PAGE Electrophoresis

The analysis of proteins by SDS-PAGE electrophoresis and the transfer to the nitrocellulose membrane is performed on a PowerEase500 protein gel electrophoresis apparatus (Fig. 4). The protein gels are stained with Coomassie Blue according to the manufacturer's recommendations.

1. 20 μL of the protein sample mixed with 20 μL of 2× Tris-Glycine-SDS sample buffer, with or without β-mercaptoethanol, and are boiled for 10 min.
2. Various amounts of samples, starting from 10 ng, are loaded in a 14 % SDS-PAGE Tris–glycine gel, and the proteins are resolved according to the manufacturer's recommendations.

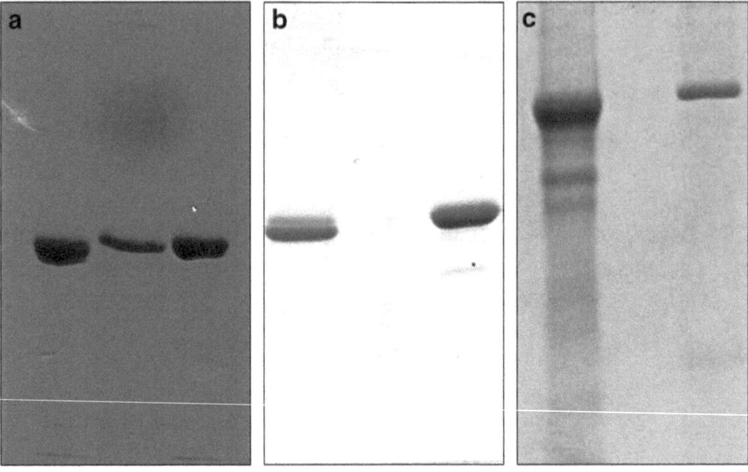

Fig. 4 SDS-PAGE gel stained with Coomassie Blue showing the folded proteins. (**a**) The catalytic domain of bovine β4Gal-T1, the *left* and the *right lanes* showing the in vitro folded protein in the absence and presence of β-mercaptoethanol, respectively, while the *middle lane* *S*-sulfonated β4Gal-T1 protein is shown. (**b**) The in vitro folded catalytic domain of *Drosophila* β4Gal-T7 protein in the absence and in the presence of β-mercaptoethanol, shown in the *left* and *right lanes*, respectively. (**c**) The *S*-sulfonated soluble domain of the ppαGalNAc-T2 protein is shown in the *left lane*, while the in vitro folded protein is shown in the *right lane* in the absence of β-mercaptoethanol

4 Notes

1. We find that purification of inclusion bodies by the repeated wash using 25 % sucrose containing Triton X100 results in relatively pure protein. However, when the yield of inclusion bodies is less than 20 mg/L of cell culture, additional purification may be required prior to *S*-sulfonation and refolding.
2. Refolding of the glycosyltransferases has been successfully performed at 4 °C inside the cold room. During the in vitro refolding protocol, the protein solution is never brought to room temperature. However, *S*-sulfonated α-lactalbumin protein is refolded in vitro at room temperature and there is no need for a low temperature at any stage of its in vitro refolding. Initially, the S-sulfonated protein that requires dissolving in 5 M guanidine HCl at 4 °C, the concentration of the protein is adjusted to 1 mg/mL. The dilution of the protein solution is done stepwise by gently mixing the dilution solution into the protein solution in equal volumes. During the dilution or refolding, the refolding yield seems to be affected by the use of magnetic stirrers. Therefore, during the glycosyltransferases, refolding protocol magnetic stirrers are used. The final refolding

concentration of guanidine HCl was determined based on the catalytic activity generated during the refolding of the catalytic domain of β4Gal-T1, and we find that the 0.5 M guanidine HCl generated maximum catalytic activity and below 0.5 M concentration, the protein precipitated during refolding. Therefore, a final concentration of 0.5 M guanidine HCl is used in all our refolding protocols (Fig. 3). Also, the time required for refolding was determined based on the catalytic activity generated at different time points. We find that nearly 80 % of catalytic activity is generated within 6 h of folding; therefore, the protein solution is kept undisturbed at 4 °C for 24–48 h to achieve maximum refolding yield.

3. The in vitro folded proteins are tested for their purity in SDS-PAGE gel with and without β-mercaptoethanol. When a protein has disulfide bonds, in the SDS-PAGE gel the non-reduced protein shows faster mobility than the same protein when reduced with β-mercaptoethanol. Therefore, the non-reduced protein shows slightly a lower molecular weight compared to the same sample when reduced (Fig. 4).
4. Using the present in vitro folding method, the catalytic domain of bovine and human β4Gal-T1 and its mutant enzymes have been produced to determine their crystal structures with various substrates (Fig. 4a). The catalytic domain of *Drosophila* β4Gal-T7 has also been successfully refolded with a 20-mg/L yield (Fig. 4b), and its crystal structure has been determined. In view of the bioconjugation applications, the soluble domain of ppαGalNAc-T2 also has been refolded using the *S*-sulfonation method (Fig. 4c). The recombinant α-lactalbumin from many species like human, bovine, rat, and mouse have been successfully refolded using the *S*-sulfonation method described here and used for the crystallization of lactose synthase complex. Also, in relation to HAMLET, the human α-lactalbumin molecule with various *C*-terminal tags has been folded using the *S*-sulfonation methods. The non-reduced form of β4Gal-T1, and β4Gal-T7 proteins show slightly high mobility in the SDS-PAGE gel, compared to their respective reduced or S-sulfonated protein, suggesting the presence of a disulfide bond in these proteins (Fig. 4).
5. Typically, when 100 mg of sulfonated protein is folded in a 1-L folding solution, it yields 10–12 mg of active, soluble, and pure β4Gal-T1. Generally, this protein does not require any further purification.
6. Typically, when 100 mg of sulfonated α-lactalbumin is folded in a 1-L folding solution, it yields 40–50 mg of active, soluble, and pure α-lactalbumin. Generally, this protein does not require any further purification.

Acknowledgments

The writing of this review was funded in whole or in part with Federal funds from the National Cancer Institute, National Institutes of Health, under contract HHSN261200800001E. The content of this publication does not necessarily reflect the view or policies of the Department of Health and Human Services, nor does mention of trade names, commercial products, or organizations imply endorsement by the US government. This research was supported [in part] by the Intramural Research Program of the NIH, National Cancer Institute, Center for Cancer Research.

References

1. Dyson MR, Shadbolt SP, Vincent KJ, Perera RL, McCafferty J (2004) Production of soluble mammalian proteins in *Escherichia coli*: identification of protein features that correlate with successful expression. BMC Biotechnol 4:32
2. Varki A, Lowe JB (2009) In: Essentials of glycobiology. Cold Spring Harbor Laboratory Press, Cold Spring Harbor, NY
3. Ramakrishnan B, Qasba PK (2002) Structure-based design of beta-1,4-galactosyltransferase-I (beta 4Gal-T1) with equally efficient N-Acetylgalactosaminyltransferase activity: point mutation broadens beta 4Gal-T1 donor specificity. J Biol Chem 277:20833–20839
4. Pasek M, Ramakrishnan B, Boeggeman E, Manzoni M, Waybright TJ, Qasba PK (2009) Bioconjugation and detection of lactosamine moiety using alpha1,3-galactosyltransferase mutants that transfer C2-modified galactose with a chemical handle. Bioconjug Chem 20:608–618
5. Khidekel N, Arndt S, Lamarre-Vincent N, Lippert A, Poulin-Kerstien KG, Ramakrishnan B, Qasba PK, Hsieh-Wilson LC (2003) A chemoenzymatic approach toward the rapid and sensitive detection of *O*-GlcNAc posttranslational modifications. J Am Chem Soc 125:16162–16163
6. Boeggeman E, Ramakrishnan B, Kilgore C, Khidekel N, Hsieh-Wilson LC, Simpson JT, Qasba PK (2007) Direct identification of nonreducing GlcNAc residues on N-glycans of glycoproteins using a novel chemoenzymatic method. Bioconjug Chem 18:806–814
7. Boeggeman E, Ramakrishnan B, Pasek M, Manzoni M, Puri A, Loomis KH, Waybright TJ, Qasba PK (2009) Site-specific conjugation of fluoroprobes to the remodeled Fc N-glycans of monoclonal antibodies using mutant glycosyltransferases: application for cell surface antigen detection. Bioconjug Chem 20:1228–1236
8. Ramakrishnan B, Boeggeman E, Manzoni M, Zhu Z, Loomis K, Puri A, Dimitrov DS, Qasba PK (2009) Multiple site-specific *in vitro* labeling of single-chain antibody. Bioconjug Chem 20:1383–1389
9. Ramakrishnan B, Boeggeman E, Pasek M, Qasba PK (2011) Bioconjugation using mutant glycosyltransferases for the site-specific labeling of biomolecules with sugars carrying chemical handles. Methods Mol Biol 751:281–296
10. Gautam S, Dubey P, Rather GM, Gupta MN (2012) Non-chromatographic strategies for protein refolding. Recent Pat Biotechnol 6:57–68
11. Thannhauser TW, Scheraga HA (1985) Reversible blocking of half-cystine residues of proteins and an irreversible specific deamidation of asparagine-67 of S-sulforibonuclease under mild conditions. Biochemistry 24:7681–7688
12. Boeggeman EE, Balaji PV, Sethi N, Masibay AS, Qasba PK (1993) Expression of deletion constructs of bovine beta-1,4-galactosyltransferase in *Escherichia coli*: importance of Cys134 for its activity. Protein Eng 6:779–785
13. Boeggeman E, Ramakrishnan B, Qasba PK (2003) The *N*-terminal stem region of bovine and human β1,4-galactosyltransferase I increases the *in vitro* folding efficiency of their catalytic domain from inclusion bodies. Protein Expr Purif 30:219–220
14. Ramakrishnan B, Shah PS, Qasba PK (2001) Alpha lactalbumin (LA) stimulates milk beta-1,4-galactosyltransferase I (4Gal-T1) to transfer glucose from UDP-glucose to N-acetylglucosamine. J Biol Chem 276:37665–37671
15. Ramakrishnan B, Boeggeman E, Qasba PK (2007) Novel method for in vitro O-glycosylation

of proteins: application for bioconjugation. Bioconjug Chem 18:1912–1918
16. Ramakrishnan B, Qasba PK (2010) Crystal structure of the catalytic domain of Drosophila beta-1,4-galactosyltransferase-7. J Biol Chem 285:15619–15626
17. Qasba PK, Kumar S (1997) Molecular divergence of lysozymes and alpha-lactalbumin. Crit Rev Biochem Mol Biol 32:255–306
18. Ramakrishnan B, Qasba PK (2001) Crystal structure of lactose synthase reveals a large conformational change in its catalytic component, the b1,4-galactosyltransferase-I. J Mol Biol 310:205–218
19. Wang M, Scott WA, Rao KR, Udey J, Conner GE, Brew K (1989) Recombinant bovine alpha-lactalbumin obtained by limited proteolysis of a fusion protein expressed at high levels in *Escherichia coli*. J Biol Chem 264:21116–21121
20. Mossberg AK, Hun Mok K, Morozova-Roche LA, Svanborg C (2010) Structure and function of human α-lactalbumin made lethal to tumor cells (HAMLET) type complexes. FEBS J 277:4614–4625
21. Mercer N, Ramakrishnan B, Boeggeman E, Qasba PK (2011) Applications of site-specific labeling to study HAMLET, a tumoricidal complex of α-lactalbumin and oleic acid. PLoS One 6:e26093
22. Thannhauser TW, Konishi Y, Scheraga HA (1984) Sensitive quantitative analysis of disulfide bonds in polypeptides and proteins. Anal Biochem 138:181–188

Chapter 25

An Assay for α1,6-Fucosyltransferase (FUT8) Activity Based on the HPLC Separation of a Reaction Product with Fluorescence Detection

Hideyuki Ihara, Hiroki Tsukamoto, Naoyuki Taniguchi, and Yoshitaka Ikeda

Abstract

N-Glycans with an α-fucose unit linked to the 6-position of the innermost GlcNAc are widely distributed among the animal kingdom, from worms and insects to human. This α1,6-linked fucosyl residue, frequently referred to as a core fucose, is formed via the action of an α1,6-fucosyltransferase, the mammalian ortholog which is systematically called FUT8. In mammals, it is well known that the extent of core-fucosylation in cellular and secreted glycoproteins varies, e.g., according to differentiation and carcinogenesis of the cells. This chapter describes a method for the sensitive and quantitative assay of FUT8 activity using a fluorescence-labeled oligosaccharyl asparagine derivative as the glycosyl acceptor substrate.

Key words Fucosyltransferase, FUT8, Fucose, Core, N-glycan, Glycoprotein, 2-Aminopyridine, Pyridylamine, Asn-linked oligosaccharide, Glycosylation

1 Introduction

Mammalian α1,6-fucosyltransferase, referred to as FUT8, catalyzes the transfer of a fucose residue from GDP-β-L-fucose to the reducing terminal GlcNAc residue of a glycan, the innermost GlcNAc, of the core structure of an Asn-linked sugar chain (Fig. 1). The structure formed by this enzyme reaction is referred to as a core fucose in the case of the transferred monosaccharide or as a core-fucosylated N-glycan in the case of the resulting oligosaccharide. This modification of an N-glycan occurs ubiquitously in wide variety of mammalian glycoproteins. The reaction does not require divalent cations or cofactors, albeit they may not or may enhance the reaction to some extent, in contrast to the absolute requirement for divalent cations in some other glycosyltransferases. The kinetic properties of FUT8 have been well characterized [1], and its crystal structure has been resolved [2]. In a previous study, we also

Inka Brockhausen (ed.), *Glycosyltransferases: Methods and Protocols*, Methods in Molecular Biology, vol. 1022, DOI 10.1007/978-1-62703-465-4_25,

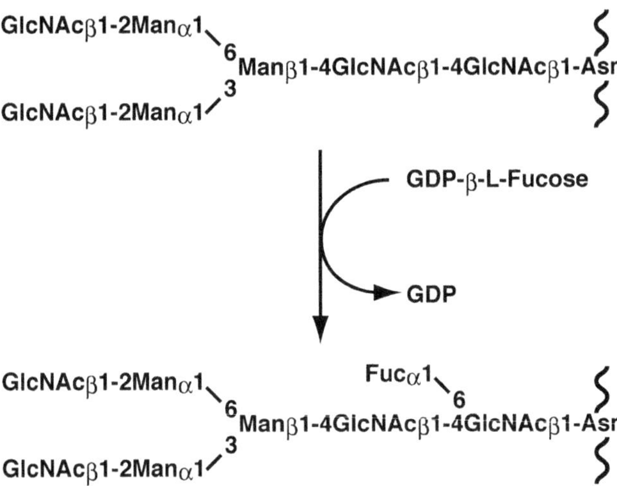

Fig. 1 A reaction catalyzed by FUT8. The *waved lines* indicate polypeptides, and the asparagine residues should be a part of the N-glycosylation site, Asn-X-Thr/Ser

reported that FUT8 is capable of catalyzing an unusual reaction in which chitooligosaccharides serve as the acceptor substrate [3].

As has been demonstrated by studies using FUT8-knockout mice, core fucosylation plays pivotal roles in permitting a variety of glycoproteins to normally function in vivo. Examples include the TGFβ1 receptor [4], the EGF receptor [5], the VEGF receptor-2 [6], LRP-1 [7], E-cadherin [8], α3β1 integrin [9], VCAM, and α4β1 integrin [10]. In addition, it has been reported that antibody-dependent cellular cytotoxicity is enhanced in the case of N-glycans that lack a core fucose in the Fc region of an IgG1 subclass [11, 12]. This discovery led to an extended application of engineering of glycosylation to antibody drug-based medicine.

α-Fetoprotein (AFP) is known to be one of the tumor markers that are most frequently used in diagnosing malignancies, particularly hepatocellular carcinoma (HCC). However, the expression and serum levels of AFP are frequently elevated in cases of benign hepatic diseases such as hepatitis and liver cirrhosis as well as HCC [13, 14]. For a better diagnosis, the AFP-L3 fraction was found to be more specific for HCC and was subsequently shown to be a core fucosylated form [15–17]. Elevated serum levels of L3 fraction were not observed in cases of benign hepatic diseases. In ovarian serous adenocarcinoma, we found that FUT8 enzyme activity and the corresponding mRNA levels are significantly elevated, compared to normal tissues and other types of ovarian carcinomas [18].

For a specific and sensitive assay of the activities of glycosyltransferases, the reactions are usually carried out using radiolabeled or fluorescence-labeled substrates, followed by appropriate separation of the substrates and products to estimate the reaction rates. While radioactive materials can be used in conjunction with both

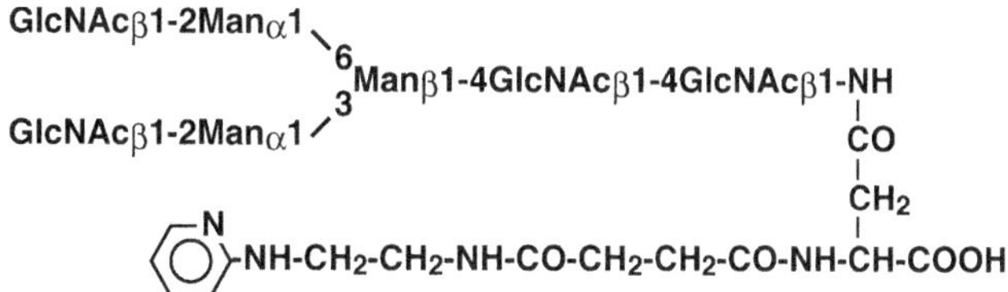

Fig. 2 An example of a fluorescence-labeled Asn-linked oligosaccharide. Preparation of the acceptor substrate is described in this chapter. The β-anomer of the cyclic (*closed ring form*) in the reducing terminal GlcNAc is essential for the transfer of fucose by FUT8

an oligosaccharide acceptor and glycosyl donor monosaccharide, fluorescence-labeling is generally used for an oligosaccharide acceptor substrate, mostly due to the bulkiness of the fluorophore. Various types of fluorophores have been examined and used in such assays, e.g., depending on the analytical method under consideration, such as high performance liquid chromatography (HPLC) and capillary electrophoresis. 2-Aminopyridine (PA) is one of the most common reagents used for the fluorescence labeling of oligosaccharide substrates and its conjugates are suitable for both normal and reversed phase modes of HPLC-based separation [19, 20].

PA is usually conjugated to the reducing end of an oligosaccharide via reductive amination involving an amino group of the fluorophore and a carbonyl group of the terminal sugar [21], and, as a result, the reducing terminal sugar is acyclic, i.e., an open chain structure. This structural alteration is tolerated by many glycosyltransferases including some types of fucosyltransferase [19, 22, 23]. On the other hand, the cyclic form of the terminal GlcNAc with a β configuration is required for determining the activity of FUT8 as it occurs in vivo, because the enzyme acts on the very sugar of the reducing end of the oligosaccharide. Therefore, the oligosaccharide needs to be prepared in an asparagine-linked form by the protease digestion of glycoproteins or glycopeptides, and is then labeled by conjugating PA-derivatives that have an amino or a carboxyl group to the asparagine via an amide bond, thus maintaining the integrity of the terminal sugar [1, 24, 25]. This chapter describes a method for the preparation of a fluorescence-labeled oligosaccharide acceptor (Fig. 2) and a reversed-phase HPLC-based assay using the substrate.

2 Materials

2.1 Preparation of Asn-Linked Oligosaccharide

1. Source of Asn-linked oligosaccharide: Sialoglycopeptide from hen's egg yolk (*see* **Note 1**).
2. Pronase: Used as a solid powder.
3. Pronase digestion buffer: 0.1 M Tris–HCl, pH 7–8.

4. Gel filtration column: Toyopearl HW-40F (Tosoh, Tokyo, Japan), the bed volume of which has an appropriate diameter × height of 70 cm or more (*see* **Note 2**).
5. Solvent for the gel filtration chromatography: 0.1 M sodium borate, pH 8.0 (*see* **Note 3**).
6. HPLC system: Including two pumps enabling gradient elution, UV detector and fluorescence detector (*see* **Note 4**).
7. Normal phase HPLC column: TSKgel Amide-80 (Tosoh) or its equivalent (*see* **Note 5**).
8. Reversed phase HPLC column: ODS column, e.g., TSKgel ODS-80TM (Tosoh) (*see* **Note 5**).
9. Solvents for the normal phase HPLC: Solvent A, 0.1 % trifluoroacetic acid (TFA)/80 % acetonitrile; Solvent B, 0.1 % TFA.
10. Solvents for the reversed phase HPLC: Solvent A, 20 mM ammonium acetate buffer, pH 4.0; Solvent B, 20 mM ammonium acetate buffer containing 1 % 1-butanol, pH 4.0 (*see* **Note 6**).

2.2 Preparation and Purification of the Fluorescence-Labeled Asn-Linked Oligosaccharide

1. Fluorescence-labeling reagent: N-[2-(2-pyridylamino)ethyl]-succinamic acid 5-norbornene-2,3-dicarboximide ester, abbreviated as PA-ES-NDCE in this chapter (*see* **Note 7**), dissolved in DMSO at a concentration of 100 mg/mL.
2. Labeling buffer: 0.1 M $NaHCO_3$, >pH 8.0 (*see* **Note 8**).
3. HPLC columns and solvents: Described above in Subheading 2.1.
4. Sialidase: neuraminidase from *Arthrobacter ureafaciens* (Nacalai Tesque, Kyoto, Japan).
5. β-Galactosidase: Jack bean β-galactosidase (Seikagaku Corp. Tokyo, Japan).
6. Glycosidase digestion buffer: 0.1 M Na-citrate buffer, pH 5.0.
7. Asn-linked oligosaccharide: Prepared according to Subheading 3.1.

2.3 FUT8 Reaction

1. 2× reaction buffer: 100 mM MES-Na buffer, 0.2 % Triton X-100, pH 7.0.
2. GDP-β-L-fucose: 5 mM solution in H_2O (*see* **Note 9**) (Wako Pure Chemical, Tokyo, Japan).
3. Fluorescence-labeled Asn-linked oligosaccharide substrate: Prepare 100 μM solution.

2.4 Separation and Quantification of the Reaction Product

1. Solvents for reversed-phase HPLC: the same as described above in Subheading 2.1.
2. HPLC system for fluorescence detection (*see* **Note 4**).
3. HPLC column: TSKgel ODS-80TM, 4.6 × 150 mm.

3 Methods

3.1 Preparation of Asn-Linked Oligosaccharide

1. Dissolve the sialoglycopeptide purified from egg yolk ([26], *see* **Note 1**) in 0.1 M Tris–HCl buffer.
2. Add Pronase powder in 1/200 to 1/20 concentrations (w/w) to the solution and dissolve gently to avoid foaming.
3. Incubate at 37 °C for 12–72 h (*see* **Note 10**).
4. Boil the sample for about 5 min to terminate the digestion. If necessary, centrifuge the sample to remove insoluble material, prior to the next step.
5. Apply the sample to an HW-40F gel filtration column (*see* **Note 11**).
6. Collect the effluent by fractionating at appropriate intervals.
7. Assess the fractions by the phenol-sulfate method to detect sugars and collect the fractions containing oligosaccharides, shown in Fig. 3 (*see* **Note 12**).
8. Load the pooled fractions to preparative normal phase HPLC to further purify (*see* **Notes 13** and **14**).

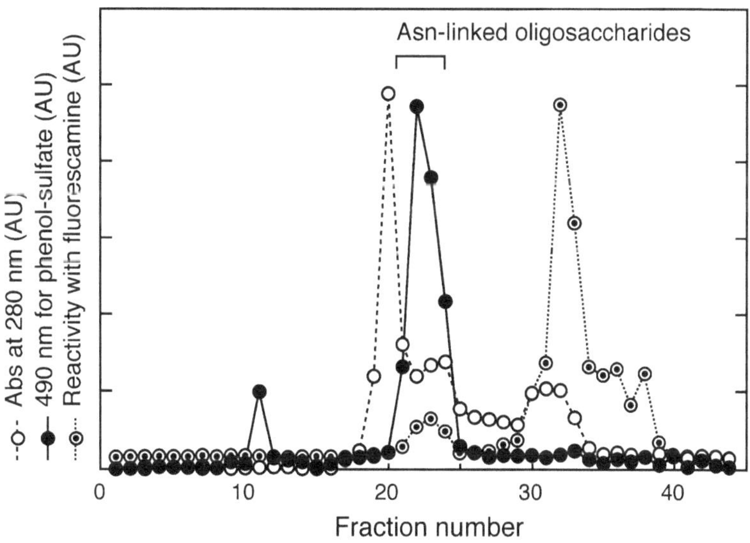

Fig. 3 Toyopearl HW40 gel filtration chromatography of a Pronase-digest of the sialoglycopeptide. *Open circles* show the undigested and/or digested peptides, as monitored by absorbance (AU) at 280 nm. *Filled circles* show the relative concentrations of sugars, as detected by the phenol-sulfate method using a 96-well plate. *Double circles* denote the fluorescence intensities as the result of the reaction of primary amines with fluorescamine, and indicate small molecules such as amino acids, oligopeptides, and buffer components

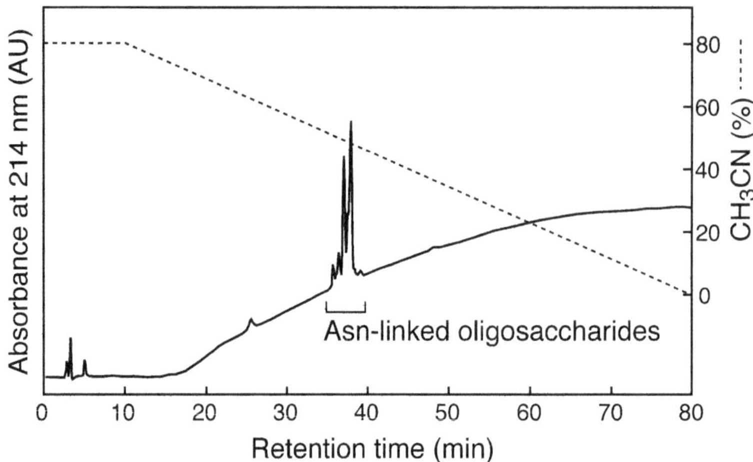

Fig. 4 An elution profile for Asn-linked oligosaccharides by normal phase HPLC. In this example, a TSKgel Amide-80, 21.5 × 300 mm, was used as a preparative scale normal phase column to separate Asn-linked oligosaccharides. Washing the column and elution were performed at a flow rate of 3.0 mL/min

9. Wash the column thoroughly with the same solvent as was used for the pre-equilibration step, monitoring by the absorbance at 214 nm based on the N-acetyl group of GlcNAc residues using a UV detector.
10. Continue to wash the column until no more obvious peaks appear (*see* **Note 15**).
11. Elute the bound oligosaccharide using a gradient of acetonitrile from 80 % down to 0 % in 0.1 % TFA, and collect all peaks observed during the gradient elution, as shown in Fig. 4 (*see* **Note 16**).
12. Evaporate the Asn-linked oligosaccharide to dryness using rotary evaporation or a similar method.

3.2 Fluorescence-Labeling, Glycosidase Digestion, and Further Purification of Asn-Linked Oligosaccharide

1. Dissolve the dried oligosaccharide in an appropriate volume of 0.1 M $NaHCO_3$ buffer, pH > 8.0.
2. Add PA-ES-NDCE, a fluorescence-labeling reagent, in DMSO to the oligosaccharide solution at a molar ratio of 2:1–10:1 to the amino group of asparagine (*see* **Note 17**).
3. Allow the coupling reaction to proceed at room temperature for 3–12 h.
4. Load the reaction mixture onto normal phase HPLC to separate the oligosaccharide from the labeling reagents, as performed for the preparation of the Asn-linked oligosaccharide described in the Subheading 3.1 (*see* **Notes 13** and **18**).
5. Wash the column thoroughly with the starting solvent, 80 % acetonitrile in a 0.1 % TFA solution.

6. Collect the oligosaccharide-containing fractions by monitoring the peak(s) that appears during elution by a gradient (*see* **Note 16**).
7. Evaporate the collected effluent to dryness using a rotary evaporator, and then dissolve the sample in the glycosidase buffer.
8. Add β-galactosidase and sialidase to the solution and carefully mix: 20 mU of each enzyme for approximately 1 μmol equivalent of asparagine.
9. Incubate the sample at 37 °C for 24 h (*see* **Note 19**).
10. Apply the resulting reaction mixture to a preparative reversed phase HPLC. Elution is performed isocratically with 0.15 % 1-butanol in 20 mM ammonium acetate buffer, pH 4.0, corresponding to a solvent B concentration of 15 % (*see* **Note 14**).
11. Carefully collect the single peak which contains the asialo-agalacto-biantennary form of the oligosaccharide (*see* **Notes 19** and **20**). Prior to this preparation, examine the elution time of the peak to be collected in your HPLC system, as performed in Fig. 5.
12. Evaporate the oligosaccharide solution to dryness on a rotary evaporator, followed by dissolution of the sample in H_2O (*see* **Note 21**).
13. Determine the concentration of the purified labeled oligosaccharide, referred to as GnGn-bi-Asn-ESPA, by comparison of its dilution with a series of PA-derivatives of known concentrations (*see* **Note 22**).

3.3 Enzyme Reaction of FUT8

1. Mix the following; 5 μL of 2× reaction buffer, 1 μL of 100 μM GnGn-bi-Asn-ESPA and 1 μL of 5 mM GDP-β-L-fucose. Make this up to a volume of 10 μL by adding appropriate volumes of H_2O and enzyme source (*see* **Note 23**).
2. Incubate the solution at 37 °C for 1–4 h, or an appropriate time, depending on the levels of the enzyme activity.
3. Terminate the reaction by rapidly boiling for 2–5 min.
4. Add 40 μL of H_2O, mix and then centrifuge at the maximum speed of a microcentrifuge, usually at 15,000 rpm (17,400 × *g*) for 10 min.
5. Carefully transfer the supernatant to a fresh microtube.

3.4 Analysis and Quantification of the Reaction Product by HPLC

1. Pre-equilibrate a TSKgel ODS-80TM column (4.6 × 150 mm) with 15 % solvent B at a flow rate of 1 mL/min, and preheat the column at 55 °C. The fluorescence detector should also be warmed up prior to use, in order to achieve a stable analysis and to obtain reproducible data.
2. Inject 20 μL of the supernatant into the HPLC column.

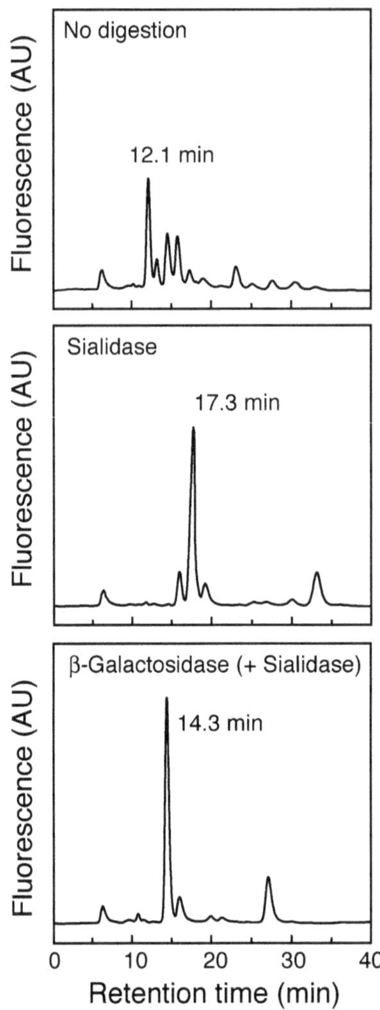

Fig. 5 Reversed phase HPLC separation of the glycosidase-digests of the labeled Asn-linked oligosaccharides. This analytical chromatography was carried out at 55 °C using a TSKgel ODS-80TM, 4.6 × 150 mm, at a flow rate of 1.0 mL/min. The oligosaccharides were eluted isocratically with 0.1 % 1-butanol in 20 mM ammonium acetate, pH 4.0, and were monitored by fluorescence at λ_{ex} 310 nm and λ_{em} 380 nm. In the profile for the galactosidase, the *main peak* indicates a degalactosylated form of the desialylated sample

3. Separate the substrate and product isocratically, and continuously monitor the effluent using a fluorescence detector at excitation and emission wavelengths of 310 nm and 380 nm, respectively.
4. Estimate the amounts of the product from the recording chart or chromatographic profiles based on fluorescence intensity (*see* **Note 24**). A typical elution profile is shown in Fig. 6.

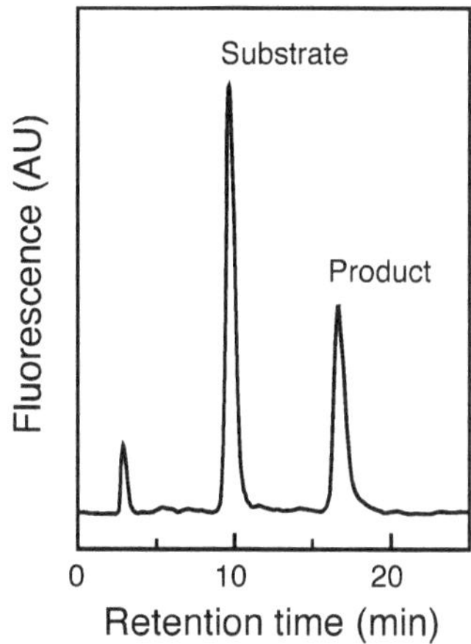

Fig. 6 A typical elution profile of the reaction product of FUT8. Separation of the substrate (GnGn-bi-Asn-ESPA) and product (core fucosylated GnGn-bi-Asn-ESPA) was performed using a TSKgel ODS-80TM, 4.6 × 150 mm. Isocratic elution was conducted at 55 °C using 0.15 % 1-butanol in 20 mM ammonium acetate, pH 4.0 at a flow rate of 1.0 mL/min. The column effluent was monitored with a fluorescence detector at λ_{ex} 310 nm and λ_{em} 380 nm

4 Notes

1. Glycoproteins such as transferrin and gamma-globulin from bovine sources are commercially available for use as the source of substrates and HPLC standards. For a relatively efficient and inexpensive large-scale preparations, however, it is much better to utilize the sialoglycopeptide as the starting material. This can be purified from hen's egg yolk according to the method of Seko et al. [26]. The relatively larger amount of oligosaccharide in the material may be advantageous for a more convenient preparation. The authors strongly recommend the use of this glycopeptide for the preparation.
2. Sephadex G-25 can also be used although the performance may not result in an efficient separation as can be obtained using Toyopearl.
3. Other types of buffers or solutions, e.g., H_2O, may be used as the solvent. In this case, borate buffer is used to detect amino groups directly in the effluents using fluorescamine without adjustment of pH.
4. It is possible to detect the oligosaccharide substrate with a UV-detector during the preparation because of relatively large amounts present; however, a fluorescence detector is desirable. For a practical analysis, as described below, a fluorescence detector is absolutely required.
5. In preparative scale separations, the use of larger bore columns, typically with diameters of 7.8–21.5 mm, are recommended. Even if preparative scale columns are not available, it is possible to process a large amount of sample by repeating

the same chromatographic separations several times using smaller bore columns.

6. Prepare by diluting 0.1 M ammonium acetate buffer, pH 4.0 (stock solution). For solvent B, further add 1/100 volume of 1-butanol to the diluted buffer.
7. Other similar 2-aminopyridine derivatives are commercially available. For example, 2-(2-pyridylamino)ethylamine can be used via the amidation of the α-carboxyl group of asparagine, similar to that reported for 4-(2-pyridylamino)butylamine [25]. The amino group of the oligosaccharyl asparagine needs to be previously modified or temporarily protected, e.g., by N-acetylation by reaction with acetic acid anhydride, and the condensation is then carried out using a water soluble carbodiimide, such as 1-ethyl-3-(3-dimethylaminopropyl) carbodiimide.
8. Other types of buffers may be used in place of $NaHCO_3$ but the components of the solution should not include ammonia or compounds that contain an amino group.
9. If it is difficult to accurately weigh dry powder due to the small amount (2 mg or less), dissolve the reagent in a sufficiently small volume of H_2O and quantify it spectrophotometrically on the basis of the UV absorption of GDP. In addition, it may be necessary to check that the pH of the solution is around neutral using pH indicator paper. It should be noted that the nucleotide sugar used for FUT8 is a β-anomer of the L-isomer, which is quite the opposite of the glycosyl donor substrates for many other glycosyltransferases.
10. The optimal incubation time depends on the amount and/or concentration of materials for digestion, but, in general, longer is better than shorter.
11. The column should be pre-equilibrated with 0.1 M Na-borate buffer. This chromatographic step is carried out in order to separate oligosaccharide fractions from amino acids produced as the result of the digestion. If necessary, for better chromatographic performance, the sample should be concentrated prior to the chromatographic step, e.g., by being lyophilized or by using a rotary evaporator. As generally carried out for gel filtration chromatography, 1/20 or less of the bed volume is suitable for this step. When a sample volume is still too large even after concentrating, repeat the chromatography as much as necessary.
12. Mix a small aliquot (50 μL) of the fractions with an equal volume of water-saturated phenol in a 96-well plate, and then add 100 μL of fresh sulfuric acid.

13. The amide column should be pre-equilibrated with 80 % (v/v) acetonitrile/0.1 % TFA, and the sample is then loaded on the column.
14. The flow rate should be appropriately varied from 0.5 to 3.0 mL/min, according to your individual system and the diameter of the column used.
15. In some cases, it may take as long as 2 h. When this step is carried out for the first time, it would be better to retain the pass-through fractions in order to check overflow or column capacity.
16. The oligosaccharides will be eluted as a large, broad peak if a large amount of oligosaccharide is loaded on the column, or otherwise is observed as a group of several closely spaced peaks, probably due to the structural heterogeneity of sugar chains (Fig. 4). The oligosaccharides are obtained by continuously collecting the effluent, monitoring the fractions by UV. Alternatively, the effluent can be fractionated at appropriate intervals, followed by the use of the phenol-sulfate method, as described above.
17. The molar concentration of the amino acid can be estimated by determining the free amino groups, e.g., using fluorescamine and *o*-phthalaldehyde. Alternatively, the concentration can be approximated by very roughly estimating the sugar content of the sample using the phenol-sulfate method using either or both of hexoses and N-acetylhexosamine as the standard. With respect to the amount of PA-ES-NDCE to be added, larger is often better for the yield than smaller.
18. At this step or later, however, monitoring by measuring the absorbance at about 310 nm may be useful for the detection of the labeled sample.
19. It is recommended to occasionally check the extent of the reaction by analytical reversed phase or normal phase HPLC (Fig. 5). As the digestion becomes complete, multiple peaks merge into a single peak that corresponds to an asialo and agalacto form of the oligosaccharide. The peak of interest is eluted later than the peak corresponding to the sialylated form but earlier than the desialylated-but-still-galactosylated form. The digestion may be stopped even if it is not complete. To save time, a 70–80 % yield, sometimes even 50 %, can be satisfactory.
20. If multiple peaks are observed due to an incomplete digestion, which occurs frequently, only the desialylated and degalactosylated oligosaccharide should be collected. However, other oligosaccharide peaks must also be kept until the correct substrate is clearly prepared, or for reuse of the undigested materials.

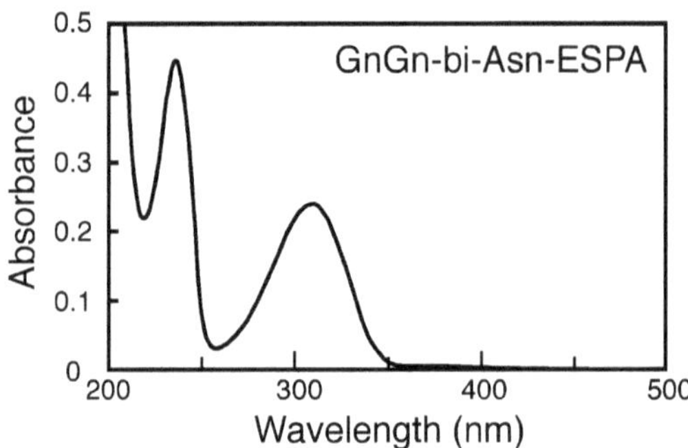

Fig. 7 UV–Visible spectrum of the purified labeled Asn-linked oligosaccharide substrate, GnGn-bi-Asn-ESPA. The concentration of the sample was 32 μM

21. If necessary, e.g., to remove contaminating peaks, further purify the sample by additional normal phase HPLC, using solvents that do not contain TFA. The absorption of the substrate to the column and washing are performed with 80 % acetonitrile in H_2O, and the substrate is then eluted by H_2O in a stepwise manner.
22. GnGn-bi-Asn-ESPA is a tentative abbreviation of GnGn-bi-Asn conjugated with a pyridylamino moiety with N-ethylsuccinamide as a linker. To determine the concentration, appropriately diluted samples are loaded together with for example PA-ES-NDCE of known concentrations on an analytical reversed phase HPLC column. The concentration can be calculated by calibration based on the areas of the peaks in the chromatogram. When the substrate solution does not contain chromophoric contaminants, the concentration can be spectrophotometrically determined using an ε value of 310 nm, approximately 6.0 mM^{-1} cm^{-1} at pH 4.0. This value was experimentally estimated by comparison with a PA-derivative. The pH of the solution is an important determinant of its absorbance and fluorescence because the protonation state of the PA affects the spectrophotometric properties of the molecule. As shown in Fig. 7, a peak at around 310 nm is characteristic of a pyridylamino moiety in the UV–Visible spectrum of the substrate at pH 4.0.
23. The total volume of the reaction mixture may be varied if a problem arises concerning a small volume, e.g., one cannot avoid drying up the solution. Alternatively, when a long incubation time is expected, it is advisable to use an oil-free thermal cycler as an incubator or to immerse the microtube completely in a water bath. The conditions used for the concentration of the substrate vary and are not universal. It is not necessary to increase the concentrations to much higher values than the K_M values.

24. In this end point assay, the amount of enzyme to be added and/or the reaction time should be varied to determine the optimal conditions. It is advisable to allow 20–30 % of the substrate to be converted for the sake of both linearity toward time and achieving reliable quantification. Much more consumption of the substrate leads to the underestimation of the enzyme activity.

References

1. Ihara H, Ikeda Y, Taniguchi N (2006) Reaction mechanism and substrate specificity for nucleotide sugar of mammalian alpha1,6-fucosyltransferase—a large-scale preparation and characterization of recombinant human FUT8. Glycobiology 16:333–342
2. Ihara H, Ikeda Y, Toma S, Wang X, Suzuki T, Gu J, Miyoshi E, Tsukihara T, Honke K, Matsumoto A, Nakagawa A, Taniguchi N (2007) Crystal structure of mammalian alpha1,6-fucosyltransferase, FUT8. Glycobiology 17:455–466
3. Ihara H, Hanashima S, Okada T, Ito R, Yamaguchi Y, Taniguchi N, Ikeda Y (2010) Fucosylation of chitooligosaccharides by human alpha1,6-fucosyltransferase requires a nonreducing terminal chitotriose unit as a minimal structure. Glycobiology 20:1021–1033
4. Wang X, Inoue S, Gu J, Miyoshi E, Noda K, Li W, Mizuno-Horikawa Y, Nakano M, Asahi M, Takahashi M, Uozumi N, Ihara S, Lee SH, Ikeda Y, Yamaguchi Y, Aze Y, Tomiyama Y, Fujii J, Suzuki K, Kondo A, Shapiro SD, Lopez-Otin C, Kuwaki T, Okabe M, Honke K, Taniguchi N (2005) Dysregulation of TGF-beta1 receptor activation leads to abnormal lung development and emphysema-like phenotype in core fucose-deficient mice. Proc Natl Acad Sci USA 102:15791–15796
5. Wang X, Gu J, Ihara H, Miyoshi E, Honke K, Taniguchi N (2006) Core fucosylation regulates epidermal growth factor receptor-mediated intracellular signaling. J Biol Chem 281:2572–2577
6. Wang X, Fukuda T, Li W, Gao CX, Kondo A, Matsumoto A, Miyoshi E, Taniguchi N, Gu J (2009) Requirement of Fut8 for the expression of vascular endothelial growth factor receptor-2: a new mechanism for the emphysema-like changes observed in Fut8-deficient mice. J Biochem 145:643–651
7. Lee SH, Takahashi M, Honke K, Miyoshi E, Osumi D, Sakiyama H, Ekuni A, Wang X, Inoue S, Gu J, Kadomatsu K, Taniguchi N (2006) Loss of core fucosylation of low-density lipoprotein receptor-related protein-1 impairs its function, leading to the upregulation of serum levels of insulin-like growth factor-binding protein 3 in Fut8-/- mice. J Biochem 139:391–398
8. Osumi D, Takahashi M, Miyoshi E, Yokoe S, Lee SH, Noda K, Nakamori S, Gu J, Ikeda Y, Kuroki Y, Sengoku K, Ishikawa M, Taniguchi N (2009) Core fucosylation of E-cadherin enhances cell-cell adhesion in human colon carcinoma WiDr cells. Cancer Sci 100:888–895
9. Zhao Y, Itoh S, Wang X, Isaji T, Miyoshi E, Kariya Y, Miyazaki K, Kawasaki N, Taniguchi N, Gu J (2006) Deletion of core fucosylation on alpha3beta1 integrin down-regulates its functions. J Biol Chem 281:38343–38350
10. Li W, Ishihara K, Yokota T, Nakagawa T, Koyama N, Jin J, Mizuno-Horikawa Y, Wang X, Miyoshi E, Taniguchi N, Kondo A (2008) Reduced alpha4beta1 integrin/VCAM-1 interactions lead to impaired pre-B cell repopulation in alpha 1,6-fucosyltransferase deficient mice. Glycobiology 18:114–124
11. Shields RL, Lai J, Keck R, O'Connell LY, Hong K, Meng YG, Weikert SH, Presta LG (2002) Lack of fucose on human IgG1 N-linked oligosaccharide improves binding to human Fcgamma RIII and antibody dependent cellular toxicity. J Biol Chem 277:26733–26740
12. Shinkawa T, Nakamura K, Yamane N, Shoji-Hosaka E, Kanda Y, Sakurada M, Uchida K, Anazawa H, Satoh M, Yamasaki M, Hanai N, Shitara K (2003) The absence of fucose but not the presence of galactose or bisecting N-acetylglucosamine of human IgG1 complex-type oligosaccharides shows the critical role of enhancing antibody-dependent cellular cytotoxicity. J Biol Chem 278:3466–3473
13. Alpert ME, Uriel J, de Nechaud B (1968) Alpha-1 fetoglobulin in the diagnosis of human hepatoma. N Engl J Med 278:984–986
14. Ruoslahti E, Salaspuro M, Pihko H, Andersson L, Seppala M (1974) Serum alpha-fetoprotein: diagnostic significance in liver disease. Br Med J 2:527–529
15. Taketa K (1990) Alpha-fetoprotein: reevaluation in hepatology. Hepatology 12:1420–1432
16. Aoyagi Y (1995) Carbohydrate-based measurements on alpha-fetoprotein in the early

diagnosis of hepatocellular carcinoma. Glycoconj J 12:194–199

17. Miyoshi E, Noda K, Yamaguchi Y, Inoue S, Ikeda Y, Wang W, Ko JH, Uozumi N, Li W, Taniguchi N (1999) The alpha1-6-fucosyltransferase gene and its biological significance. Biochim Biophys Acta 1473:9–20
18. Takahashi T, Ikeda Y, Miyoshi E, Yaginuma Y, Ishikawa M, Taniguchi N (2000) alpha1,6fucosyltransferase is highly and specifically expressed in human ovarian serous adenocarcinomas. Int J Cancer 88:914–919
19. Nishikawa A, Fujii S, Sugiyama T, Taniguchi N (1988) A method for the determination of N-acetylglucosaminyltransferase III activity in rat tissues involving HPLC. Anal Biochem 170:349–354
20. Tomiya N, Lee YC, Yoshida T, Wada Y, Awaya J, Kurono M, Takahashi N (1991) Calculated two-dimensional sugar map of pyridylaminated oligosaccharides: elucidation of the jack bean alpha-mannosidase digestion pathway of Man9GlcNAc2. Anal Biochem 193:90–100
21. Kondo A, Suzuki J, Kuraya N, Hase S, Kato I, Ikenaka T (1990) Improved method for fluorescence labeling of sugar chains with sialic acid residues. Agric Biol Chem 54:2169–2170
22. Palmerini CA, Datti A, Alunni S, VanderElst IE, Orlacchio A (1995) A fluorescent assay for the determination of UDP-GlcNAc: Gal beta 1,3GalNAc-R (GlcNAc to GalNAc) beta 1,6-N-acetylglucosaminyltransferase activity. Anal Biochem 225:315–320
23. Sasaki K, Kurata K, Funayama K, Nagata M, Watanabe E, Ohta S, Hanai N, Nishi T (1994) Expression cloning of a novel alpha 1,3-fucosyltransferase that is involved in biosynthesis of the sialyl Lewis × carbohydrate determinants in leukocytes. J Biol Chem 269:14730–14737
24. Mita Y, Aoyagi Y, Suda T, Asakura H (2000) Plasma fucosyltransferase activity in patients with hepatocellular carcinoma, with special reference to correlation with fucosylated species of alpha-fetoprotein. J Hepatol 32:946–954
25. Uozumi N, Teshima T, Yamamoto T, Nishikawa A, Gao YE, Miyoshi E, Gao CX, Noda K, Islam KN, Ihara Y, Fujii S, Shiba T, Taniguchi N (1996) A fluorescent assay method for GDP-L-Fuc:N-acetyl-beta-D-glucosaminide alpha 1-6fucosyltransferase activity, involving high performance liquid chromatography. J Biochem 120:385–392
26. Seko A, Koketsu M, Nishizono M, Enoki Y, Ibrahim HR, Juneja LR, Kim M, Yamamoto T (1997) Occurrence of a sialylglycopeptide and free sialylglycans in hen's egg yolk. Biochim Biophys Acta 1335:23–32

Chapter 26

Immunodetection of Glycosyltransferases in Gastrointestinal Tissues

Joana Gomes and Celso A. Reis

Abstract

Glycosyltransferases control the biosynthesis of glycans expressed in cells. Alterations in glycosylation in the gastrointestinal tract stem from deregulation of glycosyltransferase expression. These modifications can be detected in situ by cell and tissue immunolabelling techniques which are highly informative in physiological and pathological contexts. The protocols described here are single and double-labelling immunofluorescence techniques that allow the detection of a specific glycosyltransferase and, in the case of double labelling, the concomitant detection of the glycosyltransferase and its glycan product.

Key words Immunodetection, Glycosyltransferases, Gastrointestinal tract, Immunofluorescence, Double staining

1 Introduction

Glycosyltransferases are key enzymes controlling different steps in the biosynthesis of glycans [1]. The pattern of glycosylation in cells is crucial for several physiological and pathological mechanisms. In the gastrointestinal tract, glycans are involved in the susceptibility for infection, inflammatory processes, and neoplasia [2, 3]. Most of these modifications are generated by up- or down-regulation of the expression of glycosyltransferases involved in initiation, elongation or termination of oligosaccharides chains [4, 5]. In cancer tissues, many studies have demonstrated an association between differently expressed glycosyltransferases and carbohydrates antigens [6]. Characterization of glycosyltransferases is based on molecular, biochemical and enzymology approaches [7–10]. Cell and tissue immunolabelling techniques such as immunohistochemistry [11], immunocytochemistry [12], and immunofluorescence [6, 13, 14] are useful methods representing an important validation step in the definition of the role that glycosyltransferases play in determining the glycophenotype. These techniques require specific antibodies

Inka Brockhausen (ed.), *Glycosyltransferases: Methods and Protocols*, Methods in Molecular Biology, vol. 1022, DOI 10.1007/978-1-62703-465-4_26, © Springer Science+Business Media New York 2013

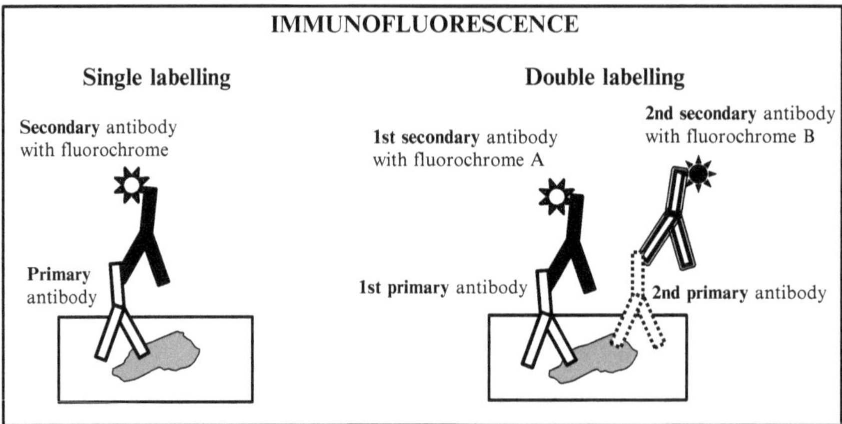

Fig. 1 Schematic representation of immunofluorescence technique for single detection of a glycosyltransferase (*left panel*) and double labelling (*right panel*) for the detection of both glycosyltransferase and its glycan product

that recognize glycosyltransferases. The majority of the antibodies used in in situ immunolabelling detects conformational epitopes and therefore requires frozen tissues [13], although few antibodies had been shown to work in paraffin sections [11, 15]. Complementary immunodetection techniques with protein extracts such as SDS-polyacrylamide gel electrophoresis (*SDS-PAGE*) and Western Blot analysis can be performed using specific antibodies directed to non conformational epitopes on glycosyltransferases [13, 16]. In this chapter we focus on the immunofluorescence method for labelling glycosyltransferases in frozen tissue sections, as well as a double immunostaining protocol used for fluorescent co-localization of glycosyltransferases and their glycan products. These techniques of immunofluorescence staining in tissue sections allow the visualization of a specific antigen using a primary antibody directed to the antigen studied followed by a specific secondary antibody covalently conjugated with a fluorochrome used to recognize the primary antibody (Fig. 1). These procedures are important tools for the identification of differentially expressed glycosyltransferases and the glycans antigens produced, which are involved in pathological conditions of the gastrointestinal tract.

2 Materials

2.1 Single Immunofluorescence

1. Adhesion Microscope Slides (Marienfeld, Lauda-Königshofen, Germany).
2. Cryostat.
3. Tris-buffer saline (TBS) 20× (*see* **Note 1**): 0.9 % NaCl (180 g), 100 mM $MgCl_2$ (9.52 g), and 200 mM Tris Base (24.22 g) in deionized water until 1 L. Adjust pH to 7.4.
4. 4 % paraformaldehyde.

5. Incubation chamber for slides (*see* **Note 2**).
6. Large glass cuvettes.
7. Hydrophobic barrier pen (Dako, Glostrup, Denmark).
8. Bovine serum albumin (BSA): 10 and 5 % in TBS.
9. Non-immune serum (Dako).
10. Primary antibodies, specific for the antigens to be detected [6, 13, 14].
11. Secondary antibodies-fluorochrome conjugated, specific to the source species of primary antibodies (Jackson ImmunoResearch Laboratories, West Grove, PA, USA).
12. Centrifuge for Eppendorf tubes.
13. DAPI (4′,6-diamidino-2-phenylindole) (Sigma, St. Louis, MO, USA) 1 μg/μL in TBS.
14. Mounting medium (Vector Laboratories, Burlingame, CA, USA).
15. Microscope Coverslips (Menzel-Glässer, Braunschweig, Germany).
16. Fluorescence microscope.

2.2 Double Immunofluorescence

1. Adhesion Microscope Slides (Marienfeld).
2. Cryostat.
3. TBS 20× (*see* **Note 1**): as described in Subheading 2.1, **item 3**.
4. 4 % paraformaldehyde.
5. Deacetylation solution: 0.1 N NaOH.
6. Neuraminidase from *Clostridium perfringens* type IV (Sigma) diluted in 0.1 M sodium acetate buffer (pH 5.5) to a final concentration of 0.1 unit/mL.
7. Incubator at 37 °C.
8. Incubation chamber for slides (*see* **Note 2**).
9. Large glass cuvettes.
10. Hydrophobic barrier pen (Dako).
11. Bovine serum albumin (BSA) 10 and 5 % in TBS.
12. Non-immune serum (Dako).
13. Primary antibodies, specific for the antigens to be detected.
14. Secondary antibodies-fluorochrome conjugated, specific to the source species of primary antibodies. (Jackson ImmunoResearch Laboratories).
15. Centrifuge for Eppendorf tubes.
16. DAPI (4′,6-diamidino-2-phenylindole) (Sigma) 1 μg/μL in TBS.
17. Mounting medium (Vector Laboratories).
18. Microscope Coverslips (Menzel-Glässer).
19. Fluorescence microscope.

3 Methods

3.1 Single Immunofluorescence

All procedures should be carried out at room temperature except when indicated.

Negative and positive controls are recommended (*see* **Note 3**).

1. Fix slides with previously cut frozen tissues (*see* **Note 4**) in freshly prepared 4 % paraformaldehyde (*see* **Note 5**) for 20 min at room temperature.
2. Wash slides twice for 5 min in TBS (*see* **Note 6**).
3. Limit the area of the tissue with a hydrophobic barrier pen (*see* **Note 7**).
4. Incubate sections (*see* **Note 8**) for 30 min with a non-immune serum (*see* **Note 9**) diluted 1:5 in TBS containing 10 % of BSA.
5. Remove non-immune serum (*see* **Note 10**) and incubate overnight at 4 °C (*see* **Note 11**) with a primary antibody (*see* **Note 12**) diluted in TBS containing 5 % of BSA, in a previously determined optimal concentration.
6. In the following day wash sections three times for 5 min in TBS.
7. Incubate fragments in the dark (*see* **Note 13**) with secondary antibody-fluorochrome conjugated specific to the source species of the primary antibody and diluted in TBS containing 5 % BSA, according to the manufacturer's instructions (*see* **Note 14**).
8. Wash sections three times for 5 min in TBS.
9. Incubate with DAPI (1 μg/μL), for 15 min (*see* **Note 15**).
10. Wash samples three times for 5 min in TBS.
11. Mount fragments in coverslips with mounting medium (*see* **Note 16**).
12. Examine immunofluorescence stained slides under a fluorescence microscope with appropriate filters for the fluorochrome used (*see* **Note 17**).

3.2 Double Immunofluorescence

All procedures should be done at room temperature except when indicated.

Negative and positive controls are recommended (*see* **Note 3**).

1. Samples designed for double-labelling immunofluorescence are fixed as described above (**step 1** of Subheading 3.1). Pretreatment of tissue samples could be necessary after fixation using for example deacetylation (saponification) protocol (*see* **Note 18**) or neuraminidase treatment (*see* **Note 19**).
2. Wash slides twice for 5 min in TBS (*see* **Note 6**).
3. Limit the area of the tissue with a hydrophobic barrier pen (*see* **Note 7**).

4. Incubate sections (*see* **Note 8**) for 30 min with a non-immune serum (*see* **Note 9**) diluted 1:5 in TBS containing 10 % of BSA.
5. Remove non-immune serum (*see* **Note 10**) and incubate overnight at 4 °C (*see* **Note 11**) with a primary antibody (*see* **Note 20**) diluted in TBS containing 5 % of BSA, in a previously determined optimal concentration.
6. In the following day wash sections three times for 5 min in TBS.
7. Incubate fragments in the dark (*see* **Note 13**) with secondary antibody with a conjugated fluorochrome (*see* **Note 21**) diluted in TBS containing 5 % BSA, according to the manufacturer's instructions (*see* **Note 14**).
8. Wash sections three times for 5 min in TBS.
9. Add to the samples non-immune serum (*see* **Note 22**) diluted 1:5 in TBS containing 10 % BSA for 20 min.
10. After remove the non-immune serum (*see* **Note 10**) incubate sections overnight at 4 °C (*see* **Note 11**) with a second primary antibody (*see* **Note 23**) diluted in PBS containing 5 % BSA.
11. Wash sections three times for 5 min with TBS.
12. Incubate with a second secondary antibody conjugated with a fluorochrome (*see* **Note 24**) and diluted in TBS containing 5 % BSA (*see* **Note 14**).
13. Wash sections three times for 5 min in TBS.
14. Incubate with DAPI (1 μg/mL), for 15 min (*see* **Note 15**).
15. Wash samples three times for 5 min in TBS.
16. Mount fragments in coverslips with mounting medium (*see* **Note 16**).
17. Observe double stained slides in fluorescence microscope with appropriate filters for all used fluorochromes (*see* **Note 17**).

4 Notes

1. A stock solution of TBS 20× may be prepared and then diluted to TBS 1× working solution.
2. Use absorbent paper soaked in water on the bottom of the incubation chamber to create a humid atmosphere during the incubation times.
3. For negative control, all the procedure are done like for the others slides but instead of primary antibody, samples are incubated with the buffer used for dilution of primary antibodies (TBS containing 5 % BSA). Positive control uses a tissue sample previously known to have a positive staining, under the same conditions of this protocol.

4. Frozen sections are obtained from tissues preserved at −80 °C or liquid nitrogen. Sections of 5 μm are acquired using a cryostat and can be stored at −80 °C.
5. Others solutions like ice-cold acetone, methanol, or ethanol 95 %, can also be used for fixation.
6. A first quick rinse step can be done just to discard the excess of reagent that remains in the section. This procedure can be done in all washing steps.
7. Hydrophobic barrier pen is used to confine the tissue zone and to limit the volume of incubating reagents. Avoid drying of specimens by adding TBS.
8. All incubation steps should use sufficient reagent to cover all the tissue. The volume could vary between 100 and 250 μL according to the size of the fragment.
9. Non-immune serum is a concentrate of immunoglobulins from the same animal species in which the secondary antibody is produced and is used to inhibit non-specific binding.
10. All excess of normal serum used for blocking should be removed cleaning around the specimen with paper without touching the tissue. No wash step is required between normal serum and primary antibody incubation.
11. An overnight incubation of approximately 16 h is done in a humid chamber at 4 °C. Some antibodies could be incubated for a shorter period of 1–3 h at room temperature according to manufacturer's instructions.
12. Primary antibodies, either monoclonal or polyclonal, can be used directed to glycosyltransferases antigens.
13. Since secondary antibodies have fluorochromes sensitive to the light all the following steps should be performed protected from light. Incubation chamber and large glass cuvettes should be covered with aluminum paper.
14. Fluorescent secondary antibody should be shortly centrifuged before use, in order to remove any precipitated material. Incubation time can vary with the antibody (45 min to 1 h 30 min) and dilutions should follow manufacturer's recommendation.
15. DAPI is used to visualize nucleus that will be stained with blue color.
16. There are some mounting media that include in their composition DAPI. If this is the case, the two previously steps can be skipped. Slides can be stored at −20 °C or 4 °C for later observation in fluorescence microscope.
17. Filters used in fluorescence microscope differ in relation to the fluorochromes.

18. In some tissues sialylated antigens can only be detected after deacetylation of sialic acid. Deacetylation of tissues is performed by treating slides after fixation with deacetylation solution for 20 min at room temperature [17].
19. Neuraminidase pretreatment is done to remove sialic acids in cases that epitope detection depends on removal of this glycan. Neuraminidase treatment can be performed using a neuraminidase from *Clostridium perfringens* type IV diluted in 0.1 M sodium acetate buffer (pH 5.5) to a final concentration of 0.1 unit/mL. The incubation is carried out for 2 h in an incubator at 37 °C. After that, two washes of 5 min are done with cold water. The control for neuraminidase treatment is incubation of the section without enzymatic pretreatment.
20. Monoclonal antibodies of specific subclasses of immunoglobulins can be used. Monoclonal and polyclonal antibodies can also be from different animal species.
21. The secondary antibody should be directed to either the immunoglobulin subclass of the primary antibody used or directed to the animal species.
22. A second incubation with non-immune serum is done if the second secondary antibody to be used was produced in a different animal species. If both secondary antibodies are from the same animal species and only differ in immunoglobulin subclass, this step is not required.
23. The second primary antibody used should be from a different subtype of immunoglobulins or from a different animal species when comparing to the first primary antibody.
24. The second secondary antibody used in double immunofluorescence should have a different fluorochrome and be directed to the animal species or immunoglobulins subclass of the second primary antibody.

Acknowledgments

This work was supported by Fundação para a Ciência e a Tecnologia—FCT: project PIC/IC/82716/2007 to CAR and grant SFRH/BD/40563/2007 to JG.

References

1. Varki A, Cummings RD, Esko JD, Freeze HH, Stanley P, Bertozzi CR, Hart GW, Etzler ME (2009) Essentials of glycobiology. Cold Spring Harbor Laboratory Press, New York
2. Fuster MM, Esko JD (2005) The sweet and sour of cancer: glycans as novel therapeutic targets. Nat Rev Cancer 5:526–542
3. Magalhães A, Reis CA (2010) *Helicobacter pylori* adhesion to gastric epithelial cells is mediated by glycan receptors. Braz J Med Biol Res 43:611–618
4. Brockhausen I (1999) Pathways of O-glycan biosynthesis in cancer cells. Biochim Biophys Acta 1473:67–95

5. Reis CA, Osorio H, Silva L, Gomes C, David L (2010) Alterations in glycosylation as biomarkers for cancer detection. J Clin Pathol 63:322–329
6. Marcos NT, Bennett EP, Gomes J, Magalhaes A, Gomes C, David L, Dar I, Jeanneau C, DeFrees S, Krustrup D, Vogel LK, Kure EH, Burchell J, Taylor-Papadimitriou J, Clausen H, Mandel U, Reis CA (2011) ST6GalNAc-I controls expression of sialyl-Tn antigen in gastrointestinal tissues. Front Biosci (Elite Ed) 3:1443–1455
7. Bennett EP, Mandel U, Clausen H, Gerken TA, Fritz TA, Tabak LA (2011) Control of mucin-type O-glycosylation—a classification of the polypeptide GalNAc-transferases gene family. Glycobiology 22:736–756
8. Carvalho AS, Harduin-Lepers A, Magalhães A, Machado E, Mendes N, Costa LT, Matthiesen R, Almeida R, Costa J, Reis CA (2010) Differential expression of alpha-2,3-sialyltransferases and alpha-1,3/4-fucosyltransferases regulates the levels of sialyl Lewis a and sialyl Lewis x in gastrointestinal carcinoma cells. Int J Biochem Cell Biol 42:80–99
9. Gram Schjoldager KT, Vester-Christensen MB, Goth CK, Petersen TN, Brunak S, Bennett EP, Levery SB, Clausen H (2011) A systematic study of site-specific GalNAc-type O-glycosylation modulating proprotein convertase processing. J Biol Chem 286:40122–40132
10. Harduin-Lepers A, Krzewinski-Recchi MA, Colomb F, Foulquier F, Groux-Degroote S, Delannoy P (2012) Sialyltransferases functions in cancers. Front Biosci (Elite Ed) 4:499–515
11. Gomes J, Marcos NT, Berois N, Osinaga E, Magalhães A, Pinto-de-Sousa J, Almeida R, Gärtner F, Reis CA (2009) Expression of UDP-N-acetyl-D-galactosamine: polypeptide N-acetylgalactosaminyltransferase-6 in gastric mucosa, intestinal metaplasia, and gastric carcinoma. J Histochem Cytochem 57:79–86
12. Marcos NT, Cruz A, Silva F, Almeida R, David L, Mandel U, Clausen H, Von Mensdorff-Pouilly S, Reis CA (2003) Polypeptide GalNAc-transferases, ST6GalNAc-transferase I, and ST3Gal-transferase I expression in gastric carcinoma cell lines. J Histochem Cytochem 51:761–771
13. Mandel U, Hassan H, Therkildsen MH, Rygaard J, Jakobsen MH, Juhl BR, Dabelsteen E, Clausen H (1999) Expression of polypeptide GalNAc-transferases in stratified epithelia and squamous cell carcinomas: immunohistological evaluation using monoclonal antibodies to three members of the GalNAc-transferase family. Glycobiology 9:43–52
14. Bennett EP, Hassan H, Mandel U, Mirgorodskaya E, Roepstorff P, Burchell J, Taylor-Papadimitriou J, Hollingsworth MA, Merkx G, van Kessel AG, Eiberg H, Steffensen R, Clausen H (1998) Cloning of a human UDP-N-acetyl-alpha-D-galactosamine: polypeptide N-acetylgalactosaminyltransferase that complements other GalNAc-transferases in complete O-glycosylation of the MUC1 tandem repeat. J Biol Chem 273: 30472–30481
15. Berois N, Mazal D, Ubillos L, Trajtenberg F, Nicolas A, Sastre-Garau X, Magdelenat H, Osinaga E (2006) UDP-N-acetyl-D-galactosamine:polypeptide N-acetylgalactosaminyltransferase-6 as a new immunohisto chemical breast cancer marker. J Histochem Cytochem 54:317–328
16. Bennett EP, Hassan H, Mandel U, Hollingsworth MA, Akisawa N, Ikematsu Y, Merkx G, van Kessel AG, Olofsson S, Clausen H (1999) Cloning and characterization of a close homologue of human UDP-N-acetyl-alpha-D-galactosamine:polypeptide N-acetylgalactosaminyltransferase-T3, designated GalNAc-T6. Evidence for genetic but not functional redundancy. J Biol Chem 274:25362–25370
17. Ogata S, Ho I, Chen A, Dubois D, Maklansky J, Singhal A, Hakomori S, Itzkowitz SH (1995) Tumor-associated sialylated antigens are constitutively expressed in normal human colonic mucosa. Cancer Res 55:1869–1874

Chapter 27

Glycosyltransferases in Chemo-enzymatic Synthesis of Oligosaccharides

Boris Tefsen and Irma van Die

Abstract

Many oligosaccharides are not commercially available, which limits studies focused on elucidation of glycan functions; therefore chemo-enzymatic methods to synthesize them can be very useful. Here, we describe the procedure to synthesize the Galα1-3GalNAcβ1-4GlcNAcβ-R (Gal-LDN) moiety, containing the Galα1-3GalNAc epitope found on the parasitic helminth *Haemonchus contortus*. An acceptor substrate providing a terminal *N*-acetylglucosamine was prepared by coupling the fluorescent hydrophobic aglycon, 2,6-diaminopyridine (DAP), to *N,N′*-diacetylchitobiose. By the subsequent action of recombinant *Caenorhabditis elegans* β1,4-*N*-acetylgalactosaminyltransferase the substrate was efficiently converted to GalNAcβ1-4GlcNAcβ-R (LDN-R). Since no recombinant α1,3-galactosyltransferase has been described that acts on terminal βGalNAc, we used bovine α1,3-galactosyltransferase to obtain a partial conversion of LDN-R to the Gal-LDN antigen. This method can be applied to synthesize any oligosaccharide, provided that specific glycosyltransferases are available, or related enzymes that can be pushed to elongate the selected acceptor.

Key words Neoglycoconjugate, 2,6-Diaminopyridine, Galactosyltransferase, Chemo-enzymatic synthesis, Gal-LDN, Helminth, Chitobiose

1 Introduction

Glycan molecules linked to proteins or lipids play important roles in cellular communication, adhesion and signalling and are key molecules in regulation of immune responses. To establish the role of individual glycans in diverse aspects of biology, the availability of oligosaccharides or neoglycoconjugates that carry defined glycan antigens is of crucial importance. Neoglycoconjugates are attractive tools to define anti-glycan responses in infection or immunization [1, 2], or to define specific carbohydrate recognition by lectin receptors that occur on many immune cells [3]. Neoglycoconjugates are also used in vaccines to elicit carbohydrate-specific antibodies that can confer protection to infection, for example to *Neisseria meningitidis and Streptococcus pneumoniae* [4, 5].

Inka Brockhausen (ed.), *Glycosyltransferases: Methods and Protocols*, Methods in Molecular Biology, vol. 1022, DOI 10.1007/978-1-62703-465-4_27, © Springer Science+Business Media New York 2013

Helminth parasites express a variety of unusual glycan antigens that are highly antigenic in infection [6]. The study of the biological properties of particular helminth glycan antigens highly depends on the availability of neoglycoconjugates, which in contrast to many mammalian-type glycans are not commercially available. A drawback for the enzymatic synthesis of unusual glycans such as helminth glycans is that not many recombinant parasite-type glycosyltransferases are available. Previously, we have synthesized LDNF antigen using β1,4-N-acetylgalactosaminyltransferase from the albumen gland of the snail *Lymnaea stagnalis* and partially purified α1,3-FucT from human milk [7] and also using the β1,4-N-acetylgalactosaminyltransferase from *C. elegans* in combination with recombinant human α1,3-fucosyltransferase VI [8]. Neoglycoconjugates containing the LDNF-moiety can be applied when antigens directly isolated from helminths are hard to obtain, as has been shown in the case of serodiagnosis of trichinellosis [9].

Recently, we showed that the Galα1-3GalNAc epitope is present on the sheep parasite *Haemonchus contortus* and that IgG antibodies against Galα1-3GalNAc were elicited during a vaccination trial with ES antigens [10]. In this chapter, we describe a chemo-enzymatic method to synthesize the neoglycoconjugate Galα1-3GalNAcβ1-4GlcNAc conjugated to DAP (Gal-LDN-DAP) in nanomolar amounts.

This synthesis starts with the chemical modification of the commercially available *N,N′*-diacetylchitobiose (chitobiose) acceptor with 2,6-diaminopyridine (DAP) to provide a fluorescent hydrophobic linker, facilitating detection and purification of the oligosaccharide products (Fig. 1), and allowing subsequent coupling to a

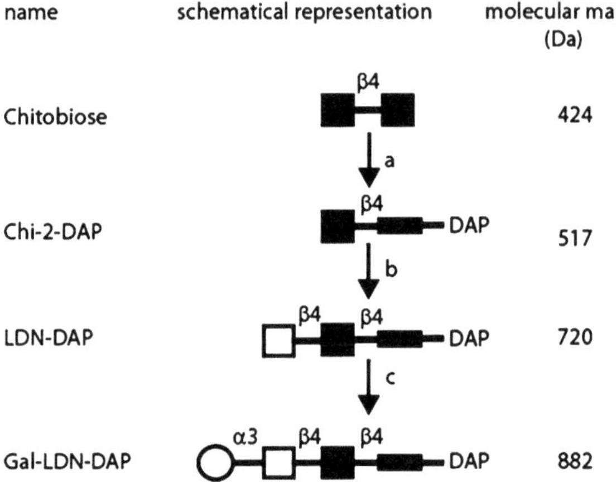

Fig. 1 Oligosaccharides used in this study. A schematic representation of the oligosaccharides mentioned in this chapter is depicted. *Filled square* (*N*-acetylglucosamine), *open square* (*N*-acetylgalactosamine), *circle* (galactose), *rectangle* (derivatized, open-chain *N*-acetylglucosamine), DAP (2,6-diaminopyridine). In addition, the abbreviations used for these oligosaccharides and their molecular masses are given

protein [11, 12]. In principle, each oligosaccharide acceptor structure can be derivatized with DAP, and after elongation by the subsequent action of appropriate glycosyltransferases, the final DAP-derivatized oligosaccharides can easily be used for different downstream applications [8, 9, 11].

Subsequently, an *N*-acetylgalactosamine residue is coupled to the DAP-derivatized chitobiose acceptor using recombinant *C. elegans* β1,4-N-acetylgalactosaminyltransferase [13] as a catalyst, resulting in a terminal GalNAcβ1-4GlcNAc moiety, which commonly is referred to as LacDiNAc (LDN). No enzymatic activity has been reported before that can catalyze the transfer of a galactose residue in α1-3 linkage to LDN. It has been described that some glycosyltransferases can be forced to act on certain acceptors that they normally would not act on [14]. Indeed, when we used bovine α1,3-galactosyltransferase [15, 16] as a catalyst on the LDN acceptor, it proved to be successful, resulting in Gal-LDN-DAP. Notably, a maximum of 10 % of LDN-DAP was transformed to Gal-LDN-DAP in a single reaction, reflecting that this is a sub-optimal reaction. The progress in synthesis of intermediate structures was monitored using normal-phase HPLC (Fig. 2). Using preparative HPLC, the synthesized oligosaccharide was purified and the structural identity of Gal-LDN-DAP was verified by ESI-MS (Fig. 3).

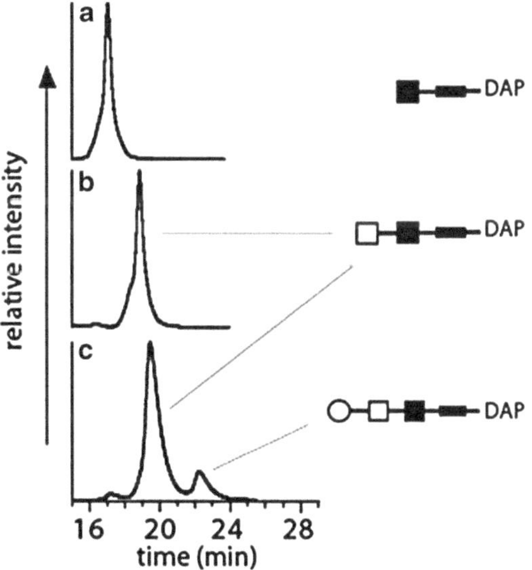

Fig. 2 Monitoring product formation of intermediate structures by HPLC. Enzymatic synthesis was monitored by analytical normal-phase HPLC. The relative intensity of DAP-derivatized glycans is plotted against time of elution from the column. The starting compound Chi-2-DAP is shown (**a**), which is converted with *C. elegans* β1,4-GalNAcT into LDN-DAP (**b**). After isolation of LDN-DAP with reversed-phase cartridges, around 10 % of this compound is converted to Gal-LDN-DAP with bovine α1,3-GalT (**c**)

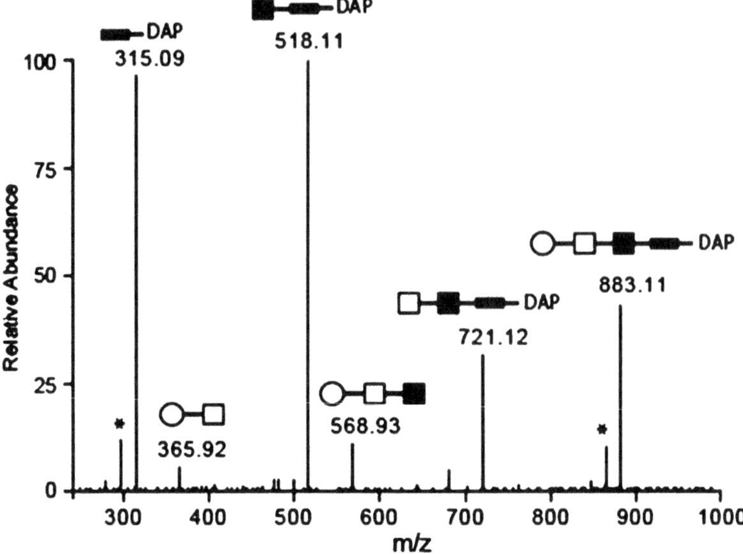

Fig. 3 Tandem mass spectrum of Gal-LDN-DAP. Characterization of Gal-LDN–DAP by tandem mass spectrometry with ESI-MS. Oligosaccharide products were characterized in the positive-ion mode. The expected molecular mass of the protonated Gal-LDN–DAP ion (*m/z* = 883.11) was found, and by fragmentation using tandem MS/MS the structure was confirmed. Schematic figures of the oligosaccharides representing the found fragments are depicted above the peaks. The *asterisks* indicate ions that are similar to the structures shown, but that lack a water molecule (–18 Da)

2 Materials

Prepare all solutions using ultrapure MilliQ water.

2.1 Derivatization of Chitobiose with 2,6-Diaminopyridine (DAP)

1. Derivatization solution: (*see* **Note 1**) mix 1,722 μL DMSO (Sigma-Aldrich, St. Louis, MO, USA) with 778 μL Acetic acid (glacial 99.99 %, Sigma-Aldrich).
2. 82.5 mg 2,6-diaminopyridine (DAP, 98 %, Sigma-Aldrich) is dissolved in 1,250 μL of the Derivatization solution.
3. 137.4 mg of sodium cyanoborohydride ($NaCNBH_3$, Sigma-Aldrich) is dissolved in 1,250 μL of the Derivatization solution (heat in heatblock at 65 °C until completely dissolved).
4. *N,N′*-Diacetylchitobiose (chitobiose, Sigma-Aldrich).
5. 12 cc reversed-phase C_{18} Sep-Pak columns (Waters, Milford, MA, USA).
6. Preparative normal-phase column (Zorbax NH_2 PrepHT, 250 × 21.2 mm, 7 μm, Agilent Technologies, Santa Clara, CA, USA).
7. Buffer A contains acetonitrile (LC-MS, Chromasolv, Riedel-de Haen, Seelze, Germany).

8. Buffer B contains 50 mM ammonium formate pH 4.4: dissolve 5 mL 10 M Ammonium formate solution and 200 μl Formic acid in 1 L water.
9. Äkta Explorer HPLC apparatus (Amersham Pharmacia Biotech; now merged with GE Healthcare Bio-Sciences AB, Uppsala, Sweden).
10. SpeedVac centrifuge.
11. 1.5-mL reaction tubes.
12. 15-mL tubes.

2.2 Production of Recombinant C. elegans UDP-GalNAc:GlcNAc β1,4-N-Acetylgalactosaminyltransferase

1. Human Embryonic Kidney 293T (HEK293T) (ATCC CRL 1573) cells.
2. Dulbecco's modified Eagle's medium (DMEM, Gibco, Life Technologies, Paisley, UK).
3. 10 % Fetal Calf Serum (FCS, Gibco).
4. Penicillin–Streptomycin (Lonza, Basel, Switzerland).
5. HEPES (Gibco).
6. 10 mM MEM nonessential amino acids (Gibco).
7. 100 mM MEM sodium pyruvate (Gibco).
8. Opti-MEM I (Gibco).
9. Lipofectamine 2000 (Invitrogen, Life Technologies, Paisley, UK).
10. Incubator at 37 °C/5 % CO_2.
11. PBS.
12. Trypsinization solution: 0.025 % trypsin (Gibco) and 0.01 % EDTA in PBS.
13. T75 culture flasks.

2.3 β1,4-GalNActransferase Reaction

1. GalNActransferase reaction buffer: mix an equivalent of 4 μL 1 M $MnCl_2$ (final concentration is 40 mM), 2.5 μL 4 M Sodium cacodylate ($C_2H_{12}AsNaO_5$), pH 7.0 (final concentration is 100 mM).
2. 100 mM UDP-GalNAc (final concentration is 20 mM).
3. Culture medium containing recombinant *C. elegans* UDP-GalNAc:GlcNAc β1,4-*N*-acetylgalactosaminyltransferase (β1, 4GalNAcT) (Subheading 3.2, **step 11**).

2.4 α1,3-Galactosyltransferase Reaction

1. Buffer for Galactosyltransferase reaction: mix an equivalent of 8 μL 1 M $MnCl_2$ (final concentration is 20 mM), 5 μL 4 M Sodium cacodylate ($C_2H_{12}AsNaO_5$) pH 7.0 (final concentration is 50 mM), 80 μL 0.5 % BSA (final concentration is 0.1 %). Add water to an end-volume of (an equivalent of) 400 μL.
2. 50 mM UDP-galactose.
3. Recombinant bovine α1,3-galactosyltransferase (*see* **Note 2**).

2.5 Monitoring of the Formation and Purification of DAP-Derivatized Glycans

1. 6 cc reversed-phase C_{18} Sep-Pak columns (Waters).
2. Methanol.
3. Normal-phase LudgerSep N1 Amide column (250×4.6 mm, Ludger, Oxfordshire, UK).
4. Surveyor HPLC (Thermo Fischer Scientific, Waltham, MA, USA).
5. Fluorimeter 470 (Waters).
6. Buffer A: Acetonitrile.
7. Buffer B: 50 mM Ammonium formate pH 4.4: dissolve 5 mL 10 M Ammonium formate solution (Sigma-Aldrich) and 200 μl Formic acid (Sigma-Aldrich) in 1 L water.

2.6 Electrospray Injection Mass Spectrometry

1. LCQ DecaXP ion trap mass spectrometer equipped with a nano-ES ionization source (Thermo Fischer Scientific).
2. 10 μL syringe.
3. Medium NanoEs spray capillary (Proxeon, Thermo Fischer Scientific).
4. 50 % acetonitrile.

3 Methods

3.1 Derivatization of Chitobiose with 2,6-Diaminopyridine (DAP)

1. Mix the derivatization solutions containing DAP and cyanoborohydride in a 1:1 ratio (*see* **Note 3**).
2. Dissolve dry, water-free pellets of chitobiose (0.4 mg, 1 μmol) (*see* **Note 4**) in 10 μL of DMSO, and subsequently add 500 μL of the mixture described in Subheading 3.1, **step 1**.
3. Incubate the reaction mixture for 18 h at 65 °C in a heat block.
4. Directly after incubation, apply the content of two tubes (containing unreacted chitobiose, Chi-2-DAP and free DAP) to one 12 cc Sep-Pak C_{18} column that was pre-activated with 5 column volumes of methanol and washed with 10 column volumes of water. Subsequently wash (*see* **Note 5**) the column with 10 column volumes of water and elute the Chi-2-DAP (*see* **Note 6**) with 5 column volumes of methanol.
5. Pool the eluted Chi-2-DAP coming from two C_{18} columns in a 15-mL tube and dry in a SpeeVac; add 0.5 mL water to the resulting pellet (contains approximately 4 μmol Chi-2-DAP and excess free DAP) and let his stand for 10 h at 4 °C (*see* **Note 7**).
6. Apply 0.5 mL from the dissolved material on the preparative normal-phase column. (*see* **Note 8**) Use a gradient that starts with 81 % acetonitrile (buffer A) and 29 % 50 mM ammonium formate, pH 4.4 (buffer B) and ends after 17 min with 30 % buffer A and 70 % buffer B on an Äkta Explorer with a column

flow of 10 mL/min. DAP-containing fractions are detected by measuring absorbance at 235 nm.

7. Collect the peak containing Chi-2-DAP (eluting at min 11.5) in a 15-mL tube (*see* **Note 9**).
8. Freeze-dry the collected Chi-2-DAP and store at −20 °C (*see* **Note 10**).

3.2 Production of Recombinant C. elegans UDP-GalNAc: GlcNAc β1,4-N-Acetylgalactosaminyltransferase (β1,4GalNAcT)

1. Culture adherent HEK293T cells in DMEM containing 10 mM HEPES, 100 μM MEM nonessential amino acids, 1 mM MEM sodium pyruvate, 10 % FCS, and 100 U/mL penicillin–streptomycin in a T75 flask at 37 °C/5 % CO_2 until the bottom is completely covered by a monolayer of cells (~6×10^6 cells/flask).
2. Prepare a transfection mixture by mixing 600 μL Opti-MEM I containing 54 μL Lipofectamine 2000 with 600 μL Opti-MEM I containing 9 μg DNA of plasmid pCMV-SH-Ceβ4GalNAcT [13].
3. Incubate the transfection mixture for 30 min at room temperature.
4. Wash the HEK293T cells twice with 9 mL Opti-MEM I.
5. Mix 4.8 mL Opti-MEM I with the transfection mixture.
6. Add this mixture to the cells and incubate the cells for 5–6 h at 37 °C/5 % CO_2.
7. Add 6 mL DMEM and culture the cells overnight at 37 °C/5 % CO_2.
8. Wash the cells with 2 mL PBS.
9. Treat the cells with 1 mL Trypsinization solution and transfer to a new T75 flask.
10. Culture the cells in 9 mL DMEM for another 3 days.
11. Harvest medium containing the active enzyme daily until 7 days after transfection.
12. Determine the activity of the enzyme by monitoring the incorporation of radiolabeled UDP-GalNAc in time [13] (*see* **Note 11**).
13. Store at −20 °C (*see* **Note 12**).

3.3 Addition of a β1-4GalNAc Residue to Chi-2-DAP to Create LDN-DAP

1. Dissolve 1 μmol dried chitobiose-DAP (final concentration is 10 mM) in 6.5 μL GalNActransferase reaction buffer.
2. Add 20 μL 100 mM UDP-GalNAc (final concentration is 20 mM) and 73.5 μL medium containing β1,4GalNAc. Transferase to a final volume of 100 μL.
3. Incubate the reaction mixture at room temperature (20–24 °C) until the Chi-2-DAP has been completely converted to the desired end-product, LDN-DAP.

3.4 Monitoring the Formation of LDN-DAP and Subsequent Purification

1. Position a LudgerSep N1 column in the HPLC.
2. Equilibrate the column with 100 % ammonium formate for 10 min, followed by 10 min of 70 % acetonitrile and 30 % 50 mM ammonium formate.
3. Dissolve 0.5 μL of the 100 μL reaction mixture of Subheading 3.3, **step 1** in 100 μL 80 % acetonitrile.
4. Inject 25 μL of this dilution onto the column.
5. Run a 30-min gradient starting with 70 % acetonitrile and 30 % 50 mM ammonium formate, pH 4.4 and ending in a 50:50 ratio at 22 °C.
6. Detect the glycan-DAP peaks by fluorescence in a fluorimeter by excitation at 345 nm and measuring the emission at 400 nm with a ten times gain of signal. The Chi-2-DAP peak elutes in this system around minute 17.8, whereas the LDN-DAP elutes around minute 19.8 (*see* Fig. 2).
7. Purify LDN-DAP when conversion is complete from the reaction mixture from Subheading 3.3, **step 2** using a 6 cc C_{18} reversed-phase Sep-Pak column as described in Subheading 3.1, **step 4**.
8. Dry the eluted LDN-DAP in the SpeedVac and store it at −20 °C.

3.5 Addition of an α1-3Galactose Residue to LDN-DAP to Create Gal-LDN-DAP

1. Dissolve 1 μmol dried LDN-DAP (final concentration is 1 mM) in 262.5 μL Galactosyltransferase buffer.
2. Add 377.5 μL water, 30 μL 50 mM UDP-galactose (final concentration is 1.5 mM) and 360 μL α1,3-galactosyltransferase to a final volume of 1 mL (*see* **Note 13**).
3. Incubate the mixture at 37 °C until no further enhancement of the desired end-product, Gal-LDN-DAP, is observed (*see* **Note 14**).

3.6 Monitoring the Formation of Gal-LDN-DAP and Subsequent Purification

1. The formation of Gal-LDN-DAP is monitored in time as described in Subheading 3.4.
2. The LDN-DAP elutes around minute 19.8 and Gal-LDN-DAP around minute 22.8 (*see* Fig. 2).
3. When the reaction from Subheading 3.5, **step 3** is stopped, purify the mixture of LDN-DAP and Gal-LDN-DAP with a 6 cc Sep-Pak as described in Subheading 3.1, **step 4**.
4. Dissolve the pellet containing this mixture in water and apply it to the preparative column, similarly as described for the purification of Chi-2-DAP in Subheading 3.1, **step 6**.
5. Collect the fractions where Gal-LDN-DAP elutes (*see* **Notes 15** and **16**). LDN-DAP starts to elute around minute 14.5 and Gal-LDN-DAP elutes around minute 15.5.
6. Dry the eluted DAP-derivatized glycans in the SpeedVac and store at −20 °C.

3.7 Confirmation of the Formation of Gal-LDN-DAP by Mass Spectrometry

1. Characterize product formation by ESI–MS on an LCQ DecaXP ion trap mass spectrometer equipped with a nano-ES ionization source.
2. Take 0.5 μL of a collected fraction from the Surveyor run, containing Gal-LDN-DAP, and dilute in 50 μL 50 % acetonitrile.
3. Inject a few μL of this sample into a medium NanoEs spray capillary (Proxeon) using the 10 μL syringe.
4. Mount the capillary in front of the ion-trap and set the capillary temperature to 200 °C.
5. Take spectra in the positive ion mode with a spray voltage of 1.0 kV and a capillary voltage of 40.1 V and subject to tandem MS to confirm the synthesis of Gal-LDN-DAP (*see* Fig. 3).

4 Notes

1. These solutions should be made fresh on the day they will be used.
2. Initially, we could obtain the recombinant bovine α1,3-galactosyltransferase from Sigma, but this product was discontinued. At that point, we kindly received the enzyme from Dr. M.M. Palcic [16].
3. The less-toxic compound picoline borate was tested instead of cyanoborohydride for the coupling of DAP to chitobiose, but it inhibited subsequent enzymatic reactions and was therefore discarded.
4. We originally started the whole procedure described here using *N*,*N*′,*N*″-triacetylchitotriose as the starting acceptor to have a greater distance between the immunogenic epitope of the neoglycoconjugate and the carrier protein. However, degradation of this compound by a β-*N*-acetylhexosaminidase present in Fetal Calf serum made this strategy impossible.
5. This step removes free chitobiose and salts.
6. Free DAP molecules will also be eluted from the column.
7. This step can be shortened, but then the amount of dissolved Chi-2-DAP will be less.
8. This step separates free DAP molecules from DAP-derivatized glycans (*see* Fig. 2 in ref. [8]).
9. Characterization by ESI-MS of the intermediate DAP-derivatized glycans, Chi-2-DAP and LDN-DAP is described in reference [8].
10. No loss of material was observed after 3 months of storage at −20 °C.

11. The enzyme-containing medium showed an activity of approximately 200 nmol/mL/h using acceptor *p*-nitrophenyl-*N*-acetyl-1-thio-β-D-glucosaminide.
12. The enzyme was stored at −20 °C for several months without loss of activity.
13. Galactosyltransferase activity of the bovine α1,3-galactosyltransferase was determined essentially as described previously [17].
14. This was after approximately 72 h. Longer incubations never yielded more than approximately 10 % of LDN-DAP to be elongated with a galactose molecule.
15. Gal-LDN-DAP purified in this way can still contain between 5 and 10 % of the starting material LDN-DAP.
16. The elutions containing LDN-DAP can be collected, dried, and reused starting from Subheading 3.5, **step 2**.

Acknowledgments

This work was supported by the Dutch Technology Foundation (STW). We thank Dr. M.M. Palcic (Carlsberg Laboratory, Denmark) for kindly providing the recombinant bovine α1,3-galactosyltransferase.

References

1. Vervelde L, Bakker N, Kooyman FN, Cornelissen AW, Bank CM, Nyame AK, Cummings RD, van Die I (2003) Vaccination-induced protection of lambs against the parasitic nematode *Haemonchus contortus* correlates with high IgG antibody responses to the LDNF glycan antigen. Glycobiology 13:795–804
2. van Remoortere A, van Dam GJ, Hokke CH, van den Eijnden DH, van Die I, Deelder AM (2001) Profiles of immunoglobulin M (IgM) and IgG antibodies against defined carbohydrate epitopes in sera of Schistosoma-infected individuals determined by surface plasmon resonance. Infect Immun 69:2396–2401
3. van Vliet SJ, Garcia-Vallejo JJ, van Kooyk Y (2008) Dendritic cells and C-type lectin receptors: coupling innate to adaptive immune responses. Immunol Cell Biol 86:580–587
4. Cuello M, Cabrera O, Martinez I, Del Campo JM, Camaraza MA, Sotolongo F, Perez O, Sierra G (2007) New meningococcal C polysaccharide-tetanus toxoid conjugate Physico-chemical and immunological characterization. Vaccine 25:1798–1805
5. Lee LH, Lee CJ, Frasch CE (2002) Development and evaluation of pneumococcal conjugate vaccines: clinical trials and control tests. Crit Rev Microbiol 28:27–41
6. Nyame AK, Kawar ZS, Cummings RD (2004) Antigenic glycans in parasitic infections: implications for vaccines and diagnostics. Arch Biochem Biophys 426:182–200
7. van Remoortere A, Hokke CH, van Dam GJ, van Die I, Deelder AM, van den Eijnden DH (2000) Various stages of schistosoma express Lewis(x), LacdiNAc, GalNAcβ1-4 (Fucα1-3) GlcNAc and GalNAcβ1-4(Fucα1-2Fucα1-3) GlcNAc carbohydrate epitopes: detection with monoclonal antibodies that are characterized by enzymatically synthesized neoglycoproteins. Glycobiology 10:601–609
8. Tefsen B, van Stijn CM, van den Broek M, Kalay H, Knol JC, Jimenez CR, van Die I (2009) Chemoenzymatic synthesis of multivalent neoglycoconjugates carrying the helminth glycan antigen LDNF. Carbohydr Res 344:1501–1507
9. Aranzamendi C, Tefsen B, Jansen M, Chiumiento L, Bruschi F, Kortbeek T, Smith

DF, Cummings RD, Pinelli E, Van Die I (2011) Glycan microarray profiling of parasite infection sera identifies the LDNF glycan as a potential antigen for serodiagnosis of trichinellosis. Exp Parasitol 129:221–226

10. van Stijn CM, van den Broek M, Vervelde L, Alvarez RA, Cummings RD, Tefsen B, Die IV (2010) Vaccination-induced IgG response to Galα1-3GalNAc glycan epitopes in lambs protected against *Haemonchus contortus* challenge infection. Int J Parasitol 40:215–222
11. Xia B, Kawar ZS, Ju T, Alvarez RA, Sachdev GP, Cummings RD (2005) Versatile fluorescent derivatization of glycans for glycomic analysis. Nat Methods 2:845–850
12. Rothenberg BE, Hayes BK, Toomre D, Manzi AE, Varki A (1993) Biotinylated diaminopyridine: an approach to tagging oligosaccharides and exploring their biology. Proc Natl Acad Sci USA 90:11939–11943
13. Kawar ZS, Van Die I, Cummings RD (2002) Molecular cloning and enzymatic characterization of a UDP-GalNAc:GlcNAcβ-R β1, 4-N-acetylgalactosaminyltransferase from *Caenorhabditis elegans*. J Biol Chem 277: 34924–34932
14. Palcic MM (2011) Glycosyltransferases as biocatalysts. Curr Opin Chem Biol 15:226–233
15. Joziasse DH, Shaper JH, Van den Eijnden DH, Van Tunen AJ, Shaper NL (1989) Bovine α1-3-galactosyltransferase: isolation and characterization of a cDNA clone. Identification of homologous sequences in human genomic DNA. J Biol Chem 264:14290–14297
16. Sujino K, Malet C, Hindsgaul O, Palcic MM (1997) Acceptor hydroxyl group mapping for calf thymus α1-3-galactosyltransferase and enzymatic synthesis of α-D-Galp-(1–3)-β-D-Galp-(1–4)-β D-GlcpNAc analogs. Carbohydr Res 305:483–489
17. Joziasse DH, Shaper NL, Salyer LS, Van den Eijnden DH, van der Spoel AC, Shaper JH (1990) α1-3-galactosyltransferase: the use of recombinant enzyme for the synthesis of α-galactosylated glycoconjugates. Eur J Biochem 191:75–83

Chapter 28

Tumor Targeting by a Carbohydrate Ligand-Mimicking Peptide

Shingo Hatakeyama, Toshiaki K. Shibata, Yuki Tobisawa, Chikara Ohyama, Kazuhiro Sugihara, and Michiko N. Fukuda

Abstract

Annexin A1 (Anxa1) is a highly specific surface marker of tumor vasculature. We used peptide-displaying phage technology to identify a carbohydrate ligand-mimicking 7-mer peptide, IFLLWQR (IF7), which can target Anxa1 in tumor vasculature. Here, we describe the binding activity of carbohydrate to Anxa1, Anxa1 to heparan sulfates, and the therapeutic potential of IF7 conjugated with anticancer drugs in tumor targeting.

Key words Annexin A1, Carbohydrate ligand-mimicking peptide, Lewis A, Heparan sulfate, Tumor targeting

1 Introduction

Although cancer malignancy is closely associated with carbohydrate structures found on tumor surfaces [1, 2], carbohydrate-based drug discovery has not been explored due to the technical difficulty involved in using chemically synthesizing complex carbohydrate structures. To improve this status, we employed peptide-displaying phage technology to identify peptides that act as carbohydrate ligands [3, 4]. During our experiments, we identified the IFLLWQR peptide (IF7) bound to annexin A1 (Anxa1) [5]. Anxa1 is a highly specific surface marker of tumor vasculature in several tumor types seen in both mice and humans [6]. Because IF7 is known to bind to Anxa1 fragments [5], we hypothesized that IF7 has potential to address the relationships between Anxa1 and carbohydrate in cancer, and it could serve as a tumor-specific drug delivery vector. We showed that IF7 mimicked the Lewis A oligosaccharide and Anxa1 bind to heparan sulfates on cell surface. Because it is well known that heparan sulfate regulates tumorigenesis, progression and metastasis [7, 8], and highly sulfated heparan

Inka Brockhausen (ed.), *Glycosyltransferases: Methods and Protocols*, Methods in Molecular Biology, vol. 1022, DOI 10.1007/978-1-62703-465-4_28, © Springer Science+Business Media New York 2013

sulfate were identified as preferential ligands of Anxa1 [9], the elongation of heparan sulfate by glycosyltransferases (EXT1/EXT2) has potential to play a key role for cancer biology though Anxa1 expression. In addition, we demonstrated that the specific Anxa1-binding peptide, IF7, is a highly efficient vector for targeted anticancer drug delivery in vivo for tumors in mice. IF7 conjugated with fluorescent Alexa Fluor 488 (IF7-A488) is known to bind to recombinant Anxa1 protein in a plate assay; in vivo imaging through a dorsal skinfold chamber window [10] showed accumulation of IF7-A488 in the tumors of mice. Moreover, tail vein injection of anticancer drugs (SN-38) conjugated with IF7 reduced tumor growth significantly. These results suggest that IF7 serves as an efficient drug delivery vector by targeting Anxa1 expressed on the surface of tumor vasculature. In this chapter we describe assays to analyze the binding activity of carbohydrate to Anxa1 and how to assess the therapeutic potential of IF7 conjugated with anticancer drugs in tumor targeting in mice.

2 Materials

2.1 Carbohydrate Binding Assays

1. Anxa1-His$_6$ stock solution: Full-length cDNA encoding Anxa1 was obtained from Invitrogen (Grand Island, NY, USA) and subcloned into the pET29a vector (Novagen, EMD Millipore Corp., Billerica, MA, USA) to produce an IF7-His$_6$ fusion protein. Recombinant proteins were purified by Ni+ affinity chromatography.
2. Superblock (10 %) (Takara Bio Company, Tokyo, JAPAN).
3. 96 well plates (Greiner bio-one North America Inc., Monroe, NC, USA).
4. 384 well plates (Greiner).
5. Blocking solution: 50 mM Tris–HCl buffer, pH 7.4 containing 0.15 M NaCl and 1 % bovine serum albumin.
6. Synthetic carbohydrates conjugated with biotinylated polyacrylamide (PAA) (Glycotech, Rockville, MD, USA).
7. TBSC solution: TBS containing 1 mM $CaCl_2$.
8. Peroxidase-conjugated streptavidin (Invitrogen, Grand Island, NY, USA).
9. Peroxidase substrate ABTS (Pierce, Rockford, IL, USA).
10. ELISA plate reader (Molecular Devices, Sunnyvale, CA, USA).
11. Peptides are synthesized by GenScript (Piscataway, NJ, USA). RQWLLFI peptide (RQ7) was used as a control as it has a sequence, which is reverse to that of IF7. To improve the solubility of the compounds, two arginine residues are added after the cysteine on IF7 and RQ7 to create IFLLWQR-C-RR [IF7(RR)] and RQWLLFI-C-RR [RQ7(RR)], respectively.

2.2 Anxa1 Binding to Heparan Sulfates

1. Mouse F2 cells, DMEM medium with 10 % fetal bovine serum (FBS) (Invitrogen, Grand Island, NY, USA).
2. 24-well dishes (Greiner).
3. Heparitinase (Seikagaku, Tokyo, Japan).
4. Bovine serum albumin (BSA), phosphate buffered saline (PBS).
5. Cell dissociation solution (EMD Millipore).
6. Anti-heparan sulfate antibody (10E4 clone, Seikagaku Corporation).
7. Rabbit anti-Anxa1 polyclonal antibodies (Invitrogen).
8. Alexa 647 conjugated anti-mouse IgM (Invitrogen).
9. Alexa 488 conjugated anti-rabbit IgG (Invitrogen).
10. FACS caliber (BD Biosciences pharmingen, San Diego, CA, USA).

2.3 IF7 And RQ7 Peptide Conjugation with Alexa Fluor 488 and an Anticancer Drug (SN-38)

1. Peptides: Peptides IF7, IF7(RR), RQ7, and RQ7(RR) were chemically synthesized as a custom order (GenScript, Piscataway, NJ, USA).
2. Solutions: Alexa Fluor 488 C5-maleimide (Invitrogen), SN-38 (Yakult Honsha, Tokyo, Japan), dimethylformamide (DMF), acetonitrile, trifluoroacetic acid (TFA), tranexamic acid, succinimidyl-4-(*N*-maleimidomethyl) cyclohexane-1-carboxylate (SMCC) (Pierce Biotechnology Inc. Rockford, IL, USA), dichloromethane (DCM), methanol, 0.1 M NaCl, 1 M Na-phosphate buffer pH 6.5, methyl tert-butyl ether (MTBE), O-(1H-6-chlorobenzotriazole-1-yl)-1, 1, 3, 3-tetramethyluronium (TCTU, Sigma-Aldrich, St. Louis, MA, USA), *N*-methyl pyrrolidone (NMP), *N,N-diisopropylethylamine* (DIPEA).
3. HPLC (Shimadzu LC-10 AD, Shimadzu Precision Instruments, Torrance, CA, USA) equipped with a C18 reverse-phase column (10 × 150 mm).
4. ESI, MALDI-TOF mass spectrometer (Waters, Micromass ZQ 2000) with MASSLYNX ver. 3.5 (Waters Corp., Milford, MA, USA).

2.4 Dorsal Skinfold Chamber Window

1. LLC cells (cultured in DME high glucose medium containing 10 % fetal bovine serum).
2. Athymic nude mice (aged 8–10 weeks, female, Balb/c) (*see* **Note 1**).
3. Anesthesia: 1.25 % 2,2,2-tribromoethanol (Sigma-Aldrich), syringes (1 mL), 26G needle, 70 % alcohol for disinfection, sterile surgical instruments [4–0 nylon strings, scissors, forceps, suture needle (size 20), and needle holder], and rubber gloves.
4. Special mounting stage: a handmade stage for holding a mouse (made from laboratory tubes or plastic instruments).

5. Dorsal skinfold chamber: Small dorsal kit (APJ Trading Co, Inc, Ventura, CA, USA), sterile coverslips (12 mm microscope glass cover, round).
6. Microscope and camera: Zeiss Axioplan fluorescence microscope and a digital camera system (DP70 and DP controller, Olympus, Centre Valley, PA, USA).
7. Antibody for inhibition assay: rabbit anti-Anxa1 antibody (N-19, Santa Cruz) and 28G syringes for tail vein injection.
8. 30G 1/2 needle, Silastic Lab tubing (0.012 × 0.025 in.) (VWR International, West Chester, PA, USA).
9. Signal intensity measurement: Image J (NIH, Bethesda, MD, USA).
10. DAPI stain (Vector laboratories, Burlingame, CA, USA).

2.5 Generation of Luciferase-Expressing HCT116-Luc Cells

1. HCT116 cells (cultured in DME high glucose medium containing 10 % fetal bovine serum) (Invitrogen).
2. Lentiviral plasmid vector CSII-Luc (constructed by Dr. Nikunj Somia, University of Minnesota, and provided by Dr. Renate Konig of Sanford-Burnham Medical Research Institute, La Jolla, CA, USA).

2.6 HCT116-Luc Tumor Monitoring by Luciferase Assay

1. Lewis lung carcinoma (LLC) and human colon carcinoma (HCT116) cells.
2. PBS, trypsin, DME medium, Matrigel (Becton-Dickinson, San Diego, CA, USA).
3. Athymic nude mice (aged 8–10 weeks, female, Balb/c).
4. Luciferin, isoflurane gas, Xenogen IVIS 200 imager (Caliper Life Sciences, Hopkinton, MA, USA).

3 Methods

3.1 Carbohydrate Binding Assay to Anxa1-His$_6$ Protein

1. Dilute Anxa1-His$_6$ stock solution (10 mg/mL, 63 μL) with 1 mL water.
2. Add 50 μL diluted Anxa1-His solution to 384-wells plate (31.25 ng/well) at 4 °C for 30 min.
3. Wash wells with water and block with 10 % superblock without Tween20. Anxa1-His$_6$ protein was coated on 96 wells plastic plate.
4. Each well is blocked with 50 mM Tris–HCl buffer, pH 7.4, containing 0.15 M NaCl (TBS) and 1 % bovine serum albumin.
5. Each synthetic carbohydrate conjugated with biotinylated polyacrylamide (PAA) is dissolved in TBS containing 1 mM $CaCl_2$ (TBSC), and is reacted with Anxa1-His$_6$ at room temperature for 15 min.

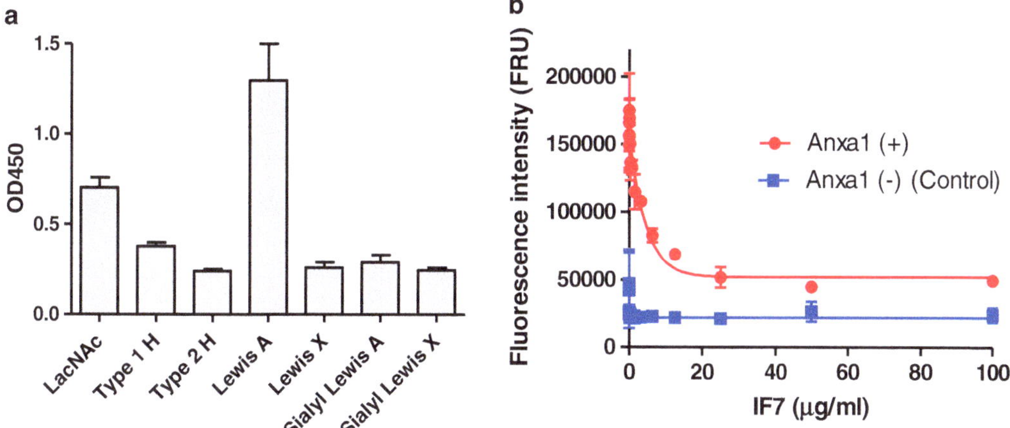

Fig. 1 Binding assay of carbohydrate antigens to Anxa1. (**a**) Binding of carbohydrate antigens conjugated to biotinylated polyacrylamide (PAA) with Anxa1-His protein. Recombinant Anxa1-His protein was coated on plastic wells and the biotinylated PAA-carbohydrate was added to the wells. *LacNAc*, Galβ1→4GlcNAcβ1→ PAA; *Type 1H*, Fucα1→2Galβ1→3GlcNAcβ1→PAA; *Type 2H*, Fucα1→2Galβ1→4GlcNAcβ1→PAA; *Lewis A*, Galβ1→3(Fucα1→4)GlcNAcβ1→PAA; *Lewis X*, Galβ1→4(Fucα1→3)GlcNAcβ1→PAA; *Sialyl Lewis A*, NeuNAcα2→3Galβ1→3(Fucα1→4)GlcNAcβ1→PAA; *Sialyl Lewis X*, NeuNAcα2→3Galβ1→4(Fucα1→3) GlcNAcβ1→PAA. (**b**) IF7C inhibition of FITC-Lewis A-PAA binding to Anxa1-His protein. Effect of IF7 (*red*) and control (*blue*) on binding of FITC-labeled polyacrylamide-Lewis A oligosaccharide to Anxa1-His protein

6. Wells are washed with TBSC, bound with peroxidase-conjugated streptavidin, and reacted with peroxidase substrate ABTS and read by an ELISA plate reader.
7. Analysis is performed in duplicate determinations. Error bars represent range (Fig. 1a). These data suggested strong candidate for anxa1 was Lewis A oligosaccharide. Wells coated with GST-His showed absorbance less than 0.113 (data not shown).

3.2 IF7C Inhibition of FITC-Lewis A-PAA Binding to Anxa1-His Protein

1. Dilute Anxa1-His_6 stock solution (10 mg/mL, 63 μL) with 1 mL water.
2. Add 50 μL diluted Anxa1-His solution to 384-wells plate (31.25 ng/well) at 4 °C for 30 min.
3. Wash wells with water and block with 10 % superblock without Tween 20.
4. Dilute 25 μL IF7C in 10 % superblock/TBSC.
5. Add 25 μL FITC-LeA-PAA (2 μg/mL) in 10 % superblock/TBSC.
6. Incubate at room temperature for 15 min.
7. Wash wells with TBSC three times and read fluorescence by plate reader (Fig. 1b). The data shown in Fig. 1b suggest that FITC-labeled polyacrylamide-Lewis A oligosaccharide binding was inhibited by the IF7 peptide.

3.3 Anxa1 Binds to Heparan Sulfates on Cell Surface

Because heparan sulfate regulates tumorigenesis, progression and metastasis [7, 8], and highly sulfated heparan sulfates were identified as preferential ligands of Anxa1 [9], the relationship between Anxa1 and heparan sulfate is determined in F2 cells.

1. F2 cells are cultured in 24-well dishes (10 % FBS-DMEM).
2. Exchange the medium to serum-free DMEM before heparitinase digestion.
3. 1 mU heparitinase is added in the wells. (BSA is used for control experiments).
4. Twenty-four hours after digestion, cells are dissociated by the cell dissociation solution.
5. The cells are washed twice with PBS.
6. The cells are reacted by anti-heparan sulfate (10E4 clone) and rabbit anti-Anxa1 antibody for 1 h on ice.
7. After washing with PBS, the cells are reacted with Alexa 647 conjugated anti-mouse IgM and Alexa 488 conjugated anti-Rabbit IgG for 1 h on ice.
8. The cells are washed twice with PBS and sequentially filtered. Cell surface Anxa1 and heparan sulfate are analyzed by FACS caliber (Fig. 2a, b) (*see* **Note 2**).

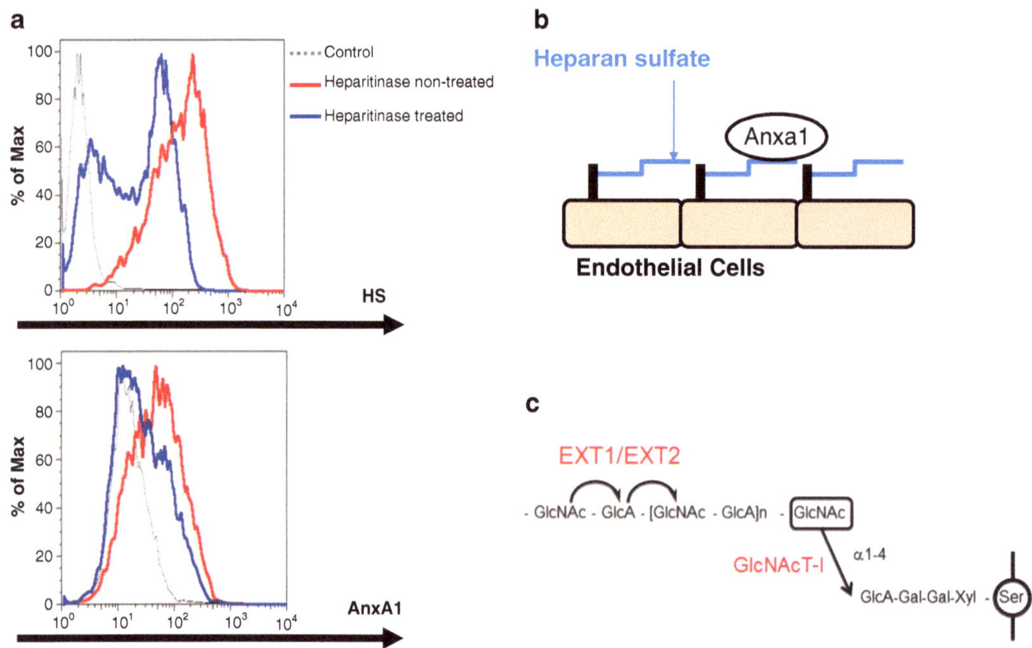

Fig. 2 Binding assay of Anxa1 to heparan sulfates on cell surface. Cell surface Anxa1 and heparan sulfate were analyzed by FACS caliber (**a**). After heparitinase digestion, the amount of cell surface heparan sulfate and Anxa1 were decreased. These data suggested the interaction between Anxa1 and heparan sulfate on cell surface (**b**), and the elongation of heparan sulfate by glycosyltransferases (EXT1/EXT2) potentially plays a key role for cancer biology (**c**)

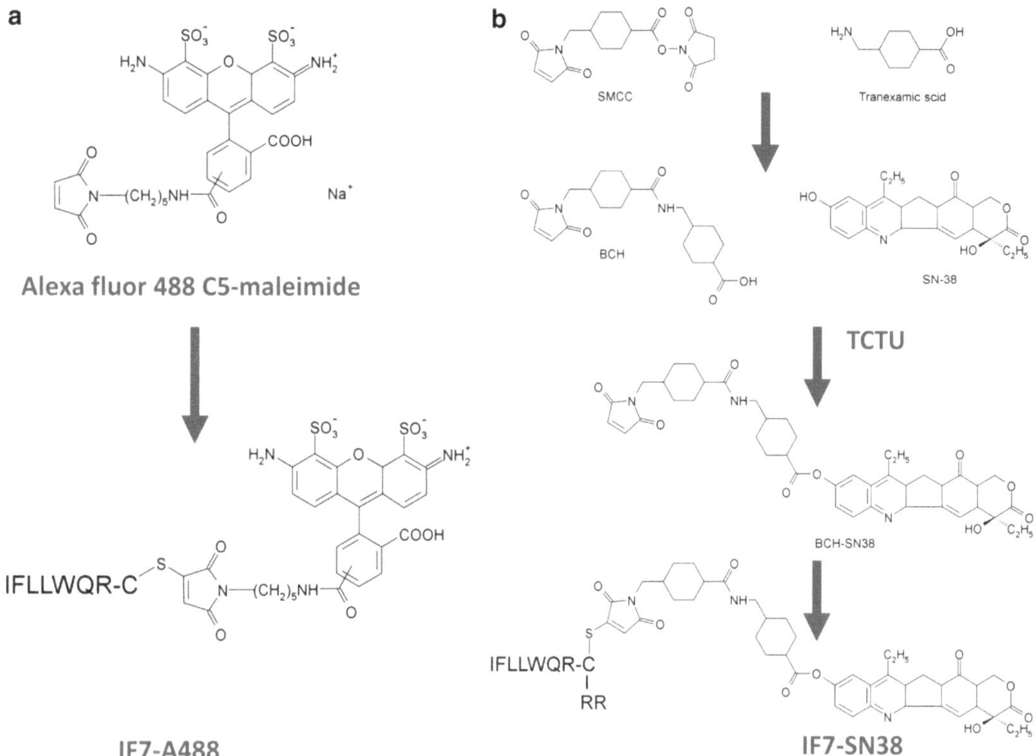

Fig. 3 IF7 peptide conjugation with Alexa Fluor 488 C5-maleimide (**a**) and IF7(RR) peptide conjugation with SN-38 (**b**). This figure shows a schematic diagram illustrating the conjugation of peptide with Alexa Fluor 488 C-5 maleimide (**a**) and a schematic diagram illustrating the conjugation of peptide with SN-38 (**b**)

3.4 IF7 Peptide Conjugation with Fluorescent Alexa 488

1. IF7 peptide (1 mg) is dissolved in DMF (0.4 mL), mixed with Alexa 488 (dissolved in 0.4 mL DMF), and incubated at room temperature for 4 h.
2. Water (2 mL) is added to the mixture and incubated at room temperature for 20 h.
3. The IF7-A488 product is purified by HPLC equipped with a C18 reverse-phase column (10 × 150 mm) using a linear gradient elution with 30–60 % acetonitrile in water containing 0.01 % TFA for 10 min at a flow rate of 2.5 mL/min.
4. The structure of the purified compound is validated by ESI mass spectrometry (Fig. 3a).

3.5 IF7 Peptide Conjugation with an Anticancer Drug (SN-38)

Peptide modification is performed in conjugation with SN-38 according to the method described by Meyer-Losic et al. [11]. Because the conjugated product (IF7-SN38) is poorly soluble in an aqueous solution, two arginine residues are added after the cysteine on IF7 to create IFLLWQR-C-RR [IF7(RR)], which thereby increased conjugate solubility.

1. Tranexamic acid (500 mg) is dissolved in water (12.5 mL), and pH adjusted by adding 1 mL 1 M Na-phosphate buffer pH 6.5.
2. SMCC (1 g) is dissolved in a mixture of acetonitrile (22 mL) and water (5.5 mL).
3. The tranexamic acid solution is then added to the SMCC solution, and the final mixture is incubated at 45 °C for 2 h.
4. Twenty milliliters of 2:1 (v/v) mixture of dichloromethane (DCM)–methanol and 60 mL DCM are added to the reaction mixture, and the organic phase is washed with water (20 mL) containing 0.1 M NaCl.
5. After removing the upper water phase, the organic phase is washed twice with water (5 mL), and evaporated using a rotary evaporator.
6. The dried material is dissolved in 9 mL of DCM–methanol (2:1) and methyl tert-butyl ether (40 mL), and left overnight at −20 °C.
7. The white precipitate 4-{4-[(*N*-maleimidomethyl)cyclohexanecarboxamido] methyl} cyclohexane-1-carboxylic acid linker (BCH) was collected by centrifugation.
8. The structure of BCH is confirmed by MALDI-TOF mass spectrometry.
9. TCTU (310 mg) is dissolved in NMP (5 mL) and DIPEA (540 μL) and added to the mixture.
10. BCH (580 mg) dissolved in NMP (5 mL) and SN-38 (450 mg) dissolved in NMP (5 mL) is then added to the mixture. The final mixture is allowed to react at room temperature for 4 h.
11. The product, BCH-SN38 conjugate, is dissolved in DCM (80 mL), which is washed thrice with 5 % aqueous citric acid and once with water.
12. The product is concentrated to 2–3 mL by a rotary evaporation, transferred to a tube to which MTBE (10 mL) is added, and precipitated at −20 °C to form a yellow precipitate.
13. BCH-SN38 (150 mg) is dissolved in DMF (1 mL) and mixed with IF7RR peptide (200 mg) dissolved in DMF (1 mL). The mixture is then incubated at 45 °C for 3 h.
14. Water (10 mL) is added to the mixture, which is then rotated at room temperature for 20 h.
15. DCM (30 mL) and 5 M NaCl (2 mL) are added to the mixture, which is then centrifuged.
16. The upper water phase is removed and the organic phase is washed thrice with water (10 mL each time).
17. IF7(RR)-BCH-SN38 forms a semi-solid interphase that is collected by decantation, suspended in water, and completely dried in vacuum.

18. Crude IF7(RR)-BCH-SN38 is dissolved in 50 % acetonitrile–water and purified by HPLC using a linear gradient elution with 50–70 % acetonitrile in water containing 0.01 % TFA over 20 min at a flow rate of 2.5 mL/min. The elution position (acetonitrile concentration) of each compound is as follows: SN-38 (50.3 %), BCH (53.5 %), BCH-SN38 (67.1 %), IF7 (48.2 %), and IF7(RR)-BCH-SN38 (59.7 %).
19. The structure of the purified compound is validated by mass spectrometry (Fig. 3b).

3.6 Surgical Preparation of the Dorsal Skinfold Chamber

The dorsal skinfold chamber is prepared according to the protocol described previously [12, 13]. We describe our protocol in brief:

1. Nude mice are anesthetized with a peritoneal injection of 1.25 % 2,2,2-tribromoethanol (25 μL/g) (*see* **Note 3**).
2. The skin on the back of each mouse is dried and disinfected with a 70 % alcohol-soaked swab (*see* **Note 4**).
3. A fold of the depilated dorsal skin parallel to the spine is then lifted up (Fig. 4a).
4. The rear of the titanium frame of the chamber is then attached to the fold on the right side of the mouse (Fig. 4b), and the mouse is temporarily fixed into position with string (Fig. 4c).
5. The mouse is then inverted (left side of the mouse facing the technician), and a resection line is marked on the skin (Fig. 4d).

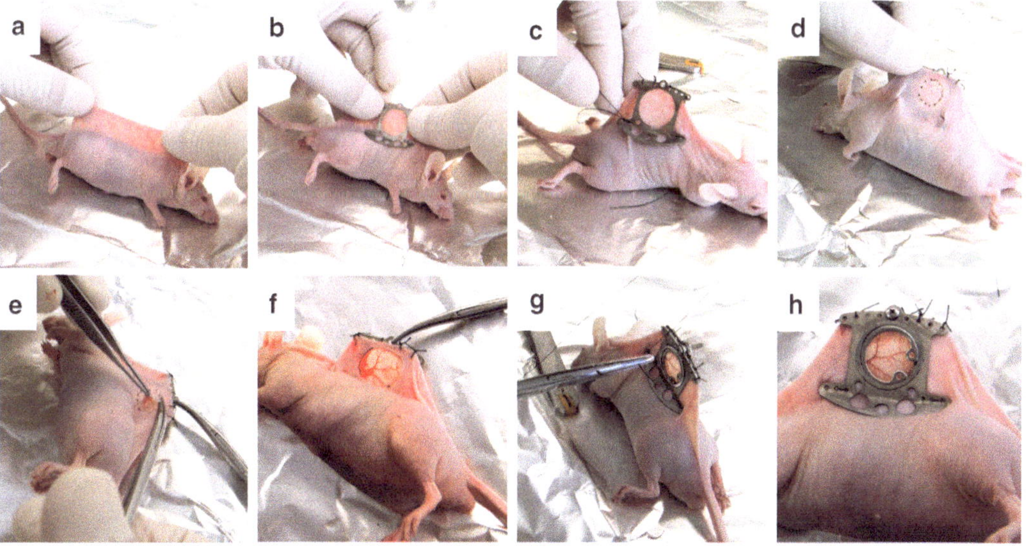

Fig. 4 Surgical preparation of the dorsal skinfold chamber. Lift up a fold of the depilated dorsal skin parallel to the spine (**a**). Place the rear part of the titanium frame of the chamber on the *right side* of the mouse (**b**), and temporarily fix it in place using string (**c**). Next, invert the mouse (*left side* of the mouse facing the technician) and mark the *resection line* (**d**). Remove the skin carefully with a pair of scissors (**e**, **f**). Place the front part of the titanium frame on the *left side* and temporarily fix it to the first frame with an upper screw and holding sutures (**g**). Hold the sandwiched skinfold together to spread out the skin and fix the edges in place using sutures (**h**)

6. The skin on the left side is then removed using a pair of scissors (Fig. 4e, f) (*see* **Note 5**).
7. The front of the titanium frame is then temporarily fixed in position, and attached to the first frame with an upper screw and holding sutures (Fig. 4g) (*see* **Note 6**).
8. Holding sutures are spread out and the edges of the sandwiched skinfold fixed in place (Fig. 4h) (*see* **Note 7**).
9. The mouse is then placed in its cage and observed until it regains consciousness. A waiting period of at least 3 days is needed before tumor implantation. The skinfold may became ischemic, and necrosis or edema occurs (*see* **Note 8**).

3.7 In Vivo Imaging in the Dorsal Skinfold Chamber Window

1. An LLC tumor is produced in a donor nude mouse by subcutaneous injection.
2. The donor mouse is then sacrificed and the LLC tumor removed under aseptic conditions using sterile surgical instruments (scissors and forceps). The LLC tumor is then placed in DMEM without serum (Fig. 5a).
3. After the tumor is removed from the donor mouse, the glass cover of the chamber window is opened and a small piece of tumor (<1 mm^3) is transplanted onto the dorsal skinfold of the recipient nude mouse (Fig. 5b) (*see* **Note 9**).

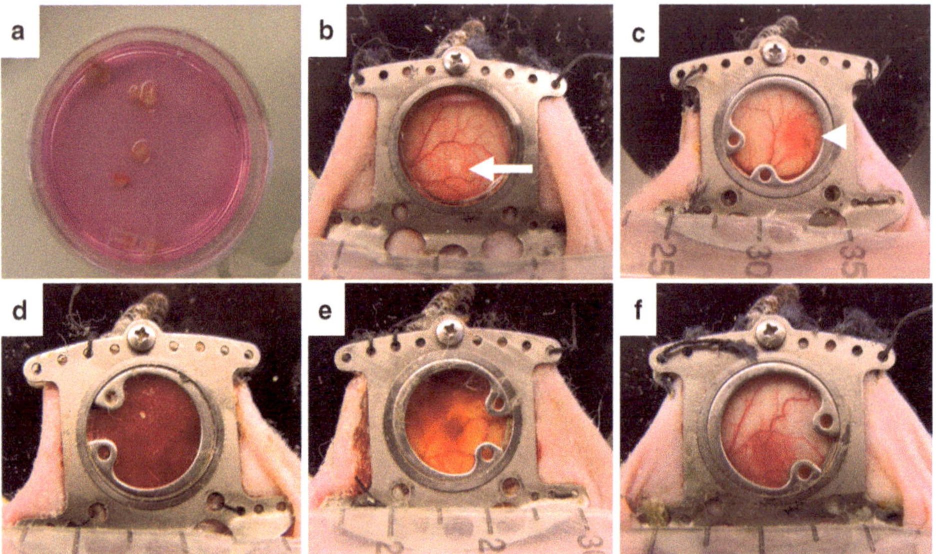

Fig. 5 Tumor transplantation from a donor mouse to the dorsal skinfold chamber. Remove a Lewis lung carcinoma (LLC) tumor produced in a donor nude mouse using subcutaneous injection and sterile techniques (**a**). Transplant a small piece of the tumor (<1 mm^3) into a dorsal skinfold chamber in a recipient nude mouse (aged 8–10 weeks, female, Balb/c) (**b**, *arrow*). Three days later, check the tumor (**c**, *arrowhead*). In some cases, bloody (**d**), edematous (**e**), or thick (**f**) tumors are observed in the skinfold chamber

4. The tumor is analyzed after 3 days (Fig. 5c). Bloody, edematous, or thick tumors are often observed due to the difficulty in creating a suitable tumor with sufficient tumor vasculature in the skinfold chamber (Fig. 5d–f, respectively).
5. The mouse is then anesthetized by peritoneal injection of 1.25 % 2,2,2-tribromoethanol (25 μL/g) and placed on the mounting stage.
6. An injection needle (30G 1/2) is then cut and connected to silastic tubing (0.012 × 0.025 in.). The other side of the needle is connected to a 26–28G needle and a 1 mL syringe filled with ultrapure water.
7. Once the mouse is placed on the mounting stage, the needle is positioned and a catheter is inserted into the tail vein (Fig. 6a, b) (*see* **Note 10**). If the technician successfully inserts the needle in the vein, back flow of blood is visible; patency is confirmed by administering 0.1–0.2 mL of the liquid in the syringe (*see* **Note 11**).
8. The tail and catheter are carefully taped to the holder (Fig. 6c, d), then the water-filled syringe is replaced with a peptide-filled syringe (IF7-A488 or RQ7-A488, 100 μL; 50 μM in 5 % glucose

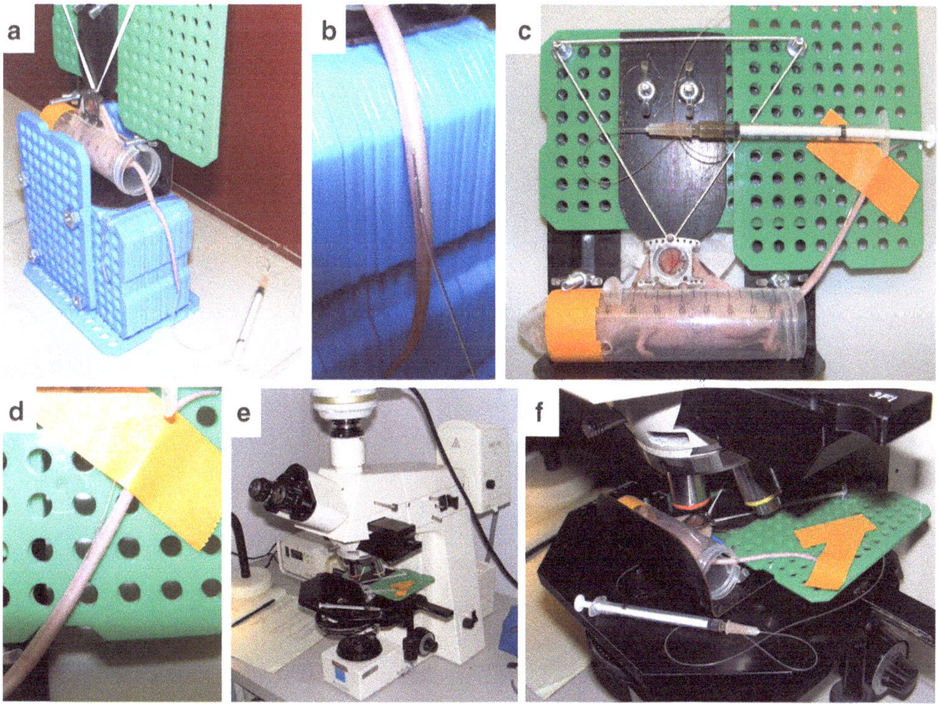

Fig. 6 The mounting stage, tail vein catheterization, and microscope settings. Insert the 30G 1/2 needle keeping the catheter in the tail vein (**a**, **b**). Tape the tail and the catheter to the holder very carefully (**c**, **d**). Next, exchange the water-filled syringe for a peptide-filled syringe (IF7-A488 or RQ7-A488, 100 μL; 50 μM in 5 % glucose solution) and place the chamber in a position for easy viewing by fluorescence microscope (**e**, **f**)

solution) and the skin chamber is assessed with the fluorescence microscope (Fig. 6e, f).

9. Tumor vasculature is initially checked by bright-field microscopy and then adjusted to zero point by fluorescence field (*see* **Note 12**).
10. IF7-A488 or RQ7-A488 is injected through the tail vein. Intravital Alexa 488 signals in the tumor are detected and recorded by a Zeiss Axioplan fluorescence microscope and a digital camera system with a quad magnification objective lens. Images are captured every 1 min from 0 to 10 min and every 5 min thereafter (Fig. 7a) (*see* **Note 13**). For the inhibition assays, 20 μg of rabbit anti-Anxa1 antibody (N-19) is injected 15 min prior to the IF7-A488 injection. Signal intensity in the tumor is measured from 0 to 40 min using Image J software. The intensity of 5–10 areas is measured and the mean calculated for each point (Fig. 7b).

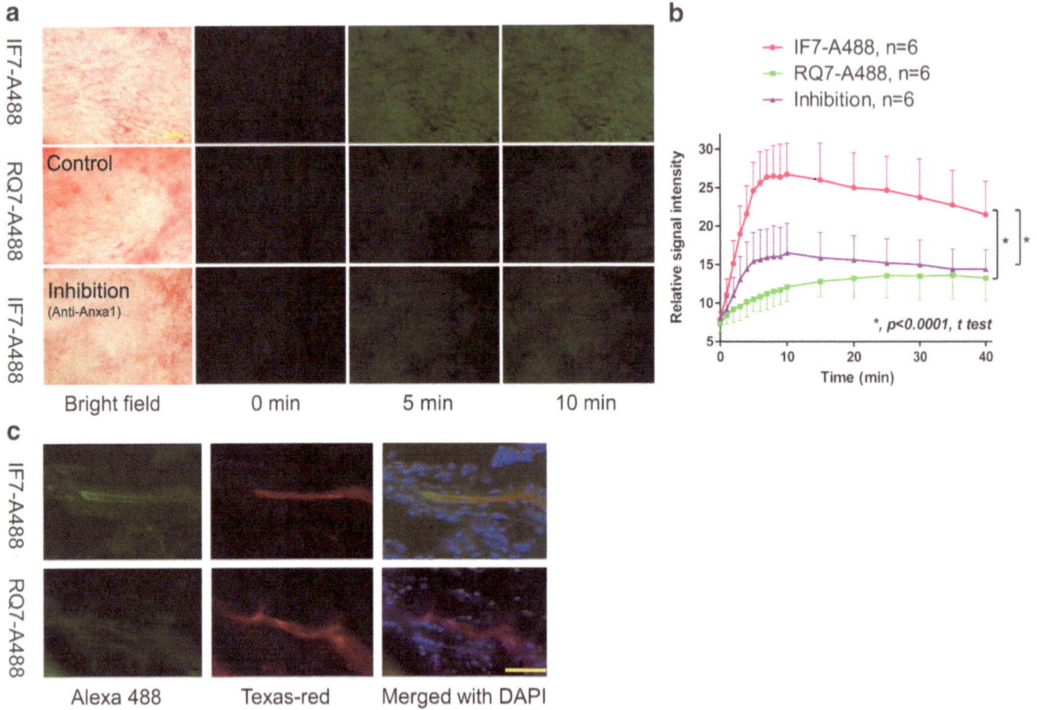

Fig. 7 In vivo imaging. When IF7-A488 was injected through the tail vein, fluorescence signals appeared in the tumor within 1 min of injection, reached a plateau after 9 min, and remained high for 40 min or until the experiment was terminated. By contrast, control peptide RQ7-A488 signals remained at background levels. When the anti-Anxa1 antibody was injected prior to IF7-A488 injection, fluorescence signals in the tumors were significantly reduced (**a**, **b**). Normal rabbit IgG (control IgG) did not alter the tumor targeting of IF7-A488 (data not shown). (**c**) Fluorescence micrographs of tumor sections 20 min after injection with IF7-A488 or RQ7-A488, coinjected with Texas-red conjugated tomato lectin. Shown are IF7-A488 or RQ7-A488 (*left*), Texas-red tomato lectin (*middle*), and the merge with DAPI nuclear staining (*right*) (scale bar; 50 μm)

11. After in vivo measurement, the tumor is isolated from the dorsal skin folder chamber, fixed with 4 % paraformaldehyde at room temperature for 15 min, immersed in Optimal Cutting Temperature (O.C.T) compound, and cryosectioned. Frozen sections are overlaid with Vectashield containing DAPI and examined under a Zeiss Axioplan fluorescence microscope (Fig. 7c).

3.8 Tumor Targeting Treatment Using IF7(RR)-SN38 for a Small Tumor

1. HCT116 cells are cultured in DME high glucose medium containing 10 % fetal bovine serum.
2. The lentiviral plasmid vector CSII-Luc and lentivirus is used to transfect HCT116 cells which are maintained as HCT116-Luc cells [14] (The lentiviral plasmid vector CSII-Luc constructed by Dr. Nikunj Somia of the University of Minnesota was provided by Dr. Renate Konig of the Sanford-Burnham Medical Research Institute).
3. HCT116-Luc cells are trypsinized and suspended in DME medium containing 1 mg/mL Matrigel placed on ice. Cell suspensions (50 μL) containing 5×10^5 cells are injected subcutaneously into 8–10-week-old female nu/nu mice.
4. After tumor injection, the mice are subjected to imaging for luciferase producing tumors; 100 μL of luciferin (30 mg/mL PBS) is injected into the peritoneal cavity of each mouse.
5. Mice are anesthetized under isoflurane gas (20 mL/min) supplemented with oxygen (1 L/min), and placed under a camera in a Xenogen IVIS 200 imager. Photons are measured (1–10 s) every 2–3 days.
6. Ten days later, the tumor-bearing mice are divided randomly into three groups and administered (1) 5 % glucose, (2) 0.81 μmol/kg (1.74 mg/kg) RQ7(RR)-SN38, or (3) 0.81 μmol/kg (1.74 mg/kg) IF7(RR)-SN38 on days 0 to 9 (10 times) (*see* **Note 14**, Fig. 8).

3.9 Tumor Targeting Treatment with IF7(RR)-SN38 for a Large Tumor

1. HCT116-Luc cells were injected subcutaneously into the dorsal flank of 8–10-week-old female nu/nu mice, and these were then subjected to imaging for luciferase producing tumors, as described previously.
2. Twenty-three days later, tumor-bearing mice were administered 6.5 μmol/kg (13.9 mg/kg) IF7(RR)-SN38. IF7(RR)-SN38 was administered on days 23–43 (20 times). Tumor size was dramatically reduced after IF7(RR)-SN38 injection (Fig. 9a, b). It is notable that the IF7-SN38 injections administered here contained 0.81 μmol/injection (1.74 mg/kg) or 6.5 μmol/kg (13.9 mg/kg), whereas the effective dose of SN-38 conjugated with a non-tumor vasculature targeting peptide was reported to be 95 mg/kg in a previous study [11] (Table 1).

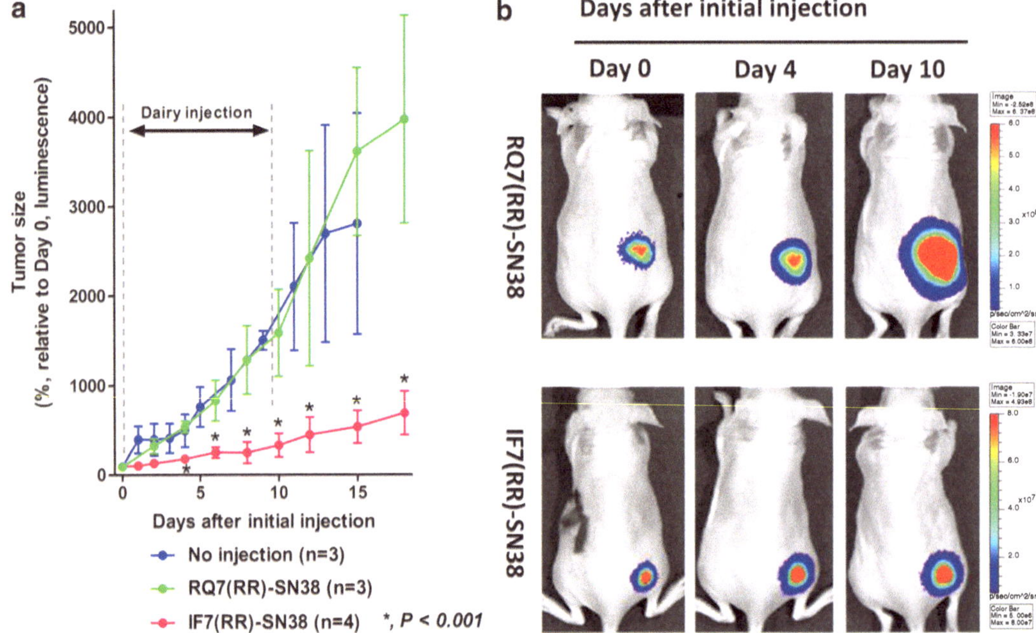

Fig. 8 Anticancer effect of IF7(RR)-SN38 on a small tumor. IF7(RR)-SN38 or RQ7(RR)-SN38 were injected into HCT116-Luc tumor bearing mice (starting from day 10 after tumor implantation) and photon numbers were measured every 2–3 days. Tumor-bearing mice were administered 0.81 μmol/kg (1.74 mg/kg) of IF7(RR)-SN38 from days 23 to 43 (20 times). In the IF7(RR)-SN38 group, tumor growth was significantly suppressed, whereas tumor growth in control mice injected with RQ7(RR)-SN38 remained unchanged (**a**). Representative photon images are shown in (**b**)

4 Notes

1. *Animal care and protocol approval.* Animal care and pain management using anesthesia and analgesia are crucial components of protocols using animals. All experiments must conform to the Principles of Laboratory Animal Care and the Guide for Care and Use of Laboratory Animals. Animal experiment protocols should be approved by the institutional animal care and use committee.

2. The data suggest the interaction between Anxa1 and heparan sulfate on cell surface (Fig. 2b). Because it is evident that heparan sulfate regulates tumorigenesis, progression and metastasis [7, 8], the elongation of heparan sulfate by glycosyltransferases (EXT1/EXT2) has potential to play a key role for cancer biology (Fig. 2c).

3. *Anesthesia in mice.* Tribromoethanol is the standard anesthetic agent used in mice. It produces short-term (15–30 min) surgical anesthesia with good muscle relaxation and moderate respiratory depression. It was once manufactured specifically for use

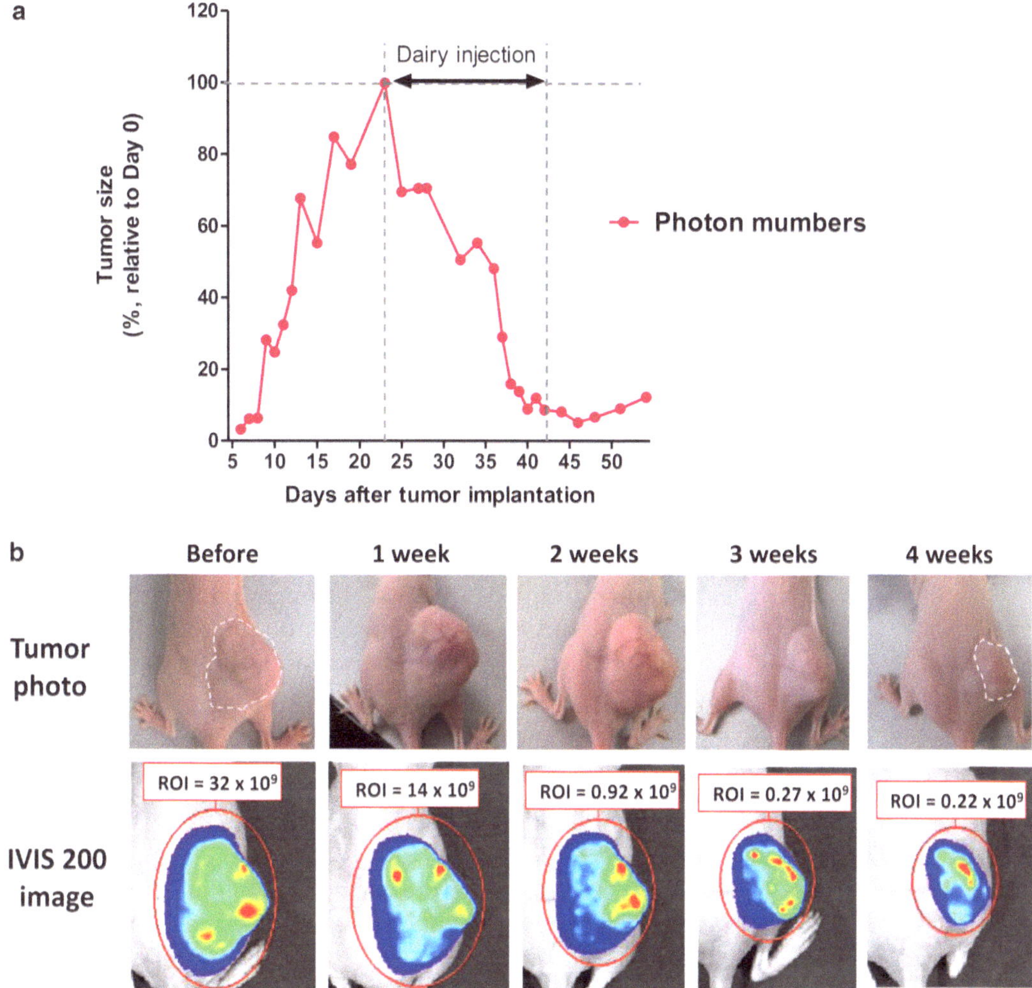

Fig. 9 Anticancer effect of IF7(RR)-SN38 on a large tumor. IF7(RR)-SN38 was injected into HCT116-Luc tumor bearing mice (starting from day 23 after tumor implantation) and photon numbers were measured every 2–3 days. Tumor-bearing mice were administered 6.5 μmol/kg (13.9 mg/kg) of IF7(RR)-SN38 from days 23 to 43 (20 times). Photon numbers and tumor size were dramatically reduced after IF7(RR)-SN38 injection (**a, b**)

Table 1
Dosage of SN-38 used in the experiments

	Anticancer drug	Dosage per single injection	Number of injection	Total dosage
Meyer-Losic et al. [8]	Irinotecan	30 μmol/kg (95 mg/kg)	3 times	90 μmol/kg
Our study	IF7(RR)-SN38	6.5 μmol/kg (13.9 mg/kg) 0.81 μmol/kg (1.74 mg/kg)	20 times for large time 10 times for small tumor	130 μmol/kg 8.1 μmol/kg

It is notable that the IF7-SN38 injections administered in this study contained 0.81 μmol/injection (1.74 mg/kg) or 6.5 μmol/kg (13.9 mg/kg), whereas the effective dose of SN-38 conjugated with a non-tumor vasculature targeting peptide was reported to be 95 mg/kg in a previous study [11]

as an anesthetic under the name Avertin®. However, this product is no longer commercially available. Investigators who wish to use tribromoethanol as an anesthetic must make their own solution. A stock solution of tribromoethanol is made by mixing equal amounts of tribromyl ethyl alcohol and tertiary amyl alcohol. This must be stored at 4 °C in the dark for no longer than 1 year. A working solution must be prepared each time by diluting the stock solution to 1.25 % using distilled water or saline. Because this agent has toxic degradation products, the technician should only use a freshly mixed solution or one that has been stored for no more than 1–2 weeks at 4 °C in the dark. Intraperitoneal injection of 400–500 μL of a 1.25 % working solution will provide adequate anesthesia for surgical experiments in mice.

4. *Disinfection of mice.* In order to prevent bacterial infection, all invasive procedures should be performed using aseptic techniques, especially when immunodeficient mice are used. Before attempting to introduce any instrument or agent into an animal's body, the injection or inoculation site should be cleansed and disinfected with 70 % ethanol or an antiseptic agent.
5. It is important to remove all thin membranes (epidermis, dermis, subcutis, and cutaneous muscle) to avoid edematous areas and to thin the skin.
6. The coverslip should be placed on the front of the frame before it is fixed to the mouse.
7. The top and bottom of the chamber should be ligated to hold the mouse in position, but care should be taken to avoid making the ligature too tight.
8. It is not easy to create suitable conditions for a skinfold chamber window. It is better to prepare 1.5–2-fold the number of mice required for the experiments.
9. It is important to take a small piece of the tumor that enables visualization of tumor vasculature.
10. Use fine forceps to hold the needle. For tail vein injection, mice should be older than 6 weeks of age because at a younger age their blood vessels are not thick enough. The most important part of the procedure is the method of holding the mouse because injection needs accurate manipulation of the needle. Three blood vessels are visible on the back of a mouse's tail: a central artery and a vein on each side. The best syringe for tail vein injection is one used for insulin injection with a 28G needle. The procedure for tail vein injection requires careful handling of the mouse and needle. Repeated practice is essential for success with this technique. For more information, please read the reference [15].

11. The catheter can easily be occluded by blood clots. To avoid clotting, flush the catheter occasionally.
12. For repeat experiments, adjust the zero point every time.
13. After 10 min, irradiation of specimens with a UV lamp is limited to only those times when photos are taken in order to avoid fluorescence bleaching.
14. The IF7(RR)-SN38 group displayed significantly suppressed tumor growth compared with the control and RQ7(RR)-SN38 group mice (Fig. 8a, b).

Acknowledgments

This study was supported by a grant-in-aid for Scientific Research 23791737 from the Japan Society for the Promotion of Science (SH) and NIH grant P01CA071932 (MNF), DoD Breast Cancer Research IDEA grant DAMD17-02-1-0311 (MNF), and a Susan Komen Breast Cancer Research grant BCTR0504175 (MNF).

References

1. Hakomori S (2002) Glycosylation defining cancer malignancy: new wine in an old bottle. Proc Natl Acad Sci USA 99:10231–10233
2. Nakamori S, Kameyama M, Imaoka S, Furukawa H, Ishikawa O, Sasaki Y, Kabuto T, Iwanaga T, Matsushita Y, Irimura T (1993) Increased expression of sialyl Lewisx antigen correlates with poor survival in patients with colorectal carcinoma: clinicopathological and immunohistochemical study. Cancer Res 53:3632–3637
3. Fukuda MN, Ohyama C, Lowitz K, Matsuo O, Pasqualini R, Ruoslahti E, Fukuda M (2000) A peptide mimic of E-selectin ligand inhibits sialyl Lewis X-dependent lung colonization of tumor cells. Cancer Res 60:450–456
4. Fukuda MN (2006) Screening of peptide-displaying phage libraries to identify short peptides mimicking carbohydrates. Methods Enzymol 416:51–60
5. Hatakeyama S, Sugihara K, Nakayama J, Akama TO, Wong SM, Kawashima H, Zhang J, Smith DF, Ohyama C, Fukuda M, Fukuda MN (2009) Identification of mRNA splicing factors as the endothelial receptor for carbohydrate-dependent lung colonization of cancer cells. Proc Natl Acad Sci USA 106:3095–3100
6. Oh P, Li Y, Yu J, Durr E, Krasinska KM, Carver LA, Testa JE, Schnitzer JE (2004) Subtractive proteomic mapping of the endothelial surface in lung and solid tumours for tissue-specific therapy. Nature 429:629–635
7. Sasisekharan R, Shriver Z, Venkataraman G, Narayanasami U (2002) Roles of heparan-sulphate glycosaminoglycans in cancer. Nat Rev Cancer 2:521–528
8. Ono K, Ishihara M, Ishikawa K, Ozeki Y, Deguchi H, Sato M, Hashimoto H, Saito Y, Yura H, Kurita A, Maehara T (2002) Periodate-treated, non anticoagulant heparin-carrying polystyrene (NAC-HCPS) affects angiogenesis and inhibits subcutaneous induced tumour growth and metastasis to the lung. Br J Cancer 86:1803–1812
9. Horlacher T, Noti C, de Paz JL, Bindschadler P, Hecht ML, Smith DF, Fukuda MN, Seeberger PH (2011) Characterization of annexin A1 glycan binding reveals binding to highly sulfated glycans with preference for highly sulfated heparan sulfate and heparin. Biochemistry 50:2650–2659
10. Lehr HA, Leunig M, Menger MD, Nolte D, Messmer K (1993) Dorsal skinfold chamber technique for intravital microscopy in nude mice. Am J Pathol 143:1055–1062
11. Meyer-Losic F, Nicolazzi C, Quinonero J, Ribes F, Michel M, Dubois V, de Coupade C, Boukaissi M, Chene AS, Tranchant I, Arranz V, Zoubaa I, Fruchart JS, Ravel D, Kearsey J (2008) DTS-108, a novel peptidic prodrug of

SN38: in vivo efficacy and toxicokinetic studies. Clin Cancer Res 14:2145–2153
12. Sckell A, Leunig M (2001) Dorsal skinfold chamber preparation in mice: studying angiogenesis by intravital microscopy. Methods Mol Med 46:95–105
13. Sckell A, Leunig M (2009) The dorsal skinfold chamber: studying angiogenesis by intravital microscopy. Methods Mol Biol 467:305–317
14. Gori JL, Podetz-Pedersen K, Swanson D, Karlen AD, Gunther R, Somia NV, McIvor RS (2007) Protection of mice from methotrexate toxicity by ex vivo transduction using lentivirus vectors expressing drug-resistant dihydrofolate reductase. J Pharmacol Exp Ther 322:989–997
15. Hatakeyama S, Yamamoto H, Ohyama C (2010) Tumor formation assays. Methods Enzymol 479:397–411

Chapter 29

Glycoengineering of Human Cell Lines Using Zinc Finger Nuclease Gene Targeting: SimpleCells with Homogeneous GalNAc O-glycosylation Allow Isolation of the O-glycoproteome by One-Step Lectin Affinity Chromatography

Catharina Steentoft, Eric Paul Bennett, and Henrik Clausen

Abstract

Lectin affinity chromatography is a powerful technique for isolation of glycoproteins carrying a specific glycan structure of interest. However, the enormous diversity of glycans present on the cell surface, as well as on individual proteins, makes it difficult to isolate an entire glycoproteome with one or even a series of lectins. Here we present a technique to generate cell lines with homogenous truncated O-glycans using zinc finger nuclease gene targeting. Because of their simplified O-glycoproteome, the cells have been named SimpleCells. Glycoproteins from SimpleCells can be isolated in a single purification step by lectin chromatography performed on a long lectin column. This protocol describes Zinc finger nuclease gene targeting of human cells to simplify the glycoproteome, as well as lectin chromatography and isolation of glycopeptides from total cell lysates of SimpleCells.

Key words Zinc finger nuclease, Glycoproteome, Lectin affinity chromatography, SimpleCell, Glycoengineering

1 Introduction

Protein glycosylation is the most abundant and diverse form of posttranslational modification [1], exhibiting enormous complexity both in regard of glycan structure as well as site of glycosylation. Although great effort has been put into the characterization of glycan structures released from the protein backbone, profiling of glycosylation-sites is hampered by technical problems. A major issue concerns mass spectrometry analysis, as complex glycans decrease ionization and fragmentation efficiency of the glycopeptides, and furthermore complicates subsequent data analysis [2, 3]. Here we present a method to genetically engineer human cell lines

Inka Brockhausen (ed.), *Glycosyltransferases: Methods and Protocols*, Methods in Molecular Biology, vol. 1022, DOI 10.1007/978-1-62703-465-4_29,

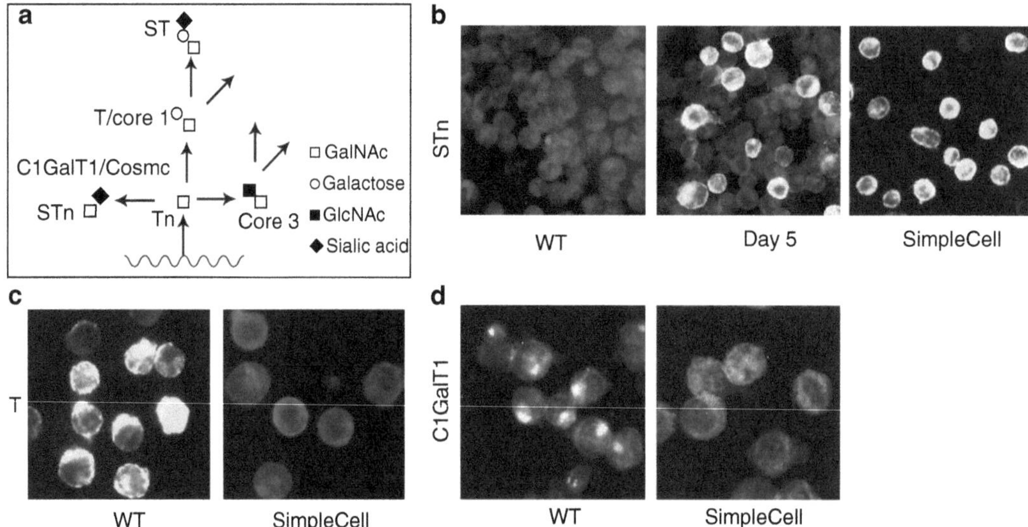

Fig. 1 Generation of SimpleCells expressing truncated homogeneous O-glycans. (**a**) O-glycosylation pathway. KO of *COSMC* results in degradation of the C1GalT1 enzyme and subsequent expression of Tn and STn glycans. The presence of core 3 glycans is still possible; however, this glycan structure is not found in most cell lines [9]. (**b**) K562 cells were screened before transfection, 5 days after transfection and after cloning with an antibody recognizing the STn antigen (MAb 3F1). Wildtype (WT) K562 cells and the final SimpleCell clone were also screened with antibodies towards the T antigen after neuraminidase treatment (MAb 3C9) (**c**) and the C1GalT1 enzyme (MAb 2G8) (**d**)

into expressing homogeneous truncated O-glycans; a goal highly desirable both in glycoproteomic studies as well as biopharmaceutical production. The method can be applied to various glycoproteomes, including those built on O-Xylose, O-Mannose, O-Glucose, and O-Fucose. Our strategy was developed for the O-GalNAc glycoproteome. We targeted *COSMC*, encoding a chaperone that is necessary for proper folding and function of the C1GalT1 enzyme [4] in mucin type O-GalNAc glycosylation. Knockout (KO) of *COSMC* prevents core 1 elongation, resulting in cell lines (designated SimpleCells) expressing only truncated Tn (GalNAcα1-O-Ser/Thr) and STn (NeuAcα2-6GalNAcα1-O-Ser/Thr) O-glycans (*see* Fig. 1).

Successful KO was achieved by the use of zinc finger nucleases (ZFNs) specifically designed to target the gene of interest. The ZFN protein consists of a DNA binding domain and a catalytic endonuclease FokI domain. When transiently expressed in the cell, heterogeneous pairs of ZFNs bind and dimerize at the specific gene target, allowing the FokI domains to introduce a double stranded break (DSB). The generated site-specific DSB is likely to be repaired by non-homologous end joining (NHEJ), which is a flawed and imprecise repair mechanism, prone to cause alterations of the DNA, e.g. loss of nucleotides. The exact size of the introduced deletions are not predetermined, but the method is efficient enough to produce stable biallelic gene knockouts in >1 % of most cell lines [5]. The exact size of the deletion created by NHEJ is subsequently validated by sequence determination of the targeted region. ZFN targeting generates stable KO cells, which is a great advantage compared to siRNA-based methods that generally have proven to be ineffective for glycoengineering, as most glycosyltransferases are stable enzymes with a long half-life. One concern regarding ZFN is off-target effects that have been recently addressed in a genome-wide study [6]. However, compared to silencing methods, the off-target effects of ZFN are better controlled for and by isolating multiple clones the issue is minimized. Recent developments in ZFN targeting design and selection methods have resulted in an increased targeting efficiency making it easier to generate multiple clones [7]. It is furthermore worth noticing that other similar methods for cell engineering are emerging such as use of chimeric transcription activator-like effector (TALE) proteins and the CRISPR/Cas9 system.

In the present chapter we describe a detailed protocol on how to generate human SimpleCell lines, and how to enrich for glycopeptides by lectin affinity chromatography by using a Tn binding *Vicia villosa* agglutinin (VVA) column. The resulting enriched glycopeptide sample may be directly subjected to mass spectrometry analysis. The strategy is shown in Fig. 2. ZFN glycoengineering and generation of SimpleCells hold great promise for further exploration of the glycoproteome, as well as investigation of the function of individual glycosyltransferases [15–17].

2 Materials

2.1 Cell Culture Materials

1. Cells: In this protocol a suspension cell line (K562) and an adherent cell line (Colo205) are used as examples of human cells. Nevertheless, the protocol can be adapted for any cell line that can be transfected with reasonable efficiency and be cloned. Cells are grown and maintained using standard cell culture methods.
2. Cell culture media: Add 10 % fetal bovine serum (FBS) (Lonza, Basel, Switzerland), and 4 mM glutamine to either RPMI (for Colo205) or DMEM (for K562). Store the glutamine stock frozen and add fresh glutamine every 10 days (In-house supplier).
3. Phosphate buffered saline (PBS) pH 7.4.
4. Trypsin: TrypLE™ Express (Gibco®, Life Technologies Europe BV Denmark).
5. Cell scraper.
6. Nucleofection solution kit: Amaxa® Cell Line Nucleofector® Kit T and Kit V (Lonza).
7. Electroporator: Nucleofector® device (Lonza).

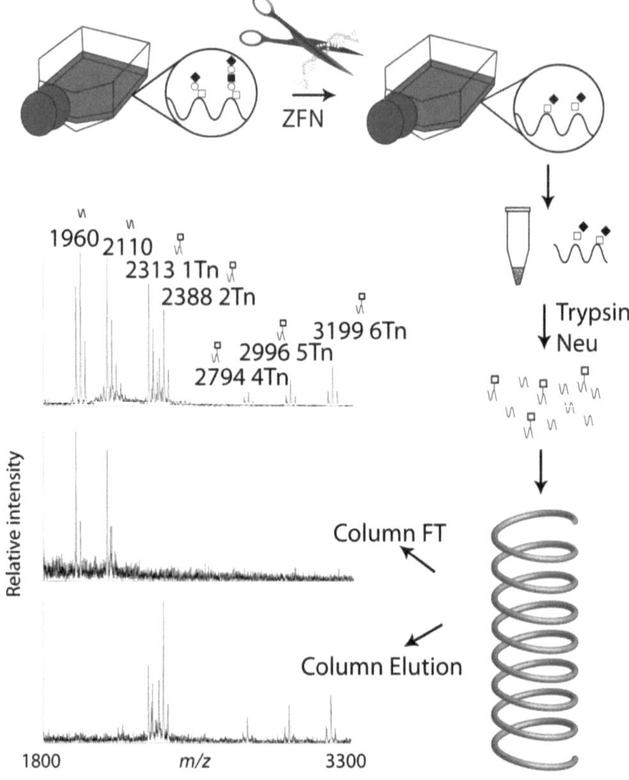

Fig. 2 An overview of the SimpleCell O-glycoproteome strategy. SimpleCells expressing homogeneous glycosylation are generated by ZFN KO. Proteins are isolated from the SimpleCells, digested with trypsin and treated with neuraminidase to remove sialic acid. The tryptic peptides and glycopeptides are loaded onto the VVA column and peptides pass through while glycopeptides are retained and subsequently eluted. In order to test the column specificity and to optimize the conditions for the lectin chromatography a peptide mixture containing 1 μg of each of two peptides and the corresponding glycopeptides were loaded onto the lectin column. MALDI-TOF spectra of the peptide mixture as well as column flow through and elution is shown in the Figure. Only the peptides pass through the column while glycopeptides with even only one and up to six Tn sites are isolated by the lectin chromatography

8. Plasmids: CompoZr® ZFN left and right plasmid (Sigma-Aldrich, St. Louis, MO, USA). The green fluorescent protein (GFP) plasmid 0.5 μg/μL included in the Nucleofector solution Kit, serves as optimization control of transfection efficiency for the protocol used.
9. Inverted fluorescent microscope: The inverted microscope is used to visualize the GFP transfected cells in culture. It can be replaced by a normal fluorescent microscope, and in such a case the cells are trypsinized and dried onto a slide (*see* Subheading 3.2) and visualized using the microscope.
10. Centrifuge: A centrifuge adaptable for Eppendorf tubes, 10 and 50 mL falcon tubes and 96-well plates are required.

Table 1
List of Monoclonal antibodies used for ICC in when generating O-GalNAc SimpleCells

Antibody	Antigen	Reference
5F4	Tn	[10]
1E3	Tn	[11]
3F1	STn	Unpublished
TKH2	STn	[12]
B72.3	STn	[13]
3C9	T	[14]
HH8	T	[14]
2G8	C1GalT1	Unpublished
5B6	C1GalT1	Unpublished
5F8	C1GalT1	Unpublished

2.2 Immunocytochemistry Materials

1. Slides: 21-well Teflon-coated diagnostic slides (ImmunoCell, Mechelen, Belgium).
2. Ice cold Acetone: Store in −20 °C freezer.
3. Primary antibody: Murine monoclonal antibodies were produced in-house as described in Chapter 30, and added to the slides as undiluted supernatant. A list of antibodies towards different O-glycan structures is provided in Table 1.
4. Secondary antibody: Fluorescein isothiocyanate (FITC)-conjugated rabbit anti-mouse immunoglobulins (Dako, Denmark). Dilute the secondary antibody 1:100 in 0.1 % Bovine serum albumin (BSA) in PBS (weigh 50 mg BSA in an 50 mL falcon tube and dissolve in 50 mL PBS).
5. Neuraminidase: *Clostridium perfringens* neuramidase Type VI (Sigma) 0.1 U/mL in 0.05 M acetate buffer pH 5.5 (mix 29.6 mL 0.1 M Acetic acid (1 mL concentrated acetic acid in 174.85 mL MQ-H_2O) with 70.4 mL 0.1 M Na-Acetate (1.3608 g in 100 mL MQ-H_2O) and add 100 mL MQ-H_2O. Check pH.).
6. Mounting media: ProLong® Gold Antifade reagent (Invitrogen, Life Technologies, Paisley, UK).
7. Fluorescence microscope.

2.3 Lectin Affinity Chromatography Materials

1. Tubing: PFA tubing 1/16″ × 50′ (Upchurch scientific, Oak Harbor, WA, USA).
2. Fittings: 2 Flangeless unions in PEEK, 1/16″, 0.5 mm (Upchurch scientific).

3. Ferrule with filter: Frit-In-A-Ferrule, PCTFE, 1/16″, 0.5 mm (Upchurch scientific). Two Frit-In-A-Ferrules are needed for one column. Take care to filter all buffers and samples to avoid clogging of the frit.
4. Endcaps: 2 flat bottom ports for sealing the column when storing (Upchurch scientific).
5. FPLC: ÄKTA FPLC system (GE Healthcare Europe GmbH, Denmark) interfaced by Unicorn 5.2 (GE Healthcare)
6. Packing column: The packing column is used as a reservoir for the agarose slurry while packing. Use XK 16 column (GE Healthcare) or similar (*see* Fig. 3).
7. VVA agarose: Use approx. 2.5 mL *Vicia villosa* Lectin (VVL, VVA) agarose slurry (3 mg/mL) (Vector Labs, Burlingame, CA, USA).

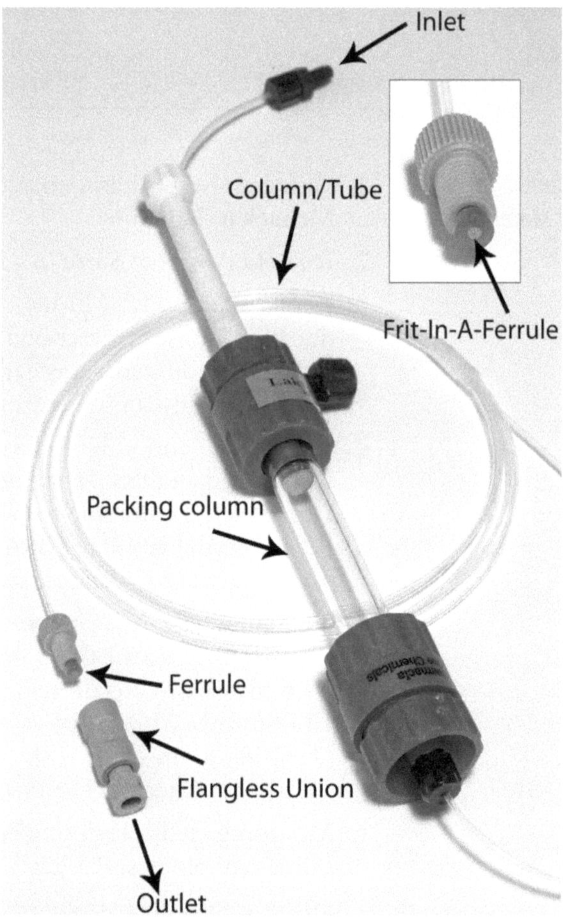

Fig. 3 Setup for lectin column packing. The start of the column tube is attached directly to the packing column, that is filled with the lectin-agarose slurry. The other end of the column tube is sealed by the end fitting with a Frit-In-A-Ferrule inserted. Once the column is packed the start of the tube is released from the packing column and a fitting with a Frit-In-A-Ferrule is attached such that the column can be used in both directions

8. LAC A buffer: 20 mM Tris–HCl pH 7.4, 150 mM NaCl, 1 M Urea, 1 mM $CaCl_2$, $MgCl_2$, $MnCl_2$, and $ZnCl_2$. Make 1 L 0.2 M Tris–HCl (31.5 g in 1 L MQ-H_2O) and 1 L 0.2 M Tris-base (24.2 g in 1 L MQ-H_2O). Titrate the Tris–HCl solution to a pH of 7.4 using the Tris-Base solution. Make 1 L 4 M NaCl, 233.8 g in 1 L MQ-H_2O. Make 50 mL 0.5 M stock of $CaCl_2$. $2H_2O$ (3.8 g in 50 mL MQ-H_2O), $MgCl_2$. $6H_2O$ (5.4 g in 50 mL MQ-H_2O), $MnCl_2$ $4H_2O$ (5.2 g in 50 mL MQ-H_2O), and $ZnCl_2$ (3.4 g in 50 mL MQ-H_2O). Weigh out 60.1 g Urea and add 100 mL 200 mM Tris–HCl pH 7.4, 37.5 mL 4 M NaCl, and 2 mL of each of the 0.5 M stock solutions with $CaCl_2$, $MgCl_2$, $MnCl_2$, and $ZnCl_2$. Add water up to 1 L. Filter the LAC A buffer through a 0.22 μm filter and degas in a sonic bath, 40 °C 10 min. Store at 4 °C.
9. Washing buffer: LAC A buffer with 0.4 M Glucose. Weigh 36.0 g of D-glucose and fill up to 500 mL with LAC A buffer. Filter and degas as per LAC A buffer in **item 8**.

2.4 Sample Preparation

1. Sep-Pak: Sep-Pak® Vac 1cc (100 mg) C18 (Waters A/S, Denmark). The following solutions are required: 100 % methanol, 50 % methanol with 0.1 % formic acid (FA) (25 mL MeOH, 25 mL MQ-H_2O, and 50 μL FA), 0.1 % FA in MQ-H_2O, and 0.1 % trifluoroacetic acid (TFA) in MQ-H_2O.
2. Low-bind tubes: Protein LoBind Tubes (Eppendorf, Hamburg, Germany).
3. 50 mM Ammonium bicarbonate (AmBic): Mix 39.5 mg with 10 mL MQ-H_2O. Make a fresh solution when preparing the cell lysate.
4. 0.1 % RapiGest (Waters), solubilize 1 vial in 1 mL 50 mM AmBic.
5. Sonicator: Probe sonicator, for cell lysate.
6. Dithiothreitol (DTT): 100 mM, mix 15.4 mg with 1 mL 50 mM AmBic, make fresh every time.
7. Iodoacetamide (IAA): 200 mM, mix 37.0 mg with 1 mL 50 mM AmBic, keep dark, make fresh every time.
8. Trypsin: 25 μg trypsin modified sequencing grade (Roche Diagnostics, Denmark).
9. Neuraminidase, 100 U neuraminidase (New England Biolabs, Ipswich, MA, USA) and 50 mM Sodium citrate pH 6.0 (provided by manufacturer).
10. SpeedVac or lyophilizer for sample concentration.
11. Filter units: ULTRAFREE ®-MC DURAPORE®-PVDF 0.22 μm (Millipore A/S Denmark).
12. Thermomixer: 37–80 °C, compatible with 2 mL Eppendorf tubes.

2.5 LAC Run

1. Elution buffer: Make 4 mL of 0.2 M (177 mg) and 2 mL of 0.4 M (177 mg) *N*-acetylgalactosamine (GalNAc) (Sigma) solubilised in LAC A buffer. In order to reduce the amount of polymers present in the elution fractions, the GalNAc solution is passed through a Sep-Pak C18 column. First wash the column with 500 μL methanol, equilibrate with 3× 500 μL 0.1 % TFA (50 μL concentrated TFA in 50 mL MQ-H_2O), pass the GalNAc solution through the column and collect the flow through.
2. Storage buffer: 0.01 M Tris–HCl pH 8, 150 mM NaCl, 0.03 % NaN_3, 20 mM Galactose.

3 Methods

3.1 Cell Culture

Perform all cell culture work under sterile conditions in a flow bench. Grow and maintain cell lines using standard cell culture techniques. Transfection is accomplished using Nucleofection with protocols provided by Lonza (http://www.lonzabio.com/resources/product-instructions/protocols/) that contain detailed information on the nucleofection solution and transfection program optimized for specific commonly used cell lines. The presented protocol explains how to generate SimpleCells from an adherent cell line (Colo205) and a suspension cell line (K562), and might need adaption for other cell lines.

1. Seed cells 2–4 days before transfection and change media the day before transfection. Cells should be 70–80 % confluent (Colo205) or at a density of $2–5 \times 10^5$ cells/mL (K562) on the day of transfection.
2. Test the transfection efficiency of a particular cell line by transfection with a GFP plasmid. For Colo205 two nucleofection programs (A-024, A-013) and two different nucleofection solutions, V and T, are suggested. In such a case add 2 μg of GFP plasmid (4 μL) in each of four Eppendorf tubes and prepare 200 μL of nucleofection solution V and 200 μL solution T in separate tubes. For K562 only one program is recommended, solution V program T-16, and thus, 1 tube with 2 μg GFP plasmid and 100 μL solution T needs to be prepared.
3. Prepare one 6-well plate with one well per transfection containing 3 mL media and place the plate in the incubator.
4. Only for adherent cells: Remove media from the cells, wash 1× in PBS, add trypsin, and incubate for approximately 5 min at 37 °C, to make the cells detach. Stop the trypsinization by adding cell culture media containing FBS.
5. Mix the cells well and take out 100 μL to count the cells using a hemocytometer (both adherent and suspension cell lines).

6. Transfer 10^6 cells to new tubes (10 or 2 mL). Make one tube per nucleofection reaction. Pellet the cells by centrifugation at $100 \times g$ for 5–10 min.
7. Perform one transfection at a time while leaving the other cell pellets standing with media in the tubes.
8. Take out the 6-well plate from the incubator.
9. Remove *all* the media and add 100 μL nucleofection solution to the pellet and mix well by pipetting up and down.
10. Transfer the mixture to the Eppendorf tube containing plasmid and mix well without making bubbles, place the solution in the cuvette, close the lid, and nucleofect with the program recommended by Lonza (*see* **Note 1** for nucleofection programs).
11. Use the Pasteur pipettes included in the nucleofection kit to add approx 0.5 mL warm media from the 6-well to the cuvette. Transfer the media with cells back into the 6-well. Repeat **step 9–10** for the other transfections and place the 6-well plate in the CO_2 incubator overnight (ON).
12. The following day remove the media (adherent cells) and wash the cells with PBS to remove dead cells and add fresh media. Alternatively spin the suspension cell line at $90 \times g$ for 10 min, remove media, add fresh media, and seed in a new 6-well plate.
13. Look at the GFP-transfected wells in an inverted fluorescent microscope. GFP transfection efficiency above 50 % is desirable, but not always possible to achieve. The GFP-transfected cells can now be discarded.
14. To do the ZFN transfection repeat **step 1–12**. However, in **step 2** prepare only one tube containing 2 μg of each PZFN plasmid (left and right). Use the nucleofection program and solution that resulted in the highest transfection efficiency with GFP. *See* **Note 2** for the use of mRNA instead of plasmid.
15. Make sure that cells do not get confluent ($>2 \times 10^6$ cells/mL for K562) in the 6-well. When cells reach around 70 %, split them and place 2/3 of the cells in an Eppendorf tube to screen for positive cells and seed the remaining cells in a new 6-well plate (the screening protocol is presented in Subheading 3.2). This is repeated approximately every third day for 2 weeks, to follow the stability and survival of positives cells (in the meantime continue to **step 16**). If positive cells remain stable after 2 weeks, the cells can be frozen.
16. Clone cells 5–7 days after transfection. The number of clones that need to be screened depends on the KO efficiency. If the KO efficiency is around 1 %, screening of 400 clones is recommended.
17. Repeat **step 4** and **5** for the transfected 6-well plate.

18. Add the amount of media needed for the 96-well plates (200 μL per well) to a petri dish, then add a number of transfected cells, so that 1–2 cells will be seeded in each well (5–10 cells/mL) (*see* **Note 3** on cloning troubleshooting). Use a multichannel pipette to transfer the cell suspension from the petri dish to the 96-well plates.
19. Change media on the 96-well once every week. Change only 50 % of the media after the first week. When the clones are ~20 % confluent in the 96-well split them for screening by removing media, washing with PBS (100 μL), adding trypsin (~20 μL) for approximately 5 min at 37 °C. Add media (80 μL) to each well and transfer 90 μL to U-shaped 96-wells. For K562 (and other suspension cell lines), remove 150 μL from each well and transfer to U-shaped 96-well plates. Refill media to the remaining cells in the old 96-wells and place in the CO_2 incubator. For the sake of simplicity, make sure to keep track of the multichannel pipette when transferring the clones, so that the new plates are exact duplicates of the old plates. It is not necessary to change tips in between each well as a second round of cloning usually is performed.
20. Screen the cells in the U-shaped 96-well plates (*see* Subheading 3.2). Make sure that the screening is finished before the cells grow to confluence.
21. Move positive clones to 24-well plates. Make sure that there are enough cells in the 96-well before you move the cells (varies from cell type to cell type). Alternatively the cells in the 96-well can be split into two 96-wells that are pooled in a 24 well once they are confluent.
22. Very often the clones are not 100 % positive, but perhaps 50 % positive. This is probably due to one positive and one negative clone being seeded in the same well during cloning. Should this be the case, clone the selected clones from the first round of cloning as in **step 17–18**, but only seed in 120 wells with 0–1 cell per well.
23. Screen the second cloning as in **step 19**, but change pipette tip between all wells when moving the cells to the U-shaped wells.
24. Once a 100 % positive clone is obtained, expand it and then freeze in aliquots for backup, meanwhile a small amount of cells can be used for verification of the KO mutation. Selected knockout clones are sequence validated for exact size determination of the ZFN mediated gene targeting. Accompanying CompoZr® ZFN primers flanking the ZFN-target locus are used to amplify the target region from the selected knockout clones, using the conditions recommended by the supplier. Amplicons are Topo-TA cloned and individual clones sequenced. In situations where mono-allelic genes are targeted, amplicons

can be sequenced directly, but for bi-allelic gene targeting (or in situations where allele copy number is uncertain) amplicons need to be cloned prior to sequencing (*see* **Note 4**). Finally the cells are expanded and used for the desired purpose.

25. For isolation and characterization of the total O-GalNAc glycoproteome in engineered cells, expand cells until they can be seeded in four T-175 flasks. Change media the following day and leave the cells for 48–72 h depending on when they get confluent.
26. Pour the media into 50 mL tubes, spin at 1,000 × *g* for 5 min, transfer to new tubes, and store at −80 °C until use.
27. For K562 cells wash the cell pellet with cold PBS and store the cell pellet in two aliquots at −80 °C until use.
28. Wash the adherent cells in cold PBS. Add new PBS and scrape the cells off the flasks. Pour the cell suspension into two 50 mL tubes, spin at 1,000 × *g*, 5 min 4 °C, remove the PBS, and freeze the cell pellets (approx. 2× 0.5 mL) at −80 °C.

3.2 Screening of Knockout Clones by Immunocytochemistry

This section presents a simple method to screen ZFN targeted cells for complete KO of *COSMC* using immunocytochemistry (ICC). PCR-based strategies can be applied as described elsewhere [7]. Cells are expected to accumulate Tn and STn on the cell surface after KO of *COSMC* (Fig. 1a). Thus, antibodies towards these glycans can be used to screen for positive KO cells in the initial screening as well as for identification of final clones (Fig. 1b). Alternatively, lack of staining (e.g. lack of stain for T in *COSMC* KO cells) can be used as well (Fig. 1c). Finally, antibodies towards the KO protein can be used. However, the latter is difficult since glycogenes often are expressed in low amounts. C1GalT1, the core 1 synthase protein dependent on cosmc, is degraded upon *COSMC* KO and can thus be used as a negative screening method (Fig. 1d). The advantage of ICC screening is that KO efficiency and complete KO clones can be visualized in the microscope. The disadvantage is that this method requires availability of appropriate antibodies to the resulting glycoform and/or the protein being targeted. Table 1 list MAbs available for O-GalNAc SimpleCell screening and the generation of MAb towards glycosyltransferases are presented in chapter 30.

This protocol is similar for the initial screening (*see* Subheading 3.1, **step 15**), as well as screening of clones from 96-wells (*see* Subheading 3.1, **step 20**).

1. Spin cells from either initial screening (Eppendorf tubes) or the cloning (96 plates) 1,000 × *g* 5 min. Remove media and wash in PBS. Use 1 mL PBS for Eppendorf tubes and 100 μL for each 96-well, spin again and remove the PBS.
2. Add sufficient amount of fresh PBS to the Eppendorf tubes so that the density of the cell slurry is almost clear, although cells

should still be visible when viewing the tube against the light, and while pipetting up and down. Resuspend the cells in the 96 wells in 15–20 μL PBS.

3. Transfer cells to 21-well Teflon-coated slides adding 15–20 μL cell slurry per well.

   For the initial screenings several wells and slides are made, so that different antibodies and conditions can be tested. Cells from the 96-well plates are transferred to one well only while keeping track of the position, so that one position on the slides correlates to the original 96-well clone in the CO_2 incubator.

4. Let the slides air-dry for 12 h until completely dry and fix in ice cold acetone (−20 °C) for 5–8 min. Let the acetone evaporate.
5. To remove sialic acid, treat the slides with 0.1 U/mL neuraminidase in 0.05 M acetate buffer pH 5.5 1–2 h 37 °C. Use 15–20 μL neuraminidase per well. Wash slides 3× 5 min in PBS. This step is optional and alternatively antibodies towards STn can be applied (*see* Table 1).
6. Add antibody to the slides ON 4 °C. Use 15–20 μL per well (undiluted supernatant).
7. Wash the slides 3× 5 min in PBS and incubate slides with the secondary antibody for 40 min at RT. Use ~350 μL pr. slide.
8. Wash the slides 3× 5 min in PBS, mount the slides and examine in the microscope.

3.3 Lectin Affinity Chromatography: Column Packing

The lectin affinity chromatography method is adopted from R. Chalkley and colleagues [8]. All lectin column work is performed at 4 °C.

1. Cut off 2.6 m tubing. Make sure to perform straight right angle cuts to avoid leakage.
2. In the end of the tube place the Frit-In-A-Ferrule and attach the fitting.
3. Attach the other end of the tube directly into the end piece of the packing column. The packing setup is shown in Fig. 3.
4. Start the ÄKTA system and fill pump B with washing buffer and pump A with LAC A buffer. Wash the system, including loop and fraction collector, with LAC A buffer.
5. Fill the packing column with approx. 2.5 mL lectin-slurry, close the packing-column, and connect it to the ÄKTA inlet. Connect the end of the tube to the ÄKTA outlet.
6. Set the flow at 0.2 mL/min to start packing. Decrease the flow to 0.1 mL/min when pressure starts to build. The pressure should never exceed 2 MPa. Flip the packing column back and forth during packing, to prevent the agarose beads sticking in the packing column, instead of moving into the long column.

The packing takes a couple of hours. The PFA tube is transparent and the packing can easily be observed. Detach the packing column when the tube is filled and attach a fitting with a Frit-In-A-Ferrule in the beginning of the long column. The column can be used in both directions.

3.4 Sample Preparation

1. Mix approximately 0.5 mL of frozen cell pellet with 1,000 μL of 0.1 % RapiGest. Keep the sample on ice for 20 min and vortex every 5 min.
2. Sonicate the sample with a sonic probe 5× 20 s, 50 W. Make sure to keep the sample on ice during sonication.
3. Spin 1,000 × *g* 5 min, and transfer the supernatant to a clean 2 mL low bind Eppendorf tube.
4. Heat to 80 °C for 10 min to further activate the RapiGest.
5. Add up to 5 mM DTT (50 μL of 100 mM stock) and heat at 60 °C for 45 min.
6. Cool to room temperature (RT) and add 10 mM IAA (50 μL 200 mM stock) and incubate at RT in the dark, 30 min.
7. Solubilize the 25 μg trypsin in 25 μL AmBic and add 20 μL to the sample ON 37°C, shaking. Freeze the remaining trypsin.
8. Add the remaining 5 μL trypsin solution to the sample and incubate for 2 h 37 °C, shaking.
9. Add 8 μL of concentrated TFA and incubate at 37 °C for 20 min. Vortex and centrifuge 10,000 × *g* 10 min.
10. Transfer the supernatant to a clean low bind tube.
11. Purify by Sep-Pak. Wash the column with 500 μL MeOH followed by 500 μL 50 % MeOH/0.1 % FA and equilibrate with 3× 500 μL 0.1 % TFA. Pass the sample through the column twice and wash with 3× 500 μL 0.1 % FA. Elute with 500 μL 50 % MeOH/0.1 % FA and 500 μL MeOH in a new low bind tube.
12. Concentrate the sample by SpeedVac or lyophilization.
13. If the cells express STn and not Tn, sialic acid is removed with neuraminidase to expose Tn. Solubilize the sample in neuraminidase buffer (G1 provided with the neuraminidase) and add 100 U of neuraminidase for 2 h 37 °C. Acidify the sample with TFA and do another Sep-Pak purification as described in **step 11–12** (Optional).
14. Solubilize the sample in 1.5 mL LAC A buffer and filter into a low bind tube using a centrifugation filter unit. Store the sample at −20 °C until use.

3.5 LAC Run

The LAC run was optimized with a glycopeptide mixture and subsequent MALDI-TOF analysis of the collected fractions as presented in Fig. 2.

1. Equilibrate the VVA column with at least 10 column volumes (CV) LAC A buffer (do the equilibration ON to save time).
2. Load the sample (either tryptic digest or glycopeptide mixture) into a 2 mL loop.
3. Start the chromatography by injecting the sample, set the flow rate to 100 μL/min, and at the same time set the fraction collector to collect 2 mL fractions in low bind tubes. Set the pressure alarm maximum at 2 MPa.
4. After 3 CV, when the flow through has passed, wash the column with 4 CV washing buffer and switch back to buffer A.
5. The glycopeptides are eluted when the column is re-equilibrated in buffer A as visualized by the conductivity. Fill the loop with 0.2 M GalNAc solution and press inject. After 20 min load 2 mL 0.2 M GalNAc more and inject. After another 20 min load 2 mL 0.4 M GalNAc and inject. Collect 1 mL fractions in low bind tubes during the elution. The fractions can now be de-salted and sent for analysis by mass spectrometry.
6. To store the column, fill one of the ÄKTA pumps with storage buffer and wash the column with two CV of storage buffer. Seal the column with end caps and keep at 4 °C.

4 Notes

1. Nucleofection program:
   Lonza provides protocols for many established cell lines (http://www.lonzabio.com/resources/product-instructions/protocols/); however, in cases where no protocol is available solution T and program T-16 as well as solution V and program H-22 have proven to be good "general" protocols with high transfection efficiency and satisfactory cell survival.
2. ZFN plasmid or mRNA:
   Sigma provides the ZFN both as separate plasmids and an mRNA pool of the left and right ZFN. It is generally believed that the use of mRNA can result in higher KO efficiency, however mRNA is unstable and care needs to be taken when performing the transfection. First of all keep the mRNA on ice right until use. Furthermore, include an extra wash step at item **step 6** in Subheading 3.1. Remove the media after centrifugation in **step 6** and add 8 mL PBS, mix by pipetting, and centrifuge again 90 × *g* 5–10 min and continue in **step 7**.
3. Cloning: Some cell lines can be very difficult to grow as single cell clones and different attempts can be made to optimize the conditions for cloning. It is strongly recommended to do a test cloning of each new cell line before transfection to establish

the optimal conditions. One possibility is to grow the cells in 50 % conditioned media. Alternatively 1–2 % of Hybridoma Cloning Supplement, (HCS) (PAA) can be added to the media. Finally for some cell lines it can be an advantage to seed 3–4 clones pr. well and then obtain a homogeneous clone by multiple cloning steps.

4. The DNA extraction, PCR amplification of the target region, Topo-TA subcloning, and DNA sequencing are carried out by standard procedures, which are not described in detail here.

Acknowledgments

This work was supported by Kirsten og Freddy Johansen Fonden, A.P. Møller og Hustru Chastine Mc-Kinney Møllers Fond til Almene Formaal, The Carlsberg Foundation, The Novo Nordisk Foundation, The Danish Research Council, a program of excellence from the University of Copenhagen, and the Danish National Research Foundation (DNRF107).

References

1. Stanley P (2011) Golgi glycosylation. Cold Spring Harb Perspect Biol 3:a005199
2. Darula Z, Medzihradszky KF (2009) Affinity enrichment and characterization of mucin core-1 type glycopeptides from bovine serum. Mol Cell Proteomics 8:2515–2526
3. Steentoft C, Vakhrushev SY, Vester-Christensen MB, Schjoldager KTBG, Kong Y, Bennett EP, Mandel U, Wandall H, Levery SB, Clausen H (2011) Mining the O-glycoproteome using zinc-finger nuclease-glycoengineered SimpleCell lines. Nat Methods 8:977–982
4. Wang Y, Ju T, Ding X, Xia B, Wang W, Xia L, He M, Cummings RD (2010) Cosmc is an essential chaperone for correct protein O-glycosylation. Proc Natl Acad Sci USA 107:9228–9233
5. Santiago Y, Chan E, Liu PQ, Orlando S, Zhang L, Urnov FD, Holmes MC, Guschin D, Waite A, Miller JC, Rebar EJ, Gregory PD, Klug A, Collingwood TN (2008) Targeted gene knockout in mammalian cells by using engineered zinc-finger nucleases. Proc Natl Acad Sci USA 105:5809–5814
6. Pattanayak V, Ramirez CL, Joung JK, Liu DR (2011) Revealing off-target cleavage specificities of zinc-finger nucleases by in vitro selection. Nat Methods 8:765–770
7. Chen FQ, Pruett-Miller SM, Huang YP, Gjoka M, Duda K, Taunton J, Collingwood TN, Frodin M, Davis GD (2011) High-frequency genome editing using ssDNA oligonucleotides with zinc-finger nucleases. Nat Methods 8:753–755
8. Chalkley RJ, Thalhammer A, Schoepfer R, Burlingame AL (2009) Identification of protein O-GlcNAcylation sites using electron transfer dissociation mass spectrometry on native peptides. Proc Natl Acad Sci USA 106:8894–8899
9. Iwai T, Kudo T, Kawamoto R, Kubota T, Togayachi A, Hiruma T, Okada T, Kawamoto T, Morozumi K, Narimatsu H (2005) Core 3 synthase is down-regulated in colon carcinoma and profoundly suppresses the metastatic potential of carcinoma cells. Proc Natl Acad Sci USA 102:4572–4577
10. Thurnher M, Clausen H, Sharon N, Berger EG (1993) Use of O-glycosylation-defective human lymphoid-cell lines and flow-cytometry to delineate the specificity of moluccella-laevis lectin and monoclonal-antibody 5F4 for the Tn antigen (galnac-alpha-L-O-Ser/Thr). Immunol Lett 36:239–243
11. Mandel U, Petersen OW, Sorensen H, Vedtofte P, Hakomori SI, Clausen H, Dabelsteen E (1991) Simple mucin-type carbohydrates in oral stratified squamous and salivary-gland epithelia. J Investig Dermatol 97:713–721
12. Kjeldsen T, Clausen H, Hirohashi S, Ogawa T, Iijima H, Hakomori S (1988) Preparation and

characterization of monoclonal-antibodies directed to the tumor-associated O-linked sialosyl-2-6 alpha-N-acetylgalactosaminyl (sialosyl-Tn)epitope. Cancer Res 48:2214–2220

13. Nuti M, Teramoto YA, Marianicostantini R, Hand PH, Colcher D, Schlom J (1982) A monoclonal-antibody (B72.3) Defines patterns of distribution of a novel tumor-associated antigen in human mammary-carcinoma cell-populations. Internat J Cancer 29:539–545
14. Clausen H, Stroud M, Parker J, Springer G, Hakomori S (1988) Monoclonal-antibodies directed to the blood group-a associated structure, galactosyl-A—specificity and relation to the thomsen-friedenreich antigen. Mol Immunol 25:199–204
15. Schjoldager KT, Vakhrushev SY, Kong Y, Steentoft C, Nudelman AS, Pedersen NB, Wandall HH, Mandel U, Bennett EP, Levery SB, Clausen H (2012) Probing isoform-specific functions of polypeptide GalNAc-transferases using zinc finger nuclease glycoengineered SimpleCells. Proc Natl Acad Sci USA 109: 9893–9898
16. Steentoft C, Vakhrushev SY, Joshi HJ, Kong Y, Vester-Christensen MB, Schjoldager KT, Lavrsen K, Dabelsteen S, Pedersen NB, Marcos-Silva L, Gupta R, Paul BE, Mandel U, Brunak S, Wandall HH, Levery SB, Clausen H (2013) Precision mapping of the human O-GalNAc glycoproteome through SimpleCell technology. EMBO J
17. Vakhrushev SY, Steentoft C, Vester-Christensen MB, Bennett EP, Clausen H, Levery SB (2013) Enhanced Mass Spectrometric Mapping of the Human GalNAc-type O-Glycoproteome with SimpleCells. Mol Cell Proteomics 12: 932–944

Chapter 30

Generation of Monoclonal Antibodies to Native Active Human Glycosyltransferases

Malene Bech Vester-Christensen, Eric Paul Bennett, Henrik Clausen, and Ulla Mandel

Abstract

Complex carbohydrates serve a wide range of biological functions in cells and tissues. Their biosynthesis involves more than 200 distinct glycosyltransferases in human cells, and the expression, properties, and topology of these enzymes regulate the glycosylation patterns of proteins and lipids. Glycosyltransferases are ER-Golgi resident enzymes with slow turnover, which makes monitoring of protein expression a method more directly linked to enzyme function, than monitoring gene expression. In situ monitoring of expression and subcellular topology of glycosyltransferase proteins by immunological techniques using monoclonal antibodies therefore provides an excellent strategy to analyze the glycosylation process in cells. A major drawback has been difficulties in generating antibodies to glycosyltransferases and validating their specificities. Here we describe a simple strategy for generating and characterizing monoclonal antibodies to human glycosyltransferases. This strategy includes a process for recombinant production and purification of enzymes for immunization, a simple selection strategy for isolation of antibodies with optimal properties for in situ detection of enzyme expression, and a comprehensive strategy for characterizing the fine specificity of such antibodies.

Key words Glycosyltransferases, Monoclonal antibodies, Purification, Immunocytology

1 Introduction

Glycans on proteins, proteoglycans, and lipids play important roles in almost every biological event at the cell, tissue, and organismal level. The secretory pathway of eukaryotic cells comprising endoplasmatic reticulum (ER), Golgi apparatus, and trans-Golgi network contains the glycosylation machinery of cells, and more than 200 distinct glycosyltransferases lined throughout this pathway work in a coordinated, albeit non-template driven fashion, to produce a large variety of oligosaccharide structures required for numerous specific functions [1]. Regulation of the glycosylation process in cells is generally thought to be controlled by the expression levels of specific enzymes. Most glycosyltransferases are type II

Inka Brockhausen (ed.), *Glycosyltransferases: Methods and Protocols*, Methods in Molecular Biology, vol. 1022, DOI 10.1007/978-1-62703-465-4_30, © Springer Science+Business Media New York 2013

transmembrane proteins with a luminal C-terminal catalytic domain, a short stalk region, an α-helix hydrophobic transmembrane segment, and a short cytosolic tail at the N-terminus [2]. The latter three domains of glycosyltransferases have been found to be important for the subcellular localization of these enzymes in cells; and proteolysis in the stalk region releases soluble active enzymes, which may be found in body fluids. Evidence of dynamic regulation through proteolytic release of these enzymes, e.g., by the furin proprotein convertase (lunatic fringe) [3] or BACE-1 (ST6Gal-I) [4], has been provided, but our understanding of this aspect is still limited. In order to analyze the glycosylation process and its regulation, it is therefore important to be able to visualize both the expression and the subcellular topology of this large group of enzymes in situ in cells and tissues. Analysis of changes in gene expression may complement this, but in itself mRNA levels may not provide sufficient information, due to long enzyme half-life and slow turnover.

Considerable efforts have in the past been devoted to the production of antibodies to glycosyltransferases, but the employed methods and choice of immunogens have generally produced antibodies lacking specificity. Initial attempts to produce antibodies utilized soluble enzymes isolated from natural sources as immunogen; however, the presence of immunogenic glycan chains on these enzymes resulted in antibodies cross-reactive with glycans in tissues [5–7]. Following isolation and sequencing of the first mammalian glycosyltransferases several investigators have used synthetic peptides and recombinantly expressed fragments derived from glycosyltransferases as immunogens. However, the generated antibodies rarely react with the native conformation of the enzymes, although they may be suitable for SDS-PAGE Western blot analysis and even in some cases immunohistochemistry after appropriate antigen retrieval methods. Surprisingly, many of these antibodies have indicated ectopic expression of glycosyltransferases, i.e., in non-Golgi localization in cell membrane or in nuclear structures [8, 9]. Other studies have shown localization to post-Golgi vesicles, which may represent accumulation of cleaved, secreted enzymes [10, 11]. However, in our hands most of these examples are related to the reagent used for detection, since monoclonal antibodies generated to the native conformation of human glycosyltransferases only show subcellular localization of immunoreactivity to ER-Golgi. However, one exception is the membrane associated glycosyltransferases in platelets [12, 13]. Furthermore, a recent interesting finding suggests that activation of Src induces relocation of polypeptide GalNAc-transferases from Golgi [14] to the ER [15]. A detailed discussion of debatable findings of ectopic expression of glycosyltransferases was previously provided by Berger [16].

Table 1
List of monoclonal antibodies against human glycosyltransferases

Antibody clone name (isotype)	Antigen	Application	Reference
UH3 (4D8) (IgG1)	GalNAc-T1	IHC, IP	[19]
UH2 (6B7) (IgG1)	GalNAc-T2	WB	[19]
UH4 (4C4) (IgG1)	GalNAc-T2	IHC, IP	[19]
2E10 (IgG1)	GalNAc-T2	IHC, IP	Unpublished
UH5 (2D10) (IgG1)	GalNAc-T3	IHC, IP	[19]
UH1 (2F6) (IgG1)	GalNAc-T3	WB	[20]
UH6 (4G2) (IgG1)	GalNAc-T4	IHC	[21]
UH7 (2F3) (IgG1)	GalNAc-T6	IHC	[22]
2E11 (IgG1)	GalNac-T6	WB	Unpublished
UH8 (1B2) (IgG1)	GalNAc-T11	IHC	[23]
1F5 (IgG1)	GalNAc-T12	WB	Unpublished
1F9 (IgG1)	GalNAc-T12	IHC	[17]
3D2 (IgG1)	GalNAc-T14	IHC	[17]
5G10 (IgG1)	GalNAc-T14	WB	Unpublished
2F5 (IgG1)	β4Gal-T1	IHC, WB	[12]
5F8	β4GalT	IHC	Unpublished
5B6	β4GalT	IHC	Unpublished
URH1 (2C3) (IgG1)	β4Gal-T7	WB	[24]
URH2 (3G6) IgG1)	β4Gal-T7	IHC	Unpublished
5E8 (IgG1)	β3Gal-T5	IHC	Unpublished
6F8 (IgG1)	β3Gal-T5	WB	Unpublished
4B10 (IgG1)	ST3Gal I	IHC, WB	[25]
2C3(IgG2A)	ST6GalNAc-I	IHC	[26]
1C9	ST6GalNAc-I	WB	[26]

We have for a number of years worked with a different strategy to produce monoclonal antibodies (MAbs) to the native conformation of enzymes, which can be used to detect expression of active enzymes in cells as well as in body fluids. We have produced a large panel of MAbs to native human glycosyltransferases using recombinant, secreted, soluble, active enzymes produced in insect cells as the immunogen (Table 1) [17]. The success of this strategy relies on a simple, yet effective screening system that allows quick selection of MAbs with strict specificity for distinct

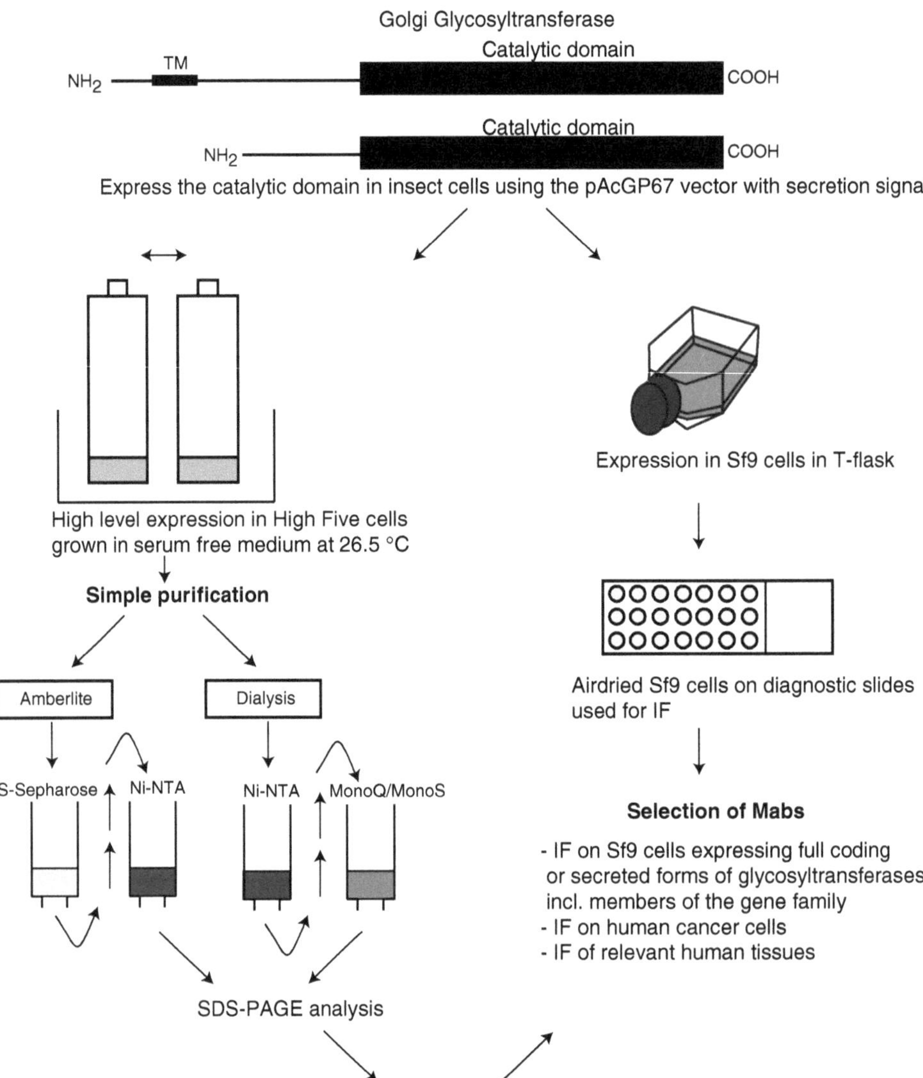

Fig. 1 The strategy for generation of monoclonal antibodies to native active human glycosyltransferases is shown here. The glycosyltransferase gene of interest is expressed in insect cells and purified using a simple two-step purification procedure. The hybridomas are tested for antibody production using IF on insect cells over-expressing the antigen, on cancer cell lines and relevant human tissue

glycosyltransferases, without cross-reactivity with carbohydrates, homologous isoenzymes, and potential affinity purification tags included in the construct used for immunization [16]. In many cases this strategy also allows for selection of MAbs to the denatured enzymes, suitable for SDS-PAGE Western blot analysis. Interestingly, however, only in few cases were antibodies isolated that reacted both with the native and the denatured enzymes.

Our strategy (depicted in Fig. 1) involves recombinant expression of secreted functionally active glycosyltransferases in insect cells;

a simple purification method suitable for mg batches; and finally the generation and selection of MAbs with defined specificities for individual members of families of homologous glycosyltransferases. The strategy is based on high level expression of soluble secreted enzymes in High Five™ cells grown in serum-free medium [18]. Purification is achieved by successive ion-exchange chromatographies and/or Ni-chromatography followed by ion-exchange for N-terminal His-tagged enzymes. For all the tested enzymes so far, these purification steps have yielded relatively pure active enzymes in large quantities, effectively serving as potent immunogens. The subsequent screening relies mainly on immunofluorescence staining of insect cells expressing the soluble, secreted constructs (or in some cases the full coding construct), while the immunogens used are soluble, secreted constructs. The specificity of selected MAbs is easily confirmed by screening a large panel of different glycosyltransferase isoforms expressed in insect cells by immunocytology. This selection strategy provides the most comprehensive analysis of specificity made to date. Selective immunoprecipitation of active enzyme is used as a second control to establish reactivity with native, non-denatured enzyme [19]. Finally, a series of human cell lines and tissues are tested to evaluate subcellular localization of immunoreactivity and correlate expression pattern with information known from other studies including those from mRNA analysis. Importantly, by using native-like enzymes (not fusion proteins or proteins tagged with an immunodominant epitope) expressed in higher eukaryote cells we obtain a high frequency of antibodies that perform well in immunohistological techniques and immuno-Electron microscopy (immuno-EM). An effort to select at least two monoclonal antibodies to different epitopes for each glycosyltransferase is made with the development of potential sandwich-capture assays in mind.

2 Materials

All cell culture reagents should be sterile.

2.1 Expression of Secreted Functionally Active Glycosyltransferases

1. Baculo-Gold DNA™ (BD Pharmingen, San Diego, CA, USA).
2. BD BaculoGold™ pAcGP67 Baculovirus Tranfer Vector (BD Pharmingen) containing gene of interest.
3. Sf9 insect cells (Life Technologies, Denmark, Europe) grown in 75 mL T-flasks.
4. Grace media (Gibco®, Life Technologies) supplemented with 10 % fetal bovine serum and 5 mM Glutamine.
5. Heating incubator set to 26.5 °C.
6. High Five insect cells (Life Technologies) grown in upright roller-bottles with 225 mL medium and split 1:1 daily.

7. Express Five® SFM (Gibco®, Life Technologies) supplemented with 5 mM Glutamine.
8. Shaking incubator set at 100 rpm and 26.5 °C.

2.2 Purification of Active Enzyme by Ion-Exchange Chromatography

This protocol is for cation-exchange chromatography purification of polypeptide GalNAc-T1, -T2, and -T3 with slightly basic pI [18]. Purification of other glycosyltransferases may require other equilibration and elution buffers or anion column, since the choice of buffer and ion-exchange material is dependent on the pI of the enzyme in question. However, the principles are the same as described here.

1. Amberlite IRA96 (Sigma-Aldrich, Denmark A/S, Denmark).
2. SP-Sepharose Fast-flow (Sigma-Aldrich).
3. Econo-Column with a length of 10 mm and inner diameter of 2.5 cm (Bio-Rad, Laboratories AB, Sweden).
4. 2 M NaCl: Weigh 467.52 g and dissolve in 4 L Milli-Q water.
5. Equilibration buffer: 25 mM Bis-Tris pH 6.0, 10 mM NaCl. Weigh 10.46 g Bis-Tris and 1.17 g NaCl and dissolve in 1.8 L Milli-Q water. Mix and adjust pH to 6.0 with 6 M HCl, add Milli-Q water until total volume reaches 2 L. Store at 4 °C.
6. Elution buffer: 25 mM Bis-Tris pH 6.0, 1 M NaCl. Weigh 5.23 g Bis-Tris and 58.44 g NaCl and dissolve in 900 mL MilliQ-water. Mix and adjust pH to 6.0 with 6 M HCl, and add Milli-Q water until total volume reaches 1 L. Store at 4 °C.
7. Whatman filter paper.
8. Equipment for SDS-PAGE and Western blotting.
9. Anti-His antibody (Santa Cruz Biotechnology, Inc., Santa Cruz, CA, USA) diluted 1:100 in 50 mM Tris pH 7.5, 150 mM NaCl, 0.05 % Tween 20 (TBS-T).

2.3 Affinity Purification of His-Tagged Enzyme Followed by Ion-Exchange Chromatography

This protocol describes Ni-NTA affinity purification followed by a high-resolution cation-exchange chromatography purification of polypeptide GalNAc-T11. The ion-exchange purification of other His-tagged glycosyltransferases may require other equilibration and elution buffers or anion column, since the choice of buffer and ion-exchange material is dependent on the pI of the enzyme. However, the principles are the same as described here.

1. Spectra/Por dialysis membrane, 10 kDa cut-off (Spectrum Laboratories Inc., CA, USA).
2. 10× dialysis buffer: 0.5 M Tris pH 8, 1.5 M NaCl. Weigh 242.28 g Trizma® base (Sigma) and 350.64 g NaCl and dissolve in 3.8 L Milli-Q water before adjusting the pH to 8.0 with 6 M HCl, add Milli-Q water until total volume reaches 4 L. Store at 4 °C.

3. 50 % Ni-NTA agarose slurry (Life Technologies).
4. Whatman filter paper.
5. Econo-Column with a length of 10 mm and inner diameter of 1 cm (Bio-Rad).
6. Ni-NTA equilibration buffer: 25 mM Tris pH 8.0, 300 mM NaCl. Weigh 242.28 g Trizma® base and 17.53 g NaCl and dissolve in 900 mL MilliQ. Mix and adjust the pH to 8.0 with 6 M HCl, then add Milli-Q water until total volume reaches 1 L. Store at 4 °C.
7. Ni-NTA wash buffer: 25 mM Tris pH 8.0, 300 mM NaCl, 10 mM imidazole. Prepare buffer as in **item 6**, but include 0.68 g imidazole.
8. Ni-NTA elution buffer: 25 mM Tris pH 8.0, 300 mM NaCl, 250 mM imidazole. Weigh 242.28 g Trizma® base, 17.53 g NaCl, and 17.02 g imidazole and dissolve in 900 mL Milli-Q water. Mix and adjust pH to 8.0 with 6 M HCl, then add Milli-Q until total volume reaches 1 L and store at 4 °C.
9. MonoS 5/50 GL, (GE Healthcare, Europe GmbH, Denmark).
10. MonoS equilibration buffer (buffer A): 25 mM Bis-Tris pH 6.5, 10 mM NaCl. Weigh 5.23 g Bis-Tris and 0.59 g NaCl and dissolve in 900 mL Milli-Q water. Mix and adjust pH to 6.5 with 6 M HCl, before adding Milli-Q water until total volume reach 1 L. Filter the buffer through a 0.45 μm filter and degas in a sonic bath, 40 °C 10 min. Store at 4 °C.
11. MonoS elution buffer: 25 mM Bis-Tris pH 6.5, 1 M NaCl. Weigh 5.23 g Bis-Tris and 58.44 g NaCl and dissolve in 900 mL Milli-Q water. Mix and adjust pH to 6.0 using 6 M HCl, add Milli-Q until total volume reaches 1 L. Filter the buffer through a 0.45 μm filter and degas in a sonic bath, 40 °C 10 min. Store at 4 °C.
12. ÄktaFPLC (GE Healthcare) interfaced by UNICORN 4.12 (GE Healthcare).
13. Equipment for SDS-PAGE and Western blotting.
14. Anti-His antibody (Santa Cruz Biotechnology, Inc) diluted 1:100 in 50 mM Tris pH 7.5, 150 mM NaCl, 0.05 % Tween 20 (TBS-T).

2.4 Immunization of Mice and Evaluation of Immune Response

1. BALBc/A mice, female, age 7–8 weeks.
2. Freund's complete adjuvant (FCA).
3. Freund's incomplete adjuvant (FIA).

2.5 Preparing NS1 Cells, Splenocytes and Thymocytes for Fusion

1. Feeder cells, i.e., thymocytes from NMR1 mice, age 3–6 weeks.
2. Fusion partner, i.e., NS1 myeloma cell line.

2.6–2.8. Fusion

1. Hybridoma Cloning Supplement, (HCS, PAA) 1 %.
2. Cell culture media: RPMI 1640 (Lonza, Basel, Switzerland) supplemented with 20 % fetal bovine serum (FBS Gold, PAA), 5 mM NaPy (5 mL), and 5 mM Glutamine, 37 °C.
3. HAT Media supplement (50×) (Sigma).
4. HT supplement (50×) (Gibco, Invitrogen).
5. PEG 1500 (Roche Diagnostics A/S, Denmark).
6. A centrifuge adaptable for 10 and 50 mL falcon tubes.
7. 96-well plates.
8. Heating incubator set to 37 °C.
9. Scissors.
10. Tweezers.
11. Strainer with a fine screen.
12. Plunge of a syringe.

2.9–2.11. Immunochemistry

1. Sf9 cells as described in Subheading 2.1.
2. Human cancer cells: cells are grown and maintained using standard cell culture methods.
3. TrypLE™ Express (Gibco®, Life Technologies).
4. 21-Well Teflon-printed diagnostic slides (Immuno-Cell, Mechelen, Belgium).
5. Ice cold acetone stored in −20 °C freezer.
6. Secondary antibody: Fluorescein isothiocyanate (FITC)-conjugated rabbit anti-mouse immunoglobulins (F 0261, Dako Denmark A/S, Denmark) or goat anti-mouse IgG (γ chain specific, 1030–02, SouthernBiotech, AL, USA). Dilute the secondary antibody 1:100/1:500 in 0.1 % Bovine serum albumin (BSA) in PBS (weigh 50 mg BSA in a 50 mL falcon tube and dissolve in 50 mL PBS).
7. Mounting media: Vectashield with DAPI (Vector labs, CA, USA).
8. Fluorescence microscope.

3 Methods

All purification steps should be carried out at 4 °C (*see* **Note 1**).

3.1 Expression of Secreted Functionally Active Glycosyltransferases in Insect Cells

1. Design an expression construct encoding the amino acid residues representing the soluble part of the glycosyltransferase and clone cDNA into the baculo-expression vector pAcGP67. Include DNA sequence encoding a His-tag in the forward primer used for amplifying the cDNA if affinity purification of the N-terminal His-tag expressed enzyme is desired.

2. Co-transfect the expression vector with Baculo-Gold DNA, and recombinant Baculovirus is obtained after two successive amplifications in Sf9 insect cells, according to the manufacturer's description. In brief, mix 0.5 μg of the pAcGP67-GalNAc-T construct with 0.7 μL Baculo-Gold DNA and co-transfect Sf9 cells in 24-well plates using BD Pharmingen's transfection kit with 100 μL of buffers A and B each. Incubate the plate at 26.5 °C for 4 h, followed by gently washing in 175 μL medium and addition of 350 μL complete Grace medium. 5–6 days post-transfection amplify the recombinant virus in 6-well plates at dilutions of 1:20 and 1:40 [18]. Make a second amplification in a T75-flask at a dilution of 1:100 and 1:200. Grow the Sf9 cells at 26.5 °C in Grace medium supplemented with 10 % fetal bovine serum.
3. Use High Five insect cells, which are adapted to grow in Express Five® SFM for large scale production of glycosyltransferases. Grow cells in roller bottles containing 225 mL Express Five® SFM each and with a cell density of approximately 2×10^5 cells/mL and split 1:1 every day. Maintain the roller bottles in upright positions at 26.5 °C and 100 rpm. Infect newly split cells with 1:500 of a stock from the second amplification. Harvest the media containing the expressed secreted enzyme 72 h post-infection by centrifugation ($2{,}000 \times g$, 4 °C, 10 min) and store it at 4 °C in the presence of 0.03 % NaN_3.

3.2 Purification of Active Enzyme by Ion-Exchange Chromatography

1. Prepare 200 mL dry Amberlite IRA96 pH 0.9 L supernatant by washing the Amberlite twice in 2 L of 2 M NaCl followed by 10 wash steps in Milli-Q water (2 L per wash). Equilibrate the Amberlite in the appropriate buffer, which is dependent on the pI of the enzyme, by incubating the resin two times 15 min in 600 mL equilibration buffer with stirring at 4 °C. In the case of GalNAc-T1, -T2, and T3 a 25 mM Bis-Tris, pH 6.0, 10 mM NaCl buffer is used.
2. Incubate the supernatant (0.9 L) with the equilibrated Amberlite for 1–1½ h with slow constant stirring at 4 °C. Before and after incubating the supernatant with the Amberlite remove a small aliquot (100 μL) of the supernatant and store it at 4 °C.
3. Filter the supernatant through a funnel with Whatman paper to get rid of the Amberlite resin.
4. Dilute the filtered supernatant threefold in the equilibration buffer. For GalNAc-T1, -T2 and T3 use a 25 mM Bis-Tris, pH 6.0, 10 mM NaCl buffer.
5. Prepare the SP-Sepharose Fast-flow column by pipetting 20 mL of the 50 % SP-Sepharose slurry into an Econo-Column with a length of 10 mm and inner diameter of 2.5 cm. This will result in a column with a bed height of approximate 2 cm and column volume (CV) of 10 mL.

6. Wash the column with 10 CV of Milli-Q water before equilibration with 10 CV of equilibration buffer, 25 mM Bis-Tris, pH 6.0, 10 mM NaCl.
7. Load the supernatant using gravity flow onto the equilibrated SP-Sepharose Fast-flow column.
8. Wash the column to remove unbound proteins in 200 mL equilibration buffer and elute the GalNAc-Ts with a linear gradient of NaCl from 10 mM to 1 M in 10 CV.
9. Analyze the purification steps using Coomassie stained gels and Western blot (anti-His for N-terminal His-tagged proteins). Load samples from the eluted fractions on SDS-PAGE gel. Include the samples from the harvest supernatant before and after Amberlite treatment, flow-through and wash.
10. For further purification of the enzyme one can either choose to use the N-terminal His-tag for Ni-NTA affinity purification or another ion-exchange chromatography step as described below.

3.3 Affinity Purification of His-Tagged Enzyme Followed by Ion-Exchange Chromatography

1. Divide the supernatant (0.9 L) into three equally sized dialysis tubes (app. 300 mL in each tube). Dialyze each tube three times against 5 L 50 mM Tris pH 8, 150 mM NaCl. Before dialysis remove a small aliquot (100 μL) of the supernatant and store it at 4 °C.
2. Prepare the Ni-NTA affinity column by pipetting 5 mL of the 50 % Ni-NTA agarose slurry into an Econo-Column with a length of 10 mm and inner diameter of 1 cm. This will result in a column with a bed height of approximate 3.2 cm and CV of 2.5 mL.
3. Wash the column with 10 CV of Milli-Q water, before equilibration with 10 CV of equilibration buffer, 25 mM Tris pH 8.0, 300 mM NaCl.
4. Pass the dialyzed supernatant over a Whatman filter paper to remove any particular matter that otherwise would clog the column.
5. Load the dialyzed and filtered supernatant (0.9 L) onto the equilibrated column using gravity flow and collect the flow-through. Before loading the supernatant onto the column remove a small aliquot (100 μL) of the supernatant and store it at 4 °C.
6. Wash the column with 25 CV of 25 mM Tris pH 8.0, 300 mM NaCl, 10 mM imidazole.
7. Elute the enzyme by adding 6 mL of 25 mM Tris pH 8.0, 300 mM NaCl, 250 mM imidazole to the column. Immediately start collecting fractions (0.5 mL) in tubes containing 4.5 mL of the appropriate buffer (*see* **step 9** and *see* **Note 2**).
8. Analyze the purification steps using Coomassie stained gels and Western blot (anti-His). Include the samples from the harvest supernatant before and after dialysis, flow-through, and wash.

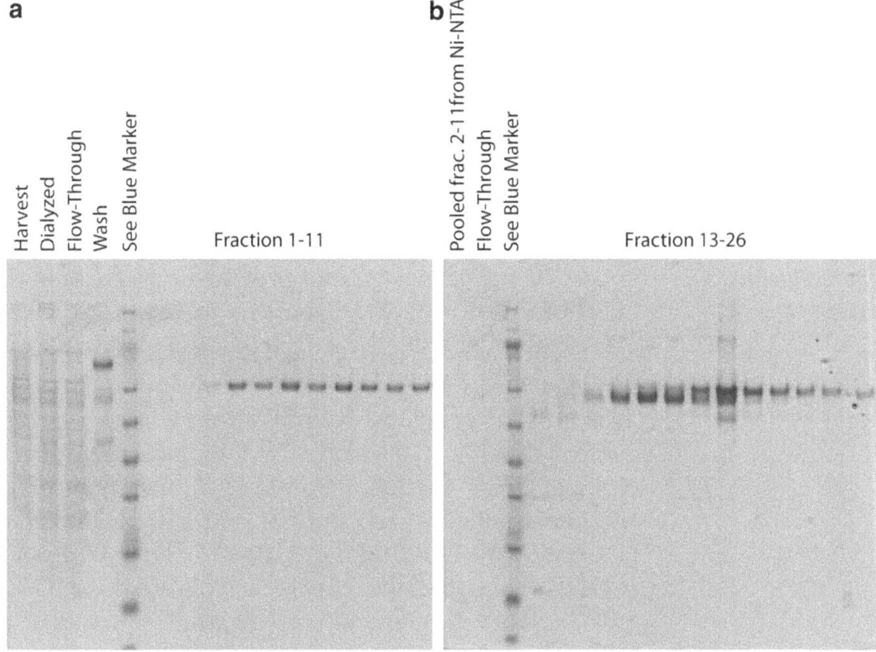

Fig. 2 Coomassie stained SDS-PAGE gel showing GalNAc-T11 purification by (**a**) Ni-NTA affinity chromatography followed by (**b**) cation exchange chromatography. The difference in band size in (**b**) is due to glycosylation of the enzyme; however, the enzyme fractions show equal activity independent on the glycosylation

9. Depending on the efficiency of the affinity purification an ion-exchange chromatography is performed. The choice of ion-exchange is depended on the pI of the enzyme. The purification of GalNAc-T11 is given here as an example. Since the theoretical pI of N-terminal His-tagged GalNAc-T11 is 8.71 use a cation-exchange column (MonoS 5/50 GL) for further purification of GalNAc-T11. In this case the appropriate buffer in **step 6** is a 10 mM Bis-Tris pH 6.5, 10 mM NaCl buffer (buffer A).
10. Pool the enzyme containing fractions from **step 6** and centrifuge at 14,000 × *g*, 4 °C, 30 min before loading on to the MonoS column. The column is pre-equilibrated in buffer A and the enzyme is loaded onto the column using the sample pump with a flow of 1 mL/min.
11. Wash the column with 15 CV of buffer A before eluting with linear salt gradient from 10 mM–1 M NaCl in 10 mM Bis-Tris pH 6.5. An ÄktaFPLC interfaced by UNICORN 4.12 control software is used for the ion exchange chromatography.
12. Analyze the purification steps using Coomassie stained gels and Western blot (anti-His). Load samples from the eluted fractions on SDS-PAGE gel, include samples from the flow-through and wash (Fig. 2).

Pool the most pure fractions containing the GalNAc-T. Quantify the protein concentration by comparative SDS-PAGE Coomassie stain gels using BSA as standard. Based on the protein quantification aliquot the enzyme with 10–20 μg for each injection (*see* **Note 3**) and store at −80 °C until it is used for immunizing the mice. Do not add sodium azide to the sample.

3.4 Immunization of Mice and Evaluation of Immune Response

1. Immunize 2–4 BALBc/A mice with one subcutaneous injection of 10–20 μg native enzyme protein in FCA, followed by 2 injections with FIA, at 2–3 weeks intervals. For each injection 100 μL of antigen solution is added slowly to 100 μL of adjuvant solution while mixing vigorously.
2. Take a blood sample (7–8 drops) from a large vein (tail/inner corner of eye) 7–10 days after third immunization. Incubate the blood at 37 °C for 1 h, followed by 2 h at room temperature (RT) or overnight (ON) at 4 °C. Add 150 μL PBS. Centrifuge blood at 3,000 × *g* RT for 5 min. Remove serum (approx. 150 μL at 1:4 dilutions) from the cell pellet and discard the cell pellet. Add 0.03 % NaN_3. Aliquot and keep frozen. Use serum at dilutions higher than 1:200 to test for induction of IgG antibodies to the immunogen by immunocytology on Sf9 cells expressing different related glycosyltransferases. Use serum also to optimize the screening protocols with human cell lines and tissues.
3. 3 weeks or later after last immunization inject final boost i.v. without adjuvant 3–4 days before the fusion.

3.5 Preparing NS1 Cells, Splenocytes, and Thymocytes for Fusion

All cell culture work is performed under sterile conditions in a flow bench without the use of antibiotics (*see* **Note 4**).

1. Thaw NS1 from stock at least 8–10 days prior to fusion. NS1 should be tested negative for mycoplasma (*see* **Note 5**). Grow cells in 2 × T75-flasks and aim for 5×10^5 cells/mL. One day before the fusion split the cells at least 50 % (*see* **Note 6**).
2. At the day of fusion transfer NS1 from 2 × T75-flasks into 2 × 50 mL tubes and centrifuge (400 × *g*). Resuspend pellet in cell culture medium without serum and transfer into 10 mL tubes. Count the cells and use 2×10^7cells for the fusion.
3. Sacrifice a BALBc/A mouse with the appropriate immune response. Remove the spleen aseptically and transfer to 10 mL tube containing cell culture medium without serum. Trim off and discard any contaminating tissue from the spleen. Likewise remove thymus from NMRI mice. Dissect the heart of the mouse with scissors and use a syringe to aspirate as much blood as possible into a small Eppendorf tube. Treat blood as in Subheading 3.4, **step 2** to obtain serum to use as a positive control.
4. Remove media from spleen and transfer spleen into the fine screen of a strainer placed in a petri dish. Carefully grind the spleen with the plunge of a 10 mL standard syringe.

5. Collect all of the ground tissue by washing plunge and strainer with 3 mL RPMI and transfer to a 10 mL tube, while discarding any contaminating tissues from the spleen.
6. Repeat the procedure with thymus from NMRI mouse.
7. Centrifuge splenocytes and thymocytes separately (800 × *g*, 5 min). Transfer pellet of thymocytes into 5 mL culture medium supplemented with HAT in a T25-flasks and transfer upright to incubator.

3.6 Fusion

1. Transfer volume of media containing 2×10^7 NS1 cells into a new 50 mL tube. Resuspend the pellet of splenocytes 2–3 times and fill into the same 50 mL tube leaving fibrous tissues behind. Resuspend NS1 and spleen cells together and centrifuge the mixture of cells (800 × *g*, 5 min). Remove and discard all the supernatant. Pellet must be completely dry (*see* **Note 7**).
2. Add PEG (800 μL, 37 °C) slowly to the cell pellet (mixture of NS1 and splenocytes) over 1 min with a 2 mL pipette with broad opening while slowly mixing.
3. Add 1 mL cell culture medium without serum (during 1 min) while shaking the tube followed by the addition of 9 mL culture medium without serum during the next 2 min.
4. Centrifuge (400 × *g*, 5 min).
5. Resuspend cell pellet slowly and carefully in 80 mL culture medium supplemented with HAT (50×) (37 °C) and transfer to a petri dish for subsequent mixing and transfer to six 96-well plates. Use only the inner 60 wells. Outer wells are filled with cell culture media without serum. Transfer plates to incubator.
6. On day 4 after the fusion add 50 μL of culture media supplemented with HAT (50×) and HCS (1 %) and thymocytes to the 200 μL of supernatant that is already in the well, i.e., prepare 20 mL of culture medium supplemented with HAT (50×) and approximately 50–60 % of the T25-flasks thymocytes prepared above and 5 % HCS.

3.7 Testing Growing Hybridomas for Appropriate Antibodies

1. At day 9–12 the clones should be ready to screen (*see* **Note 8**). Remove 50–150 μL culture medium from all the wells without disturbing the hybridomas at the bottom.
2. Transfer into round bottom 96-well plates. Add culture medium supplemented with HCS (1 %) and HT (50×) to the remaining cells and transfer plates back to the incubator (*see* **Note 9**).
3. Screen the 360 hybridoma (6 plates × 60 wells) culture supernatants by the screening methods as described in Subheading 3.9, **steps 6–8**, (*see* **Note 10**). Make sure the screening is finished before the hybridomas grow to confluence.

3.8 Single Cell Cloning

1. Transfer cells from positive 96-wells into a 24-well plate. Freeze and dilute hybridomas from the 24-wells before they become confluent.
2. Maintain original 24-wells by 1:10 dilution into new 24-wells when wells reach 50 % confluency (*see* **Note 11**).
3. Transfer 20–200 μL from the positive 96/24-well into 20 mL of culture media (20 % FBS, 1 % HCS and feeder cells) into a petri dish.
4. Seed this mixture into a 96-well plate, 200 μL in each well. For the first round of cloning aim at seeding cells in 30–100 % of a plate (*see* **Note 12**).
5. The cloning should usually be repeated 2–3 times with seeding of cells at approx. 1/3 of the 96-well plates in order to obtain a single cell clone and every well of the 96-well plate is tested positive.
6. The cloned hybridomas are then taken to T25-flasks and expanded and frozen.

3.9 Screening of Selected Clones Using Immunocytology on Insect Cells Expressing Constructs of Human Glycosyltransferases

1. Seed Sf9 cells in a T25-flask and allow to adhere for 5–10 min. Aim for almost confluent seeding, which usually requires cells detached from a T75-flask.
2. Infect cells with amplified virus (1:500) of interest and incubate for 2–3 days.
3. Harvest cells by scraping and centrifuge at 1,000 × *g* for 5 min, followed by a wash step with 5 mL PBS.
4. Resuspend washed cells in 3 mL PBS, mix, and pipette 15–20 μL cell suspension to each well of a 21 multi-well diagnostic slide (*see* **Note 13**).
5. Let the slides dry for 12–24 h and fix them in ice-cold acetone for 5–8 min and let the acetone evaporate. Additional slides should be kept frozen.
6. Add undiluted hybridoma supernatant to the slides and incubate ON at 4 °C. Use 15–20 μL per well.
7. Wash the slides 3 × 5 min in PBS and add rabbit anti-mouse Ig/goat anti-mouse IgG with FITC, 40 min RT.
8. Wash the slides 3 × 5 min in PBS; mount the slides and examine slides in a fluorescence microscope (*see* **Note 14**).

3.10 Immunocytology on Human Cancer Cell Lines That Express Glycosyltransferases Endogeneously

1. Human cancer cell lines of interest (*see* **Note 15**) are grown to subconfluency in the appropriate media as recommended.
2. Wash the cells in PBS, add TrypLE and incubate (5 min, 37 °C) until cells detach. Stop the trypsinization by adding culture medium with FBS.
3. Transfer cells to multi well Diagnostic slides and do immunocytology as described in Subheading 3.9, **steps 5–8** (*see* Fig. 3).

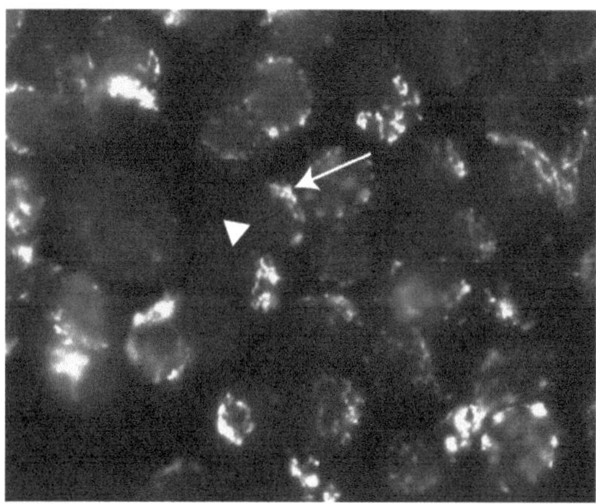

Fig. 3 Immunocytological expression of GalNAc-T2 (Mab UH4) in human pancreatic adenocarcinoma cell line Capan 1. Positive reaction is white (*arrow*) and nuclei are stained by DAPI (*arrowhead*). Note juxta-nuclear staining of the enzyme suggestive of Golgi localization

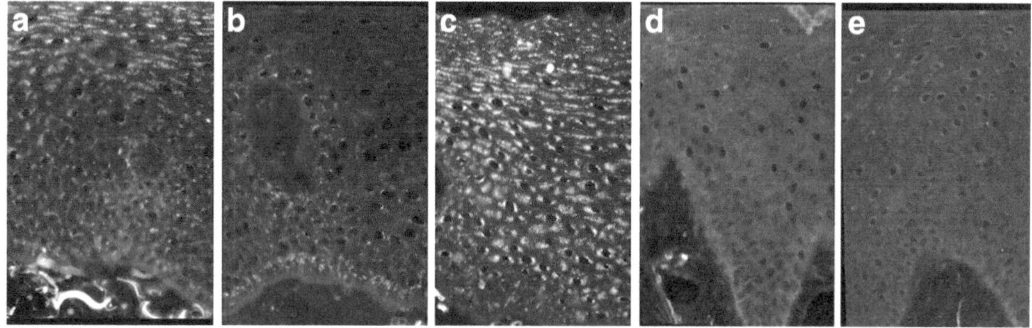

Fig. 4 Immunohistological expression of GalNAc-transferases defined by monoclonal antibodies (*see* Table 1) in neighboring sections of normal oral nonkeratinized epithelium (150×). GalNAc-T1 is strong in uttermost cell layers (**a**), GalNAc-T2 in the lower half of the epithelium (**b**) and GalNAc-T3 in all cell layers (**c**). GalNAC-T4 and T6 not expressed (**d**, **e**)

3.11 Immunohistology on Human Tissue Sections

1. Cut sections from a frozen tissue block at 5 μm and mount on gelatin coated slides. Use sections unfixed, fixed in ice cold acetone for 5–8 min or fixed in 10 % buffered neutral formalin for 10–15 min (*see* **Note 16**).
2. Add undiluted hybridoma supernatant to the slides and incubate ON at 4 °C.
3. Wash the slides 3 × 5 min in PBS and add rabbit anti-mouse Ig/goat anti-mouse IgG with FITC, 40 min RT.
4. Wash the slides 3 × 5 min in PBS; mount the slides and examine slides in a Fluorescence microscope (*see* Fig. 4).

4 Notes

1. It is important to keep the enzyme cold before, during, and after the purification, since the enzyme may lose activity at higher temperature.
2. Note that it is very important that the eluted protein is diluted immediately into a buffer without imidazole, since long time exposure to high concentration of imidazole has been shown to inactivate the GalNAc-Ts.
3. The total volume of antigen cannot exceed 100 μL. If the concentration is too low the antigen has to be concentrated before mixing with the adjuvant.
4. General use of antibiotics is not recommended as it is unnecessary and only slackens operator's attention to sterile working conditions and thereby promotes growth of mycoplasma and fungi that are resistant to antibiotics.
5. Mycoplasma contamination may inactivate thymidine in HAT by converting thymidine to thymine [27].
6. Cells must be healthy and grow rapidly.
7. Place 50 mL tube containing the pellet upside down for a few seconds and remove leftover supernatant with sterile cotton tipped applicators.
8. Note that small clones (5–20 cells) should be visible around day 4 in the microscope.
9. HT is optional.
10. Optimal antibodies for a particular purpose is best screened and selected by the assay of choice for final use, i.e., if to be used for Western blot screen with Western blot and if to be used for immunocytology screen with immunocytology.
11. Test 24 wells every 3 days to confirm that hybridomas continue to secrete antibodies.
12. Check clonings at day 4 and repeat cloning if too many (>80 %) or too few (<10 %) clones emerge.
13. The 21-wells format in each slide is a convenient number for screening a whole fusion, i.e., 20 slides is needed for screening 360 wells and there is room for controls on each slide
14. Some false positive clones with diffuse, unspecific staining will be found. Most of these will also react with untransfected cells, but not all. Selections of appropriate antibodies have to rely on the staining of correct localized protein in cells, *see* Subheadings 3.10 and 3.11.

15. If not available we usually produce a transient or stable transfected cell line with a full coding tagged construct of the enzyme in question.
16. Unfixed and acetone fixed slides can conveniently be stored at −80 °C.

Acknowledgments

This work was supported by Kirsten og Freddy Johansen Fonden, A.P. Møller og Hustru Chastine Mc-Kinney Møllers Fond til Almene Formaal, The Carlsberg Foundation, The Novo Nordisk Foundation, The Danish Research Councils, a program of excellence from the University of Copenhagen, and the Danish National Research Foundation (DNRF107).

References

1. Varki A (2009) Essentials of glycobiology. Cold Spring Harbor Laboratory Press, Cold Spring Harbor, NY
2. Paulson JC, Colley KJ (1989) Glycosyltransferases. Structure, localization, and control of cell type-specific glycosylation. J Biol Chem 264:17615–17618
3. Shifley ET, Cole SE (2008) Lunatic fringe protein processing by proprotein convertases may contribute to the short protein half-life in the segmentation clock. Biochim Biophys Acta 1783:2384–2390
4. Woodard-Grice AV, McBrayer AC, Wakefield JK, Zhuo Y, Bellis SL (2008) Proteolytic shedding of ST6Gal-I by BACE1 regulates the glycosylation and function of α4β1 integrins. J Biol Chem 283:26364–26373
5. Childs RA, Berger EG, Thorpe SJ, Aegerter E, Feizi T (1986) Blood-group-related carbohydrate antigens are expressed on human milk galactosyltransferase and are immunogenic in rabbits. Biochem J 238:605–611
6. Feizi T, Childs RA (1987) Carbohydrates as antigenic determinants of glycoproteins. Biochem J 245:1–11
7. Feizi T, Thorpe SJ, Childs RA (1987) Blood group genetic markers on human milk galactosyltransferase: relevance to the immunohistochemical approach to enzyme localization. Biochem Soc Trans 15:614–617
8. Landers KA, Burger MJ, Tebay MA, Purdie DM, Scells B, Samaratunga H, Lavin MF, Gardiner RA (2005) Use of multiple biomarkers for a molecular diagnosis of prostate cancer. Int J Cancer 114:950–956
9. Gu C, Oyama T, Osaki T, Li J, Takenoyama M, Izumi H, Sugio K, Kohno K, Yasumoto K (2004) Low expression of polypeptide GalNAc N-acetylgalactosaminyl transferase-3 in lung adenocarcinoma: impact on poor prognosis and early recurrence. Br J Cancer 90:436–442
10. Stern CA, Tiemeyer M (2001) A ganglioside-specific sialyltransferase localizes to axons and non-Golgi structures in neurons. J Neurosci 21:1434–1443
11. Taatjes DJ, Roth J, Weinstein J, Paulson JC (1988) Post-Golgi apparatus localization and regional expression of rat intestinal sialyltransferase detected by immunoelectron microscopy with polypeptide epitope-purified antibody. J Biol Chem 263:6302–6309
12. Hoffmeister KM, Josefsson EC, Isaac NA, Clausen H, Hartwig JH, Stossel TP (2003) Glycosylation restores survival of chilled blood platelets. Science 301:1531–1534
13. Wandall HH, Rumjantseva V, Sørensen AL, Patel-Hett S, Josefsson EC, Bennett EP, Italiano JE Jr, Clausen H, Hartwig JH, Hoffmeister KM (2012) The origin and function of platelet glycosyltransferases. Blood 120:626–635
14. Rottger S, White J, Wandall HH, Olivo JC, Stark A, Bennett EP, Whitehouse C, Berger EG, Clausen H, Nilsson T (1998) Localization of three human polypeptide GalNAc-transferases in HeLa cells suggests initiation of O-linked glycosylation throughout the Golgi apparatus. J Cell Sci 111(Pt 1):45–60
15. Gill DJ, Clausen H, Bard F (2011) Location, location, location: new insights into O-GalNAc

protein glycosylation. Trends Cell Biol 21: 149–158

16. Berger EG (2002) Ectopic localizations of Golgi glycosyltransferases. Glycobiology 12:29R–36R
17. Bennett EP, Mandel U, Clausen H, Gerken TA, Fritz TA, Tabak LA (2012) Control of mucin-type O-glycosylation—a classification of the polypeptide GalNAc-transferase gene family. Glycobiology 22:736–756
18. Wandall HH, Hassan H, Mirgorodskaya E, Kristensen AK, Roepstorff P, Bennett EP, Nielsen PA, Hollingsworth MA, Burchell J, Taylor-Papadimitriou J, Clausen H (1997) Substrate specificities of three members of the human UDP-N-acetyl-α-D-galactosamine: Polypeptide N-acetylgalactosaminyltransferase family, GalNAc-T1, -T2, and -T3. J Biol Chem 272:23503–23514
19. Mandel U, Hassan H, Therkildsen MH, Rygaard J, Jakobsen MH, Juhl BR, Dabelsteen E, Clausen H (1999) Expression of polypeptide GalNAc-transferases in stratified epithelia and squamous cell carcinomas: immunohistological evaluation using monoclonal antibodies to three members of the GalNAc-transferase family. Glycobiology 9:43–52
20. Sutherlin ME, Nishimori I, Caffrey T, Bennett EP, Hassan H, Mandel U, Mack D, Iwamura T, Clausen H, Hollingsworth MA (1997) Expression of three UDP-N-acetyl-α-D-galactosamine:polypeptide GalNAc N-acetylgalactosaminyltransferases in adenocarcinoma cell lines. Cancer Res 57:4744–4748
21. Bennett EP, Hassan H, Mandel U, Mirgorodskaya E, Roepstorff P, Burchell J, Taylor-Papadimitriou J, Hollingsworth MA, Merkx G, van Kessel AG, Eiberg H, Steffensen R, Clausen H (1998) Cloning of a human UDP-N-acetyl-α-D-Galactosamine: polypeptide N-acetylgalactosaminyltransferase that complements other GalNAc-transferases in complete O-glycosylation of the MUC1 tandem repeat. J Biol Chem 273:30472–30481
22. Bennett EP, Hassan H, Mandel U, Hollingsworth MA, Akisawa N, Ikematsu Y, Merkx G, van Kessel AG, Olofsson S, Clausen H (1999) Cloning and characterization of a close homologue of human UDP-N-acetyl-α-D-galactosamine:Polypeptide N-acetylgalactosaminyltransferase-T3, designated GalNAc-T6. Evidence for genetic but not functional redundancy. J Biol Chem 274:25362–25370
23. Schwientek T, Bennett EP, Flores C, Thacker J, Hollmann M, Reis CA, Behrens J, Mandel U, Keck B, Schäfer MA, Haselmann K, Zubarev R, Roepstorff P, Burchell JM, Taylor-Papadimitriou J, Hollingsworth MA, Clausen H (2002) Functional Conservation of Subfamilies of Putative UDP-N-acetylgalactosamine: PolypeptideN-Acetylgalactosaminyltransferases in Drosophila, Caenorhabditis elegans, and Mammals. J Biol Chem 277:22623–22638
24. Almeida R, Levery SB, Mandel U, Kresse H, Schwientek T, Bennett EP, Clausen H (1999) Cloning and expression of a proteoglycan UDP-galactose:β-xylose β1,4-galactosyltransferase I. A seventh member of the human β4-galactosyltransferase gene family. J Biol Chem 274:26165–26171
25. Vallejo-Ruiz V, Haque R, Mir A-M, Schwientek T, Mandel U, Cacan R, Delannoy P, Harduin-Lepers A (2001) Delineation of the minimal catalytic domain of human Galβ1-3GalNAc α2,3-sialyltransferase (hST3Gal I). Biochim Biophys Acta 1549:161–173
26. Marcos NT, Bennett EP, Gomes J, Magalhaes A, Gomes C, David L, Dar I, Jeanneau C, DeFrees S, Krustrup D, Vogel LK, Kure EH, Burchell J, Taylor-Papadimitriou J, Clausen H, Mandel U, Reis CA (2011) ST6GalNAc-I controls expression of sialyl-Tn antigen in gastrointestinal tissues. Front Biosci (Elite Ed) 3:1443–1455
27. Harlow E, Lane D (1988) Antibodies, a laboratory manual. Cold Spring Harbour Laboratory, New York, NY

ERRATUM

Study of the Biological Functions of Mucin Type Core 3 *O*-glycans

Seung Ho Lee and Minoru Fukuda

Scripps Korea Antibody Institute, Chuncheon, Gangwon, Korea;
Tumor Microenvironment Program, Cancer Center, Sanford Burnham Medical Research Institute, La Jolla, CA, USA

Inka Brockhausen (ed.), *Glycosyltransferases: Methods and Protocols*, Methods in Molecular Biology, vol. 1022, DOI 10.1007/978-1-62703-465-4, pp. 41–50, © Springer Science+Business Media New York 2013

DOI 10.1007/978-1-62703-465-4_31

The publisher regrets that in the print and online versions of this book one of three affiliations of Seung Ho Lee, author of Chapter 4 is omitted. The affiliation appears below.

Seung Ho Lee
College of Life Sciences and Bioengineering
University of Incheon
Incheon, Korea

The online version of the original chapter can be found at http://dx.doi.org/10.1007/978-1-62703-465-7_4

Index

A

B

C

D

Inka Brockhausen (ed.), *Glycosyltransferases: Methods and Protocols*, Methods in Molecular Biology, vol. 1022, DOI 10.1007/978-1-62703-465-4, © Springer Science+Business Media New York 2013

O

P

R

S

MIX
Papier aus verantwortungsvollen Quellen
Paper from responsible sources
FSC® C105338

If you have any concerns about our products,
you can contact us on
ProductSafety@springernature.com

In case Publisher is established outside the EU,
the EU authorized representative is:
Springer Nature Customer Service Center GmbH
Europaplatz 3, 69115 Heidelberg, Germany

Printed by Libri Plureos GmbH
in Hamburg, Germany